RISK ASSESSMENT

RISK ASSESSMENT

A Practical Guide to Assessing Operational Risks

Edited by

GEORGI POPOV
BRUCE K. LYON
BRUCE HOLLCROFT

Copyright © 2016 by John Wiley & Sons, Inc. All rights reserved

Published by John Wiley & Sons, Inc., Hoboken, New Jersey
Published simultaneously in Canada

No part of this publication may be reproduced, stored in a retrieval system, or transmitted in any form or by any means, electronic, mechanical, photocopying, recording, scanning, or otherwise, except as permitted under Section 107 or 108 of the 1976 United States Copyright Act, without either the prior written permission of the Publisher, or authorization through payment of the appropriate per-copy fee to the Copyright Clearance Center, Inc., 222 Rosewood Drive, Danvers, MA 01923, (978) 750-8400, fax (978) 750-4470, or on the web at www.copyright.com. Requests to the Publisher for permission should be addressed to the Permissions Department, John Wiley & Sons, Inc., 111 River Street, Hoboken, NJ 07030, (201) 748-6011, fax (201) 748-6008, or online at http://www.wiley.com/go/permission.

Limit of Liability/Disclaimer of Warranty: While the publisher and author have used their best efforts in preparing this book, they make no representations or warranties with respect to the accuracy or completeness of the contents of this book and specifically disclaim any implied warranties of merchantability or fitness for a particular purpose. No warranty may be created or extended by sales representatives or written sales materials. The advice and strategies contained herein may not be suitable for your situation. You should consult with a professional where appropriate. Neither the publisher nor author shall be liable for any loss of profit or any other commercial damages, including but not limited to special, incidental, consequential, or other damages.

For general information on our other products and services or for technical support, please contact our Customer Care Department within the United States at (800) 762-2974, outside the United States at (317) 572-3993 or fax (317) 572-4002.

Wiley also publishes its books in a variety of electronic formats. Some content that appears in print may not be available in electronic formats. For more information about Wiley products, visit our web site at www.wiley.com.

Library of Congress Cataloging-in-Publication Data:

Names: Popov, Georgi (Engineer), author. | Lyon, Bruce K. | Hollcroft, Bruce.
Title: Risk assessment : a practical guide to assessing operational risks /
   Georgi Popov, Bruce K. Lyon, Bruce Hollcroft.
Description: Hoboken : John Wiley & Sons, Inc., 2016. | Includes
   bibliographical references and index.
Identifiers: LCCN 2015047429 (print) | LCCN 2015048838 (ebook) | ISBN
   9781118911044 (hardback) | ISBN 9781119220916 (pdf) | ISBN 9781119220909
   (epub)
Subjects: LCSH: Risk assessment. | BISAC: TECHNOLOGY & ENGINEERING /
   Industrial Health & Safety.
Classification: LCC T10.68 .P66 2016 (print) | LCC T10.68 (ebook) | DDC
   658.15/5–dc23
LC record available at http://lccn.loc.gov/2015047429

Typeset in 10/12pt TimesLTStd by SPi Global, Chennai, India

Printed in the United States of America

10 9 8 7 6 5 4 3

ACKNOWLEDGEMENTS

First and foremost, the authors wish to thank Fred A. Manuele, P.E., CSP, for his vision, guidance, and contributions to the book. Without his mentorship, this text would not have materialized.

Second, we thank each of the contributing authors that helped in the development of this text. The contributing authors which are identified in each of their chapters are: Elyce Biddle, PhD; Steven Hicks, CSP, CIH; Tsvetan Popov, PhD; John Zey, EdD, CIH; Ying Zhen, CIH, CSP, and Jim Whiting.

To the wonderful and patient spouses, family, and friends of the authors, we offer many thanks.

And finally, we offer special thanks to the University of Central Missouri, the Safety Sciences program chair Dr. Leigh Ann Blunt, CSP and the Hays Companies.

During this journey, the authors have come to realize the overall magnitude and number of aspects associated with the process of operational risk assessment. No doubt, there is a vast landscape of risk concepts, methodologies, and tools to draw upon. We hope that this work will help, in some way, those that have interest in pursuing the honorable and necessary tasks of assessing and managing risks in the workplace.

CONTENTS

Preface xvii

Foreword xxi

List of Contributors xxiii

About the Companion Websites xxv

1 **Risk Assessments: Their Significance and the Role of the Safety Professional** 1
Fred A. Manuele

    1.1   Objectives, 1
    1.2   Introduction, 1
    1.3   What is a Risk Assessment? 2
    1.4   Activities at the American Society of Safety Engineers (ASSE), 2
    1.5   An Example of a Guideline that gives Risk Assessment due Recognition, 3
    1.6   ANSI/AIHA/ASSE Z10-2012: The Standard for *Occupational Health and Safety Management Systems*, 4
    1.7   ANSI/ASSE Z590.3-2011: Prevention through Design: Guidelines for Addressing Occupational Hazards and Risks in Design and Redesign Processes, 4
    1.8   THE ANSI/ASSE Z690-2011 Series, 6
    1.9   ANSI B11.0-2015: Safety of Machinery. General Safety Requirements and Risk Assessment – A Standard of Major Consequence, 7
    1.10   European Union: Risk Assessment, 8
    1.11   EN ISO 12100-2010: Safety of Machinery. General Principles for Design. Risk Assessment, and Risk Reduction, 8
    1.12   Additional European Influence, 9
    1.13   MIL-STD-882E-2012. The US Department of Defense Standard Practice for System Safety, 9
    1.14   Certain Governmental Views, 11
           1.14.1 Risk Reduction Program, 12

- 1.15 Canada, 12
- 1.16 Fire Protection, 13
- 1.17 Developments in Aviation Ground Safety, 13
- 1.18 OSHA Requirements, 14
- 1.19 EPA Requirements, 15
- 1.20 The Chemical Industry: The Extensive Body of Information, 16
- 1.21 Conclusion, 16
  Review Questions, 16
  References, 17
  Appendix 1.A: A List of Standards, Guidelines, and Initiatives That Require or Promote Making Risk Assessments: Commencing with Year 2005, 18

2 Risk Assessment Standards and Definitions   23
Bruce Hollcroft & Bruce K. Lyon

- 2.1 Objectives, 23
- 2.2 Introduction, 23
- 2.3 The Need for Risk Assessments, 24
- 2.4 Key Standards Requiring Risk Assessments, 24
- 2.5 OSHA Compliance and Risk Assessments, 24
  - 2.5.1 1910.132, Personal Protective Equipment Standard, 25
  - 2.5.2 1910.119, Process Safety Management Standard, 25
  - 2.5.3 Other OSHA Standards, 26
- 2.6 Consensus Standards Requiring Risk Assessment, 27
- 2.7 ANSI/AIHA/ASSE Z10-2012, Occupational Health and Safety Management Systems, 27
- 2.8 ISO 31000/ANSI/ASSE Z690 Risk Management Series, 28
- 2.9 ANSI/ASSE Z590.3-2011, Prevention through Design, 29
- 2.10 ANSI B11.0 Machine Safety, 30
- 2.11 NFPA 70E, 31
- 2.12 MIL-STD-882E, 11 May 2012, Department of Defense Standard Practice, System Safety, 31
- 2.13 Key Terms and Definitions, 32
- 2.14 Summary, 46
  Review Questions, 47
  References, 47

3 Risk Assessment Fundamentals   49
Bruce Hollcroft & Bruce K. Lyon

- 3.1 Objectives, 49
- 3.2 Introduction, 49
- 3.3 Risk Assessment within the Risk Management Framework, 50
- 3.4 Risk Assessments and Operational Risk Management Systems, 51
- 3.5 The Purpose of Assessing Risk, 52
- 3.6 The Risk Assessment Process, 53
- 3.7 Selecting a Risk Assessment Matrix, 53
- 3.8 Establishing Context, 55
- 3.9 The Risk Assessment Team, 57
- 3.10 Hazard/Risk Identification, 58

- 3.11 Risk Analysis, 59
  - 3.11.1 Consequence Analysis, 59
  - 3.11.2 Likelihood Analysis, 59
  - 3.11.3 Assessment of Controls, 60
- 3.12 Risk Evaluation, 60
- 3.13 Risk Treatment, 61
- 3.14 Communication, 61
- 3.15 Documentation, 62
- 3.16 Monitoring and Continuous Improvement, 63
- 3.17 Summary, 64
  - Review Questions, 64
  - References, 64

4 Defining Risk Assessment Criteria   67
Bruce K. Lyon & Bruce Hollcroft

- 4.1 Objectives, 67
- 4.2 Introduction, 67
- 4.3 Defining Risk Criteria, 68
- 4.4 Risk Scoring Systems, 69
- 4.5 Risk Assessment Matrices, 71
- 4.6 Defining Risk Values, 71
  - 4.6.1 Qualitative Risk Models, 72
  - 4.6.2 Semiquantitative Risk Models, 72
  - 4.6.3 Quantitative Risk Models, 73
- 4.7 Risk Factors, 74
- 4.8 Risk Levels, 74
- 4.9 Risk Scoring, 75
- 4.10 Severity of Consequence, 76
- 4.11 Likelihood of Occurrence, 77
- 4.12 Exposure, 79
- 4.13 Risk Reduction and the Hierarchy of Controls, 79
  - 4.13.1 Using a Protection Factor, 83
- 4.14 Acceptable and Unacceptable Risk Levels, 84
- 4.15 Documenting Risk, 85
- 4.16 Communicating Risk Criteria, 88
- 4.17 Summary, 88
  - Review Questions, 88
  - References, 89
  - Appendix 4.A, 90

5 Fundamental Techniques   91
Bruce K. Lyon

- 5.1 Objectives, 91
- 5.2 Introduction to Fundamental Hazard Analysis and Risk Assessment, 91
- 5.3 Assessments Within an Operational Risk Management System, 93
- 5.4 Hazard Analysis Versus Risk Assessment, 94
- 5.5 The Hazard Analysis and Risk Assessment Process, 96
- 5.6 Fundamental Methods, 99

- 5.7 Informal Methods, 100
- 5.8 Formal Methods, 103
  - 5.8.1 Fundamental Hazard Analysis, 103
  - 5.8.2 Pretask Hazard Analysis, 104
  - 5.8.3 Job Hazard Analysis, 104
  - 5.8.4 Fundamental Risk Assessment, 109
  - 5.8.5 Job Risk Assessment, 110
- 5.9 Conclusion, 112
- Review Questions, 112
- References, 113
- Appendix 5.A, 114
- Appendix 5.B: Common Hazards and Descriptions, 115
- Appendix 5.C: Personal Protective Equipment Hazard Assessment Form Example, 118
- Appendix 5.D: Job Hazard Analysis Form Example, 119

6 What-If Hazard Analysis 121
Bruce K. Lyon

- 6.1 Objectives, 121
- 6.2 Introduction, 121
- 6.3 Overview and Background, 121
- 6.4 Process Hazard Analysis, 122
- 6.5 Mandated Assessments, 123
- 6.6 What-If Analysis and Related Methods, 125
  - 6.6.1 Brainstorming – Structured and Unstructured, 125
  - 6.6.2 Checklist Analysis, 126
  - 6.6.3 What-If Hazard Analysis, 127
  - 6.6.4 What-If/Checklist, 130
  - 6.6.5 Structured What-If Technique (SWIFT), 131
  - 6.6.6 Hazard and Operability (HAZOP) Study, 135
- 6.7 Risk Scoring and Ranking, 137
- 6.8 Application of "What-If", 139
- 6.9 Conclusion, 143
- Review Questions, 143
- References, 144

7 Preliminary Hazard Analysis 145
Georgi Popov & Bruce K. Lyon

- 7.1 Objectives, 145
- 7.2 Introduction, 145
- 7.3 Preliminary Hazard List, 147
- 7.4 PHAs and their Application, 147
- 7.5 The Control of Hazardous Energy, 148
- 7.6 Fundamental System Safety Tenets, 149
- 7.7 Conducting a PHA, 150
- 7.8 Scoring Systems, 152
- 7.9 Practical Application, 153
- 7.10 Summary, 157

Review Questions, 157
References, 157
Practical Example, 161

8 Failure Mode and Effects Analysis 163
Georgi Popov & Bruce K. Lyon

- 8.1 Objectives, 163
- 8.2 Introduction, 163
- 8.3 Purpose and Use, 164
- 8.4 Defining Failure Modes, 166
- 8.5 Risk Description Considerations, 167
- 8.6 FMEA Process Steps, 172
- 8.7 Practical Application, 175
- 8.8 Summary, 176
  Review Questions, 179
  References, 179
  Practical Example – Assignment #2 – FMEA, 179

9 Bow-Tie Risk Assessment Methodology 181
Georgi Popov & Bruce K. Lyon

- 9.1 Objectives, 181
- 9.2 Introduction, 181
- 9.3 History, 182
- 9.4 Overview, 182
- 9.5 Bow-Tie Methodology, 184
- 9.6 Practical Application, 186
  - 9.6.1 Case Study #1: Spray Paint Operation, 186
  - 9.6.2 Case Study #2: Bhopal Disaster, 193
- 9.7 Summary, 195
  Review Questions, 195
  References, 196
  Appendix 9.A: QAP Corporation – Annual Report, 196

10 Design Safety Reviews 209
Bruce K. Lyon

- 10.1 Objectives, 209
- 10.2 Introduction, 209
- 10.3 Challenges and Obstacles to Overcome, 211
- 10.4 Standards Requiring Design Safety, 214
- 10.5 The Review of Designs, 215
- 10.6 Hazardous Energy Control, 216
- 10.7 Ergonomic Review of Designs, 217
- 10.8 Design Review Process, 218
- 10.9 Hazard Analysis and Risk Assessment in Design, 220
- 10.10 Conclusion, 224
  Review Questions, 225
  References, 225

11 Risk Assessment and the Prevention Through Design (PtD) Model — 227
Georgi Popov, Bruce K. Lyon, & John N. Zey

- 11.1 Objectives, 227
- 11.2 Introduction, 227
- 11.3 The Concept of Prevention Through Design (PtD), 229
- 11.4 Risk Assessment Process and the PtD Model, 229
- 11.5 Case Study, 234
  - 11.5.1 Methods, 234
  - 11.5.2 Results, 234
  - 11.5.3 Occupational Size-Selective Criteria and Particles Size Sampling, 237
- 11.6 PtD and the Business Process, 243
- 11.7 Summary, 244
  - Review Questions, 244
  - References, 244

12 Industrial Hygiene Risk Assessment — 247
Georgi Popov, Steven Hicks, & Tsvetan Popov

- 12.1 Objectives, 247
- 12.2 Introduction, 247
- 12.3 Fundamental Concepts, 248
- 12.4 Anticipating and Identifying Occupational Health Risks, 249
- 12.5 Determining Occupational Health Risks, 250
  - 12.5.1 Health Risk Rating Methodology, 250
  - 12.5.2 Exposure Rating Methodologies, 251
  - 12.5.3 Health Effect and Exposure Methodology, 251
  - 12.5.4 COSHH Essentials Tool, 251
  - 12.5.5 OSHA's Calculation for Mixtures, 254
  - 12.5.6 The ART Tool, 254
  - 12.5.7 Stoffenmanager, 254
- 12.6 Health Risk Assessments and Prioritization, 255
- 12.7 Modified HRR/IH FMEA Methodology, 256
  - Sampling, 257
  - Results, 257
- 12.8 Control Banding Nanotool, 261
- 12.9 Dermal Risk Assessment, 261
- 12.10 Occupational Health Risk and PTD Process Alignment, 262
- 12.11 Summary, 264
  - Review Questions, 265
  - References, 265

13 Machine Risk Assessments — 267
Bruce K. Lyon

- 13.1 Objectives, 267
- 13.2 Introduction, 267
- 13.3 Machine Safety Standards, 268
- 13.4 Machine Hazards, 270
- 13.5 Machine Safeguarding, 271
  - 13.5.1 Machine Safety Control Systems, 273

13.6	Selecting Machines for Assessment, 274	
13.7	Risk Assessment of Machines, 274	
13.8	Estimating Risk, 278	
13.9	Case Study, 279	
13.10	Assessment of Machine Maintenance and Service, 282	
	13.10.1 Risk Assessment Process, 284	
	13.10.2 Risk Reduction Process, 285	
13.11	Summary, 285	
	Review Questions, 286	
	References, 286	
	Appendix 13.A: Machine Safeguards Methods, 287	

14 Project-Oriented Risk Assessments   291

Bruce K. Lyon

14.1	Objectives, 291
14.2	Introduction, 291
14.3	Fatalities and Serious Incidents, 293
14.4	Error Traps in Nonroutine Tasks, 294
14.5	Management of Change, 294
14.6	Construction Project Work, 296
14.7	Construction Project Risk Assessment, 297
14.8	Safe Work Methods, 299
14.9	Pretask Hazard Analysis, 301
14.10	The Use of Checklists, 303
14.11	Maintenance and Service Work, 304
14.12	Operating Hazard Analysis, 305
14.13	Analyzing Specific Hazards, 308
14.14	Pre-Entry Hazard Analysis, 308
14.15	Fall Hazard Assessment, 311
14.16	Summary, 317
	Review Questions, 317
	References, 317

15 Food Processing Risk Assessments   319

Georgi Popov, Bruce K. Lyon, & Ying Zhen

15.1	Objectives, 319
15.2	Overview, 319
15.3	Introduction to Food Risk, 320
15.4	Risk Assessment Techniques in the Food Industry, 320
15.5	Food Safety-Related Hazards, 321
	15.5.1 Biological Food Hazards, 321
	15.5.2 Chemical Food Hazards, 322
	15.5.3 Physical Food Hazards, 323
15.6	Techniques for Assessing Food Risk, 323
15.7	Hazard Analysis and Critical Control Points, 324
15.8	Integration of Risk Assessment Methods, 325
15.9	PtD and HACCP Integration, 338
15.10	Conclusions, 339

16 Ergonomic Risk Assessment — 343
Bruce K. Lyon & Georgi Popov

- 16.1 Objectives, 343
- 16.2 Introduction, 343
- 16.3 Ergonomics and Design, 344
- 16.4 Ergonomic Hazards, 345
- 16.5 Ergonomic Risk Factors, 346
- 16.6 Establishing an Ergonomics Assessment Process, 346
  - 16.6.1 Scope and Context, 348
  - 16.6.2 Goals and Objectives, 348
  - 16.6.3 Responsibilities, 348
  - 16.6.4 Training, 348
  - 16.6.5 Ergonomics Team, 348
- 16.7 Assessing Ergonomic Risk, 349
- 16.8 Ergonomics Improvement Process, 350
  - 16.8.1 Identify Jobs, 350
  - 16.8.2 Assessment Tools, 351
  - 16.8.3 Assessment Team, 352
  - 16.8.4 Performing the Assessments, 352
  - 16.8.5 Identifying Corrective Measures, 353
  - 16.8.6 Implementing Measures, 353
  - 16.8.7 Verify and Refine, 353
  - 16.8.8 Communicate Results, 354
- 16.9 ERAT: A Practical Assessment Tool, 354
  - 16.9.1 ERAT Example: Pork Processing Belly Grader, 356
- 16.10 Conclusion, 359

Review Questions, 340
References, 340

Review Questions, 360
References, 360
Appendix 16.A: Sample Ergonomic Responsibilities for Involved Stakeholders, 361
Appendix 16.B: Sample Ergonomics Training for Involved Stakeholders, 363
Appendix 16.C: Ergonomic Risk Assessment Tool (ERAT) – Initial Assessment, 365
Appendix 16.D: Ergonomic Risk Assessment Tool (ERAT) – Post-Control Assessment, 366
Appendix 16.E: Hierarchy of Ergonomic Risk Controls, 367

17 Assessing Operational Risks at an Organizational Level — 369
Bruce K. Lyon

- 17.1 Objectives, 369
- 17.2 Introduction, 369
- 17.3 Risks to an Organization, 370
- 17.4 Organizational Risk Management, 371
- 17.5 Key Definitions in Organizational Risk, 372

17.6 Assessing Organizational Risk, 373
17.7 Summary, 387
Review Questions, 387
References, 387

18 Risk Assessment Applications in Lean Six Sigma and Environmental Management Systems 389

Georgi Popov

18.1 Objectives, 389
18.2 Introduction, 389
18.3 Environmental Management Systems (EMS), 390
18.4 ISO 14001 Implementation, 390
    18.4.1 Environmental Policy and Planning, 392
    18.4.2 Environmental Aspects, 393
    18.4.3 Identify Environmental Aspects, 395
    18.4.4 Identification Process, 395
    18.4.5 Location, Department, Index, and Aspect, 396
    18.4.6 Impacts to Environmental Properties, 397
    18.4.7 Impact Subtotal and Polarity Adjustment, 397
    18.4.8 Impact Severity, 398
    18.4.9 Impact Probability, 398
    18.4.10 Frequency, 400
    18.4.11 Legal Risks, 400
    18.4.12 Current Controls, 401
    18.4.13 Significance Score for Significance Scores without Controls Section, 401
    18.4.14 Personnel Risk, 401
    18.4.15 Significance Scores with Controls Section, 403
    18.4.16 Overall Significance Rating Chart, 403
18.5 EMS and Implementation of Lean Six Sigma Practices, 404
18.6 Conclusions, 407
Review Questions, 407
References, 408

19 Business Aspects of Operational Risk Assessment 409

Elyce Biddle

19.1 Objectives, 409
19.2 Introduction, 409
19.3 The Business Case Development Tool, 410
    19.3.1 Steps of the Tool, 411
19.4 Business Case Examples, 412
    19.4.1 Case Example One: Post Incident, 412
    19.4.2 Case Example Two: Regulatory Requirement, 413
    19.4.3 Case Example Three: Operational, 416
    19.4.4 Case Example Four: Postoperational, 418
19.5 Conclusion, 424
Review Questions, 424
References, 424

20 Risk Assessment: Global Perspectives   427
Jim Whiting

- 20.1 Objectives, 427
- 20.2 Introduction, 427
- 20.3 Using ISO 31000 for Maturity Assurance and Conformity, 428
- 20.4 Global Uptake of ISO 31000: International Risk Management Standard, 431
- 20.5 Global Comparison of Risk Tolerance Criteria, 432
  - 20.5.1 Individual Risk, 432
  - 20.5.2 Societal Risk, 433
- 20.6 Tolerability Criterion for Individual Risk, 433
- 20.7 Tolerability Criteria for Planning New Operations, 435
- 20.8 Investment to Prevent a Fatality, 436
- 20.9 Shifting the Paradigm from Absolute Safety to Risk Management, 438
  - 20.9.1 What Is Reasonably Practicable? 438
- 20.10 Moving Toward Risk-Based Language for more Effective Risk Conversations, 440
- 20.11 A Cautionary Concluding Note, 440
  - Review Questions, 440
  - References, 441
  - Appendix 20.A: Better Terminology and Language for Risk-Based Conversations, 442

Index   445

PREFACE

"Imagination is everything. It is the preview of life's coming attractions."
Albert Einstein <http://www.goodreads.com/author/show/9810.Albert_Einstein>

With the many titles of books on risk assessment that have been written, Fred Manuele recognized that there was a need for a fundamental guide for assessing operational risk. As a member of the Advisory Board to the Safety Sciences Program at the University of Central Missouri (UCM), Mr. Manuele challenged Dr. Georgi Popov at UCM, Bruce Lyon and Bruce Hollcroft who previously worked for Manuele to write such a text. Dr. Popov teaches risk assessment, and like Lyon and Hollcroft, performs numerous risk assessments, simple to complex, for a wide range of industries. The challenge was accepted.

This first edition of *Risk Assessment: A Practical Guide to Assessing Operational Risks* provides the fundamentals on risk assessment, with many practical applications, for undergraduate and graduate students and employed safety, health, and environmental professionals who recognize that they are expected to have risk assessment capabilities.

This book fills a void. In recent years, risk assessment has been given more prominence as an element in operational risk management systems. This text serves the needs of professors at a university level who recognize that their students have knowledge and capability with respect to risk assessment, while addressing seven of the Accreditation Board for Engineering and Technology (ABET) criteria for safety programs. In addition, the book serves as a primer for employed safety professionals who need a practical guide on various risk assessment techniques.

The authors envisioned a new format for this book: one which includes interactive exercises, links, videos, and supplemental risk assessment tools. The content of this book has been significantly impacted by events that have occurred that give greater prominence to risk assessment. Some of these include the following:

1. In 1995, the National Safety Council created an entity known as the Institute for Safety through Design. The core of safety through design is hazard identification and analysis and risk assessment.
2. In 2006, the National Institute for Occupational Safety and Health (NIOSH) began consideration of what became a major initiative on Prevention through Design (PtD). The intent of

the initiative was to encourage organizations to have processes in place to address occupational hazards and risks in the design and redesign processes. Doing so requires making risk assessments as a continuum as the design process moves forward.
3. A European led drive to have risk assessment be recognized as the cornerstone of an occupational risk management system is having an impact in the United States. In 2011, the American National Standard Institute approved a petition made by the American Society of Safety Engineers to adopt three standards on risk management developed by International Standards Organization. One of those standards has become known as *ANSI/ASSE Z690.3, Risk Assessment Techniques*. That standard is receiving broad attention.
4. A new American National Standard on PtD was adopted on September 1. 2011. A significant portion of the standard is devoted to risk assessment.
5. Educators are developing new courses related to PtD and new risk assessment tools. A chapter in this book is devoted to PtD.
6. Many industries have applied Lean Concepts to reduce waste, improve efficiency, and lower production costs. Lean Six Sigma concepts and risk assessment tools can be applied in the environmental, safety, and service fields.
7. Ergonomics-related losses account for at least 1/3 of all lost time incidents and nearly half of the insurance costs. A chapter addressing risk assessment of ergonomic risk is included.
8. For many years, businesses have been operating with tight budgets and continuously seek ways to reduce costs, among which are accident costs. One of the chapters addresses risk assessment and business aspects of safety, health, and environmental interventions.
9. In June 2013, the American Society of Safety Engineers recognized the significance of risk assessment by launching its Risk Assessment Institute, a gateway for members of the society to develop new risk assessment core competencies.
10. A risk assessment process enables the safety professional to properly deal with hazards when there are little or no applicable regulations, standards, or guidelines. It also enables organizations to make better business decisions by prioritizing hazards and their resulting risk. This makes risk assessment an "essential" skill for the safety, health, and environmental professional.

The text begins with laying the ground work in Chapters 1–4. Chapter 1. *Risk Assessments: Their Significance and the Role of the Safety Professional* presents a brief overview on risk assessment, followed by comments on the importance of the Prevention though Design standard. Chapter 2. *Risk Assessment Standards and Definitions,* Chapter 3. *Risk Assessment Fundamentals* and Chapter 4. *Defining Risk Assessment Criteria* provide the basis of the risk assessment process.

Chapters 5–8 introduce the reader to fundamental risk assessment methods beginning with hazard identification and analysis methods such as job hazard analysis, what-if analysis, preliminary hazard analysis (PHA), and fundamental risk assessment techniques such as failure mode and effects analysis (FMEA).

More specialized methods including Bow-Tie Analysis, Design Safety Reviews, PtD tools, and Industrial Hygiene (IH) methods are presented in Chapters 9–12.

Chapter 13 targets machinery risks, while Chapter 14 provides methodologies of assessing project-related risks such as construction, maintenance, and other high-risk activities. Chapter 15 provides a primer on HACCP and food safety risks, while Chapter 16 presents more advanced measures in assessing ergonomic risk factors. Chapter 17 provides a board approach to assessing risks at an organizational level.

The final three chapters provide the reader concepts and methods of incorporating risk assessment in environmental management systems (Chapter 18), the inclusion of business aspects and metrics in the risk assessment process (Chapter 19), and a view of risk from a more Global Perspective

(Chapter 20). In addition, ABET accreditation criteria state that graduates must demonstrate the application of business and risk management concepts. Therefore, Chapter 19, Business Aspects of Operational Risk Assessment is devoted to supporting this criteria requirement.

For safety students who seek employment, being able to say that they have been indoctrinated in the subjects that are currently important to management is an advantage. For employed safety professionals, being able to demonstrate that they have taken the initiative to acquire the new knowledge and skills that emerging opportunities require gives the impression of serving management needs.

This practical guide serves both the student and the safety professional in developing foundations in risk assessment. It is the authors' hope that this text will challenge the safety professional in becoming more competent and creative in their application of assessing, defining, and managing operational risks.

BRUCE K. LYON, GEORGI POPOV, AND BRUCE HOLLCROFT

Kansas City
10th July 2015

FOREWORD

BIGGEST ORGANIZATIONAL RISK!

The information and tools contained in "Risk Assessment: A Practical Guide to Assessing Operational Risks" should get into the hands and minds of every practicing environmental, health, and safety (EH&S) professional. The biggest risk to an organization is not taking a risk-based approach to protecting people, property, and the environment.

"WHAT-IF"

"What-If" the EH&S community, business leaders, and workers adopted and practiced a risk-centric approach to their work and decision-making? "What-If" a tangible face could be put on "safe work" by seeing and acting on the risk in advance of mishaps? The authors have made the business case and provided us with a road map and resources to enable these two possibilities to become a reality.

"NEW VIEW" OF SAFETY

The authors outline and take us away from the "old view" of safety (the double negative – absence and harm) to the "new view" of safety (the double positive – presence and well-being). Organizational confidence in assessing operational risks will come when adopting and building upon the principles and learnings that come from these author's body of work. Defining what "safe" looks like must become the future of our profession.

THE VOICE OF THE WORKER

It is my personal belief that if the 50,000+ workers who have died on the job over the past decade could have a collective voice today, they would clearly tell us to seek out the risk in work and processes and act on them in advance of sustaining life-ending or life-altering events. It's all about the risk!

JUST DO IT!

It is my desire and wish every EH&S professional take on a risk-based approach in their work and job role and enroll others in the process as well. Those who adopt the thinking and approach found in "Risk Assessment: A Practical Guide to Assessing Operational Risks" will enjoy a rewarding and productive career as well as leave a lasting legacy where new designs, job tasks, and process risks are free from unacceptable risk.

DAVE WALLINE, CSP

Committee Chair, ASSE Risk Assessment Institute
2013–2015

LIST OF CONTRIBUTORS

Elyce Biddle, PhD., Department of Industrial and Management Systems Engineering, West Virginia University, Morgantown, WV, USA

Steven Hicks, CIH, CSP, School of Environmental, Physical & Applied Sciences, University of Central Missouri, Warrensburg, MO, USA

Bruce Hollcroft, CSP, ARM, CHMM, Risk Control Services, Hays Companies, Lake Oswego, OR, USA

Bruce K. Lyon, CSP, P.E., ARM, CHMM, Risk Control Services, Hays Companies, Kansas City, MO, USA

Fred A. Manuele, P.E., CSP, Hazards Limited, Arlington Heights, IL, USA

Georgi Popov, PhD., QEP, CMC, School of Environmental, Physical & Applied Sciences, University of Central Missouri, Warrensburg, MO, USA

Tsvetan Popov, PhD., Inspectorate Division, Organization for the Prohibition of Chemical Weapons (OPCW), The Hague, The Netherlands

Jim Whiting, Consultant, Risk at Workplaces Pty Ltd., Indooroopilly, QLD, Australia

John N. Zey, EdD, CIH, School of Environmental, Physical & Applied Sciences, University of Central Missouri, Warrensburg, MO, USA

Ying Zhen, EHS Manager, CIH, CSP, Belzona Inc., Miami, FL, USA

ABOUT THE COMPANION WEBSITES

This book is accompanied by a companion website:

https://centralspace.ucmo.edu/handle/123456789/407

The website includes:

- PowerPoint Slides
- Supplementary Materials

* Professors who wish to use the textbook for their classes can refer to the additional "instructor specific" files.

1

RISK ASSESSMENTS: THEIR SIGNIFICANCE AND THE ROLE OF THE SAFETY PROFESSIONAL

FRED A. MANUELE
Hazards Limited, Arlington Heights, IL, USA

1.1 OBJECTIVES

- Introduce developing trends in the use of operational risk assessments
- Provide a broad overview of standards and guidelines requiring risk assessment
- Emphasize the importance of risk assessment in the safety profession

1.2 INTRODUCTION

Throughout the world, there has been a proliferation of activity resulting in provisions being included in safety standards, guidelines, and operational risk management systems requiring or promoting that risk assessments be made. This trend has had an impact on the knowledge and skills that safety professionals are required to have in their employment. It will also provide career enhancement opportunities and job satisfaction for them.

Working with design and operations personnel to assess risks and to give counsel in the decision making to achieve acceptable risk levels adds an easily recognized value. Imaginative safety professionals will recognize this opportunity to be additionally perceived as members of the management team and increase their value to their organizations.

An addendum for this chapter consists of a list of standards, guidelines, and initiatives that require or promote making risk assessments. To avoid having the list become overly lengthy, 2005 was selected as the year to commence recordings. Although data is included for 35 subjects, it is more than likely the list is not complete.

To provide guidance for safety professionals on the trending throughout the world on requirements for risk assessments and recognition of the need for safety professionals to have risk assessment capability, this chapter will comment on the following:

Risk Assessment: A Practical Guide to Assessing Operational Risks, First Edition.
Edited by Georgi Popov, Bruce K. Lyon, and Bruce Hollcroft.
© 2016 John Wiley & Sons, Inc. Published 2016 by John Wiley & Sons, Inc.

- Activities initiated by the American Society of Safety Engineers (ASSE)
- A guideline that gives risk assessment high priority within an operational risk management system
- Selected standards and guidelines to demonstrate
  - the pace and import of recent activity;
  - the variations in content for risk assessments in the standards and guidelines;
  - specificity or, lack thereof, in their content.

There are similarities and differences in the approaches taken by the drafters of these standards and guidelines. Some are industry specific. Others apply across all industries. The message they give is clear: Safety professionals will be expected to have knowledge of a variety of hazard analysis and risk assessment methods and how to apply them.

1.3 WHAT IS A RISK ASSESSMENT?

Two definitions, taken from standards, are presented here. Their interrelationship is obvious. In the introduction for ANSI/ASSE Z690.3-2011 – *Risk Assessment Techniques* (nationally adopted from IEC/ISO 31010:2009), this is the guidance given.

> *Risk assessment is that part of risk management which provides a structured process that identifies how objectives may be affected, and analyzes the risk in term of consequences and their probabilities before deciding on whether further treatment is required. Risk assessment attempts to answer the following fundamental questions*:
>
> - *What can happen and why (by risk identification)?*
> - *What are the consequences?*
> - *What is the probability of their future occurrence?*
> - *Are there any factors that mitigate the consequence of the risk or that reduce the probability of the risk?*
> - *Is the level of risk tolerable or acceptable and does it require further treatment?*

ANSI Z690.3-2011 is an adoption of IEC/ISO 31010:2009. Additional comments will be made about this standard, and of Z590.3, later.

ANSI/ASSE Z590.3-2011 is the standard *for Prevention through Design: Guidelines for Addressing Occupational Hazards and Risks in Design and Redesign Processes.* This is its definition of risk assessment:

> **Risk Assessment**. *A process that commences with hazard identification and analysis, through which the probable severity of harm or damage is established, followed by an estimate of the probability of the incident or exposure occurring, and concluding with a statement of risk.*

As described in more detail later, risk assessment is a fundamental component of the risk management process and an essential core competency for safety professionals. Examples follow that provide clear indications of the rising importance given to risk assessment.

1.4 ACTIVITIES AT THE AMERICAN SOCIETY OF SAFETY ENGINEERS (ASSE)

Several officers of ASSE had recognized that requirements for risk assessment were more frequently included in safety-related standards and guidelines and that ASSE should provide its members with educational opportunities through which the necessary skills could be acquired. A presentation on the subject was made at the February 2013 board of directors meeting, the outcome for which was the creation of the Risk Assessment Institute.

A committee was formed and its members are working on literature, videos, webinars, and other materials that could be presented at chapter meetings and at conferences. The significance of this activity is that awareness had developed among the leaders of a technical organization with an international scope that its members would be well served if they were provided means to acquire risk assessment skills. This is an important step forward for the practice of safety. The Risk Assessment Institute website can be accessed at http://www.oshrisk.org/.

1.5 AN EXAMPLE OF A GUIDELINE THAT GIVES RISK ASSESSMENT DUE RECOGNITION

Entering "ExxonMobil's OIMS" into a search engine will lead to a brochure on ExxonMobil's Operations Integrity Management System. Within that brochure, there is a depiction of its OIMS arrangement. An adaptation of it follows in Figure 1.1.

Element 1 in this 11-point outline is what would be expected – management leadership, commitment, and accountability. But note that risk assessment and management follows item 1 immediately. That is an indication of the importance given to risk assessment within ExxonMobil operations.

And facilities design and construction follows risk assessment. In the design and redesign processes, risk assessments would be made continuously as needed.

Safety professionals should not be surprised if other companies produce similar outlines as greater recognition develops that the most effective and economical method to deal with hazards and risks is to address them in the design and redesign processes.

ExxonMobil's OIMS initiative pertains to all operational risks, including occupational, environmental, product, and public safety. An example is given later in this chapter of an activity that also combines occupational and environmental safety.

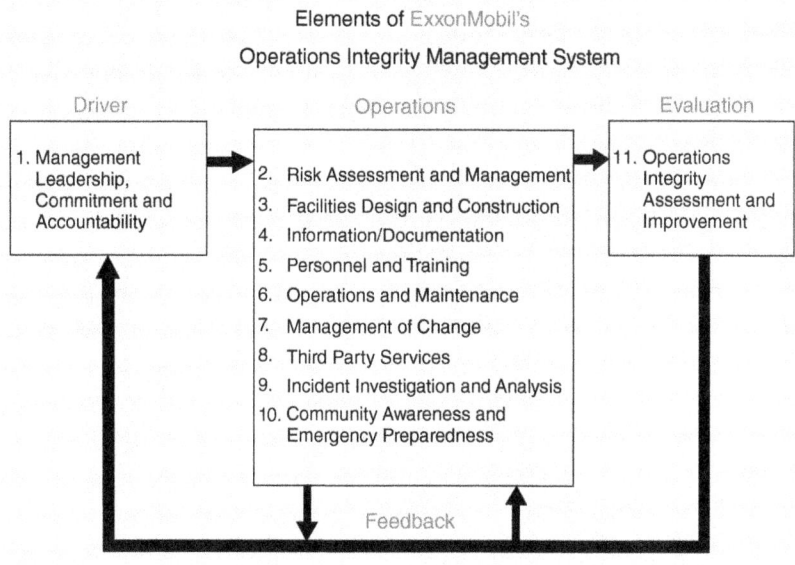

Figure 1.1   ExxonMobil's OIMS

1.6 ANSI/AIHA/ASSE Z10-2012: THE STANDARD FOR *OCCUPATIONAL HEALTH AND SAFETY MANAGEMENT SYSTEMS*

This standard continues to gain recognition as a sound outline for an occupational safety and health management system. In the first version of Z10, approved in 2005, management was required to have processes in place: "To identify and take appropriate steps to prevent or otherwise control hazards and reduce risks." While that verbiage may have implied that risk assessments were to be made, that is as close as the original version of the Z10 standard got to promoting risk assessments. A specific requirement that risk assessments be made was not included in the original standard.

Thinking changed. The 2012 version of Z10 has a "shall" provision on risk assessment at 5.1.1. It says:

The organization shall establish and implement a risk assessment process(es) appropriate to the nature of hazards and level of risk.

Safety professionals should recognize the significance of this revision. It reflects the awareness developed by the writers of the standard that making risk assessments should be an element within a safety and health management system.

1.7 ANSI/ASSE Z590.3-2011: PREVENTION THROUGH DESIGN: GUIDELINES FOR ADDRESSING OCCUPATIONAL HAZARDS AND RISKS IN DESIGN AND REDESIGN PROCESSES

This standard was approved by the American National Standards Institute on September 1, 2011. The core of Prevention through Design is risk assessment. Making risk assessments early in the design and redesign processes and continuously as needed throughout the life cycle of the system or product reduces the potential for incidents occurring. Logic in support of that premise follows:

1. Hazards and risks are most effectively and economically avoided, eliminated, or controlled in the design and redesign processes.
2. Hazard analysis is the most important safety process in that, if that fails, all other processes are likely to be ineffective (Johnson – p. 245).
3. Risk assessment should be the cornerstone of an operational risk management system.
4. If, through the hazard identification and analysis and risk assessment processes, specifications are developed that are applied in the procurement process so as to avoid bringing hazards and their accompanying risks into a workplace, the potential for injuries occurring is reduced greatly.
5. The entirety of purpose of those responsible for safety, regardless of their titles, is to manage their endeavors with respect to hazards so that the risks deriving from those hazards are acceptable.

The practice of safety is hazard based. Thus, Johnson wrote appropriately that hazard analysis is the most important safety process. Since all risks in an operational setting derive from hazards and since the intent of an operational risk management system is to achieve acceptable risk levels, it follows that risk assessment should be the cornerstone of an operational risk management system.

Figure 1.2 depicts the theoretical ideal. Prevention through Design is moved upstream in the design process. The intent is to have hazards and risks analyzed and dealt with in the Conceptual and Design steps. But, that requires unattainable perfection from the people involved. Hazards and risks will also

ADDRESSING OCCUPATIONAL HAZARDS AND RISKS IN DESIGN AND REDESIGN PROCESSES

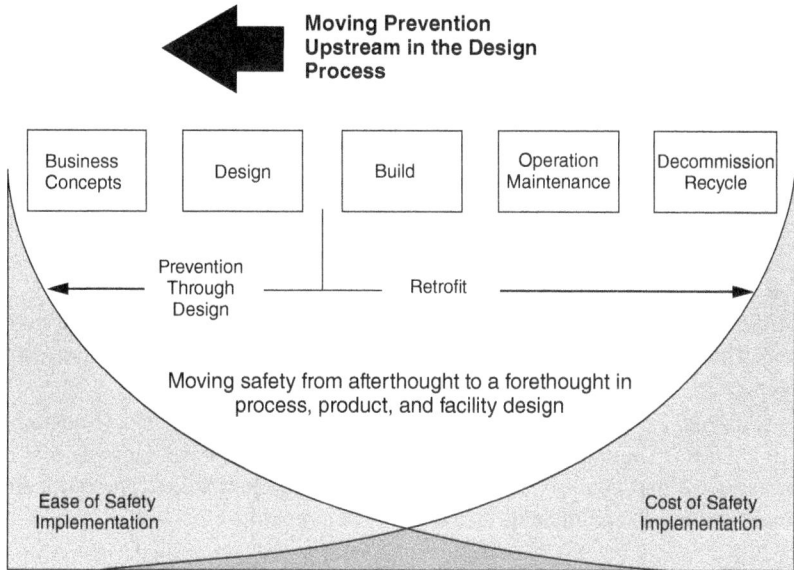

Figure 1.2 Prevention through Design. *Source:* With Permission from Christensen Consulting for Safety Excellence, Ltd

be identified in the Build and Operation and Maintenance steps for which redesign is necessary in a retrofitting process.

The hazard analysis and risk assessment process is the longest section in the Prevention through Design standard. First, an outline of the hazard analysis and risk assessment process is given. That is followed by the "how" for each of its elements. The outline follows:

- Select a risk assessment matrix.
- Establish the analysis parameters.
- Identify the hazards.
- Consider failure modes.
- Assess the severity of consequences.
- Determine occurrence probability.
- Define initial risk.
- Select and implement hazard avoidance, elimination, reduction, and control methods.
- Assess the residual risk.
- Risk acceptance decision making.
- Document the results.
- Follow up on actions taken.

For many hazards, the proper level of acceptable risk can be attained without bringing together complex teams of people. Safety and health professionals and design engineers having the experience and education can reach the proper conclusions on what constitutes acceptable risk. For the more complex risk situations, management should have processes in place to seek the counsel of experienced personnel who are particularly skilled in risk assessment for the category of the situation being considered.

Reaching group consensus is a highly desirable goal. Sometimes, for what an individual considers obvious, achieving consensus on acceptable risk levels is still desirable so that buy-in is obtained for the actions taken.

1.8 THE ANSI/ASSE Z690-2011 SERIES

Three American national standards that constitute a set should be of interest to safety generalists who want to become familiar with risk assessment techniques. The ASSE is the secretariat.

1. *ANSI/ASSE Z690.1-2011: Vocabulary for Risk Management* (National Adoption of ISO Guide 73:2009). This standard provides definitions of terms that, the originators hope, will be used in other standards.
2. *ANSI/ASSE Z690.2-2011: Risk Management Principles and Guidelines* (National Adoption of ISO 31000:2009). The intent of this standard is to provide a broad-range primer on risk management systems that could be applied in any type of organization. The requirement for risk assessments is introduced in Section 5.4: Risk Assessment.
3. *ANSI/ASSE Z 690.3-2011: Risk Assessment Techniques* (National Adoption of IEC/ISO 31010:2009). For safety generalists who want a ready reference on risk assessment concepts and methods, this standard is worth acquiring. It commences with a 15-page dissertation on risk assessment concepts and methods. Appendix A, in 5 pages, provides brief comparisons of 31 risk assessment techniques. Comments on the 31 techniques, covering Overview, Use, Inputs, Process, Strengths, and Limitations, are provided in Annex B, which covers 79 pages.

ANSI/ASSE Z 690.3-2011, particularly, is a valuable resource. A list of the 31 risk assessment techniques follows. Some could be applied only by experienced safety professionals who had knowledge of system safety concepts and techniques. Other techniques would be used by probabilistic specialists. But knowledge of a few of them will serve for a huge percentage of the needs of a safety generalist.

B01	Brainstorming
B02	Structured or semi-structured interviews
B03	Delphi
B04	Checklists
B05	Preliminary hazard analysis
B06	Hazard and operability studies
B07	Hazard analysis and critical control points
B08	Environmental risk assessment
B09	Structure – What-If analysis
B10	Scenario analysis
B11	Business impact analysis
B12	Root cause analysis
B13	Failure mode effect analysis
B14	Fault tree analysis
B15	Event tree analysis
B16	Cause–consequence analysis
B17	Cause-and-effect analysis
B18	Layer of protection analysis
B19	Decision tree

B20	Human reliability analysis
B21	Bow-Tie analysis
B22	Reliability-centered maintenance
B23	Sneak circuit analysis
B24	Markov analysis
B25	Monte Carlo simulation
B26	Bayesian statistics and Bayes nets
B27	FN curves
B28	Risk indices
B29	Consequence/probability matrix
B30	Cost–benefit analysis
B31	Multicriteria decision analysis

1.9 ANSI B11.0-2015: SAFETY OF MACHINERY. GENERAL SAFETY REQUIREMENTS AND RISK ASSESSMENT – A STANDARD OF MAJOR CONSEQUENCE

Because of the breadth of its coverage, ANSI B11.0 has major importance. This is its stated purpose: "This standard describes procedures for identifying hazards, assessing risks, and reducing risks to an acceptable level over the life cycle of machinery."

Note that its scope, as follows, has only one exclusion – portable hand tools: "This Type-A standard applies to new, existing, modified or rebuilt power driven machines, not portable by hand while working, that are used to process materials by cutting; forming; pressure; electrical, thermal or optical techniques; lamination; or a combination of these processes" (ANSI B11.0-2015).

The standard includes an explicit requirement that machinery suppliers, reconstructors, modifiers, and users achieve acceptable risk levels. ANSI B11.0 is the most comprehensive standard outlining the risk assessment process currently applicable to machinery for all of the operational categories just previously mentioned.

The foreword says "Prevention through Design or PtD is a recent term in the industry; the objectives of risk assessment, risk reduction and elimination of hazards as early as possible are integral to and not new to this standard." This objective is also taken from the foreword:

The objective of the B11 standards is to eliminate injuries to personnel from machinery or machinery systems by establishing requirements for the design, construction, reconstruction, modification, installation, set–up, operation and maintenance of machinery or machine systems. This standard should be used by suppliers and users, as well as by the appropriate authority having jurisdiction. Responsibilities have been assigned to the supplier (i.e., manufacturer, the reconstructor, and the modifier), the user, and the user personnel to implement this standard. This standard is not intended to replace good judgment and personal responsibility. Personnel skill, attitude, training and experience are safety factors that must be considered by the user.

The following sentence appears in the foreword of ANSI B11.0-2015.

This standard has been harmonized with international (ISO) and European (EN) standards by the introduction of hazard identification and risk assessment as the principal method for analyzing hazards to personnel to achieve a level of acceptable risk.

That statement presents an interesting and weighty concept. If all safety professionals accept that hazard identification and risk assessment are the first steps in preventing injuries to personnel, a major concept change in the practice of safety will have been achieved.

Adopting that premise takes the focus away from what have been called the unsafe acts of workers and redirects the emphasis to making risk assessments in the design and redesign of work systems and work methods to achieve and maintain acceptable risk levels. In this author's view, that is sound thinking.

1.10 EUROPEAN UNION: RISK ASSESSMENT

In August 2008, the European Union launched a two-year health and safety campaign focusing on risk assessment. Their bulletin (at http://osha.europa.eu/en/topics/riskassessment) says:

> *Risk assessment is the cornerstone of the European approach to prevent occupational accidents and ill health. If the risk assessment process – the start of the health and safety management approach – is not done well or not at all, the appropriate preventive measures are unlikely to be identified or put in place.*

The statement made by the European Union is seminal. Consider the significance of its campaign and its huge implications. The premise quoted recognizes the significance of risk assessment within an occupational safety and health management system, promotes the idea that the risk assessment process is where the management approach to safety should start, and specifically states that if risk assessment is not done well or not at all, the needed preventive measures are unlikely to be identified or taken.

The Europeans have been leaders in recognizing the importance of risk assessments and promoting their application. For example, employers in the United Kingdom are required to make risk assessments by law since 1999. Indications of other European involvement follow.

1.11 EN ISO 12100-2010: SAFETY OF MACHINERY. GENERAL PRINCIPLES FOR DESIGN. RISK ASSESSMENT, AND RISK REDUCTION

This standard, issued in 2010 by the International Organization for Standardization (ISO), has had an interesting history. It combines three previously issued ISO standards and replaces them. Note that "risk assessment and risk reduction" are included in the title. That is significant as it displays the status that risk assessment has attained in designing for the safety of machinery. The impact of this standard, worldwide, has been substantial.

ISO 12100-1 was titled *Safety of Machinery. Basic Concepts, General Principles for Design – Part 1*. It presented general design guidelines and required that risk assessments be made of machinery going into a workplace. ISO 12100-2 was titled *Safety of Machinery. Basic Concepts, General Principles for Design – Part 2: Technical principles*. Part 2 gave extensive detail on design specifications for the "safety of machinery." ISO 14121 was titled *Safety of Machinery. Principles of Risk Assessment*. It set forth the risk assessment concepts to be applied. EN ISO 12100-2010 combines these three standards and retains their content.

EN ISO 12100-2010 is truly an international standard and has had considerable influence worldwide. Its existence implies that a huge majority of countries agree on the principle that hazards should be identified and analyzed and their accompanying risks should be assessed in the design processes for machinery.

The EN that precedes ISO in the title indicates that the origins of the standard were in the European Community. Several standards that were applicable in the European Community that had titles commencing with the EN designation became ISO standards. Some of the relative EN standards were written in the 1990s.

The European Community standards have had considerable influence on manufacturers throughout the world. An example follows. Suppliers of products that are to go into a country that is a member of the European Community are required to place a "CE" mark on the products to indicate that all operable European Community directives have been met. Risk assessment provisions in EN ISO 12100-2010 are among those requirements.

1.12 ADDITIONAL EUROPEAN INFLUENCE

Other developments originating in Europe have also had a noteworthy impact throughout the world. Comments on one that has achieved worldwide significance follow.

BS OHSAS 18001: 2007 is the designation for a guideline titled *Occupational Health and Safety Management Systems – Requirements*. It is a British Standards Institution publication. In some contract situations, particularly in Asian countries, a bidder for a contract is required to establish that its safety management system has been "certified."

Among other things, the British Standards Institution has attained prominence as a certifying entity and 18001 is the base upon which certification is granted or withheld. In a 2007 revision of 18001, requirements for risk assessments became more explicit. The guidelines now say in 4.3.1:

The organization shall establish, implement and maintain a procedure(s) for the ongoing hazard identification, risk assessment, and determination of necessary controls.

As an indication of how broadly this guideline is known and used, Singapore adopted it fully as law in 2009.

1.13 MIL-STD-882E-2012. THE US DEPARTMENT OF DEFENSE STANDARD PRACTICE FOR SYSTEM SAFETY

The base document for the Standard Practice for System Safety, MIL-STD-882, was issued in 1969. It was a seminal document at that time and has continued to be an important reference.

MIL-STD 882 has had considerable influence on the development of hazard identification and analysis, risk assessment, risk elimination, and risk control concepts and methods. Much of the wording on risk assessments and hierarchies of control in safety standards and guidelines issued throughout the world relate to that in the several versions of 882.

Four revisions of 882 have been issued over a span of 43 years. As is said in the foreword for 882E, "This Standard is approved for use by all Military Departments and Defense Agencies within the Department of Defense." Certain contractors engaged by those departments and agencies are required to meet the requirements of the standard.

The 882 version was approved May 11, 2012. It is available at http://www.system-safety.org/. Scroll down and click on MIL-STD-882E in the right-hand column for a free copy. This author strongly recommends that safety professionals obtain a copy of this standard for informative purposes.

MIL-STD-882E extends the previous issue – 882D – considerably. For example, the 882D version, including addenda, had 26 numbered pages: the 882E version has 98 numbered pages. It replaces some of what was in 882C that was not included in 882D. In 882E:

- Achieving and maintaining acceptable risk levels dominates.
- Revisions were made in the system safety process that give additional emphasis to hazard analysis and risk assessment.
- The use of a risk assessment matrix is required.
- Noteworthy revisions are made in the design order of preference.

- Appropriate emphasis is given to managing high and serious risk levels.
- A major section is devoted to software and software assessments.

Excerpts follow, some of which are modified to avoid governmental terminology. Section 4 in 882E is titled "General Requirements." It sets forth the "requirements for an acceptable system safety effort." Section 4.3 and the following subsections of 4.3 outline and comment on the eight elements in the system safety process, as follows:

Element 1: Document the system safety approach. Describe the risk management effort and how the program is integrated into the overall business process.

Element 2: Identify and document the hazards. Hazards are identified through a systematic analysis process that includes the system hardware and software, system interfaces (to include human interfaces), and the intended use or application and operational environment.

Element 3: Assess and document risk. For each identified hazard, across all system modes, the mishap severity and probability are established in accord with the definitions given. A mishap risk assessment matrix is used to assess and display the risks.

Element 4: Identify and document risk mitigation measures. Potential risk mitigation(s) shall be identified, and the expected risk reduction(s) of the alternative(s) shall be estimated and documented. The goal should always be to eliminate the hazard as practicable.

When a hazard cannot be eliminated, the associated risk should always be reduced to the lowest practicable acceptable risk level within the constraints of cost, schedule, and performance by applying the following system safety design order of precedence in their order of effectiveness:

1. Eliminate hazards through design selection. Ideally, the hazard should be eliminated by selecting a design or material alternative that removes the hazard altogether.
2. Reduce risk through design alteration. If adopting an alternative design change or material to eliminate the hazard is not feasible, consider design changes that reduce the severity and/or the probability of the mishap potential caused by the hazard(s).
3. Incorporate engineered features or devices. If mitigation of the risk through design alteration is not feasible, reduce the severity or the probability of the mishap potential caused by the hazard(s) using engineered features or devices. In general, engineered features actively interrupt the mishap sequence and devices reduce the risk of a mishap.
4. Provide warning devices. If engineered features and devices are not feasible or do not adequately lower the severity or probability of the mishap potential caused by the hazard, include detection and warning systems to alert personnel to the presence of a hazardous condition or occurrence of a hazardous event.
5. Incorporate signage, procedures, training, and personal protective equipment (PPE). Where design alternatives, design changes, and engineered features and devices are not feasible and warning devices cannot adequately mitigate the severity or probability of the mishap potential caused by the hazard, incorporate signage, procedures, training, and PPE. Signage includes placards, labels, signs, and other visual graphics. Procedures and training should include appropriate warnings and cautions. Procedures may prescribe the use of PPE. For hazards assigned catastrophic or critical mishap severity categories, the use of signage, procedures, training, and PPE as the only risk reduction method should be avoided.

Element 5: Reduce risk. Mitigation measures are selected and implemented to achieve an acceptable risk level. Consider and evaluate the cost, feasibility, and effectiveness of candidate mitigation methods as a part of the overall operation process.

Element 6: Verify, validate, and document risk reduction. Verify the implementation and validate the effectiveness of all selected risk mitigation measures through appropriate analysis, testing, demonstration, or inspection. Document the verification and validation.

TABLE 1.1 Risk Assessment Matrix

Severity of Consequences				
Occurrence Probability	Catastrophic (1)	Critical (2)	Marginal (3)	Negligible (4)
Frequent (A)	High	High	Serious	Medium
Probable (B)	High	High	Serious	Medium
Occasional (C)	High	Serious	Medium	Low
Remote (D)	Serious	Medium	Medium	Low
Improbable (E)	Medium	Medium	Medium	Low
Eliminated (F)	This category is used only for identified hazards that are totally removed			

Source: MIL-STD-882E. Standard Practice for System Safety. Washington, DC: Department of Defense, 2012.

Element 7: Accept risk and document. Before exposing people, equipment, or the environment to known system-related hazards, the risks shall be accepted by the appropriate authority in accord with established acceptance authority levels. Definitions (in Tables and Matrices in this standard) shall be used to define the risks at the time of the acceptance decision, unless tailored alternative definitions and a tailored matrix are formally approved. The user representative shall be a part of this process and shall provide formal concurrence before all serious and high-risk acceptance decisions are made.

Element 8: Manage life-cycle risk. After the system is fielded, the system program office uses the system safety process to identify hazards, assess the risks, and maintain acceptable risk levels throughout the system's life cycle.

An instruction given in Element 7 says that "Definitions (in Tables and Matrices in this standard) shall be used to define the risks at the time of the acceptance decision, unless tailored alternative definitions and/or a tailored matrix are formally approved."

Table I presents severity categories. Table II contains probability levels. Table III in 882 is shown here as Table 1.1. It is a risk assessment matrix that combines the severity and probability categories and includes numerical and alpha indicators.

Numerical and alpha indicators are the base for expressing assessed risks in a risk assessment code (RAC), which is a combination of one severity category and one probability level. For example, a RAC of 1A is the combination of a catastrophic severity category and a frequent probability level.

For emphasis, it is said again that MIL-STD 882E is an excellent educational and resource document. Its base is hazard identification and analysis and risk assessment.

1.14 CERTAIN GOVERNMENTAL VIEWS

In a July 19, 2010, letter to the OSHA staff, Assistant Secretary David Michaels wrote on several subjects, one of which follows: "Ensuring that American workplaces are safe will require a paradigm shift, with employers going beyond simply attempting to meet OSHA standards, to implementing risk-based workplace injury and illness prevention programs" (Michaels, 2010).

If elements in injury and illness prevention programs are to be risk based, activity will be necessary to identify and assess the risks. That starts with hazard identification and analysis and, then, takes the next step to establish the risk level.

OSHA has not shown that it is adopting the concept of risk-based decision making. This statement by Dr Michaels is noteworthy because it demonstrates that the head of a major governmental entity involved in occupational safety and health has recognized that injury and illness prevention programs should be risk based. As will be seen, personnel in other governmental agencies have reached similar conclusions.

In the December 8, 2010, Federal Register, the Federal Railroad Administration issued an advance notice of proposed rulemaking for certain railroads to have a Risk Reduction Program. The Federal Register entry said "It is proposed that the Risk Reduction Program be supported by a risk analysis and a Risk Reduction Plan" (Federal Railroad Administration Risk Reduction Program, 2010). Enter "Federal Railroad Administration Risk Reduction Program" into a search engine and the following appears.

1.14.1 Risk Reduction Program

The primary mission of the Risk Reduction Program Division is ensuring the safety of the nation's railroads by evaluating safety risks and managing those risks in order to reduce the numbers and rates of accidents, incidents, injuries, and fatalities.

Our mission is accomplished by:

- Identifying, collecting, and analyzing precursor accident data to identify risks
- Developing voluntary pilot programs in cooperation with stakeholders that are designed to mitigate identified and potential risks
- Propagating and institutionalizing best practices and lessons learned to the entire rail industry
- Providing analytical support, data, and recommendations needed by stakeholders to develop strategies, plans, and processes to improve safety and promote positive organizational change
- Developing and enforcing regulations promulgated in response to the Rail Safety Improvement Act of 2008.

On October 15, 2010, the Bureau of Ocean Energy Management, Regulation and Enforcement (BOEMRE) published in the Federal Register (75 FR 63610) the Final Rule for 30 CFR Part 250 Subpart S – Safety and Environmental Management Systems (BOEMRE, 2010).

This Final Rule incorporates by reference, and makes mandatory, the American Petroleum Institute's Recommended Practice for Development of a Safety and Environmental Management Program for Offshore Operations and Facilities (API RP 75), Third Edition, May 2004, reaffirmed May 2008. BOEMRE mandated that by November 15, 2011, all operators and lessees working in the Gulf of Mexico had to submit a comprehensive SEMS plan to the regulator. This plan must address the 13 elements of API RP 75, the third of which is hazard analysis.

This development is of particular interest for two reasons. Operators and lessees affected are required by regulation to make hazard analyses (the first step in making a risk assessment). Also, the plan required is a combination that includes occupational safety, public safety, and environmental safety in one instrument. That combination deserves continual observation. Several safety directors were polled by this author to determine what proportion of the safety professionals at their locations has responsibilities for both occupational safety and environmental concerns. The range was from 50% to 90%.

Risk assessments have been made for many years in the branches of the military, the National Aeronautics and Space Administration, some chemical operations, the atomic energy field, pharmaceutical companies operating under the rules of the Food and Drug Administration, research activities pertaining to public health, traffic control studies, and other fields.

That additional Federal governmental entities have become risk conscious and are requiring that risk assessments be made is an indication of the trend.

1.15 CANADA

CSA Standard Z1000-2006 is titled *Occupational Health and Safety Management*. It was issued in the year following the first edition of Z10 and has a close relationship with respect to the content and order in the American standard.

Section 4.3.4 reads as follows: The organization shall establish and maintain a process to identify and assess hazards and risks on an ongoing basis. The results of this process shall be used to set objectives and targets and to develop preventive and protective methods. (CSA is the designation for the Canadian Standards Association.)

The excerpt above is all that is said in the standard about hazard analysis and risk assessment. The subject is dealt with further in Annex A, which is informative. But the intent of the hazard analysis and risk assessment provision is amplified in the "shall" provision of the standard at Section 4.4.7, "Management of Change."

The organization shall establish and maintain procedures to identify, assess, and eliminate or control occupational health and safety hazards and risks associated with

1. new processes or operations at the design stage
2. significant changes to its work procedures, equipment, or organizational structure and so on.

In September 2012, CSA Z1002-12: *Occupational Health and Safety – Hazard Identification and Elimination and Risk Assessment and Control* was issued. This is a major undertaking. It supports the purpose of Section 4.3.4 in Z1000-2006.

The standard relates entirely to hazards and risks in the workplace. Its issuance is another indication of the trend throughout the world whereby organizations are encouraged to have processes in place to identify and analyze hazards, to assess their accompanying risks, and to achieve acceptable risk levels.

1.16 FIRE PROTECTION

There are four entries in Addendum A for this chapter pertaining to activities of the National Fire Protection Association (NFPA) and the Society of Fire Protection Engineers (SFPE).

In 2007, NFPA issued "Guidance Document for Incorporating Risk Concepts into NFPA Codes and Standards." This is an impressive, thought-provoking, risk assessment-related document that will have a long-term effect in the fire protection field. It is available at http://www.nfpa.org/assets/files/PDF/Research/Risk-Based_Codes_and_Stds.pdf.

As an example of how risk concepts are being incorporated into NFPA standards – the 2012 edition of NFPA 70E, *Standard for Electrical Safety in the Workplace* has a new section on risk assessment (NFPA 70E, 2012).

SFPE developed an interesting course titled "Introduction to Fire Risk Assessment," which is available on the Internet (no publication date is shown, but it probably was 2006). A paraphrased and brief version of what is said about the course on the Internet follows.

This 5 h equivalent course is presented free of charge by the Society of Fire Protection Engineers. Although the course was developed primarily for fire service and fire prevention officers, it may be of value to engineers and students who would like to understand fire risk assessment. The full course consists of 19 lecture sessions each of which can be viewed in about 15 min.

This course is largely generic and deserves a look. Additional information, including the titles of the lecture sessions and how to access them, can be found at http://www.sfpe.org/SharpenYourExpertise/Education/SFPEOnlineLearning/FireRiskAssessment.aspx.

In 2006, SFPE also issued the *Engineering Guide to Fire Risk Assessment*. This is a technical book that would be of particular interest to engineers. Nevertheless, its issuance demonstrates leadership by SFPE with respect to risk assessment.

1.17 DEVELOPMENTS IN AVIATION GROUND SAFETY

One of the most interesting innovations regarding hazard analysis and risk assessment can be found in the *Safety Handbook: Aviation Ground Operation* developed by the International Air Transport

TABLE 1.2 The Risk Management Process

1. Identify the hazard
2. Assess the risk
3. Analyze risk control measures
4. Make control decisions
5. Implement risk controls
6. Supervise and review

Source: Safety Handbook: Aviation Ground Operation, Sixth Edition. Itasca, IL: National Safety Council, 2007.

TABLE 1.3 Hazard Analysis and Risk Assessment Methodologies

1. Operations analysis: Purpose – To understand the flow of events
2. Hazard analysis: Purpose – To get a quick survey of all phases of an operation. In low hazard situations, the preliminary hazard analysis may be the final hazard identification tool
3. "What-If" analysis: Purpose – To capture the input of personnel in a brainstorming-like environment
4. Scenario process tool: Purpose – To use imagination and visualizations to capture unusual hazards
5. Change analysis: Purpose – To detect the hazard implications of both planned and unplanned changes

Source: Safety Handbook: Aviation Ground Operation, Sixth Edition. Itasca, IL: National Safety Council, 2007.

Section of the National Safety Council. The Air Transport Section is truly international, having representation from all of the populated continents.

A sixth edition was published in July 2007. Chapter 2 is titled "Risk Management" (National Safety Council, 2007). The following text is taken from that chapter.

Risk management takes aviation safety to the next level. It is a six-step logic-based approach to making calculated decisions on human, material, and environmental factors before, during, and after operations.

Risk management enables senior leaders, functional managers, supervisors, and others to maximize opportunities for success while minimizing risks. Failure to successfully implement a risk management process will have a financial, legal, and social impact (p. 9).

The air transport group has outlined a way of thinking about and dealing with hazards and risks, applying a logical and sequential methodology. They have developed a "process to detect, assess, and control risk." The captions in their "six-step logic-based commonsense approach" are shown in the handbook's Table 1 (p. 11). The process is shown here as Table 1.2.

Discussions of each step in the text are extensive. Comments will be made here on the first two, only. For the first step – identify the hazards – the following hazard analysis and risk assessment methodologies, as in the handbook's Table 2, are discussed (p. 10) as Table 1.3.

For Step 2 – assess the risks – the text says: The assessment is the application of quantitative or qualitative measures to determine the level of risk associated with a specific hazard. This process defines the probability and severity of an undesirable event that could result from the hazard. The risk assessment matrix is a very useful tool in categorizing the effects of probability and severity as they relate to risk levels (p. 12).

This *Safety Handbook: Aviation Ground Operation* is a good, thought-provoking, not overly complex resource document. It is an example of what a trade group could do as a service to its members.

1.18 OSHA REQUIREMENTS

OSHA's Rule For Process Safety Management Of Highly Hazardous Chemicals, 1910.119, issued in 1992, applies to employers at about 50,000 locations, many of which are not considered chemical companies. With respect to requirements for hazard analyses being included in standards, this OSHA standard merits a review by safety practitioners. The standard requires that:

> The employer shall perform an initial hazard analysis (hazard evaluation) on processes covered by this standard. The process hazard analysis shall be appropriate to the complexity of the process and shall identify, evaluate, and control the hazards involved in the process. The employer shall use one or more of the following methodologies that are appropriate to determine and evaluate the hazards of the process being analyzed:
>
> - What-If;
> - Checklist;
> - What-If/Checklist;
> - Hazard and Operability Study (HAZOP);
> - Failure Mode and Effects Analysis (FMEA);
> - Fault Tree Analysis; or
> - An appropriate equivalent methodology.

Although affected employers are to make hazard analyses, the methodologies previously listed are risk assessment techniques. This author's recollection is that commenters on the standard prior to its promulgation expressed concern over having to use probability data – of which there is little that is statistically sound. OSHA responded favorably. This appears in the preamble to the standard.

OSHA has modified the paragraph (editorial note – paragraph on consequence analysis) to indicate that it did not intend employers to conduct probabilistic risk assessments to satisfy the requirement to perform a consequence analysis.

However, all risks are not equal. Some require attention prior to others. And managements do assess and prioritize risks in their decision making when determining which resources are to be allocated for individual projects.

1.19 EPA REQUIREMENTS

The Environmental Protection Agency (EPA) and OSHA have different legal authorities with respect to accidental releases of harmful substances. The concerns at EPA center on off-site consequences: that is, harm to the public and the environment. At OSHA, the legal authority pertains to on-site consequences.

On August 19, 1996, EPA issued Rule 40 CFR Part 68, *Risk Management Programs for Chemical Accidental Release Prevention* (EPA, 1996). Risk Management Plans required of location managements by the rule were due by June 21, 1999. Although the provisions of the rule are extensive, only the specifications for hazard analyses will be addressed here.

Processes subject to this rule are divided into three groups, labeled by EPA as Programs 1, 2, and 3. Program levels relate to the quantities and extent of exposure to toxic and flammable chemicals. For locations qualifying for program levels 1 and 2, those with lesser exposure, EPA will accept hazard reviews done by qualified personnel using suitable checklists.

Hazard reviews must be documented and show that problems have been addressed. In its literature, EPA comments on the desirability of using the "What-If" hazard identification and analysis process. EPA also proposes the use of more involved analytical techniques if findings suggest that to be desirable.

Hazard review requirement for program level 3 locations are more specific and extensive. But those locations that are compliant with the OSHA rule for Process Safety Management of Highly Hazardous Chemicals will need to do little new, although they do need to extend their hazard analyses to consider the probability of harm to the public or to the environment. As with OSHA, a team must complete the process hazard analyses required by EPA. One member of the team, at least, is to have experience with the process.

For American industry, EPA has obviously extended knowledge and skill requirements regarding hazard analysis techniques.

1.20   THE CHEMICAL INDUSTRY: THE EXTENSIVE BODY OF INFORMATION

Completing hazard analyses was a common practice in the chemical industry many years before requirements for them were established by OSHA and EPA. The body of information in the chemical industry on hazard analysis is extensive. But reference will be made here to only one publication because of its particular significant.

The Center for Chemical Process Safety is a part of the American Institute of Chemical Engineers. One of its books is titled *Guidelines for Hazard Evaluation Procedures, Second Edition with Worked Examples* (Center for Chemical Process Safety of the American Institute of Chemical Engineers, 1992). Publication of the text by a chemically oriented group should not dissuade those who want an education in the following evaluation techniques. Their descriptions are generic.

- Safety review
- Checklist Analysis
- Relative ranking
- Preliminary hazard analysis
- What-If Analysis
- What-If/Checklist Analysis
- Hazard and operability analysis
- Fault tree analysis
- Event tree analysis
- Cause–consequence analysis
- Human reliability analysis.

These techniques are dealt with broadly in the *Guidelines* within chapters titled "Overview of Hazard Evaluation Techniques" and "Using Hazard Evaluation Techniques."

1.21   CONCLUSION

The message is clear. Including provisions requiring hazard analyses and risk assessments in safety standards and guidelines is becoming ordinary. It is logical to assume that this trending will continue and that safety professionals will be expected to have the knowledge and skill necessary to give counsel on applying those provisions. Emphasis on practical applications of risk assessments is key and is the primary focus of this text.

REVIEW QUESTIONS

1. Name three new initiatives in the United States that promote and support the practice of operational risk assessment.

REFERENCES

2. How does the United States compare to other parts of the world regarding the use of risk assessments?
3. What constitutes a risk assessment? Explain the components.
4. In 2010, ANSI adopted a series of standards on risk management from ISO, which are considered fundamental to the risk profession. List these standards and describe their application.
5. Provide five other standards that have requirements for risk assessment that have been established since 2005.
6. What standard addresses risk in the life cycle of a system? Provide the stages of the life cycle of a system or product.
7. List two compliance standards that have requirements for performing hazard analyses.

REFERENCES

ANSI/AIHA Z10-2012. *Occupational Health and Safety Management Systems.* Fairfax, VA: The American Industrial Hygiene Association, 2012. ASSE is now the Secretariat. Available at: https://www.asse.org/cartpage.php?link=z10_2005 (accessed December 16, 2015).

ANSI/ASSE Z690.3. *Risk Assessment Techniques.* Des Plaines, IL: American Society of Safety Engineers, 2011.

ANSI B11.0-2015. *Safety of Machinery. General Safety Requirements and Risk Assessments.* Houston, TX: B11 Standards, Inc., 2015.

API RP 75. 2008. *Recommended Practice for Development of a Safety and Environmental Management Program for Offshore Operations and Facilities,* 3rd ed. American Petroleum Institute. Go to http://www.techstreet.com/standards/api/rp_75_r2008_?product_id=1157045 (accessed December 16, 2015).

BOEMRE – Bureau of Ocean Energy Management Regulation and Enforcement. 2010. Requirements for SEMS: Safety and Environmental Management Systems. At www.boemre.gov/semp (accessed December 16, 2015).

BS OHSAS 18001:2007. *Occupational Health and Safety Management Systems—Requirements.* London, UK: British Standards Institution (BSI), 2007.

Center for Chemical Process Safety of the American Institute of Chemical Engineers. *Guidelines for Hazard Evaluation Procedures,* 2nd ed. *With Worked Examples.* New York: Center for Chemical Process Safety of the American Institute of Chemical Engineers, 1992.

CSA Standard Z1000-06. *Occupational Health and Safety Management.* Mississauga, ON, Canada: Canadian Standards Association, 2006.

CSA Standard Z1002-12. *Occupational Health and Safety – Hazard Identification and Elimination and Risk Assessment and Control.* Mississauga, ON, Canada: Canadian Standards Association, 2012.

En ISO 12100-2010. *Safety of Machinery. General Principles for Design. Risk Assessment and Risk Reduction.* Geneva, Switzerland: International Organization for Standardization, 2010.

Environmental Protection Agency. Risk Management Programs for Chemical Accidental Release Prevention, *40 CFR Part 68.* Washington, DC: Environmental Protection Agency, 1996.

EPA. 1996. *Risk Management Programs for Chemical Accidental Release Prevention.* http://www.epa.gov/rmp.

European Union. 2008. *Risk Assessment,* http://osha.europa.eu/en/topics/riskassessment (accessed December 16, 2015).

Federal Railroad Administration Risk Reduction Program. 2010. Enter the title into a search engine to bring up information. Or go to http://www.fra.dot.gov/Page/P0049 (accessed December 16, 2015).

ISO 12100-1. *Safety of Machinery – Basic Concepts, General Principles for Design; Part 1. Basic Terminology, Methodology.* Geneva, Switzerland: International Organization for Standardization, 2003.

ISO 12100-2. *Safety of Machinery – Basic Concepts, General Principles for Design; Part 2. Technical Principles and Specifications.* Geneva, Switzerland: International Organization for Standardization, 2003.

ISO 14121. *Safety of Machinery – Principles for Risk Assessment*. Geneva, Switzerland: International Organization for Standardization, 1999.

Michaels, D. 2010. Assistant Secretary at OSHA – July 19, 2010 letter to the OSHA. At http://orc-dc.com/?q=node/3649 (accessed December 16, 2015).

MIL-STD-882E. *Standard Practice for System Safety*. Washington, DC: Department of Defense, 2012. Available at http://www.system-safety.org/. (accessed December 16, 2015) Scroll down and click on MIL-STD-882E in the right hand column for a free copy.

National Fire Protection Association. 2007. *Guidance Document for Incorporating Risk Concepts into NFPA Codes and Standards*. Available at http://www.nfpa.org/research/fire-protection-research-foundation/projects-reports-and-proceedings/current-projects/guidance-document-for-incorporating-resiliency-concepts-into-nfpa-codes-and-standards.

National Safety Council. *Safety Handbook: Aviation Ground Operation*, 6th ed. Itasca, IL: National Safety Council, 2007.

NFPA 70E. 2012. Standard for Electrical Safety in the Workplace, 2012 Edition. Begin inquiry at http://us.yhs4.search.yahoo.com/yhs/search;_ylt=A0oG7psAQy1R2CoAgLVXNyoA?p=32.%09NFPA%2070E%2C%20Standard%20for%20Electrical%20Safety%20in%20the%20Workplace%2C%202112%20Edition&fr2=sb-top&hspart=att&hsimp=yhs-att_001&type=att_lego_portal_home&type_param=att_lego_portal_home (accessed December 16, 2015).

Occupational Safety and Health Administration. *OSHA's Rule for Process Safety Management of Highly Hazardous Chemicals, 1910.119*. Washington, DC: Department of Labor, Occupational Safety and Health Administration, 1992.

Society of Fire Protection Engineers. 2006a. *Engineering Guide to Fire Assessment*. Available at http://findpdf.net/documents/SFPE-engineering-guide-to-fire-risk-assessment.html (accessed December 16, 2015).

Society of Fire Protection Engineers. 2006b *Introduction to Fire Risk Assessment*. Enter the title in a search engine for information on the course. Or go to http://www.sfpe.org/SharpenYourExpertise/Education/SFPEOnlineLearning/FireRiskAssessment.aspx (accessed December 16, 2015).

APPENDIX 1.A

A LIST OF STANDARDS, GUIDELINES, AND INITIATIVES THAT REQUIRE OR PROMOTE MAKING RISK ASSESSMENTS: COMMENCING WITH YEAR 2005

1. ANSI/AIHA Z10-2005: *Occupational Health and Safety Management Systems standard*.
   Z10 sets a benchmark provision requiring that processes be in place: to identify and take appropriate steps to prevent or otherwise control hazards and reduce risks associated with new processes or operations at the design stage.
2. *Guidance on the Principles of Safe Design for Work*. Australian Safety and Compensation Council, Australian Government, 2006.
3. In 2006, NIOSH announced a major national initiative on Prevention through Design, the core of which is risk assessment.
4. SFPE – *Engineering Guide to Fire Assessment*, 2006. This is a technical book that would be of particular interest to engineers.
5. SFPE – *Introduction to Fire Risk Assessment [Believe release date was 2006]*.
   Enter the title in a search engine for course modules on fire risk assessment.
6. CSA Z1000-2006, *Occupational Health and Safety Management Standard* issued by the Canadian Standards Association.
7. The Industrial Safety and Health Act of Japan was revised, effective in April 2006. It stipulates – without penalty – that employers should make efforts to implement risk assessment.

APPENDIX   19

8. ISO 14121-1: *Safety of Machinery. Principles for Risk Assessment.* 2007.
9. In 2007, the OSHA Alliance Construction Roundtable developed a video training program titled "Design for Construction Safety."
10. NFPA: *Guidance Document for Incorporating Risk Concepts into NFPA Codes and Standards*, 2007.
11. BS OHSAS 18001: 2007, *Occupational Health and Safety Management Systems – Requirements*, a British Standards Institution publication.

    In the 2007 revision, requirements for risk assessments are more explicit. The guidelines now say: "The organization shall establish, implement and maintain a procedure(s) for the ongoing hazard identification, risk assessment, and determination of necessary controls."
12. Nano Risk Framework, issued in June 2007 through the combined efforts of the Environmental Defense Fund and DuPont, includes a six-step guidance framework for "the responsible development of nanoscale materials."

    They are as follows: (1) Describe the material and its application; (2) profile life cycle(s); (3) evaluate risks; (4) assess risk management; (5) decide, document, and act; and (6) review and adapt.
13. ANSI B11.TR7 2007: *ANSI Technical Report for Machines – A Guide on Integrating Safety and Lean Manufacturing Principles in the Use of Machinery.*
14. China's State Administration of Work Safety published provisional regulations on risk assessment in 2008.
15. The Health and Safety Executive in the United Kingdom issued five steps to risk assessment in 2008.
16. All employers in the United Kingdom must conduct a risk assessment. An HSE bulletin says: "The law does not expect you to eliminate all risk, but you are required to protect people as far as is 'reasonably practicable'."
17. In August 2008, the European Union launched a two-year health and safety campaign focusing on risk assessment. Their bulletin says:

    Risk assessment is the cornerstone of the European approach to prevent occupational accidents and ill health. If the risk assessment process – the start of the health and safety management approach – is not done well or not at all, the appropriate preventive measures are unlikely to be identified or put in place.
18. *Machine Safety: Prevention of Mechanical Hazards*: issued by the Institute for Research for Safety and Security at Work and the Commission for Safety and Security at Work in Quebec, 2009.
19. ASSE Technical Report Z790.001: *Prevention through Design: Guidelines for Addressing Occupational Risks in the Design and Redesign Processes*, 2009.
20. Singapore Standard SS 506. *Occupational Safety and Health (OSH) Management Systems – Part 1: Requirements*, 2009.
21. ANSI-ITAA GEIA-STD-0010-2009: *Standard Best Practices for System Safety Program Development and Execution.*

    Foreword: Coupled with use of the system safety risk mitigation order of precedence, functional hazard analysis lets a program identify early in the life cycle those risks which can be eliminated by design, and those which must undergo mitigation by other controls in order to reduce risk to an acceptable level.
22. ExxonMobil issued its *Operations Integrity Management System* in July 2009. It pertains to safety, health, the environment, and product safety. The first 4 of 11 elements in this management system are:

1. Management leadership, commitment, and accountability
2. Risk assessment and management
3. Facilities design and construction
4. Information and documentation

23. ISO/IEC 31000: 2009 – *Risk Management – Principles and Guidelines* and ISO/IEC 31010: 2009 – *Risk Assessment Techniques.*

24. EN ISO 12100-2010: *Safety of Machinery. General Principles for Design. Risk Assessment and Risk Reduction.*

    This standard combines three previously issued ISO standards (including item 8 in this listing) and replaces them. Risk assessments are explicitly required.

25. In a July 19, 2010, letter to the OSHA staff, Assistant Secretary David Michaels wrote on several subjects, one of which follows:

    Ensuring that American workplaces are safe will require a paradigm shift, with employers going beyond simply attempting to meet OSHA standards, to implementing risk-based workplace injury and illness prevention programs.

26. ANSI B11.0: *Safety of Machinery. General Safety Requirements and Risk Assessments*, December, 2010.

    Purpose: This standard describes procedures for identifying hazards, assessing risks, and reducing risks to an acceptable level over the life cycle of machinery.

27. In the December 8, 2010, Federal Register, the Federal Railroad Administration issued an advance notice of proposed rulemaking for certain railroads to have a risk reduction program.

    It is proposed that the Risk Reduction Program be supported by a risk analysis and a Risk Reduction Plan.

28. ANSI/PMMI B155.1 – March 2, 2011: *Safety Requirements for Packaging Machinery and Packaging-Related Converting Machinery.*

    Foreword: This standard has been harmonized with international (ISO) and European (EN) standards by the introduction of hazard identification and risk assessment as the principal method for analyzing hazards to personnel to achieve a level of acceptable risk.

29. Pipeline and Hazardous Materials Safety Administration, March 11, 2011.

    Hazardous Materials Regulations are to be modified to require that risk assessments be made of loading and unloading operations.

30. *OSH Management System: A Tool for Continual Improvement* – issued by the International Labour Organization, Geneva. April 28, 2011.

    Hazard and risk assessments have to be carried out to identify what could cause harm to workers as well as property so that appropriate preventive and protective measures can be developed and implemented.

31. ANSI-ASSE Z590.3 – September 1, 2011. *Prevention through Design: Guidelines for Addressing Occupational Hazards and Risks in Design and Redesign Processes.*

    The core of Z590.3 is risk assessment to be performed as a continuum in the design and redesign processes.

32. NFPA 70E, *Standard for Electrical Safety in the Workplace*, 2112 Edition, has a new section on risk assessment.

33. MIL-STD 882E. *US Department of Defense Standard Practice for System Safety*, approved May 11, 2012. It is available at http://www.system-safety.org/links/. Click on Home. Click on 882E for a free download.

34. ANSI/AIHA Z10-2012: *Occupational Health and Safety Management Systems* standard.

APPENDIX

The second version of Z10, approved in June 2012, now contains a specific requirement for a risk assessment process to be in place.

5.1.1 Risk Assessment.

The organization shall establish and implement a risk assessment process(es) appropriate to the nature of hazards and level of risk.

35. CSA Z1002-12: *Occupational Health and Safety – Hazard Identification and Elimination and Risk Assessment and Control.* Canadian Standards Association. September 2012.

2

RISK ASSESSMENT STANDARDS AND DEFINITIONS

BRUCE HOLLCROFT
Risk Control Services, Hays Companies, Lake Oswego, OR, USA

BRUCE K. LYON
Risk Control Services, Hays Companies, Kansas City, MO, USA

2.1 OBJECTIVES

- Introduce the need for risk assessments
- Review compliance standards that require hazard determination and analysis
- Review consensus standards that prescribe risk assessment
- Define key risk assessment terminology

2.2 INTRODUCTION

The single most important component within operational risk management is the process of identification, analysis, and assessment of risk. Without this element, there is little hope of managing risk. The purpose of risk assessment is to identify and analyze hazards and assess their risks so that management can properly avoid, mitigate, and manage operational risks. This is no small endeavor. Risk assessment is a discipline that must be learned, practiced, and executed on a consistent basis to effectively anticipate, recognize, and manage risks. As further explained in this text, risk assessment is an essential component of an effective operational risk management system used to identify, analyze, and evaluate risks in the workplace.

In Chapter 1, Manuele describes the growing significance of risk assessments and development of standards, guidelines, and practices requiring or advocating formal risk assessments. The focus of Chapter 2 is to provide the reader with an understanding of key standards and guidelines prescribing operational risk assessment for organizations and their importance in shaping the safety, health, and environmental professional's role in occupational safety and health. Key terms and definitions for

---

Risk Assessment: A Practical Guide to Assessing Operational Risks, First Edition.
Edited by Georgi Popov, Bruce K. Lyon, and Bruce Hollcroft.
© 2016 John Wiley & Sons, Inc. Published 2016 by John Wiley & Sons, Inc.

risk assessment and management are reviewed and explained in context of the safety, health, and environmental professional's responsibilities.

2.3 THE NEED FOR RISK ASSESSMENTS

Fundamentally, the safety profession has long recognized the importance of managing hazards and risk through proper identification, assessment, and control. This is reflected in the standards, guidelines, and best practices found in various industries and business segments such as environmental, financial, medical, and nuclear, among others that require or recommend risk assessment. Some of these business segments are covered by regulatory compliance requirements, specific to their industry.

In the first chapter, an addendum can be found listing 35 standards, guidelines, and initiatives that have been established since 2005, which require or promote the use of formal risk assessments. A review of these reveals that there are similarities and differences in the approaches taken by the drafters of these standards and guidelines. Some are industry specific, while others apply across all industries. But as Manuele declares the message they give is clear: Risk assessment is important, and safety professionals are expected to be knowledgeable and skilled in their application.

2.4 KEY STANDARDS REQUIRING RISK ASSESSMENTS

In Europe, Australia, New Zealand, Canada, and other parts of the world, occupational risk assessments have become a common practice. This is largely due to the fact that these countries have national standards requiring risk assessments in the workplace. For instance, in the United Kingdom, the Health and Safety Executive (H&SE) has legally required all employers with five or more employees to perform risk assessments since 1999. Similar requirements are found in other countries such as Australia and New Zealand.

However, few risk assessments are mandated in the United States, with the exception of the Occupational Safety and Health Administration (OSHA) *29 CFR 1910.119 Process Safety Management of Highly Hazardous Chemicals* standard and the Environmental Protection Agency (EPA) *40 CFR Part 68, Risk Management Plan* standard. Other OSHA standards contain requirements for hazard determination and interpretations that suggest analysis and in some cases, "assessment" such as *1910.132, Personal Protective Equipment* and hazard assessments; *1910.178, Control of Hazardous Energy Sources*; and *1910.146, Permit-required Confined Space Entry*.

2.5 OSHA COMPLIANCE AND RISK ASSESSMENTS

In the United States, OSHA is the safety compliance standard minimum for employers. In 1970, the OSH Act was formed, making employers responsible for providing a safe and healthful workplace. OSHA's stated mission is to assure safe and healthful workplaces by setting and enforcing standards and by providing training, outreach, education, and assistance. The following description is provided by the OSHA website:

> *The OSH Act covers most private sector employers and their workers, in addition to some public sector employers and workers in the 50 states and certain territories and jurisdictions under federal authority. OSHA covers most private sector employers and their workers in all 50 states, the District of Columbia, and other U.S. jurisdictions either directly through Federal OSHA or through an OSHA-approved state program. Workers at state and local government agencies are not covered by Federal OSHA, but have OSH Act protections if they work in those states that have an OSHA-approved state program. Four additional*

OSHA COMPLIANCE AND RISK ASSESSMENTS 25

states and one U.S. territory have OSHA-approved plans that cover public sector workers only. State-run health and safety programs must be at least as effective as the Federal OSHA program. OSHA standards are rules that describe the methods that employers must use to protect their employees from hazards.

There are OSHA standards for Construction work, Maritime operations, Agriculture, and General Industry, which is the set that applies to most worksites.

Employers must also comply with the General Duty Clause of the OSH Act, which requires employers to keep their workplace free of serious recognized hazards. This clause is generally cited when no OSHA standard applies to the hazard.

Before OSHA can issue a standard, it must go through an extensive and lengthy process that includes substantial public engagement, notice, and comment periods. This is known as OSHA's "rulemaking process." According to the "OSHA's rulemaking process" flowchart, the process begins with an identified health or safety hazard and includes hazard analyses and risk assessments. In fact, many of the standards are in direct response to a serious accident or large loss such as the *1910.119, Process Safety Management of Highly Hazardous Chemicals* standard and the Union Carbide, Bhopal, methyl isocyanate (MIC) gas release tragedy of 1984.

2.5.1 1910.132, Personal Protective Equipment Standard

Hazard determination, analysis, and assessment are touched upon in some OSHA regulations but not formally referred to as "risk assessment." The most basic is OSHA *29 CFR 1910.132, Personal Protective Equipment (PPE) General Requirements*. This requires employers to assess the workplace to determine if hazards are present, or likely to be present, which will necessitate the use of PPE. It applies to types of PPE that are not required by another specific OSHA regulation covering hazards such as noise or respiratory contaminants. This regulation requires employers to perform and document a hazard assessment of workplace hazards requiring PPE. Generally, it requires employers to identify types of hazards and select appropriate PPE as stated in 1910.132 (d) below:

Hazard assessment and equipment selection – *The employer shall assess the workplace to determine if hazards are present, or are likely to be present, which necessitate the use of personal protective equipment (PPE). If such hazards are present, or likely to be present, the employer shall:*

- *Select, and have each affected employee use, the types of PPE that will protect the affected employee from the hazards identified in the hazard assessment;*
- *Communicate selection decisions to each affected employee; and,*
- *Select PPE that properly fits each affected employee. Note: Non-mandatory Appendix B contains an example of procedures that would comply with the requirement for a hazard assessment.*

The employer shall verify that the required workplace hazard assessment has been performed through a written certification that identifies the workplace evaluated; the person certifying that the evaluation has been performed; the date(s) of the hazard assessment; and, which identifies the document as a certification of hazard assessment.

An example of a PPE hazard assessment can be found in Chapter 5, "Fundamental Techniques."

2.5.2 1910.119, Process Safety Management Standard

As indicated earlier, OSHA *29 CFR 1910.119, Process Safety Management (PSM) of Highly Hazardous Chemicals* is the most prominent standard containing requirements for risk assessment. The Process Safety Management standard was established in 1992 and requires process hazard analyses

(PHAs) for regulated industrial processes containing 10,000 lb or more of a hazardous chemical for the purpose of protecting the employees working in and around such processes. 1910.119(e) requires an initial PHA be conducted and revalidated every five years with a follow-up PHA. While this requirement is called a process hazard analysis, a PHA is similar to a risk assessment in that it identifies and analyzes the hazards and existing controls to determine if additional controls are needed (OSHA, 1992; Highly Hazardous Chemicals, 1910.119).

The PHA must be conducted on a prioritized basis and consider the extent of the hazard, number of potentially affected people, age of the process, and operating history of the process. Some of the same risk assessment techniques mentioned in this book and various assessment standards are referenced by the OSHA PSM regulation including those below.

- What-If Analysis
- Checklist Analysis
- What-If/Checklist Analysis
- Hazard and Operability Study (HAZOP)
- Failure mode and effects analysis (FMEA)
- Fault tree analysis (FTA).

The PHA must address the potential consequences of past incidents and the potential consequences of the failure of engineering and administrative controls. It must also consider human factors and include a qualitative evaluation of safety and health effects of the failure of controls on employees in the workplace. Other notable sections of the Process Safety Management standard include requirements for Pre-Start-up Safety Reviews (PSSR) and Management of Change (MOC) procedures.

The references to prioritization and consequences in the PSM regulation are consistent with a risk assessment. All of these PHA activities must be formal and documented. Further discussion can be found on PHAs in Chapter 6 "What-If Hazard Analysis."

2.5.3 Other OSHA Standards

Other OSHA standards require that a hazard or exposure determination be made and that existing hazards are controlled. Examples include the *1910.146, Permit-required Confined Space* standard, which requires an initial evaluation of the workplace to identify confined spaces and determine if they are "permit-required" spaces and the identification and evaluation of hazards prior to entry into permit spaces and the *1910.1200, Hazard Communication* standard, which requires an inventory of hazardous chemicals and a hazard determination be made. Thus, risk assessments, while not specifically required, are a technique that should be used to comply with these requirements.

Also consider the OSHA General Duty Clause requirement that all employers must provide their employees a workplace that is free of recognized hazards. The General Duty Clause is intended to apply to all hazards that the employer should be aware of but that are not covered by a specific OSHA regulation. A logical approach to this requirement is to perform hazard identification and analysis and risk assessment. Again, risk assessment is not specifically required by the General Duty Clause, but a thorough risk assessment would enable employers to comply with this OSHA requirement.

It is the authors' opinion that the United States has fallen behind other parts of the world in the use of formal risk assessments, partially due to the fact that most OSHA standards do not include specific requirements to do so. However, there is a growing interest in the use of risk assessment taking shape in consensus standards and industry practices.

2.6 CONSENSUS STANDARDS REQUIRING RISK ASSESSMENT

Even though compliance standards are not the current driving force for risk assessment, there is a movement toward formal risk assessments in several key consensus standards and technical reports. These include the American National Standard, *ANSI/AIHA/ASSE Z10, Occupational Health and Safety Management Systems*; the *ANSI/ASSE Z590.3, Prevention through Design*; the *ISO 31000/ANSI/ASSE Z690 Risk Management series*; and the *ANSI B11.0 Machine Safety* standard and *B11.TR3 Technical Report*.

2.7 ANSI/AIHA/ASSE Z10-2012, OCCUPATIONAL HEALTH AND SAFETY MANAGEMENT SYSTEMS

As the standard states, risk assessment is essential to an organization's occupation health and safety management system. To begin, a clear understanding of occupational health and safety management systems is necessary. The American National Standard, *ANSI/AIHA/ASSE Z10-2012, Occupational Health and Safety Management Systems*, defines such a systems as "a set of interrelated elements that establish and/or support occupational health and safety policy and objectives, and mechanisms to achieve those objectives in order to continually improve occupational health and safety." In essence, it is a business system employed by the organization to effectively manage operational risks to enable the organization to achieve its business objectives.

The ANSI Z10 standard was originally approved and published in 2005 and revised in 2012. It is fairly consistent with other international health and safety standards such as the British Standards Institution's (BSI) OHSAS 18001 and International Labour Organization's (ILO) Guidelines on Occupational Health and Safety Management Systems, 2001. They all, in varying degrees, contain requirements for identifying and assessing risk, designing in safety, and managing change. Several examples include:

- ANSI/AIHA Z10-2012, the Occupational Health and Safety Management Systems standard
  - Assessment and Prioritization – Section 4.2
  - Risk Assessment – Section 5.1.1 and Appendix F
  - Design Review and Management of Change – Section 5.1.3 and Appendix H
- BSI OHSAS 18001-2007, Occupational Health and Safety Management Systems
  - Hazard Identification, Risk Assessment and Determination of Controls – Section 4.3.1
- ILO OHSMS-2001
  - Initial Review – Section 3.7
  - Continual Improvement – Section 3.16
  - Management of Change – Section 3.10.2
- OSHA VPP – 2008
  - Hazard Analysis of Routine Jobs, Task, and Processes
  - Worksite Analysis
  - Hazard Analysis of Significant Changes
  - Pre-use Analysis.

The most recent version of ANSI Z10-2012 dedicates significant emphasis to risk assessment in Sections 4.2 Assessment and Prioritization and 5.1.1. Risk Assessment and Appendix F. Risk Assessment (informative). The standard is formatted into two columns to distinguish requirements (left column) from recommended practices and explanatory information (right column). Requirements are identified by the word "shall," while the recommended practices or explanatory notes are described by the word "should." In Section 5.1.1 of ANSI Z10, it states that organizations *shall* establish and implement a risk assessment process as part of the Occupational Health and Safety Management Systems. In doing so, the organization *should* select the most appropriate methods to address the types of hazards and risks facing the organization.

Risk assessments are a requirement of Z10. Appendix F provides a very informative section on the risk assessment process and select risk assessment methods. It states that no single risk assessment technique will apply in all situations and describes the importance of risk assessments within a management system. The standard cites examples of risk assessments ranging from overall organizational reviews, design safety reviews, change management process, to procurement and use of contractors.

ANSI Z10 is a consensus standard, meaning that ANSI has verified that the requirements for due process, consensus, and approval criteria have been met in the development of this standard. The use of consensus standards is voluntary; however, they are often incorporated into regulatory compliance standards and citations. Other safety management system guidelines have been issued; however, only Z10 is approved by an accrediting organization.

ANSI Z10 should be a primary reference for the safety professional. It defines the minimum requirements for a health and safety management system, and its primary purpose is to reduce the risk of occupational injuries, illnesses, and fatalities.

All health and safety management system standards will likely be considered when the International Organization for Standardization (ISO) finalizes ISO 45001, Occupational Health and Safety Management Systems.

2.8 ISO 31000/ANSI/ASSE Z690 RISK MANAGEMENT SERIES

The American National Standard, ANSI/ASSE Z690-2011 Risk Management standard series are three fundamental consensus standards for the practice of managing risk. Specifically, *ANSI/ASSE Z690.3*-2011, *Risk Assessment Techniques*, nationally adopted from ISO 31010:2009, is a standard for current good practices in the selection and use of risk assessment techniques. The series contains the following:

- ANSI/ASSE Z690.1-2011, Vocabulary for Risk Management (National Adoption of ISO Guide 73:2009)
- ANSI/ASSE Z690.2-2011, Risk Management Principles and Guidelines (National Adoption of ISO 31000:2009)
- ANSI/ASSE Z690.3-2011, Risk Assessment Techniques (National Adoption of IEC/ISO 31010:2009).

The Risk Assessment Techniques standard, ISO 31010/ANSI Z690.3, was approved in 2011 by the American Society of Safety Engineers (ASSE) in the United States. It is the official adoption of International Electrotechnical Commission (IEC)/ISO 31010:2009 by the United States.

ISO 31010/ANSI Z690.3 is exclusively about the risk assessment process from the approach of "risk" management. The risk management perspective on risk assessment is generally broader and higher level that the occupational health and safety perspective. However, the process is essentially the same. ISO 31010/ANSI Z690.3 provides guidance on the risk assessment phase of the overall risk

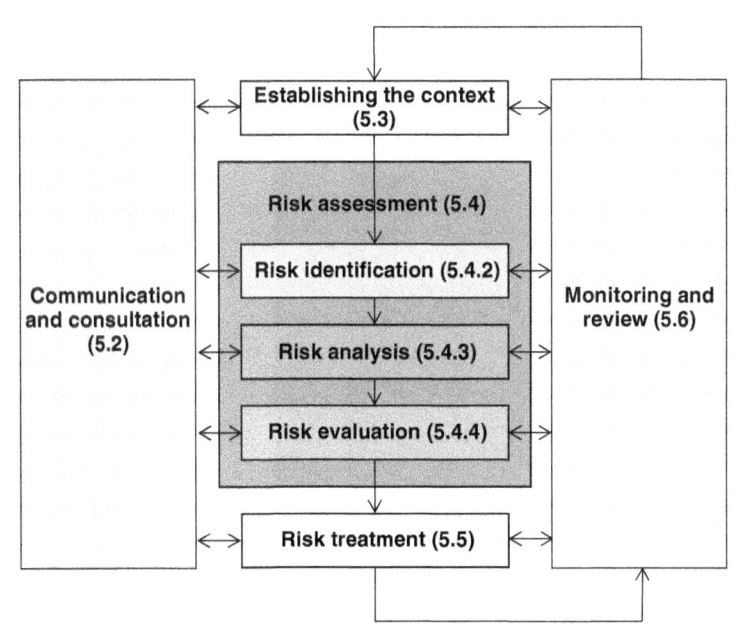

Figure 2.1 Risk Management Process. *Source:* Reprinted with permission from ISO 31000/ANSI/ASSE Z690.2-2011 (Courtesy of the American Society of Safety Engineers)

management process. The purpose is to provide evidence-based information and analysis to make informed decisions on how to treat particular risks.

Risk assessment is defined as the process of risk identification, risk analysis, and risk evaluation. Figure 2.1 illustrates the relationship risk assessment has within the risk management process reprinted with permission from ISO 31000/ANSI/ASSE Z690.2.

ISO 31010/ANSI Z690.3 states that the risk assessment process provides decision makers and stakeholders a better understanding of risks that could impact an organization's business objective and the efficacy of controls in place, so that the organization can better manage its operational risks. In essence, the risk assessment process provides a basis for decisions to be made regarding the most appropriate risk control measures to achieve an acceptable risk level. Without proper risk assessment, risks remain unknown and cannot be adequately managed.

The ISO 31000/ANSI Z690 series should also be an important reference for the safety professional, especially those who work in the risk management and insurance industry.

2.9 ANSI/ASSE Z590.3-2011, PREVENTION THROUGH DESIGN

The *ANSI/ASSE Z590.3*-2011, *Prevention through Design, Guidelines for Addressing Occupational Hazards and Risks in the Design and Redesign Processes* was published in 2011. This was after a number of years of efforts by the ASSE, National Safety Council (NSC) and National Institute for Occupational Safety and Health (NIOSH). The standard was developed to provide consistent procedures for conducting hazard analysis and risk assessment in the design and redesign process.

Risk assessment is considered to be a major component of the Prevention through Design process by ANSI Z590-3-2011. A significant portion of the standard is dedicated to risk assessment. Section 7, "The Hazard Analysis and Risk Assessment Process" covers the following steps on risk assessment:

- Select a risk assessment matrix.
- Establish the analysis parameters.
- Identify the hazards.
- Consider the failure modes.
- Assess the severity of consequences.
- Determine occurrence probability.
- Define initial risk.
- Select and implement control methods.
- Assess the residual risk.
- Risk acceptance decision making.
- Document the results.
- Follow up on actions taken.

The standard also presents several risk assessment techniques and provides samples and templates in the appendices. ANSI Z590.3 probably dedicates more emphasis to risk assessment than any other aspect of Prevention through Design. This standard provides guidance on the avoidance, elimination, reduction, and control of occupational safety and health hazards and risks in the design or redesign process.

This standard is unique in that it is intended to be applied at pre-operation, operational, post incident, and/or postoperational stages of products, processes, and operations. Identifying hazards, assessing risks, and determining controls during the design or redesign process are the most reliable and cost-effective times to perform these activities. Thus, this standard should be a primary reference for the safety professional.

2.10 ANSI B11.0 MACHINE SAFETY

ANSI B11.0-2015, Safety of Machinery, General Requirements and Risk Assessment is considered the centerpiece for machine safety and risk assessments. It is one of more than 30 standards and technical reports for metal working machinery published by ANSI and B11 Standards, Inc., known as the B11 series. In the absence of machine-specific standards, ANSI B11.0 and *ANSI B11.19-2010, Performance Criteria for Safeguarding* combine to form the foundation for the B11 series of machine-specific (Type C) standards and for other industrial machinery lacking a machine-specific safety standard. The B11 standards and technical reports are organized with the ISO A–B–C level structure briefly summarized below:

- **Type A standards** (basis standards) provide basic concepts, principles for design, and general aspects that can be applied to machinery.
- **Type B standards** (generic safety standards) address one or more safety aspects or one or more types of safeguards that can be used across a range of machinery.
- **Type C standards** (machinery-specific safety standards) address detailed safety requirements for a particular machine or group of machines.

Since it applies to an array of machines and contains general requirements, ANSI B11.0 is considered a Type A standard. Its applies to new, existing, modified, and rebuilt power-driven stationary machines used to shape or form materials through cutting, impact, pressure, electrical, and other processing methods. ANSI B11.0 also states that machinery suppliers and users are responsible for defining and achieving acceptable risk and that any risks associated with the operation, maintenance, dismantling, and disposal of machinery shall be reduced to an acceptable level. The standard includes a formal method to conduct and document the risk assessment process and also identifies some preparations that must be made before a risk assessment begins. The standard presents the basic risk assessment process in a step-by-step approach to assist in achieving this goal.

The *ANSI B11.TR3*-2000, *Risk Assessment and Risk Reduction – A Guide to Estimate, Evaluate and Reduce Risks Associated with Machine Tools* is an ANSI Technical Report. As part of the ANSI B11 series of technical reports and standards, TR3 pertains to the design, construction, care, and use of machine tools. The report defines a method for identifying hazards associated with a particular machine or system when used as intended and provides a procedure to estimate, evaluate, and reduce the risks of harm to individuals associated with these hazards under the various conditions of use of that machine or system. Examples of tasks and machine hazards, as well as risk reduction methods, are included in the technical report.

Finally, ANSI has also adopted *ISO 12100, Safety of Machinery – General Principles for Design – Risk Assessment and Risk Reduction*. This standard specifies the basic terminology, principles, and a methodology for achieving safety in the design or machinery. It specifies principles of risk assessment and risk reduction to help designers in achieving this objective. While this is an international standard, it is an ISO standard that safety professionals in the United States should be aware of even if they do not have international responsibilities or involvement.

2.11 NFPA 70E

The 2015 edition reflects a major shift in how stakeholders evaluate electrical risk. The National Fire Protection Association's 2015 *NFPA 70E: Standard for Electrical Safety in the Workplace* is the latest edition of the voluntary consensus standard. The standard addresses workplace electrical safety and includes guidance for performing risk assessments of arc flash, shock, and electrical hazard exposures.

The 2015 edition has several significant changes over previous editions. Where an arc flash, shock, or electrical "hazard analysis" was required, the 2015 edition now requires a "risk assessment" be performed. Further, the standard has expanded the "risk assessment procedure" to include the requirement of hazard identification, assessment of risks, and the implementation of risk controls according to hierarchy of risk control methods specified in ANSI/AIHA Z10. The addition of the Informative Annex F, Risk Assessment Procedure, provides information regarding the process steps, risk estimation, risk reduction, and risk evaluation. This shift in the 2015 edition from "hazard analysis" to "risk assessment" reflects a change in awareness about the potential for failure, moving from a hazard-based to a risk-based focus. In support of this shift, new definitions for *hazard*, *hazardous*, *risk*, and *risk assessment* are now included.

2.12 MIL-STD-882E, 11 MAY 2012, DEPARTMENT OF DEFENSE STANDARD PRACTICE, SYSTEM SAFETY

This official Department of Defense (DoD) standard, originally published in 1969, provides detailed guidance on system safety including the risk assessment process. It has been updated numerous times since and MIL-STD 882E, 11 May 2012, is the current version at the time of this writing. The standard is approved for use by all departments and agencies within the DoD, and it is used by many contractors and others outside the DoD.

MIL-STD-882 is one of the earliest US standards on risk assessment that was publically available. General requirements of the standard include the identification and documentation of hazards and assessment and documentation of risk. The standard defines severity categories and probability levels for application in a risk matrix. Below are all eight system safety elements of MIL-STD-882E.

1. Document the system safety approach.
2. Identify and document hazards.
3. Assess and document risk.
4. Identify and document risk mitigation measures.
5. Reduce risk.
6. Verify, validate, and document risk reduction.
7. Accept risk and document.
8. Manage life-cycle risk.

This standard clearly includes a significant risk assessment provision, and it is one of the earliest US standards to do so as mentioned previously. It has been referred to by safety professionals for decades.

2.13 KEY TERMS AND DEFINITIONS

Found within the aforementioned standards and other related works are a number of important terms and definitions related to the practice of hazard analysis, risk assessments, and risk management. The following are selected risk-related terms and how they are defined in their context. Many of the definitions are taken from referenced standards, which in some cases have multiple variations. These variations have been provided to illustrate subtle differences and similarities for certain terms used in the standards.

Acceptable Risk:
- The risk level that is considered by the organization to be acceptable in its current context. This level of risk is generally lowered as the organization matures and the control technologies improve (Authors).
- That risk for which the probability of an incident or exposure occurring and the severity of harm or damage that may result are as low as reasonably practicable (ALARP) in the setting being considered (ANSI/ASSE Z590.3-2011).
- Knowledge that risk is low enough or adequately controlled to take action. A risk level achieved after risk reduction measures have been applied. It is a risk level that is accepted for a given task (hazardous situation) or hazard. For the purpose of this standard, the terms acceptable risk and tolerable risk are considered to be synonymous (ANSI B11.0-2015).
- Risk that has been reduced to a level that can be tolerated by the organization having regard to its legal obligations and its own OH&S policy (BS OHSAS 18001-2007).
- Risk that the appropriate acceptance authority (as defined in DoDI 5000.02) is willing to accept without additional mitigation (MIL-STD-882E-2012).

As Low As Reasonably Achievable (ALARA, 2015):
- ALARA is an acronym for "as low as (is) reasonably achievable," which means "making every reasonable effort to maintain exposures to ionizing radiation as far below the dose limits as practical, consistent with the purpose for which the licensed activity is undertaken, taking into account the state of technology, the economics of improvements in relation to state of technology, the economics of improvements in relation to benefits to the public health and safety, and other societal and socioeconomic considerations, and in relation to utilization of nuclear energy and licensed materials in the public interest" (Title 10, Section 20.1003, of the Code of Federal Regulations).

KEY TERMS AND DEFINITIONS

As Low As Reasonably Practicable (ALARP):

- That level of risk that can be further lowered only by an increase in resource expenditure that is disproportionate in relation to the resulting decrease in risk (ANSI/ASSE Z590.3-2011).

Asset:

- Something valuable that an entity owns, benefits from, or has use of in generating income or to provide a service to society. Examples include employees and management, customers and vendors, property and buildings, liability, income, technology and information, and reputation (adapted from Businessdictionary.com).

Audit:

- A systematic, independent, documented process for obtaining information and data and evaluating it objectively to determine the extent to which defined audit criteria are fulfilled (ANSI/AIHA/ASSE Z10-2012).

Barrier:

- Physical or procedural control measures that are put in place to prevent or reduce likelihood of risk exposure (proactive) and/or reduce severity of impact/consequence (reactive) resulting from a hazardous event. A fixed machine guard is an example of a "proactive barrier." An example of a "reactive barrier" might be an air bag that is deployed in a car crash (Authors).

Causal Factor:

- One or several mechanisms that trigger the hazard that may result in a mishap (MIL-STD-882E-2012).

Compliance:

- Meeting the requirements of local, state, or federal statutes, standards, or regulations (ANSI/AIHA/ASSE Z10-2012).

Conformance:

- Meeting the requirements of the organization's OHSMS and this standard (ANSI/AIHA/ASSE Z10-2012).

Consequence:

- Outcome of an event affecting objectives (ISO Guide 73/ANSI/ASSE Z690.1-2011).

Continual Improvement:

- The process of enhancing the OHSMS to achieve ongoing improvement in overall health and safety performance in line with the organization's health and safety policy and performance objectives (ANSI/AIHA/ASSE Z10-2012).

Contractor:

- A person or organization providing services to another organization in accordance with agreed-upon specifications, terms, and conditions (ANSI/AIHA/ASSE Z10-2012).
- An entity in private industry that enters into contracts with the government to provide goods or services (MIL-STD-882E-2012).

Control:

- Measure that is modifying risk (ISO 31000/ANSI/ASSE Z690.2-2011).

Corrective Action:

- Action taken to eliminate or mitigate the cause of a system deficiency, hazard, or risk (e.g., fix an existing problem) (ANSI/AIHA/ASSE Z10-2012).

Critical Control Point (CCP) Decision Tree:

- A decision-making flowchart tool used to identify "high severity with low probability" risks that require additional control measures to reduce severity (Authors).

Current State Risk:

- For present conditions, a risk assessment that takes into consideration existing controls and their effects on risk is referred to as a "current state" risk level (Authors).

Design:

- The process of converting an idea or market need into the detailed information from which a product, process, or technical system can be produced (ANSI/ASSE Z590.3-2011).
- To plan and develop the machine to meet the intended purpose and function during its life cycle (ANSI B11.0-2015).

Design Safety Review:

- An important management process tool for integrating safety and health into the design process for new facilities, processes, or operations and for changes in existing operations (ANSI/ASSE Z590.3-2011).

Ergonomics:

- The scientific discipline concerned with the understanding of interactions among humans and other elements of a system and the profession that applies theory, principles, data, and other methods to design in order to optimize human well-being and overall system performance (Human Factors and Ergonomics Society).
- Occupational ergonomics is the applied science of designing workplace demands and environment to accommodate human capabilities and limitation for well-being and optimum performance (Authors).

Event:

- Occurrence or change of a particular set of circumstances (ISO Guide 73/ANSI/ASSE Z690.1-2011).

Exposure:

- Extent to which an organization and/or stakeholder is subject to an event (ISO Guide 73/ANSI/ASSE 690.1-2011).
- Contact with or proximity to a hazard, taking into account duration and intensity (ANSI/AIHA/ASSE Z10-2012).

KEY TERMS AND DEFINITIONS

- Exposure includes the frequency and duration of a hazard coming into contact with the population or assets at risk. Frequency of exposure describes how often an event might take place over a specified time period. Duration of exposure is the length of time that a single exposure occurs (Authors).
- State or condition of being unprotected and open to damage, danger, risk of suffering a loss in a transaction, or uncertainty. Examples of exposure to hazards include natural hazards, fire and explosion, spills or releases, process breakdowns, utility failures, transportation or distribution disruption, human error, intentional acts, lack of regulatory compliance, and liability (adapted from Businessdictionary.com).

Exposure Assessment:

- For occupational health and environmental purposes, exposure assessment is the multidisciplinary field that identifies and characterizes workplace exposures, develops estimates of exposure response and makes risk assessment studies, and evaluates the significance of exposures and effectiveness of intervention strategies (ANSI/ASSE Z590.3-2011).
- The process of measuring or estimating the exposure profiles of workers, including the relevant characteristics of the exposures such as the duration and intensity (ANSI/AIHA/ASSE Z10-2012).

Failure Mode:

- What is observed to fail or to perform incorrectly (ISO 31010/ANSI Z690.3-2011).
- A failure mode is the manner in which the item or operation potentially fails to meet or deliver the intended function and associated requirements. Failure modes may include functions that fail to perform within defined limits, inadequate or poor performance, intermittent performance, or performing an unintended or undesired function (Carlson, 2012).

Financial Risks:

- Risks that arise from the effect of market forces on financial assets or liabilities and include market risk, credit risk, liquidity risk, and price risk (The Institutes).

Frequency:

- Number of events or outcomes per defined unit of time (ISO Guide 73/ANSI/ASSE Z690.1-2011).
- "Frequency" is a rate measure and needs a denominator such as a unit of time (e.g., per hour/per year), the number of tasks performed (e.g., per 1000 miles driven/per 100 drill holes completed), units produced (e.g., per million tons of coal), and life cycle of equipment, process, or product. *Note: "Likelihood" is a general term that can be expressed specifically as a "frequency" or "probability" of the future occurrence of the chosen consequence scenario* (Whiting).

Future State Risk:

- When additional control measures are proposed, a "future state" risk level is estimated considering their effect in reducing risk. Future state residual risk assessments are performed to also validate and support the proposed risk reduction measures (Authors).

Harm:

- Physical injury or damage to health of people (ANSI B11.0-2015; ANSI B11.TR3-2000).

Hazard:

- The potential for harm (ANSI/ASSE Z590.3-2011).
- Source of potential harm (ISO Guide 73/ANSI/ASSE Z690.1-2011).
- A condition, set of circumstances, or inherent property that can cause injury, illness, or death (ANSI/AIHA/ASSE Z10-2012).
- A potential source of harm (ANSI B11.0-2015; ANSI B11.TR3-2000).
- A source of possible injury or damage to health (2015 NFPA 70E).
- Any real or potential condition that can cause injury, illness, or death to personnel; damage to or loss of a system, equipment, or property; or damage to the environment (MIL-STD-882E -2012).

Hazard: Insurance Context:

- Condition or situation that creates or increases chance of loss in an insured risk, separated into two kinds: (1) Physical hazard: Physical environment that could increase or decrease the probability or severity of a loss. It can be managed through risk improvement, insurance policy terms, and premium rates. (2) Moral hazard: Attitude and ethical conduct of the insured. It cannot be managed but can be avoided by declining to insure the risk (adapted from Businessdictionary.com).

Hazard Analysis:

- A process that commences with the identification of a hazard or hazards and proceeds into an estimate of the severity of harm or damage that could result if the potential of an incident or exposure occurs (ANSI/ASSE Z590.3-2011).
- It includes an analysis of severity level but does not consider probability of occurrence. Examples might include maximum foreseeable loss or maximum probable loss used by insurance underwriting practices (Manuele, 2013).

Hazard Area (Zone):

- An area or space that poses an immediate or impending hazard (ANSI B11.TR3-2000).
- Any space within and/or around a machine in which a person can be exposed to a hazard, also referred to as hazard area or hazard space (ANSI B11.0-2015).

Hazard Identification:

- Process of recognizing that a hazard exists and defining its characteristics (BS OHSAS 18001-2007).
- The act of anticipating and recognizing existing and potential hazards and their characteristics (Authors).

Hazard Risks:

- Risks arising from property, liability, or personnel loss exposures, which are generally the subject of insurance (The Institutes).

Hazard/Risk Avoidance:

- Prevent the entry of hazards into a workplace by selecting and incorporating appropriate technology and work method criteria during the design processes (ANSI/ASSE Z590.3-2011).

KEY TERMS AND DEFINITIONS

Hazard/Risk Elimination:

- Eliminate workplace and work method risks that have been discovered (ANSI/ASSE Z590.3-2011).

  Author's Cautionary Note: In risk management, care must be taken when using absolute terms such as "eliminate" or "stop" or "prevent" as they imply the false confidence belief that zero risk is achievable. Risk management can never reduce risk to zero. Usually, only some risk factors which form parts of a risk scenario can be absolutely removed or eliminated. More often than not, risk treatment is usually about substituting a lower risk – risk factor rather than completely eliminating a risk. For example, a highly toxic solvent may be replaced by a less toxic solvent. So in some respects the exposure to the highly toxic solvent has been eliminated but there has been essentially a reduction in risk by substitution not elimination. Also be careful with believing that simply declaring a rule that a particular solvent is not to be used, is no guarantee that the chances of further use and exposure has been completely eliminated (Whiting).

Hazardous:

- Involving exposure to at least one hazard (2015 NFPA 70E).

Hazardous Event:

- An event that can cause harm (ANSI B11.0-2015).

Hazardous Situation:

- A circumstance in which a person is exposed to a hazard (ANSI B11.TR3-2000).
- Circumstance in which a person is exposed to at least one hazard (ANSI B11.0-2015).

Hierarchy of Controls:

- A systematic approach to avoiding, eliminating, controlling, and reducing risks, considering steps in a ranked and sequential order, beginning with avoidance, elimination, and substitution (ANSI/ASSE Z590.3-2011).

Human Factors:

- Human factors are concerned with the application of what we know about people, their abilities, characteristics, and limitations to the design of equipment they use, environments in which they function, and jobs they perform (Human Factors and Ergonomics Society).
- Applied human factors engineering is "the designing of systems with the user in mind" (Authors).

Incident:

- An event in which a work-related injury or illness (regardless of severity) or fatality occurred or could have occurred (commonly referred to as a "close call" or "near miss") (ANSI/AIHA/ASSE Z10-2012).

Initial Risk:

- The first assessment of the potential risk of an identified hazard. Initial risk establishes a fixed baseline for the hazard (MIL-STD-882E-2012).

Level of Risk:

- Magnitude of a risk or combination of risks, expressed in terms of the combination of consequences and their likelihood (ISO Guide 73/ANSI/ASSE Z690.1-2011).

Life Cycle:

- The phases of design, construction, operation, maintenance, and disposal for a facility, equipment, process, and material (ANSI/ASSE Z590.3-2011).
- The phases of a machine including design and construction; transport and commissioning; reassembly, installation, initial adjustment, and relocation; use (e.g., setting, teaching/programming or process changeover, operation) and care (cleaning, trouble shooting [fault finding], maintenance [planned and unplanned]); and decommissioning, dismantling, and, as far as safely is concerned, disposal (ANSI B11.TR3-2000).
- The phases of a machine including but not necessarily limited to planning and specification; acquisition and contracting; design and construction; transport and commissioning, reassembly, installation, initial adjustment, and relocation; operation and maintenance (e.g., setting, teaching/programming, process changeover, cleaning, planned or unplanned maintenance, troubleshooting); modification; and decommissioning, dismantling, and, as far as safety is concerned, disposal (ANSI B11.0-2015).
- All phases of the system's life, including design, research, development, test and evaluation, production, deployment (inventory), operations and support, and disposal (MIL-STD-882E-2012).

Likelihood:

- Chance of something happening (ISO Guide 73/ANSI/ASSE Z690.1-2011).
- "Likelihood" is a general term that can be expressed specifically as a "frequency" or "probability" of the future occurrence of the chosen consequence scenario. The likelihood of the complete scenario needs to include estimates of the likelihoods of each scenario event and circumstance so as to obtain an overall likelihood. This compounding of contributing likelihoods can be qualitative or quantitative (Whiting).
- *Author's Cautionary Note: It is very useful for novice risk assessors to clearly distinguish between the terms "likelihood" and "probability" and "frequency.". The effective way of avoiding confusion of terms is to use "likelihood" as the generic term with "probability" and "frequency" as being alternative ways or subsets of expressing likelihood. This can be illustrated by improving the way in which the likelihood scale can better provide guidance of how to estimate likelihood more reliably and consistently.*

Mishap:

- An event or series of events resulting in unintentional death, injury, occupational illness, damage to or loss of equipment or property, or damage to the environment. For the purposes of this standard, the term "mishap" includes negative environmental impacts from planned events (MIL-STD-882E-2012).

Mitigation Measure:

- Action required to eliminate the hazard or, when a hazard cannot be eliminated, reduce the associated risk by lessening the severity of the resulting mishap or lowering the likelihood that a mishap will occur (MIL-STD-882E-2012).

KEY TERMS AND DEFINITIONS

Occupational Health and Safety Management Systems (OHSMS):
- A set of interrelated elements that establish and/or support occupational health and safety policy and objectives and mechanisms to achieve those objectives in order to continually improve occupational health and safety (ANSI/AIHA/ASSE Z10-2012).

OHSMS Issues:
- Hazards, risks, management system deficiencies, and opportunities for improvement (ANSI/AIHA/ASSE Z10-2012).

Operational Risk:
- Risks that are generated from work-related hazards (Authors).
- Risk of loss resulting from inadequate or failed processes, people, and systems or from external events. This definition includes legal risk but excludes strategic and reputational risk (Global Association of Risk Professionals, 2011).
- Arise from people or a failure in processes, systems, or controls, including those involving information technologies (The Institutes, 2012).

Operational Risk Management System:
- A management system that encompasses all operational risks such as occupational safety, health, environmental, liability, and other risks that must be managed to achieve and sustain the organization's business objectives through a continually improve process (Authors).

Organization:
- A public or private company, corporation, firm, enterprise, authority, or institution or part or combination thereof, whether incorporated or not, that has its own management functions. This can consist of one or many sites or facilities (ANSI/AIHA/ASSE Z10-2012).

Prevention through Design:
- Addressing occupational safety and health needs in the design and redesign process to prevent or minimize the work-related hazards and risks associated with the construction, manufacture, use, maintenance, retrofitting, and disposal of facilities, processes, materials, and equipment (ANSI/ASSE Z590.3-2011).

Preventive Action:
- Action taken to reduce the likelihood an underlying system deficiency or hazard will occur or recur in another similar process (ANSI/AIHA/ASSE Z10-2012).

Probability:
- An estimate of the likelihood of an incident or exposure occurring that could result in harm or damage for a selected unit of time, events, population, items, or activity being considered (ANSI/ASSE Z590.3-2011).
- Measure of the chance of occurrence expressed as a number between 0 and 1, where 0 is impossibility and 1 is absolute certainty (ISO Guide 73/ANSI/ASSE Z690.1-2011).
- An expression of the likelihood of occurrence of a mishap (MIL-STD-882E-2012).

- "Probability" is an expression of chances or odds and is pure number with no units usually expressed in a variety of ways such as 1 chance in 100 or 1% of the occurrences or 0.01 or even 1E-02. The first of these expressions is recommended as the best, most easily interpreted way to describe a probability (Whiting).

Process:

- A series of progressive and interrelated steps by which an end is attained; continuous action, operation, or a series of changes taking place in a definite manner; the action of going forward (ANSI/ASSE Z590.3-2011).

Protective Device:

- Device (other than a guard) that reduces a risk, either alone or associated with a guard. Note: This does not include PPE (e.g., hand tools, safety glasses/goggles, face shields, safety shoes) (ANSI B11.TR3-2000).

Protective Measures:

- Design, safeguarding, administrative controls, warnings, and training or PPE used to eliminate hazards or reduce risks (ANSI B11.TR3-2000).
- Any action or means used to eliminate or control access to hazards and/or reduce risks (ANSI B11.19-2010).

Qualitative Risk Assessment:

- A risk assessment based on subjective ratings (Authors).

Quantitative Risk Assessment:

- A risk assessment based on numerical ratings (Authors).

Raw Risk:

- The initial risk assessed assuming no risk reduction methods are in place. Raw risk serves as a baseline for the measurement of further risk reduction. Raw risk estimations may also be applicable to situations where the existing controls are considered very low on the hierarchy of controls and/or unreliable such as warnings, administrative level, or PPE-type controls (Authors).

Reasonable Foreseeable Misuse:

- The predictable use of facilities, equipment, or materials in a way not intended in the original design (ANSI/ASSE Z590.3-2011).
- The predictable use of a machine in a way not intended by the supplier or user, but which may result from human behavior (ANSI B11.TR3-2000).

Redesign:

- A design activity that includes all retrofitting and altering activities affecting existing facilities, equipment, technologies, materials, and processes and the work methods (ANSI/ASSE Z590.3-2011).

Residual Risk:

- The risk remaining after risk reduction measures have been taken (ANSI/ASSE Z590.3-2011).
- Risk remaining after risk treatment (ISO Guide 73/ANSI/ASSE Z690.1-2011).

KEY TERMS AND DEFINITIONS

- Risk remaining after protective measures have been taken (ANSI B11.TR3-2000).
- The risk remaining after risk reduction measures (protective measures) are taken (ANSI B11.0-2015).

Risk:

- An estimate of the probability of a hazard-related incident or exposure occurring and the severity of harm or damage that could result (ANSI/ASSE Z590.3-2011).
- Effect of uncertainty on objectives (ISO Guide 73/ANSI/ASSE Z690.1-2011).
- An estimate of the combination of the likelihood of an occurrence of a hazardous event or exposure and the severity of injury or illness that may be caused by the event or exposures (ANSI/AIHA/ASSE Z10-2012).
- A combination of the probability of occurrence of harm and the severity of that harm (ANSI B11.0-2015; ANSI B11.TR3-2000).
- A combination of the likelihood of occurrence of injury or damage to health and the severity of injury or damage to health that results from a hazard (2015 NFPA 70E).
- A combination of the severity of the mishap and the probability that the mishap will occur (MIL-STD-882E-2012).

Risk Acceptance:

- Informed decision to take a particular risk (ISO Guide 73/ANSI/ASSE Z690.1-2011).

Risk Analysis:

- Process to comprehend the nature of risk and to determine the level of risk (ISO Guide 73/ANSI/ASSE Z690.1-2011).

Risk Assessment:

- A process that commences with hazard identification and analysis, through which the probable severity of harm or damage is established, followed by an estimate of the probability of the incident or exposure occurring, and concluding with a statement of risk (ANSI/ASSE Z590.3-2011).
- Overall process of risk identification, risk analysis, and risk evaluation (ISO Guide 73/ANSI/ASSE Z690.1-2011).
- Process(es) used to evaluate the level of risk associated with hazards and system issues (ANSI/AIHA/ASSE Z10-2012).
- The process by which the intended use of the machine, the tasks and hazards, and the level of risk are determined (ANSI B11.0-2015; ANSI B11.TR3-2000).
- An overall process that identifies hazards, estimates the potential severity of injury or damage to health, estimates the likelihood of occurrence of injury or damage to health, and determines if protective measures are required (2015 NFPA 70E).
- The scientific process that characterizes the nature and magnitude of health risks to humans and ecological receptors from chemical contaminants and other stressors that may be present in the environment (Environmental Protection Agency).

Risk Assessment Process:

- The entire process of identifying hazards, assessing risk, reducing risk, and documenting the results (ANSI B11.0-2015).

Risk Avoidance:

- Informed decision not to be involved in, or to withdraw from, an activity in order not to be exposed to a particular risk (ISO Guide 73/ANSI/ASSE Z690.1-2011).

Risk Centric:

- The state when an organization gains a sense of urgency around a fatal or serious injury/illness level risk as an actual catastrophic event; seeing risk of harm as actual harm itself resulting in the action of mitigating risk in advance of mishaps (Walline).
- The mind-set of acting upon risk rather than hazards (Authors).

Risk Criteria:

- Terms of reference against which the significance of a risk is evaluated (ISO Guide 73/ANSI/ASSE Z690.1-2011).

Risk Description:

- Structured statement of risk usually containing four elements: sources, events, causes, and consequences (ISO 31010/ANSI/ASSE Z690.3-2011).

Risk Evaluation:

- Process of comparing the results of risk analysis with risk criteria to determine whether the risk and/or its magnitude is acceptable or tolerable (ISO Guide 73/ANSI/ASSE Z690.1-2011).

Risk Factor:

- A component of risk derived from an identified hazard used to estimate and measure a risk level. The primary risk factors used in risk assessments are severity of consequence (S) and likelihood (L) or probability (P) of occurrence. Other risk factors used include exposure (E), frequency of exposure (F), detection of failure (D), control reliability (CR), and prevention effectiveness (PE) (Authors).

Risk Identification:

- Process of finding, recognizing, and describing risks (ISO Guide 73/ANSI/ASSE Z690.1-2011).

Risk Level:

- The characterization of risk as either high, serious, medium, or low (MIL-STD-882E-2012).

Risk Management:

- Coordinated activities to direct and control an organization with regard to risk (ISO Guide 73/ANSI/ASSE Z690.1-2011).

KEY TERMS AND DEFINITIONS

Risk Management Plan:

- Scheme within the risk management framework specifying the approach, the management components, and resources to be applied to the management of risk (ISO 31000/ANSI/ASSE Z690.2-2011).

Risk Management Process:

- Systematic application of management policies, procedures, and practices to the activities of communicating, consulting, and establishing the context and identifying, analyzing, evaluating, treating, monitoring, and reviewing risk (ISO 31000/ANSI/ASSE Z690.2-2011).

Risk Matrix:

- Tool for ranking and displaying risks by defining ranges for consequence and likelihood (ISO Guide 73/ANSI/ASSE Z690.1-2011).

Risk Priority Number:

- A combined risk score of three or more risk factors such as severity, probability, and exposure or prevention effectiveness to produce a priority number used in risk ranking (Authors).

Risk Professional:

- A person skilled, knowledgeable, and experienced in the risk assessment and management process (Authors).

Risk Profile:

- Description of any set of risks (ISO Guide 73/ANSI/ASSE Z690.1-2011).

Risk Reduction:

- That part of the risk assessment process involving the elimination of hazards or selection of other appropriate risk reduction measures (protective measures) to reduce the probability of harm or its severity (ANSI B11.0-2015).

Risk Reduction Measures:

- Any action or means used to eliminate hazards and/or reduce risks (ANSI B11.0-2015).

Risk Register:

- Record of information about identified risks (ISO Guide 73/ANSI/ASSE Z690.1-2011).

Risk Retention:

- Acceptance of the potential benefit of gain, or burden of loss, from a particular risk (ISO Guide 73/ANSI/ASSE Z690.1-2011).

Risk Source:

- Element that alone or in combination has the intrinsic potential to give rise to risk (ISO Guide 73/ANSI/ASSE Z690.1-2011).

Risk Tolerance:

- Organization's or stakeholder's readiness to bear the risk after risk treatment in order to achieve its objectives (ISO Guide 73/ANSI/ASSE Z690.1-2011).

Risk Treatment:

- Process to modify risk (ISO 31000/ANSI/ASSE Z690.2-2011).

Safe:

- "Safe" is viewed as having reached a level of acceptable or minimal residual risk (ANSI/AIHA/ASSE Z10-2012).
- Deciding that a thing is safe or not safe requires judgments of whether the probability of an undesired incident occurring and the severity of its outcome are acceptable (Manuele, 2013).
- The state of being protected from recognized hazards that are likely to cause physical harm. There is no such thing as being absolutely safe, that is, a complete absence of risk (Whiting).

Safe Work Procedure(s):

- Formal written documentation developed by the user that describes steps that are to be taken to safely complete tasks where hazardous situations may be present or hazardous events are likely to occur (ANSI B11.0-2015).

Safeguarding:

- Guards, safeguarding devices, awareness devices, safeguarding methods, and safe work procedures (ANSI B11.TR3-2000).
- Protection of personnel from hazards by the use of guards, safeguarding devices, awareness devices, and safeguarding measures (ANSI B11.0-2015).

Safety:

- Freedom from unacceptable risk (ANSI/ASSE Z590.3-2011).

Safety Function:

- Function of a machine, the malfunction of which would increase the risk of harm (ANSI B11.TR3-2000).

Safety Professional:

- Trained individual dedicated to the protection of people, assets, and the environment (Authors).

Severity (of Consequence):

- An estimate of the magnitude of harm or damage that could reasonably result from a hazard-related incident or exposure (ANSI/ASSE Z590.3-2011).
- The magnitude of potential consequences of a mishap to include death, injury, occupational illness, damage to or loss of equipment or property, damage to the environment, or monetary loss (MIL-STD-882E-2012).

- *Author's Cautionary Note: The author recommends not using emotive subjective and judgmental descriptors for severities of consequences. It is appropriate to use such terms for sizes or levels of risk when deciding priorities for risk treatment, for example, if desired, the term "catastrophic" could be used instead of "high" in the A1 cell of the matrix. But if used with consequence severities, there is an expectation to consider tolerability based on consequence alone. Using them with severities encourages the restrictive and inappropriate practice of decision making based ONLY on consequence severity instead of risk [both likelihood AND consequence $R = L \times C$] (Whiting).*

Stakeholder:

- Person or organization that can affect, be affected by, or perceive themselves to be affected by a decision or activity (ISO Guide 73/ANSI/ASSE Z690.1-2011).

Standard:

- A required or recommended practice endorsed by a credible organization (Authors).

Strategic Risks:

- Risks that arise from trends in the economy and society, including changes in the economic, political, and competitive environments, as well as from demographic shifts (The Institutes).

Supplier:

- Any entity that provides or makes available equipment, material, or professional services (ANSI/ASSE Z590.3-2011).
- An entity that provides or makes available for use all or part of a [machine] or [system] (ANSI B11.TR3-2000).

System:

- An integrated composite of people, products, and processes that provide a capability to satisfy a stated need or objective (ANSI/ASSE Z590.3-2011).
- The organization of hardware, software, material, facilities, personnel, data, and services needed to perform a designated function within a stated environment with specified results (MIL-STD-882E-2012).

System Safety:

- The application of engineering and management principles, criteria, and techniques to achieve acceptable risk within the constraints of operational effectiveness and suitability, time, and cost throughout all phases of the system life cycle (MIL-STD-882E-2012).

Task:

- Any specific activity that is done on or around the machine during its life cycle (ANSI B11.0-2015; ANSI B11.TR3-2000).

Threat:

- Often used in security-related concerns, a threat is an indication of an approaching or imminent menace. A threat is a negative event that can cause a risk to become a loss, expressed as an aggregate of risk, consequences of risk, and the likelihood of the occurrence of the event.

A threat may be a natural phenomenon such as an earthquake, flood, or storm or a man-made incident such as fire, power failure, sabotage, etc. (adapted from Businessdictionary.com). Anything that might exploit a vulnerability of an asset. Examples include arson, theft, cyber-attacks, sabotage, and terrorism (Rausand, 2011).

Tolerable Risk:

- Risk that is accepted for a given task and hazard combination [hazardous situation] (ANSI B11.TR3-2000).

Trigger:

- An event or action that initiates the exposure to a hazard allowing a pathway to a mishap (Authors).

User:

- An entity that utilizes the [machine], [system], or related equipment. Note: Under certain circumstances (i.e., acting as a builder, modifier, integrator), the user may act as a supplier (ANSI B11.TR3-2000).

Vulnerability:

- Degree to which an asset is susceptible to harm, degradation, or destruction by being exposed to a hazard (adapted from Businessdictionary.com). A weakness of an asset that can be exploited by one or more threat agents. Vulnerability refers to the security flaws in a system that allow an attack to be successful (Rausand, 2011).

Warning:

- A means used to call attention to a hazard (ANSI B11.0-2015).

Worst Conceivable Risk:

- The worst conceivable consequence from an incident that could occur, but probably will not occur, within the lifetime of the system (ANSI/ASSE Z590.3-2011).

Worst Credible Consequence:

- The worst credible consequence from an incident that has the potential to occur within the lifetime of the system (ANSI/ASSE Z590.3-2011).

2.14 SUMMARY

Safety, health, and environmental professionals should develop an understanding of standards and guidelines prescribing operational risk assessment for organizations and their importance in shaping the SH&E professional's role in occupational safety and health. The key terms and definitions for operational risk assessment and management reviewed in this chapter are fundamental and should be in every SH&E professional's vocabulary. As the operational risk profession matures, reference to these standards and future trends in standard development regarding the practice of risk assessment will be required.

REVIEW QUESTIONS

1. Explain the difference between a consensus standard and a compliance standard. Provide examples of each.
2. Provide five examples of where OSHA standards require hazard determination and analysis.
3. Summarize the main requirements of operational risk management systems and list three standards for such systems.
4. List the name and purpose of the ANSI standard series for risk management.
5. List the steps of a hazard analysis and risk assessment as outlined in ANSI Z590.3 Prevention through Design standard.
6. Which ANSI standard series addresses machine safety and provides the primary standard for machine risk assessment.
7. Provide examples that can be found in the workplace for acceptable risk level.

REFERENCES

ANSI/AIHA Z10-2012. *Occupational Health and Safety Management Systems.* Fairfax, VA: The American Industrial Hygiene Association, 2012. ASSE is now the Secretariat. Available at: https://www.asse.org/cartpage.php?link=z10_2005.

ANSI/ASSE Z590.3-2011. *Prevention through Design: Guidelines for Addressing Occupational Hazards and Risks in Design and Redesign Processes.* Des Plaines, IL: American Society of Safety Engineers, 2011.

ANSI/ASSE Z690.1. *Vocabulary for Risk Management.* Des Plaines, IL: American Society of Safety Engineers, 2011.

ANSI/ASSE Z690.2. *Risk Management Principles and Guidelines.* Des Plaines, IL: American Society of Safety Engineers, 2011.

ANSI/ASSE Z690.3. *Risk Assessment Techniques.* Des Plaines, IL: American Society of Safety Engineers, 2011.

ANSI B11.0-2015. *Safety of Machinery – General Safety Requirements and Risk Assessments.* Houston, TX: B11 Standards, Inc., 2015.

ANSI B11.TR3-2000. *Risk Assessment and Risk Reduction – A Guide To Estimate, Evaluate, and Reduce Risks Associated with Machine Tools.* McLean, VA: The Association for Manufacturing Technology, 2000.

BS OHSAS 18001:2007. *Occupational Health and Safety Management Systems – Requirements.* London, UK: British Standards Institution (BSI), 2007.

BusinessDictionary.com. 2015. Available at: http://www.businessdictionary.com/ (accessed December 15, 2015).

Carlson, C. S. *Effective FMEAs – Achieving Safe, Reliable, and Economical Products and Processes using Failure Mode and Effects Analysis.* Hoboken, NJ: John Wiley & Sons, Inc., 2012.

EPA. 1990. *Risk Management Programs for Chemical Accidental Release Prevention,* http://www.epa.gov/rmp.

Global Association of Risk Professionals. 2011. *Operational Risk Management. Chapter 1: Operational Risk.* Available at http://www.garp.org/media/665968/icbrr-operational0711preview.pdf (accessed December 15, 2015).

Manuele, F. A., *On the Practice of Safety.* Hoboken, NJ: Wiley, 2013.

MIL-STD-882E. *Standard Practice for System Safety.* Washington, DC: Department of Defense, 2012. Available at http://www.system-safety.org/. Scroll down and click on MIL-STD-882E in the right hand column for a free copy.

NFPA 70E. Standard for Electrical Safety in the Workplace. 2015 Edition. Quincy, MA: National Fire Protection Association, 2015.

Occupational Safety and Health Administration. *OSHA's Rule for Process Safety Management of Highly Hazardous Chemicals, 1910.119*. Washington, DC: Department of Labor, Occupational Safety and Health Administration, 1992.

OSH Act of 1970, *Section 5, Duties, General Duty Clause*. Washington, DC: Department of Labor, Occupational Safety and Health Administration, 1970.

OSHA 29 CFR 1910.132, *Personal Protective Equipment, General Requirements*. Washington, DC: Department of Labor, Occupational Safety and Health Administration, 2011.

OSHA Law & Regulations. 2012 found at https://www.osha.gov/law-regs.html (accessed December 16, 2015).

Rausand, M., *Risk Assessment: Theory, Methods, and Applications*. Hoboken, NJ: John Wiley & Sons, Inc., 2011.

The Institutes, 2012. *Risk Management Principles and Practices*, *Quadrants of Risk: Hazard, Operational, Financial, and Strategic*. American Institute for Chartered Property Casualty Underwriters. Retrieved March 18, 2015 from: http://www.theinstitutes.org/comet/programs/arm/assets/arm54-chapter.pdf (accessed December 16, 2015).

U.S.NRC. ALARA. United States Nuclear Regulatory Commission Library. 2015. Available at http://www.nrc.gov/reading-rm/basic-ref/glossary/alara.html (accessed December 16, 2015).

3

RISK ASSESSMENT FUNDAMENTALS

BRUCE HOLLCROFT
Risk Control Services, Hays Companies, Lake Oswego, OR, USA

BRUCE K. LYON
Risk Control Services, Hays Companies, Kansas City, MO, USA

3.1 OBJECTIVES

- Describe the fundamentals of the risk assessment process
- Introduce the steps in the process
- Describe how to complete the steps successfully.

3.2 INTRODUCTION

Organizations, whether they realize it or not, are exposed to hazards and their risks each day capable of significantly affecting the ability to achieve important business goals or even remain in business. Risk assessment is an important and sophisticated tool used to assess an organization's operational risks so that proper decisions can be made to avoid or effectively mitigate and manage risks to an acceptable level. It is considered the cornerstone of risk management and the basis for the practice of safety.

In Europe, the importance of operational risk assessment is well known and publicized as indicated in the following statement from the European Agency for Safety and Health at Work (EU-OSHA) website.

> "If the risk assessment process – the start of the health and safety management approach – is not done well or not done at all, the appropriate preventive measures are unlikely to be identified or put in place." (EU-OSHA, 2015)

In fact, risk assessments are a common practice in Europe, Australia, New Zealand, Canada, and other parts of the world. In the United Kingdom, risk assessments have been legally required since

---

Risk Assessment: A Practical Guide to Assessing Operational Risks, First Edition.
Edited by Georgi Popov, Bruce K. Lyon, and Bruce Hollcroft.
© 2016 John Wiley & Sons, Inc. Published 2016 by John Wiley & Sons, Inc.

1999 by the Health and Safety Executive (H&SE). As stated in the opening chapter by Fred Manuele, the United States is behind in the use of risk assessment; however, there is significant momentum being generated by recent standards, risk-centric organizations and their leaders.

3.3 RISK ASSESSMENT WITHIN THE RISK MANAGEMENT FRAMEWORK

A central theme in this text is the concept of assessing risk within the framework of risk management. According to the American National Standard Institute's ANSI/ASSE Z690.1-2011, *Vocabulary for Risk Management* standard (national adoption of ISO Guide 73:2009), risk management is defined as "coordinated activities to direct and control an organization with regard to risk." In other words, it is the process of making management decisions based on known risks and the organization's acceptance for those risks.

The term "risk assessment" is often misused. It's the authors' experience that some organizations (and even some safety professionals) refer to hazard inspections, surveys, and compliance audits as "risk assessments." Thus, a clear understanding of the term is necessary. ISO Guide 73/ANSI Z690.1-2011 states that there are three distinct components to the act of "risk assessment" which are given as follows:

- Risk identification – finding, recognizing, and recording hazards
- Risk analysis – understanding consequences and probabilities and existing controls
- Risk evaluation – comparing levels of risk and considering additional controls.

Consequences are the potential outcomes of an undesirable event, which is measured by severity. Probability or likelihood is an estimation of the chances of the undesirable event occurring over a unit of time or for a specific activity. Risk assessment is an attempt to "predict" the worst event that could reasonably happen as a result of the hazard or operation, and how likely it is to occur. This estimation is often qualitative in nature; however some are semiquantitative or quantitative based. It's important to remember that the risk level relates to uncertainty and its effect on an organization's ability to achieve its objectives.

Within the risk management process, risk assessment is the primary component. This is illustrated in Figure 3.1 reprinted with permission from the ANSI/ASSE Z690.2-2011 *Risk Management Principles and Guidelines* consensus standard, nationally adopted from ISO 31000:2009.

Unfortunately, risk assessments have not been a common practice in the United States. One example is the April 20, 2010 Deepwater Horizon incident. According to estimates, the losses from the offshore oil rig accident resulted in 11 lives lost, $40 billion, and 4.9 million barrels of oil released in the Gulf during the 87-day incident. BP's internal investigation team of the Deepwater Horizon accident (i.e., "Deepwater Horizon Accident Investigation Report" September 8, 2010; p. 36) concluded that one of the eight key causes to the accident was that no risk assessment was performed of the cement slurry barrier application. The report stated "the investigation team has not seen evidence of a documented risk assessment regarding annulus barriers." The accuracy of cement slurry barriers was described as "critical" in the report, yet no formal risk assessment was performed.

Other examples indicate risk assessments are inconsistently performed. In a webinar hosted by the American Society of Safety Engineers (ASSE), "Prevention through Design: Guidelines for Addressing Occupational Hazards and Risks in Design and Redesign Processes" November 30, 2011, one of the webinar facilitators, Bruce Main, quoted a study conducted by a Fortune 500 company indicating that 65% of serious incidents had no previous risk assessment. This number may be indicative of other

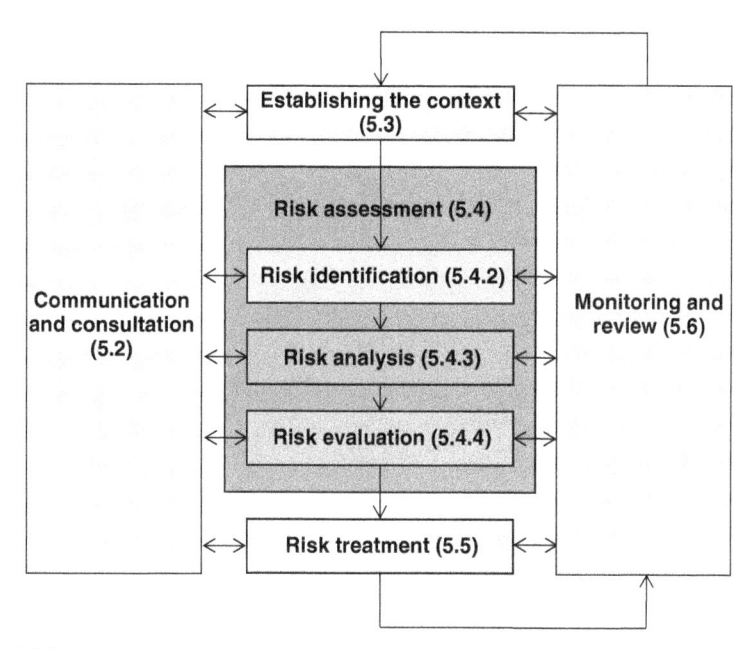

Figure 3.1 Risk Management Process. *Source:* Reprinted with permission from ISO 31000/ANSI/ASSEZ690.2-2011 (Courtesy of the American Society of Safety Engineers)

Fortune 500 companies and supports the authors' experience that many smaller companies perform very few if any risk assessments.

The take away message here is that organizations should establish a strategy for determining when and how risks should be assessed. Basic criteria for a written policy for conducting risk assessments and when assessments are needed might include some of the following:

- Projects or tasks that have not had a formal risk assessment
- New facilities, processes, or equipment
- When there are a number of risks present or introduced that make it necessary to apply risk priorities in an organized way
- When there is a risk that could have serious consequences, and where control measures are unclear
- Where there is a planned change to equipment, machinery, or a particular process (as outlined in ANSI Z10 5.1.2 Design Review and Management of Change).

3.4 RISK ASSESSMENTS AND OPERATIONAL RISK MANAGEMENT SYSTEMS

As within the risk management framework, risk assessment is central to an operational risk management system. The ultimate goal of an operational risk management system is to effectively manage risks and associated costs of occupational incidents through a "management-lead" continual improvement process that involves its employees. This is evidenced in the ANSI/AIHA Z10-2012 *Occupational Health and Safety Management Systems* standard and other management system standards and guidelines. For instance, the process of hazard analysis and risk assessment is a "required" core element in the following standards and guidelines:

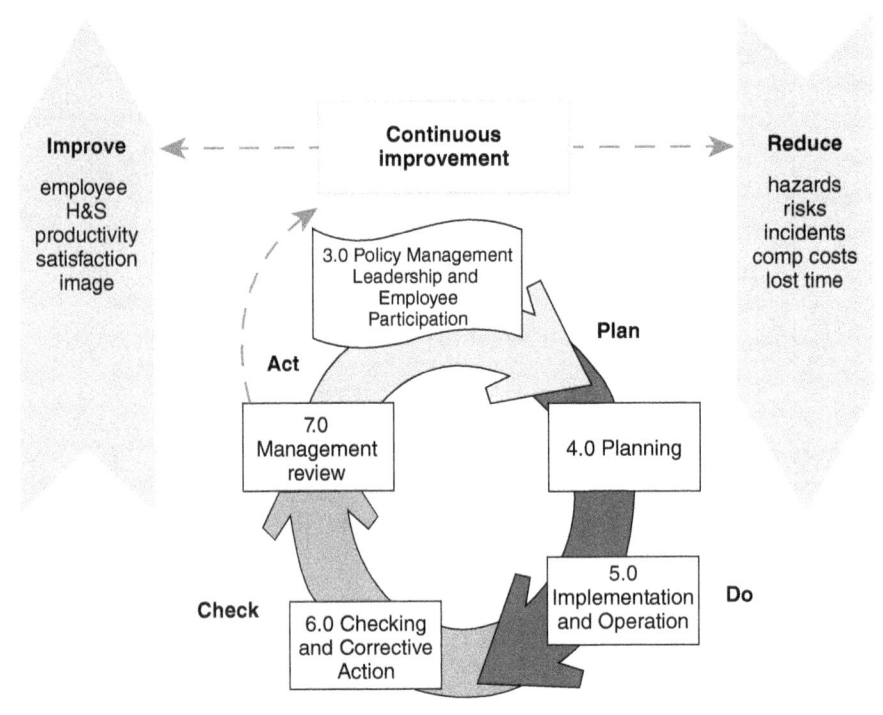

Figure 3.2 The OHSMS Process. *Source:* Reprinted with permission from ANSI/ASSE Z10-2010 (Courtesy of the American Society of Safety Engineers)

- Occupational Safety and Health Administration's (OSHA, 2003) *Voluntary Protection Program* (VPP)
- ANSI/AIHA/ASSE Z10-2012, *Occupational Health and Safety Management Systems*
- BS OHSAS 18001-2007, *Occupational Health and Safety Management*
- International Labor Office ILO-OSH 2001 "*Guidelines on occupational safety and health management systems*"
- ISO 14001-2004, *Environmental management systems – Requirements with guidance for use*
- ISO 45001-2015-16, *Occupational Health and Safety Management Systems.*

Operational risk management systems are based on a continual improvement process using the plan, do, check, act (PDCA) model promoted by Dr. Edwards Deming, known for his efforts in continuous improvement and quality initiatives. The OHSMS continual improvement process is illustrated in Figure 3.2 reprinted with permission from ANSI/ASSE Z10-2010.

The effectiveness of an operational risk management system requires the continual identification, analysis, and evaluation of risks to understand their magnitude of loss, and potential of occurring, as well as adequacy of existing control measures and needed improvements within the organization. Therefore, the risk assessment process is crucial to understanding and managing risks to an acceptable level within an operational risk management system.

3.5 THE PURPOSE OF ASSESSING RISK

Risk assessments can be performed for many reasons and have many purposes. According to the ANSI/ASSE Z690.3-2011, *Risk Assessment Techniques* standard (national adoption of IEC/ISO

SELECTING A RISK ASSESSMENT MATRIX 53

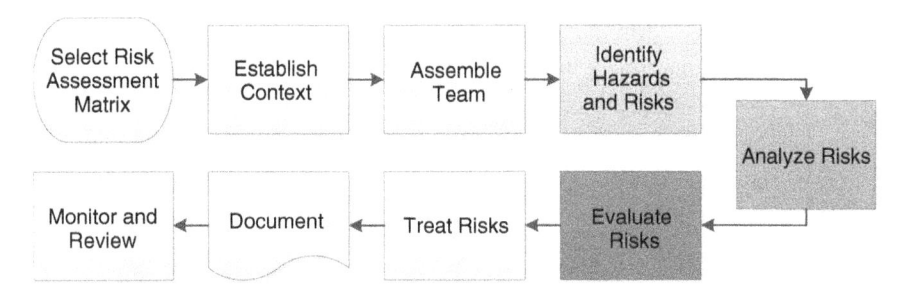

Figure 3.3 Simplified Steps of a Risk Assessment

31010:2009), the purpose of risk assessment is to "provide evidence-based information and analysis to make informed decisions on how to treat particular risks and how to select between options." The use of a consistent risk assessment process allows an organization to understand risk levels, compare those risks, and address those with the greatest risk first.

Generally, risk assessments are performed by safety professionals to determine the risk level resulting from a risk source (hazard or operation) and apply appropriate risk control measures according to the hierarchy of controls to reduce risk to an acceptable level. Other forms of risk treatment are available to risk management through insurance or other risk financing mechanisms to cover incidents that are not prevented.

Through the use of risk assessment, an organization is able to make better decisions regarding risk and achieve its business objectives. Removing uncertainly by assessing risk allows an organization to manage with a certain degree of confidence.

3.6 THE RISK ASSESSMENT PROCESS

The fundamental process of identifying, analyzing, and evaluating risk is necessary in providing those responsible for making business decision an understanding of the risk. This understanding allows decisions to be made regarding whether the identified risk is acceptable, and what control measures are most appropriate. Ultimately, the "output" of risk assessment is an "input" to the decision-making processes (ISO 31010/ANSI Z690.3-2011).

The following sections describe the sequence of components that take place within a risk assessment process. This is illustrated in Figure 3.3.

3.7 SELECTING A RISK ASSESSMENT MATRIX

In the ANSI/ASSE Z590.3-2011, *Prevention through Design: Guidelines for Addressing Occupational Hazards and Risks in Design and Redesign Process*es standard, the initial steps outlined in (Section 7) "The Hazard Analysis and Risk Assessment Process" are (7.1) "Management Direction" followed by (7.2) "Selecting a Risk Assessment Matrix." In the ISO 31010/ANSI Z690.3 standard, the process of defining and establishing risk criteria is included in Section 4.3.3, "Establishing the Context"; however, the selection of a matrix is not specifically mentioned. The authors think the selection of a risk assessment matrix is significant and should be included at the beginning of the process, either prior to or during the establishment of the context.

An organization should select or develop a risk assessment matrix that the stakeholders broadly agree upon to be used in the risk assessment process. This key component is used to define and

TABLE 3.1 Example of a Risk Assessment Matrix

Likelihood of Occurrence or Exposure for Select Unit of Time or Activity	←Severity of Injury or Illness Consequence→			
	Negligible	Marginal	Critical	Catastrophic
Frequent	Medium	Serious	High	High
Probable	Medium	Serious	High	High
Occasional	Low	Medium	Serious	High
Remote	Low	Medium	Medium	Serious
Improbable	Low	Low	Low	Medium

Source: Reprinted with permission from ANSI/AIHA/ASSE Z10-2012 (Courtesy of the American Society of Safety Engineers)

determine risk levels within an organization. An example of a risk assessment matrix reprinted with permission from ANSI/AIHA/ASSE Z10-2012 is provided in Table 3.1.

The purpose of the risk assessment matrix is to provide "a method to categorize combinations of probability of occurrence and severity of harm, thus establishing risk levels." (ANSI/ASSE Z590.3-2011) In essence, it is a risk "measuring stick" and communication tool used to help categorize and prioritize risks within the organization so that decision makers can take the most appropriate action in regards to risks and their treatment. There are a number of sources from which to select a risk assessment matrix. Notable examples are provided in the ANSI Z590.3 Prevention through Design standard, the US Military standard MIL-STD-882E, and the ANSI B11.TR3-2000 technical report, among others.

It is important that the risk rating criteria and matrix used by an organization are consistent. When developing or selecting a risk assessment matrix that expresses numerical values, rating criteria should be standardized so that a lower risk score or risk priority number (RPN) value indicates a lower risk level. Thus, on a 10-point risk scale, a risk score of 1 is considered the lowest level, while a 10 is considered the highest risk. For example, Table 3.2 provides an example of risk criteria reprinted with permission from ANSI/ASSE Z590.3. The example uses severity of consequence and occurrence probability factors, which are multiplied to determine risk level. Risk levels are defined as very high (15 or greater), high (10–14), moderate (6–9), and low (1–5). The corresponding *Risk Scoring Levels and Actions Required* shown in Table 3.3 are reprinted with permission from ANSI/ASSE Z590.3. The standards states that these numbers are judgmentally determined, are qualitative, and only have value in relation to each other.

TABLE 3.2 Example 5-Point Risk Assessment Matrix

	Occurrence Probabilities and Values				
Severity Levels and Values	Unlikely (1)	Seldom (2)	Occasional (3)	Likely (4)	Frequent (5)
Catastrophic (5)	5	10	15	20	25
Critical (4)	4	8	12	16	20
Marginal (3)	3	6	9	12	15
Negligible (2)	2	4	6	8	10
Insignificant (1)	1	2	3	4	5
Very high risk: 15 or greater; High risk: 10–14; Moderate risk: 6–9; Low risk: 1–5					

Source: Reprinted with permission from ANSI/ASSE Z590.3-2011 (Courtesy of the American Society of Safety Engineers)

ESTABLISHING CONTEXT

TABLE 3.3 Risk Scoring Levels and Action Required

Category	Risk Score	Action
Very high risk	15 or greater	Operation not permissible. Immediate action necessary
High risk	10–14	Remedial actions to be given high priority
Moderate risk	6–9	Remedial action to be taken at appropriate time
Low risk	1–5	Remedial action discretionary

Source: Reprinted with permission from ANSI/ASSE Z590.3-2011 (Courtesy of the American Society of Safety Engineers)

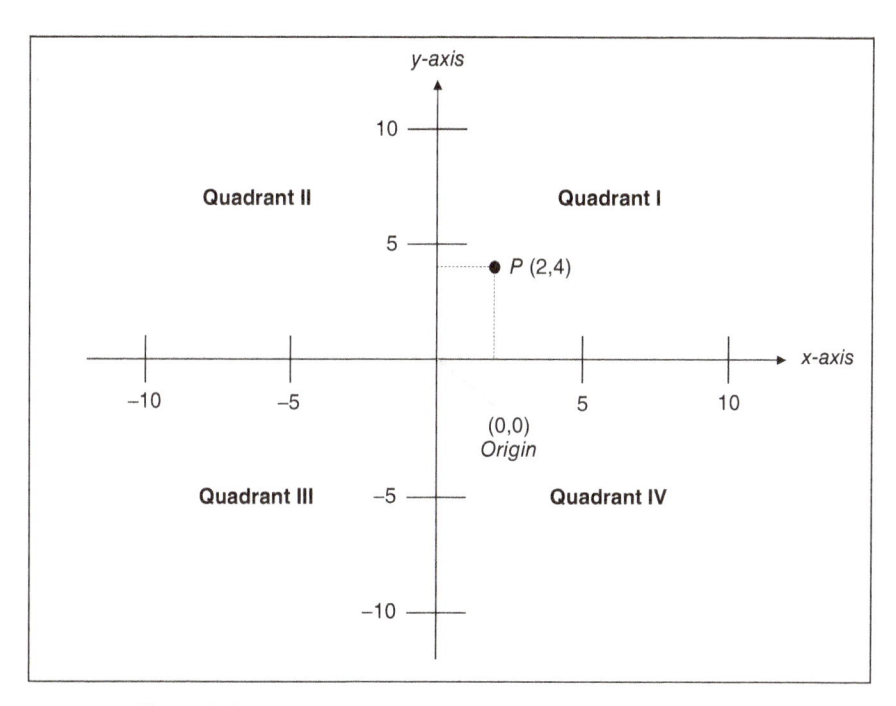

Figure 3.4 The Four Quadrants of a Cartesian Coordinate System

Notice that both matrix examples are in the upper right-hand quadrant known as Quadrant I of the Cartesian coordinate system. It is the authors' opinion that a risk assessment matrix should be positioned in the first quadrant (Quadrant I). This allows the user to read the matrix from left to right/bottom to top, with risk levels progressing from low to high accordingly, as indicated in Figure 3.4.

At this stage, an organization should consider their acceptance risk level in terms of the planned risk assessment exercise. Management and other involved stakeholders should agree on the levels that require "stop work," "immediate action required," "remedial action," down to "acceptable level – no further action." As stated in ANSI/ASSE Z590.3, "Personnel who craft risk assessment matrices may have differing ideas about acceptable risk levels and the management actions that should be taken in a given risk situation, and those differences must be resolved so that all personnel understand the process."

3.8 ESTABLISHING CONTEXT

The purpose and scope known as the context of the risk assessment must be established. This step is considered critical since it sets the direction, tone, and expectations for the project

(Lyon & Hollcroft, 2012). Within the organization's risk management process, the context should define the purpose and scope of the assessment; the stakeholders/team members responsibilities and accountabilities; the degree, extent, or rigor of the assessment; the risk assessment methodologies; the risk criteria; and resources available. Ultimately, the context defines the parameters for managing the risk throughout the process as stated in ISO 31010/ANSI Z690.3-2011.

The context for the risk assessment should be clear, concise, and well understood by all stakeholders. Every successful risk assessment needs a tightly defined beginning and end so the assessment team is not tempted to make it more complicated than needed or take it further than it is intended. The context should set the boundaries for the assessment with internal (resources, knowledge, culture, and values among others) and external (legal, regulatory, economy, perceptions of external stakeholders, etc.) parameters in mind.

To ensure the risk assessment team is focused on the correct elements, the context should "clearly explain" what is expected. For example:

- If the hazards associated with the job of cleaning windows were to be assessed, would it just cover the actual cleaning or would it include associated hazards like setting up the ladders, transferring the cleaning chemicals, or even driving to and from the location?
- If a facility was to be assessed, would it include just the plant operations on site, or would it include items like transportation, utilities, waste, and emissions to and from the facility?
- For a risk assessment of potential emergencies/disasters, should the assessment be limited to emergencies/disasters at facility sites or include events off site? Should it include natural, man-made, or technological emergencies/disasters or all of them?

Setting the scope too narrow might prevent a hazard and the resulting risk from being identified and assessed, or making it too broad could prevent the risk assessment from getting to the real purpose. It is important to get input from those who will be using the risk assessment to make decisions.

A well written and understood context is important as a guide that can be checked against frequently to keep the risk assessment team from drifting off course. The authors recommend that a concise "Purpose and Scope" statement be produced and written at the beginning of each risk assessment. Two simple examples are provided in the following:

1. *The purpose and scope of the risk assessment for the robotic welding cell # 214 is to determine the risk level to operators when entering the cell to change welding tips on the robot considering existing controls using a preliminary hazard analysis method by a cross-functional team of certified assessors.*
2. *The purpose and scope of the risk assessment for the Westfall Plant is to determine the risk level that natural hazards (including wind, severe rain, hail, tornado, earthquake, lightning, hurricane, and flood) present to the plant considering existing controls. The assessment team consisting of risk management, insurance professionals, and an outside consultant will use an organizational risk assessment methodology to identify and quantify risk levels that may need additional risk treatment or risk financing.*

Prior to conducting a risk assessment, an organization should clearly define and communicate its own risk criteria and "acceptable risk" level. Risk criteria should correspond to the selected risk assessment matrix and take into account the type of consequences expected, how likelihood or probabilities will be determined and expressed, the specific risk level classifications to be used, at what level of risk will corrective action be required, and the organization's established acceptable risk level.

The criteria used to determine risk criteria and acceptable risk level should include the organization's safety and health goals and the use of cost/benefit analyses of risks and their treatment and will be influenced by its culture and industry setting (ISO 31010/ANSI Z690.3, p. 21). Typically, as an organization matures and improves its risk control measures, the acceptable risk level will move closer to the negligible risk level. Further discussion on risk criteria and acceptable risk are provided in Chapter 4.

3.9 THE RISK ASSESSMENT TEAM

The context of the risk assessment assignment will determine the size and makeup of the team needed. The risk assessment team should include a cross-functional group of individuals who are familiar and knowledgeable with the hazards and operations being assessed. In many cases, the team will be comprised of several members on a consistent basis, while others are likely to come and go based on their ability to contribute to the particular risk assessment. For example, a risk assessment for a production operation might include engineering, maintenance, quality, production, safety, and an operator. A risk assessment for a transportation operation might include a driver, mechanic, and a dispatcher, along with the safety professional. A risk assessment for a product might include a product designer, engineer, legal counsel, production employee, marketing, customer service representative, insurance representative, and safety.

It is important to include members from within as well as outside of the organization that are knowledgeable about the hazards or operations and can make a positive contribution to the risk assessment. In some situations, external members are crucial to an assessment. During the "Occupy Movement" in 2011, a risk assessment was being performed for emergency planning purposes at a company near the Port of Oakland, California. The company did not have a security person, and it was recommended that the security firm for the complex be invited to participate. As a result, the outside security professional identified that the port was subject to civil unrest during the risk assessment. Recommendations from the assessment were addressed. Soon after "Occupy Oakland" occurred disrupting operations, but the company was prepared as a result of the input from the external security person.

A facilitator is also an important member of the risk assessment team. The facilitator could be an additional member or one of the members who is knowledgeable about the hazard or operation. The facilitator must know the risk assessment process and techniques. They must also understand the purpose and scope and be able to communicate this to the other members in order to keep them focused and on track. The facilitator should be confident but not controlling. Confidence is important to keep the risk assessment progressing in the right direction, staying objective, and on a reasonable schedule. However, the facilitator must avoid being controlling or dominating and should prevent other team members from dominating the risk assessment. Each member should feel comfortable contributing their ideas to the risk assessment and be assured that their contributions will be valued. The facilitator is key to the success of a good risk assessment.

A risk assessment team of about 3–10 members seems to work well. Less than three members may not provide enough perspectives or insights into the risk assessment. Risk assessment teams with more than 10 members may be difficult to control and keep focused. At some point a large risk assessment team will likely be dominated by some of the stronger personalities. The facilitator must function so as to avoid this situation.

A team that is committed performs the best and most thorough risk assessments. To be committed, the team must understand the purpose, scope, and the value of the risk assessment. They must be informed and understand how the risk assessment will contribute to the organization's goals and objectives. They should have the support of their manager and understand that a certain amount of prestige is associated with being on the team. A good facilitator can solicit a great deal of commitment out of team members though good communication on these matters.

The risk assessment team should have the appropriate skills to perform a successful risk assessment. They should understand the basic risk assessment concepts and methods. They can learn these concepts and methods through experience over time, or they can be provided basic instruction immediately prior to the risk assessment. They should be clear on the purpose and scope of each individual risk assessment. This includes knowing why the risk assessment is being performed and how the information will be applied and used. This communication should be clear and concise, and the facilitator should provide it or make sure it has previously been provided. Good communication and skills are essential.

3.10 HAZARD/RISK IDENTIFICATION

Hazards are the source of risk. Thus, if risks are to be assessed, hazards must first be identified. Risk identification is defined as the process of finding, recognizing, and recording risks. Its purpose is to identify what might happen and/or the situations that could impact the system or organization. The key components of risk identification are the identification of the causes and source of the risk (hazard in the context of physical harm), events, situations, or circumstances, which could have a material impact upon objectives and the nature of that impact. Once identified, existing controls for the risk should also be identified (ISO 31010/ANSI Z690.3-2011).

Methods for identifying existing and potential hazards/risks in the workplace are many. A likely starting point might begin with collecting available information about the operations to be assessed. Some hazards are easily identified intuitively or through recent or past experience, while others require more systematic methods of identification. Depending on the operation or subject of the assessment, the level of effort will vary accordingly. A simple job or task may only require a job hazard analysis, while a more complex system may require a series of methods to identify existing and potential hazards.

There are many ways to go about identifying hazards and operations for assessment, but a systematic approach will likely be more thorough and reliable. Some of the more common methods and sources used by safety professionals to identify hazards are listed in the following:

- Brainstorming
- Checklists
- Regulations (OSHA, EPA, DOT, etc.)
- Consensus industry standards (ANSI, ASTM, NFPA, etc.)
- Experts (external or internal)
- Job hazard analyses/job safety analyses
- Accident/incident investigations
- OSHA injury and illness records
- Insurance claims
- Formal hazard/risk identification techniques (31 listed in ISO 31010/ANSI Z690.3-2011).

Each of these hazard identification techniques listed previously and those listed in ISO 31010/ANSI Z690.3 have their own strengths, weaknesses, and limitations. They vary in their complexity, required capabilities and resources, and ability to provide quantitative results. In some cases, more than one technique may be used to identify the hazard. All of these techniques will be used in conjunction with a risk assessment matrix that will be discussed more in the next chapter.

3.11 RISK ANALYSIS

Upon identifying hazards, the team will analyze the potential risk. As stated by ISO 31010/ANSI Z690.3, risk analysis involves developing an "understanding" of the risk. This analysis of each hazard/risk includes the following tasks:

- Determining the severity of consequences
- Estimating the likelihood of occurrence
- Assessment of the effectiveness of existing controls
- An estimation of the risk level.

The level of risk takes into consideration a combination of the possible consequences and likelihood. A single event or task can have many possible consequences and impact multiple assets.

Risk analysis can be qualitative, semiquantitative, or quantitative in nature depending upon the context of the assessment and available data. Qualitative analyses are the most common and use descriptors such as "high," "serious," "medium," and "low" for degrees of severity of consequence, likelihood of occurrence, and risk level. Semiquantitative methods use numerical ratings for consequence and likelihood to produce a level of risk, which are based on qualitative descriptive criteria rather than quantitative data. And finally, quantitative analyses, which are not as common, use estimated values for consequences and their likelihood producing numerical values of risk in specific units defined in the context. As stated by ISO 31010/ANSI Z690.3, full quantitative analysis may not always be possible or desirable due to insufficient information or the needs of the assessment. In many cases, a comparative semiquantitative or qualitative ranking of risks by qualified assessors is desired for the assessment.

3.11.1 Consequence Analysis

The assessment team determines the nature and type of consequences that could result for exposure to a particular hazard or event. A single hazard or event may produce a number of impacts with various magnitudes (levels of severity) and could affect multiple assets or stakeholders. The assessment's context determines the types of consequence analyzed and stakeholders affected.

Consideration and direction should be given to how certain impacts will be handled. Some hazards may present a low severity level of consequence but a high likelihood of occurrence, while others may present a high severity and a low likelihood of occurrence. An organization's acceptable risk level will help determine the priorities for these types of risk. As mentioned in ISO 31010/ANSI Z690.3, it may be appropriate to focus on risks with potentially very large outcomes such as fatal and serious incident (FSI) type consequences, as these are often of greatest concern to managers. In other situations, it may be important to analyze both high and low consequence risks separately. Guidance should be established during the development of the context for such decisions.

3.11.2 Likelihood Analysis

Determining probability or likelihood generally involves (1) a review of relevant historical data to identify events or situations that have occurred; (2) predictive-type techniques such as fault tree analysis and event tree analysis; and (3) a structured systematic process guided by a qualified, knowledgeable expert(s). Any available data used should be relevant to the focus of the assessment. Where historical data shows a very low frequency of occurrence, it may be difficult to properly estimate probability. Therefore, it may be necessary to consider exposure frequency, time, and duration to a certain hazard or event in the likelihood analysis.

3.11.3 Assessment of Controls

The adequacy and effectiveness of existing control measures greatly affect the level of risk and must be assessed. This assessment of controls should include determining the type of controls for each specific risk and a judgment of their effectiveness based on the hierarchy of controls. For instance, controls such as permanent or fixed guards (engineering controls) are considered more effective than employee training, warnings (administrative controls), or personal protective equipment. The assessment should ensure existing controls are being applied/operated as intended and that their effectiveness can be demonstrated and verified. The effectiveness for a single control, or combination of related controls, can be expressed in qualitative, semiquantitative, or quantitative terms. The main focus of the control assessment should be on determining whether existing controls are adequate in reducing risk to an acceptable level or whether improved control measures are needed.

3.12 RISK EVALUATION

Risk evaluation involves comparing the estimated risk levels with the defined risk criteria to determine the significance of the level and type of risk. It is based on the combination of estimated consequences and likelihood and uses information from the hazard/risk identification and risk analysis phases to make recommendations for decision makers. These decisions may include implementing further controls, other forms of risk treatment, or avoiding the hazard or operation all together. Additional inputs to the decision-making process include legal, financial, ethical, and other considerations. This process may also be used to prioritize possible actions should more than one possible action be feasible.

Methods for defining risk criteria can range from a single level dividing risks that require treatment from those that do not to multiple levels of risk requiring graduated degrees of actions. Decisions on treating a risk will likely depend on the costs and benefits of risk and the costs and benefits of implementing improved controls. The "as low as reasonably practicable" or ALARP criteria is used to determine when the cost of further reduction is disproportionate to the benefits gained in risk reduction and safety (Manuele, 2008). Figure 3.5 provides an illustration of the ALARP principle reprinted with permission from ANSI/ASSE Z690.3 (Courtesy of ASSE) and Fred Manuele.

The selected risk assessment matrix is used to consider both the consequence and likelihood risk levels for each risk. The risk assessment matrix example provided in Table 3.1 is qualitative in nature with risks ranging from "high" to "low." In the example matrix, risks that fall in the "low" category would most likely be considered acceptable by an organization, while those in the "medium" category may be considered acceptable with some additional controls. Risks in the "serious" category would require immediate action, and those in the "high" category are considered the highest risk and would be unacceptable to an organization, requiring immediate action to avoid or reduce the risk to acceptable levels. In each case, the criteria for severity of consequence and likelihood of occurrence will need to be customized and defined by the organization's stakeholders.

In Manuele's (2010) article "Acceptable Risk: Time for SH&E professionals to adopt the concept" published in Professional Safety, May 2010, he suggests that safety professional have yet to fully embraced the concept of "acceptable risk." The fact is that there will always be some residual risk.

The ultimate goal is to reduce the risk of the hazard or operation to an acceptable level so as to feel confident in engaging in the activity. This may be accomplished though the implementation of additional or better controls according to the hierarchy of controls concept mentioned throughout this book, or it may involve other forms of risk treatment. Avoiding the risk by deciding not to engage in the hazard is always an option if the risk of engaging in it cannot be reduced to an acceptable level. A personal example might include a homeowner that decides not to clean their gutters because it is too dangerous or risky to work at heights. Instead they hire a contractor who has better equipment and is better trained to perform the work. The homeowner decided the risk was not acceptable so they avoided it by having a contractor perform the work.

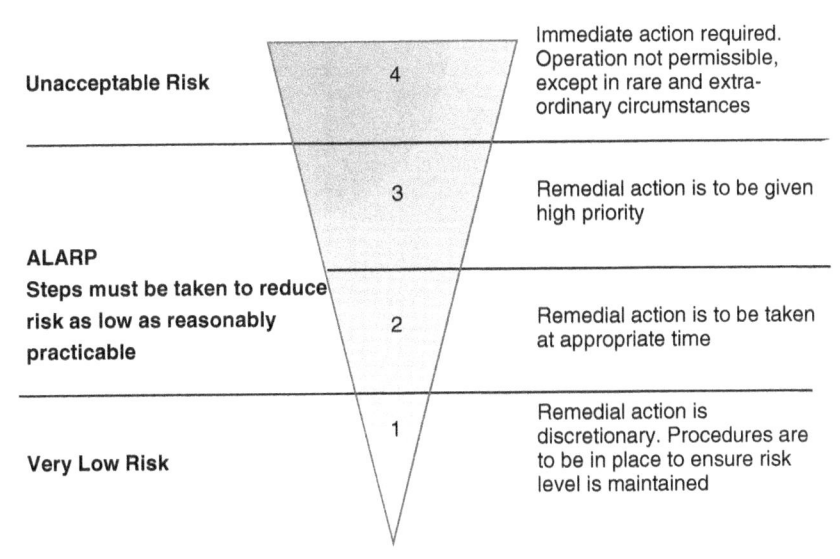

Unacceptable Risk	4	Immediate action required. Operation not permissible, except in rare and extraordinary circumstances
ALARP Steps must be taken to reduce risk as low as reasonably practicable	3	Remedial action is to be given high priority
	2	Remedial action is to be taken at appropriate time
Very Low Risk	1	Remedial action is discretionary. Procedures are to be in place to ensure risk level is maintained

Figure 3.5 The ALARP Model. *Source:* Reprinted with permission from ANSI/ASSE Z690.3 (Courtesy of the American Society of Safety Engineers) and Fred Manuele

It should also be mentioned that risk assessments are a process of continuous improvement. Risks that are estimated to be acceptable today will likely not be acceptable in the future. As an organization's operational risk management system matures, and new technologies are incorporated, the levels of acceptable risk tend to be further reduced through continuous improvement. The term "acceptable" refers to a point in time and will not likely be true in the future as expectations rise and what is considered acceptable in terms of risk lowers. As higher risks are controlled, the next ranking of risks are addressed until all risks are reduced to an acceptable level by the organization.

3.13 RISK TREATMENT

Risk treatment is the process of modifying risk. As mentioned previously, risks that are judged unacceptable to an organization must be "treated" to reduce risk through the use of risk controls. Risk treatment generally involves the selection and implementation of one or more risk control measures or enhancements to existing controls. The risk treatment process involves (1) the assessment of a risk treatment, (2) determining if residual risk levels are acceptable, (3) selecting new risk treatments for those residual risks that are not acceptable, (4) and assessing the effectiveness of any new control measure. Selection of control options should be made using the hierarchy of controls model. Figure 3.6 illustrates the hierarchy of controls model reprinted with permission from ANSI/ASSE Z590.3.

As indicated by ISO 31000/ANSI Z690.2-2011, risk treatment options are not always mutually exclusive or appropriate for all situations. Treatment options include (1) avoiding the risk by deciding not to start or continue with the activity that gives rise to the risk, (2) removing the risk source, (3) changing the likelihood, (4) changing the consequences, (5) sharing the risk with another party such as insurance contracts and risk financing, and (6) retaining the risk by informed decision.

3.14 COMMUNICATION

Successful risk assessment processes are dependent on effective communication among stakeholders prior to, during, and after the process. Without proper communication, severe consequences can occur.

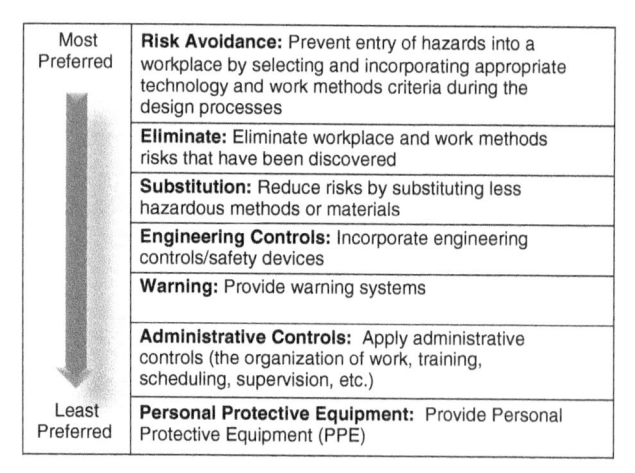

Most Preferred ↓ Least Preferred	**Risk Avoidance:** Prevent entry of hazards into a workplace by selecting and incorporating appropriate technology and work methods criteria during the design processes
	Eliminate: Eliminate workplace and work methods risks that have been discovered
	Substitution: Reduce risks by substituting less hazardous methods or materials
	Engineering Controls: Incorporate engineering controls/safety devices
	Warning: Provide warning systems
	Administrative Controls: Apply administrative controls (the organization of work, training, scheduling, supervision, etc.)
	Personal Protective Equipment: Provide Personal Protective Equipment (PPE)

Figure 3.6 Risk Reduction Hierarchy of Controls. *Source*: Reprinted with permission from ANSI/ASSE Z590.3 (Courtesy of the American Society of Safety Engineers)

Take for instance, the NASA's Space Shuttle Columbia explosion, which occurred February 1, 2003, claiming seven lives. The investigation that followed determined that a significant root cause to the incident was a lack of effective communication of critical safety information. The synopsis of the report of the Columbia Accident Investigation Board concluded that organizational causes including lack of communication contributed to the incident. "Cultural traits and organizational practices detrimental to safety were allowed to develop, including: reliance on past success as a substitute for sound engineering practices … , organizational barriers that prevented effective communication of critical safety information and stifled professional differences of opinion; lack of integrated management across program elements; and the evolution of an informal chain of command and decision-making processes that operated outside the organization's rules. (p. 9)"

Communication is a provision of both ISO 31010/ANSI Z690.3 and ANSI Z590.3 and is also required by virtually all of the national and international health and safety management standards such as ANSI Z10, OHSAS 18001, and OSHA VPP. However, ineffective communication continues to be a leading cause to poor outcomes such as fatal and serious incidents.

As with many other functions in organizations, it should be made a priority to communicate effectively when performing risk assessments. Those involved in the risk assessments should think about who could help them do the risk assessment more effectively. For example, they could ask others within their own departments for input. Alternatively, they should think about who might be interested and benefit from the risk assessment that is being performed and let them know the outcome. An effective process will involve stakeholders throughout the process and seek their input. Internal personnel such as the assessment team, management, affected operators, and uses of the systems as well as external stakeholders such as customers, investors, partners, suppliers, and vendors should be included in the communication process. In rare situation an exception may exist where there are legal or Fair Trade Commission implications or other reasons where communication about the risk assessment may need to be limited.

3.15 DOCUMENTATION

Documentation is an important form of verification and communication. Virtually all aspects of the risk assessment process discussed in this chapter should be documented including details on the following:

MONITORING AND CONTINUOUS IMPROVEMENT

- Selecting the risk assessment matrix
- Determining the purpose and scope (context)
- Selecting the team
- Identifying the hazards or operations to be assessed
- Hazard/risk identification
- Risk analysis
- Risk evaluation
- Communication and documentation
- Monitoring and continuous improvement.

Documenting the risk assessment serves to record the efforts of the assessment process, as well as the resulting risk findings and recommendations. It also allows those who were not directly involved to have some understanding of the risks and risk reduction measures as well as an organization to prove and demonstrate their efforts and actions. Documentation allows future reviewers to recreate and understand the thought process behind the previous risk assessment. It enables teams conducting future risk assessments to build upon and further improve the work that has already been done. Possibly most importantly, a good risk assessment that is well documented might serve to limit an organization's potential liability at some point in the future by showing their decisions and actions that were well thought-out by a qualified team to the best of their ability at the time.

Communication, documentation, and employee involvement are components of virtually all standards relating to risk assessment including ANSI Z10-2012, ANSI Z590.3-2011, and ISO 31010/ANSI Z690.3-2011 to mention a few. Furthermore, risk assessments would be more successful if those performing them would take a little time to think about who might be able to contribute and who might benefit from the risk assessment and engage those parties where possible.

3.16 MONITORING AND CONTINUOUS IMPROVEMENT

They say that only one thing is certain and that is there will always be change. With that in mind, it is important to monitor risk assessments and the hazards and operations they cover. Hazards and operations continuously change and with these changes come new and different risks. Examples of these might include different equipment, processes, operating environments, production rates, etc. Each of these changes could have an effect on the existing controls and their effectiveness. Thus, it may be appropriate to update risk assessments to consider these possible changes.

New technologies and controls are also being created, and monitoring these may offer opportunities to improve controls and further reduce the risk. New controls that are being introduced may be higher in the hierarchy of controls and more reliable. An example might be the Global Positioning Systems (GPS)/telematics available for vehicles today. In the past, driver training, instruction, and possibly their logs were manually used to monitor their location, speed, and driving performance – controls consider low on the hierarchy of controls. The GPS/telematics of today are much more accurate, factual, and reliable and might be consider higher on the scale.

Monitoring, updating, and further reducing of risk are continuous improvement. In fact, risk assessment done properly is the perfect example of a continuous improvement process. Certain risks that are accepted today will not likely be accepted in the future. There are a variety of reasons for this condition. Personal expectations are always increasing. People increasingly expect working environments that are more comfortable, convenient, and safe. This expectation comes with the improvement in technology and working conditions, as well as enhanced knowledge of risks. This is especially true as more people work for someone else as opposed to being self-employed and willing to take more risk. Not too long ago many people would have been satisfied just working indoors at a desk. Now the expectation is that the room is properly air conditioned, and more and more people are expecting good ergonomics like sit/stand workstations. Expectations will continue to rise as will technology resulting in continuous improvement.

Regulations and standards also continue to have higher requirements to keep up with technology and expectations. The methods of improving the safety of hazards and operations and reducing the resulting risk are getting dramatically better. Thus, regulations and standards are being amended and developed to require these improvements. This is continuous improvement and risk assessment can lead the way. The term ALARP or as low as reasonably practicable is used in this book and what is reasonably practicable is increasing, continuously enabling risk to be further reduced going forward. Think about fire, smoke, and carbon dioxide (CO_2) detectors. At one time they had not been invented. Then, they were invented but expensive and only used for major hazards and operations. Now, they are simple, inexpensive, and installed in most homes.

3.17 SUMMARY

Safety professionals must understand the risk assessment process and be able to complete the steps in the risk assessment process competently if they are to fulfill their roles successfully. Risk assessments will be the norm in the future as they are required and referenced by more regulations and standards going forward. Just as safety professionals have been expected to be competent in OSHA regulations and their compliance in the past, they will be expected to lead hazard analyses and risk assessments. This is the advancement of the safety profession.

REVIEW QUESTIONS

1. Describe risk assessment's role within the framework of risk management.
2. Name three standards or guidelines that have the process of hazard analysis and risk assessment as a "required" core element.
3. State the primary purpose of an operational risk assessment.
4. Explain the difference between a hazard and a risk.
5. Within a risk assessment process, explain how "establishing the context" affects the process. When is the context established and what should it contain?
6. Provide a brief description of the three types of risk analysis values used.
7. Explain the concept of ALARP.

REFERENCES

ANSI/AIHA/ASSE Z10-2012. *American National Standard – Occupational Health and Safety Management Systems.* Fairfax, VA: American Industrial Hygiene Association, 2012.

ANSI/ASSE Z590.3-2011. *Prevention Through Design: Guidelines for Addressing Occupational Hazards and Risks in Design and Redesign Processes.* Des Plaines, IL: American Society of Safety Engineers, 2011.

ANSI/ASSE Z690.1-2011. *American National Standard – Vocabulary for Risk Management.* Des Plaines, IL: American Society of Safety Engineers, 2011.

ANSI/ASSE Z690.2-2011. *American National Standard – Risk Management Principles and Guidelines.* Des Plaines, IL: American Society of Safety Engineers, 2011.

REFERENCES

ANSI/ASSE Z690.3-2011. *American National Standard – Risk Assessment Techniques*. Des Plains, IL: The American Society of Safety Engineers, 2011.

ANSI B11.TR3-2000. *Risk Assessment and Risk Reduction – A Guide to Estimate, Evaluate and Reduce Risks Associated with Machine Tools.* New York: ANSI, 2000.

ASSE. 2011. Prevention Through Design: Guidelines for Addressing Occupational Hazards and Risks in Design and Redesign Processes [Webinar]. Retrieved from http://eo2.commpartners.com/users/asse/session.php?id=7823 (accessed December 17, 2015).

BP. 2010. Deepwater Horizon Accident Investigation Report. Retrieved from www.bp.com/liveassets/bp_internet/globalbp/globalbp_uk_english/incident_response/STAGING/local_assets/downloads_pdfs/Deepwater_Horizon_Accident_Investigation_Report.pdf (accessed December 17, 2015).

BS OHSAS 18001:2007. *Occupational Health and Safety Management Systems – Requirements.* London, UK: British Standards Institution (BSI), 2007.

Columbia Accident Investigation Board (CAIB). 2003. *Columbia Accident Investigation Board Report (Vol. 1).* Washington, DC: NASA. Retrieved from www.nasa.gov/columbia/home/CAIB_Vol1.html.

EU-OSHA. 2015. Risk Assessment – Safety and Health at Work. European Agency for Safety and Health at Work. Retrieved April 2, 2015 at: https://osha.europa.eu/en/topics/riskassessment/index_html.

ILO-OSH 2001. *Guidelines on Occupational Safety and Health Management Systems.* Geneva, Switzerland. International Labour Office, 2001.

ISO 14001:2004(E). *Environmental Management Systems – Requirements with Guidance for Use.* Geneva, Switzerland. International Organization for Standardization, 2004.

Lyon, B. K. and Bruce H., 2012. "Risk assessments – Top 10 pitfalls & tips for improvement." *Professional Safety,* 57(12), 28–34.

Manuele, F. A., *Advanced Safety Management: Focusing on Z10 and Serious Injury Prevention.* Hoboken, NJ: Wiley, 2008.

Manuele, F.A., 2010. "Acceptable risk: Time for SH&E professionals to adopt the concept." *Professional Safety,* 55(5), 30–38.

MIL-STD-882E. *Standard Practice for System Safety.* Washington, DC: Department of Defense, 2012.

OSHA. 2003. *Voluntary Protection Programs: Policies and Procedures Manual.* Washington, DC: U.S. Department of Labor, OSHA. Retrieved from www.osha.gov/OshDoc/Directive_pdf/CSP_03-01-003.pdf (accessed December 17, 2015).

Wikipedia. 2015. Quadrants (Plane Geometry). Retrieved April 18, 2015 from: http://en.wikipedia.org/wiki/Quadrant_%28plane_geometry%29 (accessed December 17, 2015).

4

DEFINING RISK ASSESSMENT CRITERIA

BRUCE K. LYON
Risk Control Services, Hays Companies, Kansas City, MO, USA

BRUCE HOLLCROFT
Risk Control Services, Hays Companies, Lake Oswego, OR, USA

4.1 OBJECTIVES

- Introduce risk assessment criteria
- Discuss risk scoring systems and their components
- Describe how to select, develop, and apply risk criteria
- Discuss risk reduction and the hierarchy of controls in risk estimation
- Review acceptable risk level criteria

4.2 INTRODUCTION

Before operational risks can be assessed, an organization must first establish and define the risk criteria from which to measure and score. Such criteria must be clearly defined and communicated by an organization to consistently evaluate operational risks and make proper risk-based decisions. A number of existing risk criteria measures are available; however, it is essential that an organization carefully selects and/or develops its own risk criteria to reflect its values, goals, industry setting, and overall culture.

In the *risk evaluation* stage, established *risk criteria* are compared to the results of the *risk analysis* to determine if the risks are acceptable. The following terms and their definitions taken from the *ISO Guide 73/ANSI Z690.1-2011, Vocabulary for Risk Management*, standard are presented to provide an understanding of their use (ANSI/ASSE Z690.1-2011):

Risk Assessment: A Practical Guide to Assessing Operational Risks, First Edition.
Edited by Georgi Popov, Bruce K. Lyon, and Bruce Hollcroft.
© 2016 John Wiley & Sons, Inc. Published 2016 by John Wiley & Sons, Inc.

3.3.1.3 Risk criteria. Terms of reference against which the significance of a risk (1.1) is evaluated.

3.6.1 Risk analysis. Process to comprehend the nature of risk (1.1) and to determine the level of risk (3.6.1.8).

3.7.1 Risk evaluation. Process of comparing the results of risk analysis (3.6.1) with risk criteria (3.3.1.3) to determine whether the risk (1.1) and/or its magnitude is acceptable or tolerable. NOTE: Risk evaluation assists in the decision about risk treatment (3.8.1).

In his book *Risk Assessment: Challenges and Opportunities*, Bruce Main states that "assessing risk is perhaps the most controversial part of the risk assessment process" (Main, 2012). This, in part, is due to the "great diversity of opinion" that exists as evidenced by the many variations of risk scoring methods available ranging from very basic to complex. The challenge for the safety professional is selecting the "right" risk criteria that will serve the needs and context of the endeavor and yield effective results. For certain, there is no single risk scoring system that suits all industries, applications, or situations.

Thoughtful consideration should be given in obtaining an effective risk criteria system. However, caution is advised in placing too much emphasis in finding the "perfect" risk scoring system. Rather, selecting a method that allows the user to effectively assess, measure, and achieve acceptable risk levels should be the primary focus (Main, 2012). Safety professionals are advised to educate themselves in the fundamentals of risk criteria so as to appropriately select and develop specific risk scoring methods within their organization. This chapter will provide an overview of the concepts and applications of risk criteria used in operational risk assessments.

4.3 DEFINING RISK CRITERIA

Risk criteria are the reference points against which the significance of risk are evaluated and measured. Such criteria are derived from the organization's culture and industry, external and internal context, applicable laws, standards, and other requirements. In general, risk criteria should include a risk scoring system that includes risk factors, defined scales of risk levels, and a risk matrix for an organization to measure risk for the purpose of prioritizing and making proper decisions.

Risk is defined as a hazard's estimated likelihood (or probability) of occurrence and the resulting severity of consequences. Thus, risk criteria must begin with these two *risk factors*: *likelihood* and *severity*. Several other risk factors can be added to further define risk such as exposure or frequency of exposure, duration of exposure, vulnerability, failure detectability, control reliability (CR), and prevention effectiveness (PE).

In *ISO 31010/ANSI Z690.3-2011, Risk Assessment Techniques*, the following guidance is provided regarding selecting and defining risk criteria in *Section 4.3.3, Establishing the Context (ANSI/ASSE Z690.3-2011)*:

(d) Defining risk criteria involves deciding:

- *the nature and types of consequences to be included and how they will be measured,*
- *the way in which probabilities are to be expressed,*
- *how a level of risk will be determined,*
- *the criteria by which it will be decided when a risk needs treatment,*
- *the criteria for deciding when a risk is acceptable and/or tolerable,*
- *whether and how combinations of risks will be taken into account.*

Criteria can be based on sources such as:

- *agreed process objectives,*
- *criteria identified in specifications,*
- *general data sources,*
- *generally accepted industry criteria such as safety integrity levels,*
- *organizational risk appetite,*
- *legal and other requirements for specific equipment or applications.*

As indicated in this guidance, stakeholders charged with determining risk criteria must consider the overall context of the assessment and any applicable factors such as operational, technical, financial, legal, social, or environmental. The assessment context, including the setting, situation, and perspective of the external and internal environment and stakeholders, plays a major role in the development of risk criteria. The "internal" risk criteria may be derived from an organization's standards, policies, procedures, culture, and an organization's overall tolerance for risk. Outside influences affecting risk criteria may include laws, regulations, contractual requirements, and political, stakeholder, and societal expectations. To be effective, risk criteria must correspond to the type of risks and the way in which risk levels are expressed. Further development or refinement of criteria for particular risks identified during the process may also be required.

Much of the risk assessment criteria are related to defining the scoring or ranking of risk including the severity of consequences and likelihood of occurrence as part of the overall risk assessment process. When defining the context and criteria of the risk assessment process, the following elements should be considered:

- The purpose, goals, and objectives of the risk assessment
- Responsibilities in the risk assessment process
- Specific inclusions and exclusions
- Resources, time, and location
- The risk assessment methodologies to be used
- The risk scoring system to be used in the assessment.

4.4 RISK SCORING SYSTEMS

The primary purpose of risk assessment is to identify hazards and assess and reduce their risk to an acceptable level. To achieve this, a measurement system that includes a baseline (an organization's acceptable risk level) and a method of scoring (a risk scoring system) must be established.

Two of the better references on risk scoring systems can be found in Fred Manuele's *Innovations in Safety Management* (2001) and Bruce Main's *Risk Assessment: Challenges and Opportunities* (2012). Manuele was the first to use the term "risk scoring system" for operational risk assessments in his 2001 book. He states that two-dimensional risk assessment matrices using likelihood (L) of event occurrence and severity (S) of consequence have been commonly used in risk assessment exercises. However, risk scoring systems with three or four risk factors are becoming more common, adding a third or fourth factor such as failure detectability, control effectiveness, vulnerability, or others.

What is meant by the term *risk scoring system*? The short answer is a means or method to consistently measure and compare risks. In the American National Standard, *ANSI B11.0-2015*, *Safety of Machinery – General Requirements and Risk Assessment*, a section is devoted to risk scoring systems. *Section 6.4.1*, *Selecting a Risk Scoring System*, states that risks shall be assessed using a scoring system that combines risk factors to produce a risk level. The system selected may be specified by the organization, stakeholder, or industry and should allow prioritized action for risk reduction. A review of standard risk scoring systems reveals several common components in their makeup including:

1. **Risk factors**. Specific *risk factors* used to measure risk such as severity, likelihood or probability, frequency, duration, failure detectability, control effectiveness, vulnerability, or other risk measures
2. **Risk levels**. Specified *risk levels or categories* for each risk factor (typically 3–6 levels)
3. **Risk values**. Defined *qualitative, quantitative, or semiquantitative values* for risk levels
4. **Risk actions**. Decision guidelines or *action required* for each risk level
5. **Risk screening and communication tools**. A risk assessment *matrix, graph, or risk priority numbers* (RPNs) used to measure, screen, compare, and prioritize as well as communicate risk within an organization.

TABLE 4.1 Variables in Select Risk Scoring Systems

Standard/System	Values	Risk Factors	Matrix Type	Risk Levels/Categories
ANSI B11.0-2015	Qualitative	Probability (P) of occurrence × severity (S) of harm	4 × 4	Four risk levels: • High • Medium • Low • Negligible
ANSI Z10-2012	Qualitative	Likelihood (L) of exposure × severity (S) of injury or illness	5 × 4	Four risk levels with actions required: • High • Serious • Medium • Low
ISO 31010/ANSI Z690.3-2012	Semiquantitative	Likelihood (L) × consequence (C)	5 × 6	Five risk levels: • I (highest) • II • III • IV • V (lowest)
MIL-STD 882E	Qualitative	Probability (P) × severity (S)	6 × 4	Five risk levels: • High • Serious • Medium • Low • Eliminated
ANSI Z590.3 PtD	Semiquantitative	Severity (S) × probability (P)	5 × 5	Four descriptive risk levels: • Very high risk • High risk • Moderate risk • Low risk

In Table 4.1, five risk scoring systems from key standards are highlighted with specific variables for values, risk factors used to derive risk, number of risk levels and correlated actions required, and matrix type available.

For most risk assessments, the criteria categories used are either (1) adopted from a variety of the standards such as those referred to in this book, (2) previously established by an organization or industry group, or (3) created to meet the needs of the particular risk assessment. The safety professional should be competent at selecting or developing a risk scoring system to meet the needs of the risk assessment.

4.5 RISK ASSESSMENT MATRICES

A key part of a risk scoring system is the risk assessment matrix. A matrix helps visualize and communicate risk levels to decision makers by providing a means for categorizing combinations of likelihood and severity and their risk levels. They are often used as a screening tool when there are many risks to evaluate. The selection of a matrix is generally made before or during the development of the risk assessment's context, as discussed in Chapter 3 of this text.

As mentioned in Chapter 3, several best practices are offered when selecting or developing risk scoring systems. First, the risk rating criteria and matrix used within an organization should remain consistent. Second, it is the authors' opinion that risk scoring numerical values should intuitively correspond with risk levels. In other words, higher risk scores indicate higher risk levels, while lower risk scores represent lower risks. (Most risk assessment models follow this guideline; however, a couple of exceptions are found in ISO 31010/ANSI Z690.3, the MIL-STD 882E standards.) Third, the risk matrix should be positioned in the first quadrant of the Cartesian coordinate system – presenting risk levels from low to high, reading left to right and from bottom to top (refer to Chapter 3).

Many examples are available ranging from three to six levels or magnitudes of severity (S) of consequences and likelihood (L) of occurrence. The following are several recommended references for risk assessment matrices to be considered:

- *ANSI Z10-2012, Occupational Health and Safety Management Systems*
- *ANSI Z590.3-2011, Prevention through Design*
- *ISO 31010/ANSI Z690.3-2011, Risk Assessment Techniques*
- *ANSI B11.0-2015, Safety of Machinery, General Requirements and Risk Assessment*
- *MIL-STD 882E, 11 May 2012, Department of Defense Standard Practice, System Safety*
- *Risk Assessment: Challenges and Opportunities* (2012), Bruce W. Main
- *Advanced Safety Management: Focusing on Z10 and Serious Injury Prevention* (2014) and *Innovation in Safety Management* (2001), Fred A. Manuele
- *ASSE's Risk Assessment Institute – www.oshrisk.org/.*

4.6 DEFINING RISK VALUES

Risk assessment models and their matrices may be classified as qualitative, semiquantitative, or quantitative. When defining the risk criteria and risk scoring system to be used, stakeholders must take into consideration the level of detail desired and data and resources available. Most matrices have qualitative or semiquantitative values using descriptive or subjective measures.

TABLE 4.2 Qualitative Risk Assessment (5 × 4) Matrix

	←Severity of Injury or Illness Consequence→			
Likelihood of Occurrence or Exposure for Select Unit of Time or Activity	Negligible	Marginal	Critical	Catastrophic
Frequent	Medium	Serious	High	High
Probable	Medium	Serious	High	High
Occasional	Low	Medium	Serious	High
Remote	Low	Medium	Medium	Serious
Improbable	Low	Low	Low	Medium

Source: Reprinted with Permission from ANSI/AIHA/ASSE Z10-2012 (Courtesy of the American Society of Safety Engineers).

4.6.1 Qualitative Risk Models

Qualitative risk models are based on qualitative or subjective descriptions rather than numerical or statistical data and require less precise information to be developed and used. Qualitative risk models define severity of consequence, likelihood, and level of risk using descriptive words such as "high," "medium," and "low," which are evaluated according to qualitative criteria. An example of a qualitative risk assessment matrix reprinted with permission from ANSI/AIHA/ASSE Z10-2012 is presented later in Table 4.2.

4.6.2 Semiquantitative Risk Models

Semiquantitative risk models use qualitative data; however the values are expressed as numerical risk ratings using a formula to produce a risk level or score. Risk level scores produced can be linear or logarithmic based on the formula selected. One advantage of a semiquantitative model is that more precision can be given by adding definitions that include some numerical ranges for the severity of consequences and likelihood of occurrence. The addition of numerical ranges often helps in comparing and communicating risk levels. In Tables 4.3–4.6, an example of a semiquantitative model, which includes a risk matrix, descriptions for risk factors, and risk scoring, is provided. It should be noted that the numbers used in semiquantitative models are typically determined by "qualified" judgments and experience, largely without quantitative data, and only have value in relation to each other.

TABLE 4.3 Semiquantitative Risk Assessment (5 × 4) Matrix Example

	←Severity of Injury or Illness Consequence→			
Likelihood of Occurrence or Exposure for Select Unit of Time or Activity	Negligible (1)	Marginal (2)	Critical (3)	Catastrophic (4)
Frequent (5)	5	10	15	20
Probable (4)	4	8	12	16
Occasional (3)	3	6	9	12
Remote (2)	2	4	6	8
Improbable (1)	1	2	3	4

DEFINING RISK VALUES

TABLE 4.4 Semiquantitative Descriptions for Likelihood of Occurrence Example

Risk Level	Likelihood (L) of Occurrence	Description
5	Frequent	Almost certain to occur. Has occurred more than once within the last 12 months. Conditions exist for it to occur
4	Probable	Very likely to occur. Has occurred once within the last 12 months. Conditions often exist for it to occur
3	Occasional	Likely to occur if conditions exist. Has occurred within the last 24 months. Conditions can exist for it to occur
2	Moderate	May occur if conditions exist. Has occurred within the last 36 months. Conditions sometimes exist for it to occur
1	Unlikely	Unlikely to occur. Has not occurred within last 5 years. Conditions rarely exist for it to occur

TABLE 4.5 Semiquantitative Descriptions for Severity of Consequence Example

Risk Level	Severity (S) of Consequence	Description
4	Catastrophic	One or more fatalities; multiple serious hospitalizations; incident resulting in more than $250K
3	Critical	Disabling injury or illness; permanent impairment; incident resulting in more than $50K
2	Marginal	Medical treatment or restricted work; recordable incidents; incident resulting in more than $1K
1	Low	First aid or nontreatment incidents; incident resulting in less than $1K

TABLE 4.6 Risk Scoring Levels and Action Example

Risk Level	Risk Score	Action
Very high	12 or greater	Operation not permissible; immediate action required
High	8–10	Remedial action required; high priority
Moderate	4–6	Remedial action suggested
Low	1–3	Remedial action discretionary

4.6.3 Quantitative Risk Models

Quantitative risk models use data to define values for severity of consequences and likelihood of occurrence and produce risk level values in specific numerical units. As described in ISO 31010/ANSI Z690.3, "full quantitative analysis" may not be possible or desired if there is insufficient information or data available about the system or activity to be analyzed or the efforts required exceed the needs of the assessment. It should be recognized that even with fully quantitative values, the levels of risk calculated are still "estimates." However, a comparative ranking of numerical risk ratings by stakeholders is oftentimes useful and necessary for the context of the assessment.

4.7 RISK FACTORS

Risk factors are the components of risk derived from an identified hazard that are estimated and measured to produce a risk score. Risk factors can be identified by a simple risk assessment code (RAC), generally with capitalized initials within parentheses to denote its identity such as (S) for severity of consequence or (L) for likelihood of occurrence.

Risk assessments generally have two-dimensional risk scoring systems, which use two risk factors such as severity (S) of consequence and likelihood (L) or probability (P) of occurrence. As an example, the *ANSI Z590.3-2011, Prevention through Design* standard presents a two-factor scoring system with risk codes for the severity (S) of consequences and occurrence (O) probability. Severity of consequence rating (S) is described as an estimate of the magnitude of harm or damage that could reasonably result from a hazard-related incident or exposure. The occurrence (O) probability rating is based on the likelihood of the potential failure or hazardous event occurring. The ANSI Z590.3 standard defines probability as "an estimate of the likelihood of an incident or exposure occurring that could result in harm or damage for a selected unit of time, events, population, items, or activity being considered" (ANSI/ASSE Z590.3-2011).

The following are risk factors and their RACs used in various three- and four-dimensional models:

- **Exposure (E)** is used as a general measure of exposure events/units.
- **Frequency of exposure (F)** is used as a number of exposure events for a unit of time.
- **Time duration of exposure (T)** is used as a time period that a single exposure occurs.
- **Vulnerability (V)** is sometimes used in security threat analyses and generally refers to weaknesses in a system that are factored into the risk estimation.
- **Detection of failure (D)** is used in many FMEA models as a third risk factor in the risk level scoring system. The detection rating is based on an estimate of how easily the potential failure could be detected prior to its occurrence.
- **Control reliability (CR)** is used in machine risk assessments and factors the reliability of a selected control into the risk estimation.
- **Prevention effectiveness (PE)** is a risk factor sometimes used in FMEA and other methods to evaluate a control effectiveness in preventing a failure from occurring.

Note: When formulating the risk scoring system, it is advised to keep the number of different risk factors to the minimum necessary that best estimate risk for the application.

4.8 RISK LEVELS

Risk levels or categories are the defined graduated levels of increasing risk established for each risk factor and for risk scores produced from combined risk factors used in the risk assessment matrix. For each risk factor, defined risk levels ranging from low to high are developed as shown in Tables 4.4 and 4.5. The combined risk factors produce a risk score or rating that are sometimes accompanied with required action such as immediate work stoppage or required controls to reduce risk to an acceptable level as shown in Table 4.6. At the low end, the acceptable risk level is defined. These risk levels are incorporated into the risk assessment matrix and are used to determine, rank, and compare risks.

Decisions for selecting or creating a risk scoring system and its matrix will include determining the magnitude and number of categories for severity (S) of consequence and likelihood (L) of occurrence desired. Matrix configurations often have three to six level structures; however some models use varying combinations of severity and likelihood or probability levels such as the 4×5 matrix found in ANSI Z10 or the 4×6 matrix in MIL-STD 882E (MIL-STD-882E, 2012).

The number of graduated risk levels to be used in the risk assessment matrix should be considered. Too few levels and the resulting risks may not be adequately separated, while too many levels can lead to confusion of ranking and wasted resources. The assessment team should agree on the number of risk levels that create a useful scale for distinguishing important differences in risk so that decisions can be made. Some stakeholders prefer four levels for each risk factor (4 × 4) so that no middle exists; risks are either ranked below or above the middle in the matrix. Others use five categories (5 × 5) so that there is a midlevel option of risk. In either case, safety professionals should be capable of modifying or creating a custom risk scoring system and matrix to meet the needs of the risk assessment and decision makers in the organization.

4.9 RISK SCORING

In many risk assessment models such as failure mode and effect analysis (FMEA), a third factor, "detection of failure" (D), is used to score risk. The detection rating is based on an estimate of how easily the potential failure can be detected prior to its occurrence. In other applications, where hazard control measures are analyzed, a "prevention effectiveness" (PE) factor is used. Prevention effectiveness is an estimate of a control's efficacy in preventing the failure and its effects.

Risk scores are produced by combining risk factors. When three or more risk factors are used, an RPN is produced. As described by ANSI Z590.3, the RPN is a semiquantitative measure of criticality obtained by multiplying numbers from rating scales (usually between 1 and 10) for consequence of failure, likelihood of failure, and ability to detect the problem. (A failure is given a higher priority if it is difficult to detect.) For instance, a three-dimensional risk score formula using "detection of failure" would be represented as follows:

$$Risk\ Priority\ Number = Severity \times Likelihood \times Detection$$

There are other models suggesting a four variable scoring system including severity, probability, frequency of exposure and detection, or other risk factors. The use of three and four risk factor systems should be carefully examined. As indicated by Manuele, a four-factor risk scoring system can distort or dilute severity level of a particular risk if all four factors are given equal weight. For instance, Manuele presents a hypothetical scenario of a fatality that is obviously an unacceptable risk; however, when applying a four-factor risk score, it is rated as acceptable. This occurs due to a dilution of severity by the other three factors through the mathematical scoring giving each risk factor a weighting of 25%. In a three-factor risk scoring system where risk is estimated by multiplying the results of severity, probability, and frequency of exposure, severity is also discounted in the final score (Manuele, 2014). With three factors in the equation, each has a weighting of 33% of the final risk score as shown in the following.

$$Severity \times Probability \times Frequency\ of\ Exposure = Risk$$

To more accurately score risk levels, Manuele proposes that severity receive proper weighting to reflect the impact severity has on incident outcomes. In the following equation, the rating for occurrence probability and rating for frequency of exposure are *added* together and then multiplied with severity:

$$Severity \times (Probability + Frequency\ of\ Exposure) = Risk$$

As a final note, hazards that present high severity and low probability risks, especially those that have a potential for fatalities and serious incidents should be reviewed very carefully.

4.10 SEVERITY OF CONSEQUENCE

Consequences are the results, outcomes, or losses of an event caused by a hazard(s). Consequences most often refer to the damage or harm caused to people, assets/property, or the environment. As a primary risk factor, the "severity levels" of consequences to be used in an assessment must be determined upfront, during the development of the context. This will include the types of consequences and levels of severity.

Different types of consequences may be necessary for various types of risk assessments. For example, when conducting a product safety risk assessment, descriptions of product failure types or levels might be used. For a motor vehicle risk assessment, varying degrees of severity of vehicle accidents might be appropriate, while an assessment of health risk may use the spread and severity of disease as severity risk levels. Such data may exist in governmental agencies such as workplace injuries and fatalities from the Department or Labor or Bureau of Labor Statistics or motor vehicle accidents and fatality rates per miles driven from the Department of Transportation. Other sources may include insurance or industry groups.

Hazard events may produce a range of impacts with different magnitudes or severity levels and may affect a number of different assets, objectives, or stakeholders. It is possible, and sometimes common, to have more than one consequence from a particular hazard or risk. For example, a large flammable chemical tank explosion might destroy a major portion of a plant or factory, but it might also result in toxic releases that could harm the health of employees or others.

Risk should be evaluated for the *worst credible case* rather than worst *conceivable* risk (Clemens & Simmons, 1998). A worst credible consequence is something that has a reasonable potential of occurring within the lifetime of the system, while a worst conceivable risk could occur, but would not likely occur, within the system's lifetime (ISO Guide 73/ANSI Z690.1-2011). It has been the authors' experience that on occasion overly ambitious team members are tempted to get creative and propose unrealistic consequences such as "being struck by an asteroid." While worst *conceivable* events may be possible, they are highly unlikely and would not warrant consideration.

Historical incident data can be of significant value in forming a baseline when defining severity levels. As described by Fred Manuele, "informed speculations are made to establish the consequences of an incident or exposure." When developing the range of severity levels to be used in an assessment, certain values can be used such as the number and severity of injuries and illnesses, value of property or assets damaged, extent of environmental impact, or loss of production or business.

Severity of consequence levels may range from a basic description of outcomes to a very detailed quantitative model depending upon the context of the risk assessment. As suggested by ISO 31010/ANSI Z690.3, the assessment team should answer some key questions as to how severity will be assessed such as the following:

TABLE 4.7 Severity of Consequence Categories Example

Severity Category	Injury/Illness Levels	Financial Loss Levels
Catastrophic (4)	Fatality(ies) or permanent total disability	More than $1 M
Critical (3)	Hospitalizations, permanent–partial or temporary disability in excess of three months	$100K–$1 M
Marginal (2)	Recordable injury/illness, minor injury, lost workday incident	$10–$100K
Negligible (1)	First aid or minor medical treatment	$0–$10K

LIKELIHOOD OF OCCURRENCE

- Will the analysis consider the effect of existing controls on consequences, or will "raw risk" be initially determined?
- In addition to direct/immediate consequences, will the analysis include delayed or long-tail consequences that may later arise?
- Will secondary consequences that may affect outside or adjacent elements, systems, or activities be considered?
- How will events with an estimated low severity but high likelihood of occurrence be treated?
- Will high severity and low likelihood events such as those with a *fatality or serious incident* (FSI) potential be given more weight? Such cases are of greater concern by management and may warrant greater attention.
- How will frequently occurring low-impact or chronic problem with larger long-term effects be treated?
- What type severity values will best communicate severity levels to management: dollar amounts, production rates, injury or illness measures, or by other values?

Although the safety professional generally facilitates this process, risk criteria decisions are determined by the risk assessment team. Input from end users, management, and other affected stakeholders should be solicited. For example, the organization's chief financial officer (CFO), treasurer, risk manager, etc. could be queried about financial impact when using dollar figures to describe consequences with monetary values. Common questions such as "how much financial loss could be tolerated without concern" to identify a low level consequence and "what dollar amount would be considered devastating to the organization" to determine a high level consequence. Gaining input from all affected stakeholders improves communication and the overall credibility of the risk assessment.

The completed severity categories and their descriptions will be used in the risk matrix. For example, the risk assessment matrix provided in ANSI Z10 uses four categories of severity: (1) negligible, (2) marginal, (3) critical, and (4) catastrophic. It is important that severity categories are clearly defined so that consequences can be consistently ranked or scored by the risk assessment team. Table 4.7 provides some basic examples of severity categories and simple definitions using injury/illness severity and dollar losses (as do previous tables in this chapter).

It should be noted that in a risk assessment, severity of consequence is estimated first, followed by an estimation of its likelihood of occurring.

4.11 LIKELIHOOD OF OCCURRENCE

Risk is the product of a hazard's severity of consequence and its likelihood of occurrence. Likelihood is sometimes referred to as probability in risk management terminology. Although these terms are often used interchangeably, there are distinctions to take into consideration. *Likelihood* is the chance of an event or something happening, generally expressed qualitatively. *Probability* is a quantitative or numerical measure of the chance of something happening expressed as a percentage. Both can be used successfully. Definitions for likelihood or probability must be selected or developed and provide a clear understanding of their meaning for each level.

Three methods are used to estimate likelihood or probability for risk assessments. These include the use of (1) historical data, (2) predictive techniques, and (3) expert opinion. If adequate data of past events and losses exist, such information can be used to help project future probabilities of occurrence. This method requires sufficient data relevant to the type of system or activity being considered. When such data are unavailable, probability forecasts using methods such as fault tree analysis or event tree analysis can be applied to arrive at probability estimates. In many cases, experienced safety professionals or risk experts are used to further analyze and estimate probability. All three methods can be used individually or jointly depending on the context and the resources available.

Likelihood of occurrence level categories can range from three to six different levels but most often in four or five such as the following descriptive categories:

1. Improbable/very unlikely
2. Remote/seldom/rare
3. Occasional/possible
4. Probable/likely
5. Frequent/very likely.

These likelihood categories will often be further defined. This should be done by the risk assessment team with input from the organization and users of the risk assessment information. A simple set of definitions are provided in Table 4.8.

Some risk assessment likelihood criteria can be fairly quantitative. For example, a natural hazard risk assessment could consider statistics on weather events such floods and other natural hazards such as earthquakes. Information is available on the frequency and magnitude of these types of events. As previously mentioned, injury, illness, and fatality rates exist and as do accident and fatality rates for motor vehicle miles. International SEMATECH is a consortium of 13 semiconductor manufacturers from seven countries, which have developed industry guidelines for risk assessment. Table 4.9 taken from the *Failure Mode and Effects Analysis (FMEA): A Guide for Continuous Improvement for the*

TABLE 4.8 Likelihood Categories Example

Category Description	Time Period	Frequency
Improbable	Century	Every 100 years or more
Remote	Decade	Every 10–100 years
Occasional	Annually	Every 1–10 years
Probable	Monthly	Every 1–12 months
Frequent	Weekly	Every 1–4 weeks

TABLE 4.9 Scoring Table for Occurrence Ranking

Occurrence Ranking Criteria	
Rank	Description
1	*An unlikely probability of occurrence* during the item operating time interval. Unlikely is defined as a single failure mode (FM) probability < 0.001 of the overall probability of failure during the time operating interval
2–3	*A remote probability of occurrence* during the item operating time interval (i.e., once every two months), defined as single FM probability > 0.001 but <0.01
4–6	*An occasional probability of occurrence* during the item operating time interval (i.e., once a month), defined as a single FM probability > 0.01 but <0.10
7–9	*A moderate probability of occurrence* during the item operating interval (i.e., once every two weeks), defined as a single FM probability > 0.10 but <0.20
10	*A high probability of occurrence* during the item time operating interval (i.e., once a week), defined as a single FM probability > 0.20

Source: Failure Mode and Effects Analysis (FMEA): A Guide for Continuous Improvement for the Semiconductor Equipment Industry, SEMATECH Technology Transfer #92020963B-ENG. Reprinted with permission.

Semiconductor Equipment Industry document published by SEMATECH provides more quantitative descriptions for likelihood (International SEMATECH, 1992).

Once the credible consequence and likelihood criteria are established, the risk assessment team can make the estimates, chart them on the risk assessment matrix, and have a high degree of confidence in their assessment of the risk being considered.

4.12 EXPOSURE

Exposure is an indication of the extent to which the organization is subject to the consequences based of the amount of exposure in numbers. According to the "law of large numbers," the greater the number of exposures, the greater the likelihood or probability of an occurrence. Some risk assessments include exposure as a third risk factor to severity and likelihood. Exposure can be measured as the frequency of an event or exposure, its duration, and/or the assets exposed to risk. *Frequency of exposure* describes how often an event might take place over a specified time period. *Duration of exposure* is the length of time that a single exposure occurs. Some of the variables for exposure might include the following:

- The number of employees or people exposed
- How frequent an activity is performed
- The miles driven or number of vehicles used in transportation
- The number of customers or products for a product risk assessment
- The number of locations or facilities for a property risk assessment.

There are a number of variations in such risk factors used to score risk. Considering "exposure" in a risk assessment can lead to a deeper, more thorough risk assessment process, however, it also increases complexity and consuming time. The risk assessment team should evaluate the need for additional risk factors to fit the needs of the context. In many cases, a separate "exposure" risk factor may not be necessary since "exposure level" is normally considered in likelihood and probability ratings.

4.13 RISK REDUCTION AND THE HIERARCHY OF CONTROLS

Dave Walline states that a major misconception currently held by many in operational risk is the belief that low level controls such as administrative controls, warnings, and personal protective equipment mitigate severity of risk. He states that low level controls only have the ability to reduce the "likelihood" of an event (Walline, 2014). While the argument can be made that personal protective equipment can reduce the "impact" of certain consequences (i.e., fall protection harness, or hard hat) such controls are considered the least reliable and least effective. Unfortunately, such controls are often the first and only choice. To compound problems, many times organizations view layers of protection or controls such as machine guards, training, and PPE as each having the same mitigation weight. These misconceptions point to the importance of the concept of the hierarchy of controls and its application in selecting risk reduction measures.

Assessing risk requires an evaluation of existing controls. During the risk assessment process, risk is assessed before and after control measures are applied providing an initial risk level and a residual risk level. In many cases, there are existing mitigation measures in the system or process being assessed. During the development of the risk assessment context, decisions regarding existing risk controls and their consideration in risk assessment should be made. Will initial risk estimations include the effect of existing controls? Or will raw risk be first measured? The following are several terms and their uses in risk estimation:

Initial risk (*raw risk*) – Initial risk is assessed by assuming no risk reduction methods are in place. This is sometimes referred to as the "raw risk." Initial risk serves as a baseline for the measurement of further risk reduction. Raw risk estimations may also be applicable to situations where the existing controls are considered very low on the hierarchy of controls and/or unreliable such as warnings and administrative-level or PPE-type controls.

Residual risk – Risk cannot be completely eliminated. Therefore, the risk that remains after risk reduction measures are put into place is called residual risk. Following an initial risk estimate, a residual risk assessment is performed to include existing controls and their effect on risk level. Residual risk assessments occur for "current state" (present conditions) as well as "future state" (projected risk level) for anticipated additional controls. Residual risk assessments should be conducted to validate the effectiveness of selected risk reduction measures.

Current state risk – For present conditions, a risk assessment that takes into consideration existing controls and their effects on risk is referred to as a "current state" risk level.

Future state risk – When additional control measures are proposed, a "future state" risk level is estimated considering their effect in reducing risk. Future state residual risk assessments are performed to also validate and support the proposed risk reduction measures.

In ISO 31010/ANSI Z690.3, Risk Assessment Techniques, the standard provides guidance on the assessment of controls in Section 5.3.2, Controls Assessment, as part of the risk analysis. Assessment of controls for a particular risk should determine and assess several key elements:

- The type of control(s) for a specific hazard
- The adequacy of control level as it relates to the hierarchy of controls to reduce risk to an acceptable level
- Demonstrated effectiveness of existing control
- Reliability of existing control.

The hierarchy of controls is a model for identifying the risk reduction effectiveness of control types. It is defined by ANSI Z590.3, Prevention through Design, as:

A systematic approach to avoiding, eliminating, controlling, and reducing risks, considering steps in a ranked and sequential order, beginning with avoidance, elimination, and substitution. Residual risks are controlled using engineering controls, warning systems, administrative controls, and personal protective equipment.

As indicated in the definition, the most effective risk reduction is achieved through avoidance of the risk or elimination by design or redesign. This requires risk reduction efforts in the design phase, which are further discussed in Chapters 10 and 11. The remaining "residual" risks in the system are then controlled through substitution for lesser hazardous components, engineering applications to control existing hazards, warnings and administrative-type controls to manage hazards, and finally PPE to create a final barrier between the hazard and the individual. Figure 4.1 illustrates the hierarchy of control model reprinted with permission from ANSI/ASSE Z590.3-2011, Prevention through Design.

The hierarchy of controls is used in the assessment of existing controls and the selection of new risk reduction measures. If the risk is considered unacceptable after the initial or preliminary risk assessment, then the hierarchy of controls is applied to the consideration of additional or improved controls.

RISK REDUCTION AND THE HIERARCHY OF CONTROLS   81

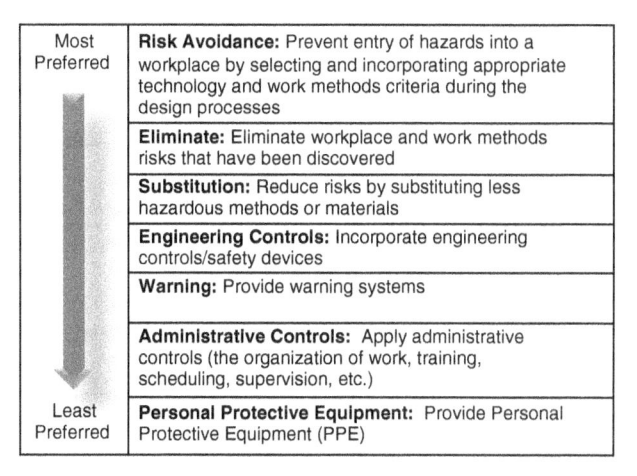

Most Preferred ↓ Least Preferred	
	Risk Avoidance: Prevent entry of hazards into a workplace by selecting and incorporating appropriate technology and work methods criteria during the design processes
	Eliminate: Eliminate workplace and work methods risks that have been discovered
	Substitution: Reduce risks by substituting less hazardous methods or materials
	Engineering Controls: Incorporate engineering controls/safety devices
	Warning: Provide warning systems
	Administrative Controls: Apply administrative controls (the organization of work, training, scheduling, supervision, etc.)
	Personal Protective Equipment: Provide Personal Protective Equipment (PPE)

Figure 4.1 Risk Reduction Hierarchy of Controls. *Source:* Reprinted with Permission from ANSI/ASSE Z590.3-2011 (Courtesy of the American Society of Safety Engineers)

In certain Risk Assessment Techniques such as FMEA, "prevention effectiveness" is a risk factor considered in the estimation of risk. Prevention effectiveness is an estimate of a control measure's efficacy in controlling the failure and its effects and is determined according to the hierarchy of control model. The prevention effectiveness (PE) risk factor may be preferred to other factors such as "detection of failure" since PE aligns closely with the "Prevention through Design" concept found in the ANSI Z590.3 standard.

The following progression of controls should be considered and applied in the risk control process:

1. **Risk avoidance or elimination** should be the first choice when in the design or redesign stage, since they have the greatest risk reduction and reliability factors and generally are the most cost-effective. If a hazard is avoided or eliminated in its design, the hazard and its risk will not exist in the system. This should be the safety professional's goal in all cases possible.
2. **Substitution** of a less hazardous component or element for a more hazardous one such as a nonflammable chemical for a highly flammable chemical is considered very effective. By changing the nature of the hazard, risk is reduced.
3. **Engineering controls** and design changes to existing systems can be applied to contain or prevent exposure to a hazard. Engineering controls are considered effective and reliable when the application is functioning as intended. These types of controls often require ongoing maintenance and testing to verify their effectiveness. An example might be a well-guarded machine with interlocks that prevent access to moving parts that could cause harm.
4. Controls below this line are considered less reliable since they rely on human behavior to a large degree. **Warning systems** require the individual to recognize and obey the warnings. Warnings can be compromised, obstructed, and ignored if false alarms or warning are given or may fail to operate due other factors.
5. **Administrative controls** such as policies, procedures, and training are only effective if they are known, reinforced, and followed. These measures rely on the individual to understand their intent and the adherence by the individual to these protocols. Communication, training, retraining, supervision, reinforcement, and discipline are required to "manage" employee practices to ensure they are in line with the administrative controls.
6. **PPE** is commonly used as a last defense against hazard exposures in the workplace. Their effectiveness is limited and dependent upon the proper selection, use, and care of equipment.

Figure 4.2 The Hazard Control Hierarchy. *Source:* Reprinted with Permission from ANSI B11.0-2015 – Safety of Machinery

Risk Reduction Measures	Examples	Influence on Risk Factors	Classification
Elimination or Substitution	• Eliminate pinch points (increase clearance) • Intrinsically safe (energy containment) • Automated material handling (robots, conveyors, etc.) • Redesign the process to eliminate or reduce human interaction • Reduced energy • Substitute less hazardous chemicals	• Impact on overall risk (elimination) by affecting severity and probability of harm • May affect severity of harm, frequency of exposure to the hazard under consideration, and/or the possibility of avoiding or limiting harm depending on which method of substitution is applied	Design Out
Guards, Safeguarding Devices, and Complimentary Measures	• Barriers • Interlocks • Presence sensing devices (light curtains, safety mats, area scanners, etc.) • Two-hand control and two hand trip devices	• Greatest impact on the probability of harm (occurrence of hazardous events under certain circumstance) • Minimal if any impact on severity of harm	Engineering Controls
Awareness Devices	• Lights, beacons, and strobes • Computer warnings • Signs and labels • Beepers, horns, and sirens	• Potential impact on the probability of harm (avoidance) • No impact on severity of harm	Administrative Controls
Training and Procedures	• Safe work procedures • Safety equipment inspections • Training • Lockout / tagout / verify	• Potential impact on the probability of harm (avoidance and/or exposure) • No impact on severity of harm	
Personal Protective Equipment (PPE)	• Safety glasses and face shields • Ear plugs • Gloves • Protective footwear • Respirators	• Potential impact on the probability of harm (avoidance) • No impact on severity of harm	

(Most Preferred → Least Preferred)

As mentioned in ANSI Z590.3, a combination of risk reduction measures is many times required to achieve acceptable risk levels. Lower levels in the hierarchy of controls should only be selected after practical applications of higher-level controls are considered. For these reasons, controls lower on the hierarchy should not be given much of any credit during the initial or preliminary risk assessment. ANSI Z590.3, Addendum I – The Logic Supporting the Hierarchy of Controls (Informative), provides additional information on the hierarchy model.

A second source for the hierarchy of control concept is presented in the ANSI B11.0-2015, *Safety of Machinery – General Requirements and Risk Assessment*, standard. B11 refers to this as the "hazard control hierarchy" and states that risk reduction can be achieved by lowering the severity of harm, improving the likelihood of avoidance and reducing the need for exposure to the hazard. In selecting the most appropriate risk reduction measures, the hazard control hierarchy principles should be applied in the order they appear in Figure 4.2, reprinted with permission from ANSI B11.0-2015 – Safety of Machinery.

> Note: A significant element added to the ANSI B11.0 hierarchy is the column for 'Influence on Risk Factors.' Notice that lower level 'Administrative Controls' have little or no impact on severity of harm, and are considered the least effective and least reliable.

In ANSI B11.0, *Annex E – Approaches to Risk Reduction*, the importance of higher-level controls is emphasized in the two-stage approach that is introduced. In its approach, it states that it is necessary to implement the hierarchy of controls in two stages. First, eliminate the hazard entirely or substitute hazardous elements with those that are less hazardous where possible. Second, apply the remaining hazard control options (engineering controls, awareness systems, procedures, training, and PPE) in a balanced approach to reduce risk to an acceptable level. This is illustrated in Figure 4.3, reprinted with permission from ANSI B11.0-2015 – Safety of Machinery.

RISK REDUCTION AND THE HIERARCHY OF CONTROLS 83

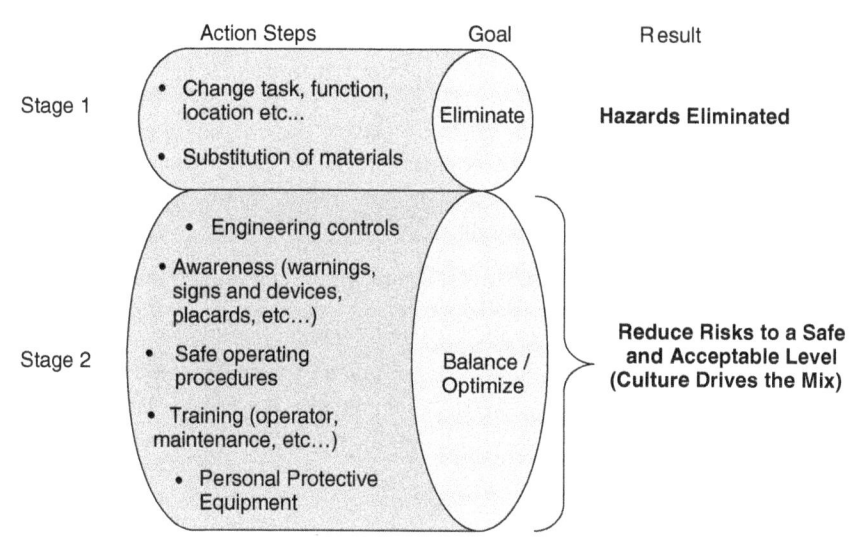

Figure 4.3 Two-Stage Iterative Approach to the Hierarchy of Controls and Risk Reduction. *Source:* Reprinted with Permission from ANSI B11.0-2015 – Safety of Machinery

This two-stage iterative approach is significant in that it requires risk professionals to first look for ways to eliminate hazards or substitute components or materials that are less hazardous. The importance of applying the hazard control hierarchy concept when selecting controls and pursuing "high-level" controls cannot be overstated and should be well understood by SH&E professionals.

4.13.1 Using a Protection Factor

Protection factors provided by both existing controls and those being proposed can be selected and used in a risk assessment. When incorporating a control or protection factor, the risk estimate is adjusted to reflect the level of protection provided. An example protection factor formula and scale are provided in Table 4.10.

Using the formula earlier, the level of risk reduction effectiveness is factored into the risk level estimate. Higher-end risk reduction controls such as elimination, substitution, and multiple

TABLE 4.10 Risk Formula with Protection Factor Multipliers

Risk Formula	
Severity × (Likelihood × Protection Factor) = Risk	
Protection Factor (PF)	Multiplier
Elimination	0.1
Substitution	0.4
Engineering – multiple	0.6
Engineering – single	0.7
Warning	0.8
Administrative	0.9
PPE	0.95
No controls	1

engineering controls have a greater reduction of risk since they impact the severity of harm as well as the probability of occurrence. Lower-level controls such as warning systems, administrative, and PPE have a lower reduction of risk due to their lack of impact on severity. The following example demonstrates how it is used. The risk criteria definitions and matrix used in this example are found in Appendix 4.A.

Example

A metal fabricator has a parts washer used to wash parts with a hot caustic solution. The parts must be then transferred to a drying oven.

Task: An operator hand loads parts into the washer and then removes and transfers washed parts to the oven. The number of parts handled by the operator is 240 parts per hour or 4 parts per minute. Parts weigh approximately 20 lb each and are 5 ft long.

Hazards: The primary hazards are repetitive motion, contact with hot caustic residue and high temperature parts, and cuts from sharp edges.

Existing controls: Operators are trained in proper lifting and handling and provided heat and cut-resistant gloves. No other controls are identified.

Hazard/risk assessed: Repetitive handling risk represents the greatest potential for risk by the team.

Severity (S) level: The risk assessment team determines that severity level is rated *Marginal* (3) since exposure can result in lost time or redistricted duty injuries (see Appendix 4.A). In addition, a previous case was reported in a similar operation.

Likelihood (L) level: Likelihood is determined to be *frequent* (5) or very high since it occurs throughout the work day, each day; and the exposure time and duration are significant.

Risk level (RL): The formula used to calculate risk level is as follows:

$$Severity \times (Likelihood \times Protection\ Factor) = Risk\ Level$$

Current protection factor (PF): Since only training, an administrative-level control, is in place for the identified hazard, a multiplier of 0.9 is assigned.

Current state risk level: The risk level for repetitive handling exposure is calculated as *high* (13.5) using the S × (L × PF) = RL formula as follows: **3 × (5 × 0.9) = 13.5**

Additional controls proposed: Risk levels calculated as *high* require further action to reduce risk. Using the hierarchy of controls, the team determines that a conveyor system is needed to transfer parts from the washer to the oven, eliminating a manual handling step.

Future protection factor: Elimination controls are given a protection factor multiplier of 0.1.

Future state risk level: The new risk level is estimated to be greatly reduced through the elimination of the hazard to 1.5 using the following formula: **3 × (5 × 0.1) = 1.5**

4.14 ACCEPTABLE AND UNACCEPTABLE RISK LEVELS

The concept of acceptable risk can be difficult to accept, until it is realized that there is always some level of risk in all operations, systems, and activities. The residual risks – those that remain in the system – are "accepted," until they are recognized as unacceptable. Organizations generally know what they will not accept in terms of losses and exposures. Those organizations with a more mature operational risk management system tend to have defined criteria for unacceptable and acceptable risks.

DOCUMENTING RISK

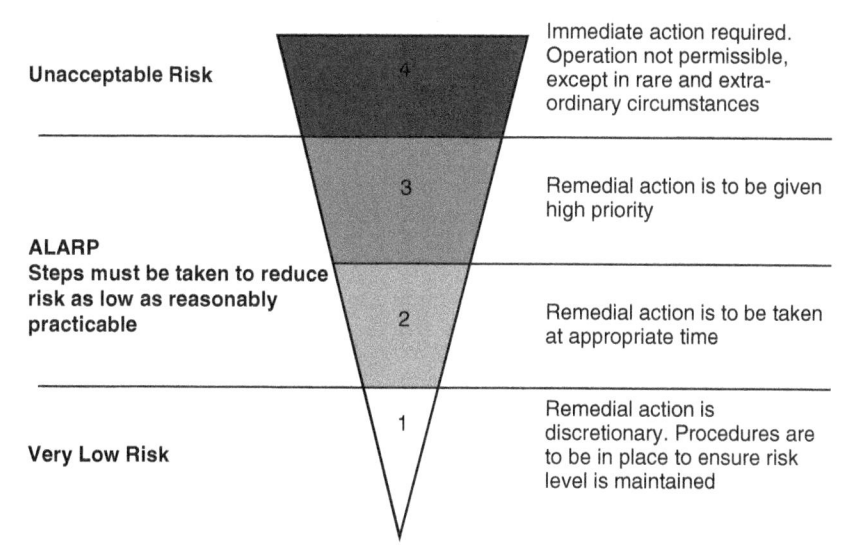

Risk Level	Label	Required Action
4	Unacceptable Risk	Immediate action required. Operation not permissible, except in rare and extraordinary circumstances
3	ALARP — Steps must be taken to reduce risk as low as reasonably practicable	Remedial action is to be given high priority
2	ALARP	Remedial action is to be taken at appropriate time
1	Very Low Risk	Remedial action is discretionary. Procedures are to be in place to ensure risk level is maintained

Figure 4.4 Risk Levels and Their Required Actions. *Source:* Reprinted with Permission from ANSI/ASSE Z690.3 (Courtesy of the American Society of Safety Engineers) and Manuele

Acceptable risk level can be defined as the risk level an organization is willing to accept in its current context. Acceptable risk levels as well as unacceptable levels tend to be lowered as an organization becomes more effective in their risk management efforts, reducing risk and improving control technologies.

ANSI Z590.3 defines acceptable risk as follows:

Acceptable Risk. That risk for which the probability of an incident or exposure occurring and the severity of harm or damage that may result are as low as reasonably practicable (ALARP) in the setting being considered.

Risk acceptance is a function of many factors and varies greatly across industries, organizations, and even operations within an organization. The culture of an organization ultimately defines to a great extent the acceptable risk level.

The concept of as low as reasonably practicable (ALARP) is the practical application of risk reduction using a cost/benefit analysis-type approach. ALARP is the level of risk that is acceptable which cannot be reduced further without expenditures that exceed the benefit (Manuele, 2014). Figure 4.4 reprinted with permission from ANSI/ASSE Z690.3 with modifications from Fred Manuele provides a visual illustration of how operational risk levels can be stratified and prioritized from "unacceptable" to "acceptable" within an organization.

4.15 DOCUMENTING RISK

Risk assessment results should be well documented to demonstrate the methods, to communicate the results, and to be referred to and understood by different people at different times. In ANSI Z590.3-2011, 7.12, *Document the Results*, it suggests documenting the names, titles, and qualifications of the risk assessment team and the methods, hazards identified, risks, controls, and follow-up actions. ANSI Z590.3 comments that activities should be traceable and provide the foundation for improvement and that the risk management process and results should be clearly expressed.

TABLE 4.11 Example of a Risk Register

					Risk Register			
Case #	Location	Task	Hazard #	Hazard	Current State Risk Level	Additional Controls	Completion Date	Future State Risk Level
1	QC lab	Plasma cutter	1.1	Electrical shock	14.00	Adm.	2/20/15	12.00
1	QC lab	Plasma cutter	1.2	Burns	15.20	Adm., PPE	3/15/15	12.00
1	QC lab	Plasma cutter	1.3	Arc flash	11.20	Adm.	2/20/15	9.80
1	QC lab	Plasma cutter	1.4	Noise	19.00	Eng.	3/15/15	8.40
1	QC lab	Plasma cutter	1.5	Fire	14.00	Adm.	3/15/15	12.00
1	QC lab	Plasma cutter	1.6	Dust	11.20	Adm.	3/15/15	9.60
2	QC lab	Weld destruct	2.1	Ergo strains	14.00	Adm.	4/15/15	12.00
2	QC lab	Weld destruct	2.2	Vibration	19.00	Elim.	4/15/15	4.80
2	QC lab	Weld destruct	2.3	Noise	11.20	PPE	4/15/15	10.80
2	QC lab	Weld destruct	2.4	Struck by	15.20	PPE	2/20/15	14.40
2	QC lab	Weld destruct	2.5	Dust	16.00	Multi-Eng.	4/15/15	8.40
2	QC lab	Weld destruct	2.6	Struck against	11.40	Multi-Eng.	3/15/15	6.30
2	QC lab	Weld destruct	2.7	Falls same level	16.00	Eng.	3/15/15	11.20
3	Finishing	Wash station	3.1	Hot liquid	9.00	Sub.	4/15/15	6.30
3	Finishing	Wash station	3.2	Struck against	14.25	Elim.	4/15/15	0.20
3	Finishing	Wash station	3.3	Chem-corrosive	11.20	Sub.	4/15/15	4.20
3	Finishing	Wash station	3.4	Hot surfaces	14.25	Elim.	4/15/15	2.10
3	Finishing	Wash station	3.5	Mechanical	9.60	Multi-Eng.	3/15/15	4.80
3	Finishing	Wash station	3.6	Ergo-strains	11.20	Elim.	4/15/15	0.20

A risk register is one well-accepted method of documenting the risk assessment and its results. Depending on the assessment and its complexity, the risk register may need to be broadened with an introduction and/or conclusion. An example of a risk register is provided in Table 4.11.

There are many different examples of risk registers from the very simple to the very complex, but all serve to document the risk assessment process. They often take the form of spreadsheets and cover items such as those as follows:

- Operation or process
- Hazard(s)
- Exposure
- Existing controls
- Consequence(s)
- Likelihood
- Risk rank or score
- Additional controls desired
- Responsible party
- Time line
- Residual risk rank or score.

A risk register with the appropriate heading coupled with a brief introduction or summary is a common and effective method of documenting a risk assessment in a very consistent manner. Sometimes

TABLE 4.12 Triggers for Risk Assessment

Organizational	• Mergers and acquisitions • Divestitures • Expansions
Design and redesign	• New facilities, processes, systems • New methods • New products or services • Major capital projects or expenditures
Change management	• Modification to existing facilities, processes, or systems • Additions or expansions to existing operations • Changes in methods and materials • Change in setting or environment
Procurement	• New chemicals, substances, or materials • New equipment, machinery, and tools • Other physical assets
Third-party interactions	• Contractors, vendors, and suppliers • Temporary workers • Customers and visitors • Rental/leases • Multiple employer worksites
Nonroutine activities	• Construction • Maintenance and repair • Installations • Debugging and adjustments • Decommission • Demolition and disposal
High-risk activities	• Elevated work • Confined space entry • Electrical or other high-energy works • Line breaking and energy isolation • Hot work • Work around moving objects • Hazardous chemical processes
Incidents	• Fatalities • Serious mishaps • Serious near hits • At-risk observations • Complaints and concerns • Incident trends • Environmental releases • Emergencies • Upsets and breakdowns
External requirements	• Existing regulations • New regulations • Customers

a narrative risk assessment report is used to document the risk assessment. Both can serve to meet the documentation requirements and needs.

4.16 COMMUNICATING RISK CRITERIA

Defining risk criteria has little value if it is not effectively communicated to all affected stakeholders. Communication must start from the beginning during the context phase throughout the process and include monitoring and verifying risk reduction results.

To begin with, an organization must define when, where, and how risk assessments will be performed. An excellent resource can be found on ASSE's Risk Assessment Institute website, which provides a series of videos at http://www.oshrisk.org/videos/. The Institute's video entitled "Triggers for Risk Assessment" explains that a "universal trigger" is any risk sources or events that give rise to risk as defined in ISO 31000/ANSI Z690.2. A list of "triggers" for conducting risk assessments is presented in Table 4.12.

The organization and management should establish policies and procedures about risk assessments that are communicated and well known to the entire organization. It should be understood that the goal is to achieve acceptable levels of risks for the activities the organization engages in.

Communication is important with internal and external stakeholders to ensure that their interests are understood and considered. It also allows them to contribute their views and expertise to the risk management process. The ANSI Z690.2-2011, Risk Management Principles and Guidelines, standard states that "*communication and consultation with the external and internal stakeholders should take place at all stages of the risk management process*" (ANSI/ASSE Z690.2-2011). Stephen Covey, the author of *The Seven Habit of Highly Effective People*, said, "*Communication is the most important skill,*" and this holds very true in the risk management and risk assessment processes.

4.17 SUMMARY

At its core, risk assessment is governed by the specific risk criteria established by its stakeholders. The importance of well-defined risk criteria within a risk scoring system cannot be overemphasized. However, SH&E professionals should always keep in mind the ultimate purpose of risk assessment, which is to "reduce risk to an acceptable level."

"Perfect" can be the enemy of "good." Delaying and prolonging the process to acquire the "perfect" risk assessment criteria or scoring system can undermine the ultimate goal of reducing risk. Remember, a "good" risk assessment conducted is better than a "perfect" risk assessment planned. To quote Dave Walline, "Just jump in and make a difference." Truer words have not been spoken.

REVIEW QUESTIONS

1. Describe the concept of risk criteria and provide a summary of what makes up risk criteria.

2. Describe a risk scoring system. List common components found in risk scoring systems.

3. Explain the concept of the hierarchy of controls and how it is applied. What are the progressive levels from "most preferred" to "least preferred" found in ANSI Z590.3, Prevention through Design Standard? What are the levels that are listed in ANSI B11.0?

4. Explain how high-level risk reduction measures such as substitution and engineering controls affect severity of consequences and likelihood of occurrence and how lower-level measures such as warning systems, administrative controls, and PPE affect severity and likelihood.

5. Risk assessments generally have a two-dimensional scoring system. What are these two risk factors? List additional risk factors that are used in three- and four-dimensional scoring systems.

6. Most risk assessment matrices have qualitative or semiquantitative values using descriptive or subjective measures. Explain what is meant by a qualitative and semiquantitative matrix. Where can examples for each be found?

7. List eight "triggers" for conducting a risk assessment.

8. Describe how "communication" should occur in the risk assessment process and why it is important.

REFERENCES

ANSI/AIHA/ASSE Z10-2012. *American National Standard – Occupational Health and Safety Management Systems*. Fairfax, VA: American Industrial Hygiene Association, 2012.

ANSI/ASSE Z590.3-2011. *Prevention Through Design: Guidelines for Addressing Occupational Hazards and Risks in Design and Redesign Processes*. Des Plaines, IL: American Society of Safety Engineers, 2011.

ANSI/ASSE Z690.1-2011. *American National Standard – Vocabulary for Risk Management*. Des Plaines, IL: American Society of Safety Engineers, 2011.

ANSI/ASSE Z690.2-2011. *American National Standard – Risk Management Principles and Guidelines*. Des Plaines, IL: American Society of Safety Engineers, 2011.

ANSI/ASSE Z690.3-2011. *American National Standard – Risk Assessment Techniques*. Des Plains, IL: The American Society of Safety Engineers, 2011.

ANSI B11.0-2015. *Safety of Machinery – General Safety Requirements and Risk Assessments*. Houston, TX: B11 Standards, Inc., 2015.

Clemens, P.L. & Simmons, R.J. *System Safety and Risk Management – A Guide for Engineering Educators*. U.S. Department of Health and Human Services, National Institute for Occupational Safety and Health, 1998.

International SEMATECH. *Failure Mode and Effects Analysis (FMEA): A Guide for Continuous Improvement for the Semiconductor Equipment Industry*. Technology Transfer #92020963B-ENG. Albany, NY: SEMATECH, 1992. Also available at http://www.sematech.org/docubase/document/0963beng.pdf (accessed December 23, 2015).

Main, B. W., *Risk Assessment: Challenges and Opportunities*. Ann Harbor, MI: Design Safety Engineering, Inc., 2012.

Manuele, F. A., *Innovations in Safety Management – Addressing Career Knowledge Needs*. Hoboken, NJ: John Wiley & Sons, Inc., 2001.

Manuele, F. A., *Advanced Safety Management: Focusing on Z10 and Serious Injury Prevention* (2nd ed). Hoboken, NJ: Wiley, 2014.

MIL-STD-882E. *Standard Practice for System Safety*. Washington, DC: Department of Defense, 2012.

Walline, D. L., 2014 "Prevention through design: Proven solutions from the field." *Professional Safety,* 59(11), 43–49.

APPENDIX 4.A

Risk Matrix (adapted/modified from ANSI Z10)

		Severity (S)				
		Insignificant (1) Inconsequential with respect to injuries or illnesses, system loss or downtime, or environmental release.	**Negligible (2)** First aid or minor medical treatment only, non-serious equipment or facility damage, chemical release requiring routine cleanup without reporting.	**Marginal (3)** Medical treatment or restricted work, minor subsystem loss or damage, chemical release triggering external reporting requirements.	**Critical (4)** Disabling injury or illness, major property damage and business downtime, chemical release with temporary environmental or public health impact.	**Catastrophic (5)** One or more fatalities, total system loss, chemical release with lasting environmental or public health impact.
		1	2	3	4	5
Likelihood (L)	**Frequent (5)** Likely to occur repeatedly. 5	5	10	15	20	25
	Likely (4) Probably will occur several times. 4	4	8	12	16	20
	Occasional (3) Could occur intermittently. 3	3	6	9	12	15
	Seldom (2) Could occur, but hardly ever. 2	2	4	6	8	10
	Unlikely (1) Improbable, may assume incident or exposure will not occur. 1	1	2	3	4	5

Risk Criteria

Risk Level (RL)	Risk Scores	Actions
Very High	15 to 25	Operation not premissible. Immediate action required.
High	9 to 14	Remedial action to be given high priority.
Moderate	5 to 8	Remedial action to be taken at appropriate time.
Low	1 to 4	Remedial action discretionary.

Protection Factor (PF)	Multiplier
Elimination	0.1
Substitution	0.4
Engineering - Multiple	0.6
Engineering - Single	0.7
Warning	0.8

Risk Formula

Severity x (Likelihood x Protection Factor) = Risk

5

FUNDAMENTAL TECHNIQUES

Bruce K. Lyon
Risk Control Services, Hays Companies, Kansas City, MO, USA

5.1 OBJECTIVES

- Introduce fundamental hazard analysis and risk assessment
- Review hazard analysis and risk assessment process
- Review informal techniques
- Review formal techniques
- Examine the use of techniques and their strengths and limitations
- Provide guidance on the use of fundamental risk assessment techniques

5.2 INTRODUCTION TO FUNDAMENTAL HAZARD ANALYSIS AND RISK ASSESSMENT

The act of identifying operational hazards, assessing their risk exposures, and applying appropriate control measures is the key to risk management and one of the most important functions performed within an operational risk management system (ORMS). The American National Standard *ANSI/ASSE Z690.3*-2011 *Risk Assessment Techniques*, nationally adopted from *IEC/ISO 31010:2009*, is a consensus standard for current good practices in the selection and use of risk assessment techniques. ISO 31010/ANSI Z690.3 states that the risk assessment process provides decision makers and stakeholders a better understanding of the risks that could impact an organization's business objective, and the efficacy of controls in place, so that the organization can better manage its operational risks. In essence, the risk assessment process provides a basis for decisions regarding the most appropriate risk control measures to take to achieve an "*acceptable*" risk level in an organization. Without proper risk assessment, risks cannot be adequately managed.

Within the risk management process, risk assessment is the central component, as depicted in Figure 5.1 reprinted with permission from the *ANSI/ASSE Z690.2*-2011, *Risk Management Principles and Guidelines* consensus standard, nationally adopted from *ISO 31000:2009*. The standard defines

---

Risk Assessment: A Practical Guide to Assessing Operational Risks, First Edition.
Edited by Georgi Popov, Bruce K. Lyon, and Bruce Hollcroft.
© 2016 John Wiley & Sons, Inc. Published 2016 by John Wiley & Sons, Inc.

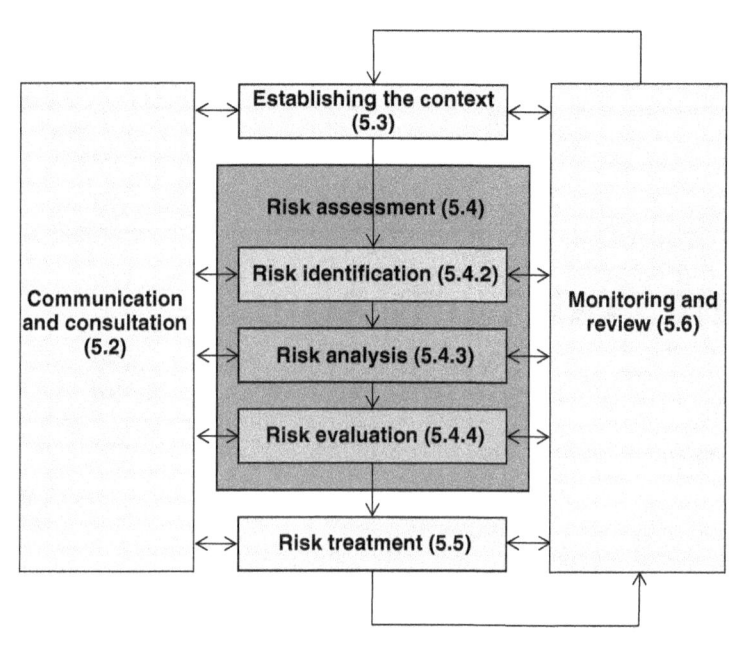

Figure 5.1 Risk Management Process. *Source:* Reprinted with Permission from ISO 31000/ANSI/ASSE Z690.2-2011 (Courtesy of the American Society of Safety Engineers)

risk assessment as a three-phase process of (1) risk identification, (2) risk analysis, and (3) risk evaluation. ISO 31010/ANSI Z690.3, *Risk Assessment Techniques*, says in its introduction the following about risk assessment and its importance in risk management:

> *Risk assessment is that part of risk management which provides a structured process that identifies how objectives may be affected, and analyzes the risk in term of consequences and their probabilities before deciding on whether further treatment is required.*
>
> *Risk assessment attempts to answer the following fundamental questions:*
>
> - *What can happen and why (by risk identification)?*
> - *What are the consequences?*
> - *What is the probability of their future occurrence?*
> - *Are there any factors that mitigate the consequence of the risk or that reduce the probability of the risk?*
> - *Is the level of risk acceptable and does it require further treatment?*

There are many different risk assessment techniques and variations of techniques to choose from. ISO 31010/ANSI Z690.3 describes 31 specific techniques, while *ANSI/ASSE Z590.3*-2011, *Prevention through Design Guidelines for Addressing Occupational Hazards and Risks in Design and Redesign Processes*, covers eight different methods. A table listing specific techniques from these standards is included in Appendix 5.A. Each risk assessment technique is designed to provide a general or specific level of information, analysis, and assessment for its selected application in order to provide adequate information for decision making on the treatment or reduction of risk. Since there are many different types of risk exposures and levels of complexities in organizations, it is rare that a single method of risk assessment would adequately address every type of risk in a workplace. However, as a general rule, when selecting a risk assessment tool, the simplest method that provides sufficient information to make an appropriate risk management decision is advised.

Even though there are numerous methods and variations of techniques for assessing operational risks, all are based on the same fundamental process. The risk assessment process is outlined and described in ISO 31010/ANSI Z690.3 and includes a review of the targeted activity/subject, identification of hazards and risks, analysis of risks produced from hazards, and evaluation of the risks compared to other risks according to defined risk tolerability criteria [see Chapter 4].

To manage operational risk, an organization must evaluate and prioritize tasks, jobs, and processes within an operation according to risk level so that necessary controls can be applied to reduce risk to an acceptable level. This effort requires practical methods that can be consistently and readily applied by all stakeholders from skilled safety practitioners to managers, supervisors, and workers at the ground level.

This chapter provides a review and outline of fundamental analysis and assessment methods commonly used by organizations, frontline workers, supervisors, and safety, health, and environmental professionals. Examples of highlighted methods and case studies are included in this chapter to provide useful perspective and practical application of these tools and techniques. These methods include the following:

- Informal pretask hazard analysis
- Task analysis/job hazard analysis (JHA)
- Job risk assessment.

With the increasing number of safety regulations, consensus standards, and guidelines that include risk assessment requirements, organizations are seeking out safety professionals skilled in performing and facilitating hazard analyses and risk assessments, with the ultimate goal of achieving acceptable risk levels in their workplace. Safety professionals that are proficient in fundamental risk assessment techniques will be better positioned to take advantage of these future opportunities.

5.3 ASSESSMENTS WITHIN AN OPERATIONAL RISK MANAGEMENT SYSTEM

Hazard analysis and risk assessments are stated core elements in standards and guidelines addressing occupational health, safety, and environmental management systems and are a recognized best practice for managing risk. Such models should be extended to encompass all operational risks to an organization such as liability, business interruption, employment practices, business practices, and compliance, among others – referred to as an ORMS. Several notable standards and guidelines for management systems that include hazard analysis and risk assessment requirements are listed:

- Occupational Safety and Health Administration's (OSHA) Voluntary Protection Program (VPP)
- ANSI/AIHA/ASSE Z10-2012 Occupational Health and Safety Management Systems
- BS OHSAS 18001-2007, Occupational Health and Safety Management
- International Labor Office ILO-OSH 2001, "Guidelines on Occupational Safety and Health Management Systems"
- ISO 14001-2004, Environmental Management Systems
- ISO 45001-2016, Occupational Health and Safety Management Systems

ORMS are designed to continually improve risk management performance and are aligned with the Plan, Do, Check, Act (PDCA) model made popular by Dr. Edwards Deming who championed continuous improvement and quality initiatives. ORMS require multiple levels of implementation beginning with performance-oriented strategic policies and procedures, followed by specific actions and duties performed by each level of management and employee, monitoring and measuring performance, modifying and adjusting where necessary, and reinforcement of the process. It is up to the stakeholders to develop the necessary processes, job instructions, and documents (including fundamental task/job analysis and risk assessments) to support the ORMS.

Several key components commonly found in ORMS include (1) management commitment and employee involvement, (2) hazard analysis and risk assessment, and (3) risk management and control (ANSI/AIHA/ASSE Z10-2012). Fundamental risk assessment methods and practices play a large role in these components.

5.4 HAZARD ANALYSIS VERSUS RISK ASSESSMENT

There are subtle but significant differences between hazards and risks. Hazards can be defined as having the "potential for harm" and include aspects of technology and activity that, if left uncontrolled, can create risk. Hazards are produced by the use of equipment, technology, energy sources, substances and chemicals, and materials and by human actions and inactions. Basic workplace hazard classifications include physical and mechanical, chemical, biological, ergonomic, and psychosocial. A list of common hazards is provided in Appendix 5.B.

Risks are derived or produced from hazards when their exposures to people and/or assets pose a chance for loss. This chance for loss or "risk" is measured by the likelihood of the event occurring and the resulting severity or consequence of the loss. If the risk level exceeds the organization's acceptable risk level, risk control and management efforts are applied.

The fact that a hazard exists does not necessarily mean that a risk is produced. For a risk to exist, the exposure to a hazard must pose a severity of loss or harm and a likelihood of occurring. If no person or no asset is exposed to the hazard, then there is no risk. For a hazard to create a risk there needs to be exposure – the forgotten term in risk communication (Whiting, 2013). The following illustration provides an example of when a hazard produces a risk:

A manufacturing operation has an in-house maintenance department responsible for performing preventive maintenance, equipment service and general repair. Sometimes the maintenance work involves welding, cutting or hotwork in various areas of the plant. The open flames and sparks (ignition sources) generated during welding and cutting operations present a "hazard". If no person or asset is exposed to the flames and sparks then there is no risk. Also a hazard can exist but the level of risk it creates can be controlled to a low level by the use of administrative control measures which may include a Hotwork Permit system. Welding, cutting and other hotwork activities (producing ignition source hazards) properly performed in a controlled environment such as a 'Hotwork Permit' area free of combustible materials with a fire watch and fire suppression equipment can adequately control the hazard and reduces the risk of fire. However, if the same 'hotwork' hazard is introduced into a flammable liquid storage room, a significant risk of fire or explosion is produced.

It is important to realize the difference between hazards and risks. And it is equally important to understand the differences between hazard-based and risk-based control efforts. The following provides some guidance on these distinctions:

Hazard-based efforts – Control measures that are developed based on the existence of an identified hazard are done so from a hazard-based effort. In a hazard-based analysis, identification of hazard classifications that exist or have a potential of existing in an operation leads to prescribed hazard controls using the hierarchy of hazard controls. Hazard-based decisions are made solely on the existence of a hazard classification without a detailed description and consideration of a scenario for *how* the hazard can produce the unwanted harm and without an assessment of the risk from the scenario. Some regulatory standards in the United States such as the OSHA's 1910.1200, Hazard Communication, are hazard based and "prescriptive" in nature. The scope of the Hazard Communication, which indicates by its very title that it is a hazard-based standard, applies to any employer that has chemicals (a hazard classification) in its workplace. No risk assessment is required to determine the application of the standard requirements, only a determination that a chemical hazard exists. If chemicals exist in the workplace, compliance with the Hazard Communication standard is required.

Risk-based efforts – Risk-based efforts take a more comprehensive approach. They typically begin with identifying the hazard classification, followed by an estimate of the consequence or severity of the harm and an estimate of the likelihood of the hazard causing harm. Based on the resulting risk level, control measures are then implemented to reduce risk to an acceptable level. An example of a risk-based standard is the OSHA's 1910.119, Process Safety Management of Highly Hazardous Chemicals. The standard applies to operations that use a listed highly hazardous chemical at or above the standards specified threshold quantity and requires covered operations to perform a process hazard analysis to assess the risk.

As one works in the field of operational risk management, it is very likely that they will hear the terms "analysis" and "assessment" used to describe the same process. This raises the question, "Is there a difference between an analysis and an assessment?" It has been the authors' experience that these terms are often used interchangeably by stakeholders (including safety practitioners, risk managers, insurance professionals, etc.) tasked with operational risk management responsibilities. According to the Merriam-Webster's dictionary, these terms are defined as follows:

> <u>Analysis</u> – *a careful study of something to learn about its parts, what they do, and how they are related to each other: an explanation of the nature and meaning of something.*
>
> <u>Assessment</u> – *the act of making a judgment about something: the act of assessing something.*

Looking at these definitions reveals that there are differences. An analysis requires the "study" of the subject to "understand" its nature and relationship with its surroundings, while an assessment requires not only a "study" but also a "judgment" or rating of the subject in comparison with its surroundings.

To put these terms in the context of operational risk management, *ANSI/ASSE Z690.1*-2011, *Vocabulary for Risk Management* (nationally adopted from ISO Guide 73:2009), provides these definitions:

> <u>Risk analysis</u> – *process to comprehend the nature of risk and to determine the level of risk.*
>
> <u>Risk assessment</u> – *overall process of risk identification, risk analysis, and risk evaluation.*

To further clarify, *ANSI Z590.3-2011, Prevention through Design*, provides the following definitions within the context of the hazard analysis and risk assessment process:

> <u>Hazard analysis</u> – *a process that commences with the identification of a hazard or hazards and proceeds into an estimate of the severity of harm or damage that could result if the potential of an incident or exposure occurs.*
>
> <u>Risk assessment</u> – *a process that commences with hazard identification and analysis, through which the probable severity of harm or damage is established, followed by an estimate of the probability of the incident or exposure occurring, and concluding with a statement of risk.*

Therefore, the distinguishing differences between a "risk assessment" and a "hazard analysis" are that a risk assessment (1) encompasses risk identification, risk analysis, and risk evaluation and (2) includes the act of making a judgment or an estimation of risk levels. Figure 5.2 provides a comparison of the basic elements that are comprised in a hazard analysis and a risk assessment. Understanding these differences between hazards and risks and analyses and assessments is important if safety professionals are to be consistent and effective in their efforts.

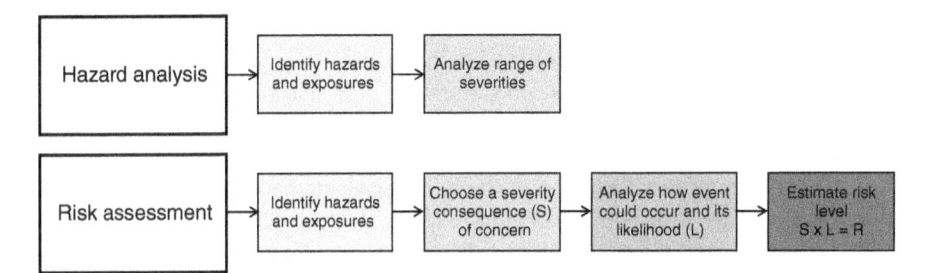

Figure 5.2 Comparisons of Hazard Analysis and Risk Assessment Steps

Risk is an estimate of the likelihood of a *hazard* exposure occurring and the severity of harm that could result. As mentioned earlier, in *ISO 31000/ANSI Z690.2-2011, Risk Management Principles and Guidelines*, "risk assessment" is defined as having three essential components: (1) identification, (2) analysis, and (3) evaluation. In many safety activities, hazards are identified and analyzed by various means such as checklists, inspections, and audits. However, without the evaluation of the resulting risk's probability and severity and a comparison to established risk criteria, only a "hazard-based" analysis is achieved.

It should be noted that different exposures to a hazard can produce consequences of different severities by slightly different scenarios of events and circumstances. Different consequences require different hazard exposure scenarios. Each chosen consequence of interest or concern and its corresponding hazard exposure scenario need to be assessed separately. There is no need to argue about which consequences can eventuate from different hazard exposure scenarios. Rather a risk assessment work group can choose and agree on which consequence is of interest or concern and assess them separately.

5.5 THE HAZARD ANALYSIS AND RISK ASSESSMENT PROCESS

For any analysis and assessment method used, the fundamentals remain the same. Guidance in the hazard analysis and risk assessment process is provided in *ISO 31010/ANSI Z690.3, Risk Assessment Techniques*, and *ANSI Z590.3, Prevention through Design*. The following is a summary of the fundamental process in performing a *hazard analysis and risk assessment* adapted from ISO 31010/ANSI Z690.3, ANSI Z590.3, and Manuele (2012):

1. *Establish purpose and scope.*

    As with any risk management effort, a clear and defined purpose and scope should be established to provide direction and boundaries. This purpose and scope should be effectively communicated to all stakeholders. An effective purpose and scope statement defines the basic parameters for managing risk and sets the scope and criteria for the risk assessment process. This includes considering internal and external parameters, as well as the background to the particular risks being assessed. Once the context has been defined, the task, job, process, scenario, or system to be analyzed is selected. Selection criteria should be based on the purpose, scope, and parameters of the assessment and will include tasks with higher perceived risk and/or loss experience. Determine the scope of analysis/assessment regarding exposures affected such as employees, contractors, the public, environment, property, assets, productivity, quality, or other elements. Establish responsibilities and accountabilities for stakeholders involved in the process. Needed training and resources should be determined and provided. Other parameters to consider are the operating phase, such as whether the task is a standard operation, maintenance and service activity, new equipment start-up, teardown, or other, and any interface with other operations.

THE HAZARD ANALYSIS AND RISK ASSESSMENT PROCESS 97

2. *Identify hazards and risks.*

   Select a formal method or technique of analysis and assessment; review and identify sources of existing and potential hazards and risks (i.e., technology, activities, materials, etc.) associated with the task; determine the characteristics of hazards and their affects. Consider whole system risk and combined risk as well as individual risk.

3. *Analyze potential failure modes.*

   Determine how and why failures might occur in the task; determine the conditions and causal factors that can lead to failure modes; identify the affected parties, property, or assets impacted by potential failures.

4. *Analyze existing controls.*

   The level of risk will depend on the efficacy of existing controls; identify and list controls in place for each existing or potential hazard; test and verify existing controls to determine if they are missing, inadequate, or ineffective.

5. *Analyze exposure frequency and duration.*

   Determine the frequency of task, activity, or process; determine the exposure period of task, activity, or process; determine the number of people, property, or assets exposed.

6. *Analyze and choose consequence.*

   Consider the range of "credible" scenarios and different consequences associated with the hazard exposures identified. Determine and select the credible worst-case scenario that is of most concern. Other scenarios can be analyzed separately later. (Hazard analysis and risk analysis cannot be completed simultaneously for multiple scenarios.) The severities in different risk domains can include number of injuries and illnesses and their severity, number of fatalities, estimate of cost from damage to property or assets, business interruption or lost productivity estimate, and extent of environmental impact.

   *Upon completion of the severity of consequences estimate of a hazard-related incident, a hazard analysis has been completed.

7. *Analyze and estimate likelihood.*

   Upon completing the hazard analysis, the risk is assessed with the following steps. The estimation of likelihood needs to be as informed as possible with inputs from knowledgeable experts but typically can be a subjective process unless empirical data are available. Estimating the probability of a hazardous event or exposure occurring, complex events, or incidents may require brainstorming among a team of knowledgeable people. "Likelihood" is a general term that can be expressed specifically as a "frequency" or "probability" of the future occurrence of the chosen consequence scenario. "Frequency" is a rate measure and needs a denominator such as a unit of time (e.g., per hour/per year), the number of tasks performed (e.g., per 1000 miles driven/per 100 drill holes completed), units produced (e.g., per million tons of coal), or life cycle of equipment, process, or product. "Probability" is an expression of chances or odds and is a pure number with no units such as 1 chance in 100 or 1% of the occurrences or 0.01. The likelihood of the complete scenario needs to include estimates of the likelihoods of each scenario event and circumstance so as to obtain an overall likelihood. This compounding of contributing likelihoods can be qualitative or quantitative.

8. *Define the risk.*

   Develop a statement of risk for the chosen hazard-related scenario and consequence. The statement of risk needs to include the chosen severity of consequence of the hazard-related scenario, the estimated likelihood, and a risk level or category rating (high, moderate, low, etc.) that can be obtained by using a risk assessment matrix. Placing individual risks in risk level categories assists in communicating with decision makers on risk levels.

9. *Evaluate and prioritize risks.*

   Establish the risk ranking or risk scoring system – chosen in Part 8 – for comparing and ranking risks. Evaluate and compare the risk levels against legal and corporate risk acceptability criteria such as the, as low as reasonably practicable, (ALARP) concept. Use risk rankings to prioritize risks for their control and management.

10. *Formulate additional control measures.*

    When evaluation of risk levels indicates the moral, legal, and commercial need to reduce risk, development and choice of risk control options are required. The formulation of any modified or new controls should include the use of the "hierarchy of controls" to select and implement the best options for additional control measures to reduce risk level to an acceptable level.

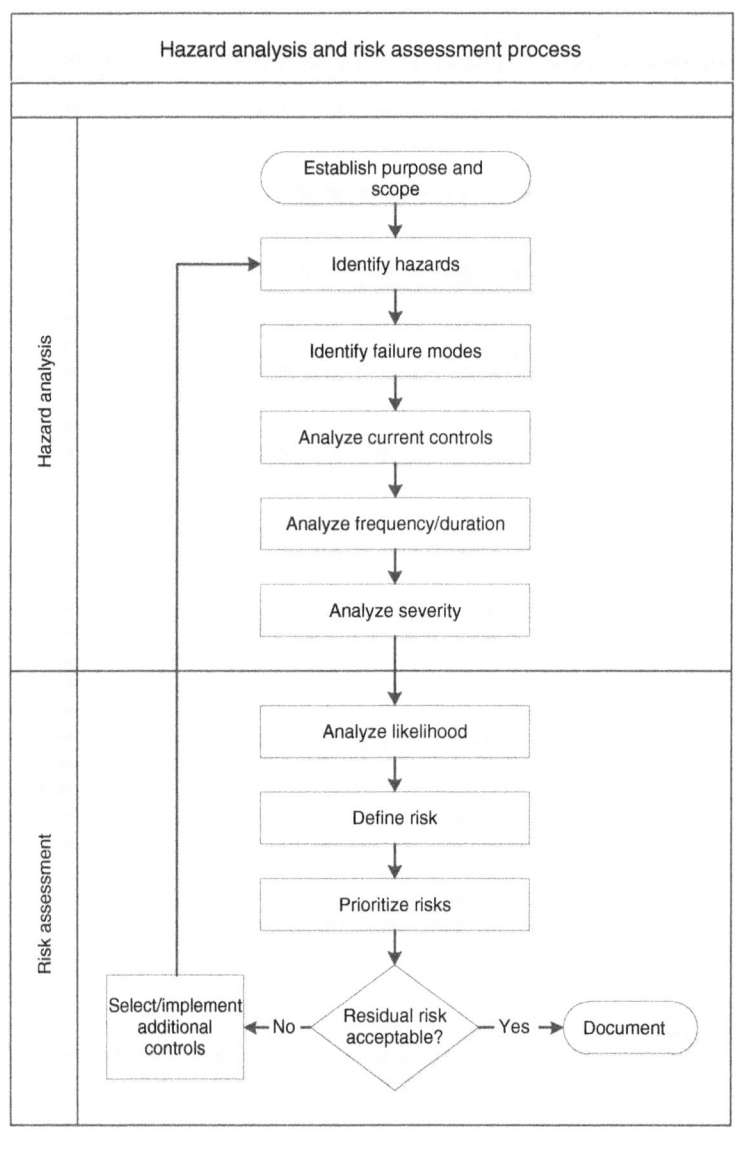

Figure 5.3 Hazard Analysis and Risk Assessment Process

FUNDAMENTAL METHODS

The following flowchart in Figure 5.3 illustrates the sequential steps taken in conducting hazard analysis and risk assessment and the dividing point between the two components.

5.6 FUNDAMENTAL METHODS

Fundamental hazard analysis and risk assessment techniques include informal pretask hazard analysis techniques and more formal methods such as job hazard analysis (JHA)/job safety analysis (JSA) and job risk assessment. These basic methods are considered an essential component to managing risk within an operation because of their ease of use and adaptability, especially where nonroutine tasks occur and where tasks have potential changes or variables. For these reasons, these analysis and assessment techniques have become some of the most widely used in the workplace.

However, a word of caution is advised. Traditional task analyses/JHA are not risk assessments. Severity of harm and likelihood of occurrence are not estimated in their use. This can lead organizations and those performing the job to believe the job is "safe" or of an acceptable risk level. JHA/JSA and safe work methods do not equate to low risk, unless documented as such. A modified JHA with a risk assessment component is provided in this chapter.

Table 5.1 provides a summary of commonly used fundamental hazard analyses and risk assessment techniques covered in this textbook.

Frontline workers and supervisors, field personnel, service and installation workers, construction workers, and other remote operators use fundamental techniques to help make proper decisions regarding the job and its safe execution. In addition to the benefits of assessing and managing risks, fundamental hazard analyses and risk assessments offer a meaningful and effective way for employees to contribute and participate in the safety process. Employee involvement in the safety and health process is a required element in ORMS standards and guidelines such as OSHA VPP, ANSI Z10, and BSI OHSAS 18001 and is essential to an effective and sustainable process.

TABLE 5.1 Types of Fundamental Methods

Method	Application	Description	Type
Pretask hazard analysis	Prior to nonroutine, remote, or unusual tasks often used in mining, construction, field operations, and oil and gas industries	Simple methods that require training in the practice of steps to analyze tasks and hazards prior to performing job	Informal methods involving a structured mental exercise, pretask cards, and checklists
Task/job hazard analysis	New, existing, or modified tasks/jobs	Identifies steps, associated hazards, and control measures	Formal methods using task cards and three-column spreadsheets
Personal protective equipment (PPE) hazard assessment	Work areas or activities where personal protective equipment is required	Identifies hazards and appropriate personal protective equipment	Formal method using a form or spreadsheet listing types of hazards, the body part affected, and the specific PPE required
Task/job risk assessment	New, existing, or modified tasks/jobs	Identifies steps, associated hazards, control measures, and risk level	Formal methods using task cards and multiple-column spreadsheet with risk estimation and rating

Typically, hazard analyses and risk assessments at the task and job level are performed as a result of a larger, more comprehensive risk assessment covering an entire process or operation. Effective "micro-assessments" of an operation or process are necessary to fully understand specific steps of an activity, the potential hazards associated with each step, and the required control measures to ensure the risks remain acceptable and ALARP as defined. Organizations base many workplace decisions on the identified, perceived operational risks and rely on the expertise of the safety practitioner to provide accurate and reliable information. These fundamental methods are vital in providing risk-based information that can be used to make proper decisions in managing operational risks. Every discussion about safety risks and selecting control options significantly benefits from risk assessments. Subjective disagreements about "safe" and "at risk" are replaced by considering the realistic ALARP criterion.

In the United States, only a few analyses/assessments are required by regulatory standards. One is the OSHA's 1910.132, Personal Protective Equipment (PPE) standard, which requires employers to perform a "personal protective equipment hazard assessment" to identify existing and potential hazards in a job and the appropriate PPE to be used. Here is what OSHA 1910.132 says regarding these required assessments:

> *1910.132 (d)(1): The employer shall assess the workplace to determine if hazards are present, or are likely to be present, which necessitate the use of personal protective equipment (PPE).*
>
> *1910.132(d)(2): The employer shall verify that the required workplace hazard assessment has been performed through a written certification that identifies the workplace evaluated; the person certifying that the evaluation has been performed; the date(s) of the hazard assessment; and, which identifies the document as a certification of hazard assessment.*

Often times, PPE hazard assessments are incorporated into a JHA/JSA or job risk assessment as a means of providing more complete safety-related information for the operator performing the job. Note that OSHA requires PPE hazard assessments to include the name of the assessor(s), date(s) of assessment, identity of evaluated workplace, and certification of the hazard assessment in the document. An example of a PPE hazard assessment is included in the Addendum 5.C. *[Note: Most PPE is "protective" rather than "preventive" and only reduces the impact of the consequence. A risk should always be managed by a mix of both preventive (reduce "likelihood") and protective (reduce "impact" of consequence) controls.]*

5.7 INFORMAL METHODS

The most fundamental hazard analysis and assessment method is the informal technique. Basically, informal methods are practiced when employees are asked to "think" about the steps and potential hazards of a task and properly prepare for the job. Informal analyses and assessments of hazards are made consciously and subconsciously by stakeholders in many situations every day. For example, an individual crossing a busy intersection performs a mental assessment of the traffic, conditions, distance, and speed of vehicles, among other variables, in order to cross the street safely. The same mental exercise is used in the workplace. Informal methods have always been an important part of hazard and risk management. In his 2008 book *Advanced Safety Management: Focusing on Z10 and Serious Injury Prevention*, Manuele (2008) states the following regarding informal methods:

> *When a safety professional identifies a hazard and its potential for harm or damage and decides on the probability that an injurious or damaging incident can occur, a risk assessment has been subjectively made. In doing so, for the simpler and less complex hazards and risks, the assessment may be based entirely on a prior knowledge and experience, without documentation. Making informal risk assessments has been an integral part of the practice of safety and health professionals from time immemorial.*

Informal pretask hazard analyses, sometimes referred to as "field-level" analyses, are typically used by frontline workers, supervisors, remote operators, and workers that perform high-risk, nonroutine activities (Hudson and Smith, 1998). These brief but effective methods are applied by the stakeholder prior to initiating a task or job, oftentimes in the field away from supervision, and when there are anticipated changes or deviations that can increase the risk levels of routine tasks or jobs.

Field-level analysis practices are considered informal since they are performed as a mental exercise (rather than on paper) by the individual or team prior to the job and are designed to improve the affected stakeholder's situational awareness and understanding of the hazards and risks. Even though field-level analyses do not document the anticipated tasks, hazards, and controls, they are sometimes the most practical way of helping stakeholders mentally prepare and safely perform their tasks. To perform these steps correctly, training is required for the workers. Like formal methods, the ultimate purpose of performing informal techniques is to enable stakeholders to work safely by making proper decisions in controlling hazards of the task.

Pretask analyses are particularly useful in mentally preparing the worker for tasks that may have variations or changes to an activity and for nonroutine jobs such as maintenance, service, installation, setup, and teardown. There are many variations used in industries such as mining, construction, oil and gas extraction, and other operations where individuals perform certain tasks in the field. Australia, Canada, the United Kingdom, and other countries have made extensive use of these informal techniques. In the United States, industries such as mining and petrochemical have incorporated informal methods in their ORMS. For instance, the US Mine Safety and Health Administration (MSHA) Safety and Health Outreach website provides information and training material on the use of techniques such as Stop–Look–Analyze–Manage (SLAM) (MSHA, 2008).

Although the actual assessment is considered informal, a structured process is required that defines the scope and context, the procedure, stakeholders involved, training and coaching requirements, supervision, and management. Once trained, stakeholders are skilled in identifying and addressing hazards involving unacceptable risks in their jobs.

The concept is based on a set of sequential steps taken by the stakeholder to identify hazards and determine their significance in how they can cause harm. Ultimately, the process must lead to taking the appropriate actions to mitigate identified risks prior to beginning the task. The steps typically include a checklist or mental review of the job steps, visual inspection for hazards and potential hazards, consideration for the development of hazards during the job steps, and a check of required safety equipment and safety procedures. These process steps normally include the following:

- Pause before starting task.
- Look for hazards and how they can harm.
- Consider the risks.
- Act to reduce unacceptable risk (or do not proceed with the task).
- Report any unacceptable risk that needs further action.

The information from these processes should be recorded to ensure required action is taken and feedback provided to the initiator. Some examples of informal methods include:

SLAM: A mental process that requires the worker to stop and consider the work process before beginning and during the task, examine the work environment, analyze the work process, and manage the risk. Often used by miners, a 2004 webcast presented by the US MSHA provides the following description of the SLAM risk concept (MSHA, 2004):
  (a) The first step in this initiative is STOP. It is important for each stakeholder to stop and consider the work involved. Take time to consider the task at hand and ensure the job can be done safely. Stakeholders need to consider if they have been properly trained and have

the knowledge and skills to perform the job safely. Estimate the manpower that is needed to complete the job successfully and if assistance is needed. Consider the consequences of stakeholder actions on fellow workers. Make sure that any work performance will not have serious consequences on coworkers and others.

(b) The second step is LOOK. Upon starting the job, take a moment; look for and identify the hazards associated with this job. This process of looking for hazards should not only be done prior to starting the job but continues throughout the task. In many cases new hazards may appear during the job. Be alert and be prepared to address them. Ask if these new hazards should have been identified: Are there any new hazards that could occur?

(c) The third step is ANALYZE. Take time to think through the procedures necessary to accomplish this task with minimal risk. Question whether following the normal procedure would actually involve a higher risk than using an approved variation. Contemplate the worst-case scenario of performing a task incorrectly.

   (a) Does the stakeholder have everything needed to accomplish the task safely?
   (b) Are the tools the proper ones to use in the particular case?
   (c) Is PPE appropriate for hazards involved in the task being undertaking?
   (d) Does everyone understand the task and their role to finish the task safely?
   (e) Is there someone who needs to be notified or made aware of the task to be performed?
   (f) Is the likelihood of an incident small or if one were to occur would any injury result?
   (g) If the hazard(s) cannot be eliminated for this task, talk with the supervisor prior to starting the task. Take measures to lessen both the potential severity of an injury and the likelihood one could occur. Decide if these measures are sufficient to permit the task to be performed.

(d) The final step is MANAGE. Safety can be managed by developing and implementing controls. Eliminate hazards where possible or reduce exposure to the risk and take the necessary action to see the job is done safely. Some items that the stakeholder can do to control hazards are as follows:

   (a) Following procedures or approved variations
   (b) Eliminating hazards
   (c) Using proper PPE
   (d) Assuring proper safeguards are in place
   (e) Reassessing the tasks from the perspective of safety
   (f) Asking for additional help.

Take 5 for Safety: A brief (5-minutes) mental process sometimes supported with a pretask checklist that requires the stakeholder to stop, think through the job steps, identify hazards, and ensure safety procedures are in place before commencing and during the work. There are variations of the method used by construction, field work, and other similar industries with the following basic elements:

1. Stop and think through the job and its individual steps, noting any changes or possible variations of the task that may be needed.
2. Inspect the work area, equipment, materials, and conditions for potential hazards that could result in the task. Some methods provide sample hazard categories for human factors, environmental factors, use of equipment, energy sources, etc. to aid the stakeholder in analyzing common hazards.
3. Plan for safe work and perform the following:

(a) Ensure procedures are in place (i.e., emergency action plans, standard operating procedures (SOPs), approved variations, roles and responsibilities, etc.).
(b) Conduct a prework safety briefing.
(c) Provide appropriate communication.
(d) Secure the work area free from known hazards.
(e) Ensure controls, safeguards, and PPE are in use.
4. Monitor, supervise, and manage the task to ensure it is performed safely.

Step Back 5 × 5: A simple mental exercise method that requires the worker/work group to "step back" five steps from the job and spend 5 minutes reviewing and planning the tasks to be performed. During this exercise, the worker/work group identifies existing and potential hazards associated with the tasks and takes measures to prevent hazards from creating a risk. The basic steps are as follows:

1. Take five steps back from the job to think about the tasks, and observe the surroundings.
2. Take 5 minutes to discuss with coworkers what the known hazards are associated with each task and what risk control measures are needed.
3. Review task steps to be performed, any changes required, known or potential hazards, and necessary control measures.
4. Verify that necessary equipment, tools, and PPE are available to perform job safely.
5. Monitor and review work being performed. If a problem arises, the job is stopped to determine what measures are needed to proceed safely.

To be effective in reducing risk, the Step Back 5 × 5 method is intended to be used by workers before and during each task throughout the day, each day, as part of their daily routine.

5.8 FORMAL METHODS

To properly manage risk, more formal techniques in hazard analysis and risk assessment are required for many types of jobs and activities. This section provides a review of fundamental methods of hazard analysis and risk assessment commonly used in the workplace.

5.8.1 Fundamental Hazard Analysis

A formal hazard analysis is defined for the purpose of this text as one that is conducted according to a defined methodology to identify and analyze hazards and existing controls and recorded in a prescribed document format to be used for decision making, training, and managing operational risk. Fundamental methods of hazard analysis common in the workplace include task analysis and JHA and PPE hazard analysis or assessment. Formal analysis methods involve a standardized, systematic process of identification and analysis of hazards recorded in a document or interactive spreadsheet. They are common in many industries and most useful in new employee orientation, job training, PPE selection, and incident investigation and analysis.

Formal analysis of a task becomes necessary when there are unknown hazards, numerous potential hazards or more complex hazards, and higher risk hazards, as well as when there are changes in the work environment, work methods, materials, or equipment. By formalizing and documenting the hazard analysis and risk assessment process, the assessor gains a better understanding of the hazards and risks and the necessary control measures and is able to communicate this information to decision makers and other affected stakeholders.

It is advisable that organizations develop guidelines defining where, when, and how jobs and tasks are analyzed and assessed from an operational risk standpoint. ISO 31010/ANSI Z690.3, *Risk Assessment Techniques*, provides guidance on how risk assessments should be included in the overall risk management process.

5.8.2 Pretask Hazard Analysis

Similar to an informal analysis, a pretask hazard analysis is performed in the field by the individual prior to conducting the task. However, unlike informal techniques, the analysis is documented on a small card that is kept by the individual throughout the day. This method is used in high-hazard environments where small crews or individuals perform various tasks in the field, such as construction, repair, maintenance, and service work. Variations of this technique include preshift safety task analysis, daily hazard analysis, safety task analysis card (STAC), and TASK assessment card. As a hazard recognition and safety preparation tool, the pretask hazard analysis is used to:

- Identify the anticipated task steps, hazards, and needed controls in new tasks.
- Reinforce workers' knowledge of the specific task steps, hazards, and necessary controls to perform the task safely in routine tasks.
- Identify task scope changes, new hazards, and needed controls when tasks or conditions change.

Pretask hazard analyses are performed in the field and documented, oftentimes on a preprinted card that can be folded and kept in the individual's shirt pocket. Similar to a JHA, the pretask hazard analysis requires the worker to fill out three sections including (1) task step descriptions, (2) potential hazards, and (3) necessary control measures. However, the pretask hazard analysis differs from a JHA in that it is:

- A more pinpointed analysis of a specific task rather than an established, standardized job analysis
- Performed by the worker in the field prior to the task
- Dynamic in nature to address the various changes that occur in daily tasks
- A daily awareness and analysis tool
- Kept with the worker throughout the day to review and update as necessary.

One of the key benefits of pretask hazard analysis is that it directly involves the worker in hazard recognition and control efforts in their daily tasks. The worker is required to fill out the card prior to performing the task and update the analysis card whenever changes in tasks, hazards, or conditions occur. To be effective, the individual must list the steps, potential hazards, and controls briefly, but in sufficient detail to provide adequate information for controlling hazards and making proper decisions. An example of a Pretask Hazard Analysis Card taken from a case study posted on the OSHA website is displayed in Figures 5.4 and 5.5 (OSHA, 2008).

5.8.3 Job Hazard Analysis

Likely the most common hazard-based analysis technique used in operational risk management is the Job Hazard Analysis (JHA) or sometimes referred to as a Job Safety Analysis (JSA). JHA is a simple hazard analysis tool that is used to help stakeholders identify, analyze, and manage existing and potential hazards in the tasks they perform. These methods are often used to review job tasks and identify hazards that may have been overlooked in the design or redesign of the task.

The technique centers on defining the sequential job tasks and the associated hazards for each step along with needed control measures. Typically, JHA's are documented on a spreadsheet with three

FORMAL METHODS 105

Types of Hazards	Permit Requirements	Pretask Hazard Analysis Card
Access Congested area, uneven ground, confined space, overhead obstructions, objects in walkways, unsecured decking, clutter	**Y N/A** ( ) ( ) Analysis Card ( ) ( ) Safe Work ( ) ( ) Hot Work ( ) ( ) Excavation Work ( ) ( ) Confined Space Entry	**Personal Work Permit** Today's Date ____ Time ____ Name _____ Company _____
Caught-In/Struck-By Sharp objects, pinch points, hot/cold surfaces, open holes, overhead workers, struck-by objects, strike against objects, fire/spark	( ) ( ) Line Entry ( ) ( ) Critical Lift ( ) ( ) Scaffold Inspection ( ) ( ) Other _____ **Y N/A Process Requirements** ( ) ( ) Job scope understood	Foreman _____ Job Location _____ Job Description _____
Environment Noise, dust, weather, lightning, heat, wet areas, wind, plant processes, lead, asbestos, hot/cold surface, heat stress	( ) ( ) Orientation ( ) ( ) LO/TO verified ( ) ( ) SDS reviewed ( ) ( ) Lines drained/purged ( ) ( ) Low points checked	I have been employed less than 90 days and I am considered an at-risk employee Yes ___ No ___
Ergonomic Bad body position, improper or static body position (awkward objects or work position), excessive force (heavy objects, torque), excessive repetition, excessive duration, over reaching	( ) ( ) Oxygen/flammability check ( ) ( ) Line identification confirmed ( ) ( ) Close drain/vents ( ) ( ) Initial entry procedure review ( ) ( ) Confined space procedure ( ) ( ) CSE Rescue Plan reviewed ( ) ( ) Proper Tool/Equipment ( ) ( ) Communication ( ) ( ) Other	The name of my assigned "Buddy" is _____ List Crew _____
Falls To a lower level, at the same height, slippery surface, floor or wall opening, unprotected perimeter, climbing, relocating corrosion	**PPE Hazard Assessment** ( ) Hardhat ( ) Goggles ( ) Gloves: _____	**Emergency Information**
Hazardous Material Burns, exposure, inhalation, splashing, fumes, spills, airborne particles, trapped substances, lead, asbestos, radiation	( ) Safety Glasses ( ) Face shield ( ) Clothing: _____ ( ) Safety shoes ( ) Hearing protection ( ) Air Monitors ( ) Respiratory Protection ( ) Welding Shield ( ) Other: _____	( ) Wind direction? _____ ( ) Emergency alarms/phone numbers reviewed ( ) My escape route(s): _____ ( ) My evacuation assembly point is: _____ ( ) Location of eye wash/shower station: _____
Stored Energy Pressure, tension, electrical, combustible, flammable/explosion, static electricity		
Tools/Equipment Airborne particles, fumes, arc flashes, sharp edges, line of fire, wrong tool for job, broken tools, rotating parts, vibration, shock	**Y N/A Job Completion Review** ( ) ( ) Work area cleaned up ( ) ( ) All red tags signed off ( ) ( ) Permit turned in ( ) ( ) Job status communicated ( ) ( ) Customer: _____	Foreman/Coach Signature: _____

Figure 5.4 Pretask Hazard Analysis Card – Front Side

columns consisting of (1) the task or step, (2) existing or potential hazards, and (3) control measures. An example is provided in Addendum 5.D. The OSHA's booklet entitled "Job Hazard Analysis" (OSHA 3071, 2002) defines JHA this way:

A job hazard analysis is a technique that focuses on job tasks as a way to identify hazards before they occur. It focuses on the relationship between the work, the task, the tools, and the work environment. Ideally, after you identify uncontrolled hazards, you will take steps to eliminate or reduce them to an acceptable risk level.

In the United States, OSHA, the National Safety Council, the American Society of Safety Engineers (ASSE), and others consider JHA and JSA to be the same technique. In *The Dictionary of Terms*

List of Task(s)	List Hazard(s) for Each Task	How Can the Hazard Be Eliminated or Controlled?
1. _____	1a. _____	E1a. _____ C1a. _____ E1b. _____ C1b. _____ E1c. _____ C1c. _____
	1b. _____	
	1c. _____	
2. _____	2a. _____	E2a. _____ C2a. _____ E2b. _____ C2b. _____ E2c. _____ C2c. _____
	2b. _____	
	2c. _____	
3. _____	3a. _____	E3a. _____ C3a. _____ E3b. _____ C3b. _____ E3c. _____ C3c. _____
	3b. _____	
	3c. _____	
4. _____	4a. _____	E4a. _____ C4a. _____ E4b. _____ C4b. _____ E4c. _____ C4c. _____
	4b. _____	
	4c. _____	
		E = Eliminated C = Controlled

Figure 5.5 Pretask Hazard Analysis Card – Back Side

Used in the Safety Profession published by ASSE, JHA and JSA are defined as the same method (Lack, 2001, p. 58). Most safety practitioners in the United States use the terms interchangeably.

However, some argue there are differences between JHA and JSA. For example, in Australia, JHA and JSA are considered to be two separate techniques. Jim Whiting, a Australian risk engineer and leader in the risk assessment field, indicates that *JHA* most often does not include a risk analysis, only a hazard analysis, while a *JSA* usually includes some form of risk analysis such as determination of risk levels or scores to be used for decision making and prioritizing risk control resources (Whiting, 2013).

For the purposes of this text, *JHA* and *JSA* are viewed as a "hazard analysis," while *job risk assessment* is considered a "hazard analysis and risk assessment," which is covered in this chapter. Since JHA focuses on the specific hazards of the job rather than the risks, it is considered a hazard-based approach.

The JHA process was derived from the scientific management practice of job analysis. Job analysis was used in the early 1900s to develop standardized work instructions or "SOPs" for specific jobs. As the use of job analysis progressed, hazards and preventive measures associated with the job began to be identified and added in varying degrees. It was later referred to as JHA or JSA. More recently, the term job safety analysis or JSA has been used by some to denote a more positive view of the process. Some of the benefits derived from conducting an effective JHA include:

- Defined and improved job methods that can be used for employee orientation and training and post-incident analysis
- reduced exposure to hazards leading to fewer employee injuries and illnesses
- reduced costs resulting from employee absenteeism and injuries
- Increased productivity and quality.

5.8.3.1 Conducting a JHA

A JHA involves careful study and recording of each job step, identifying existing or potential job hazards (both safety and health) associated with each step, and determining the best way to perform the job and reduce or eliminate hazards. When performing JHA, there is a great opportunity to engage employees in the process and leverage their unique and critical job knowledge. It is important to clearly communicate the purpose of the analysis and that it is not an evaluation of employee performance. The following steps provide guidance on conducting a JHA:

1. *Select job for analysis.*

   A JHA is a "second-tier analysis," generally following a broader, more comprehensive assessment of the workplace. The National Safety Council defines a job as a "sequence of separate steps or activities that together accomplish a work goal" and states that jobs suitable for analysis are those that are neither too broad nor too narrow (National Safety Counicl, 1992). Generally, an inventory of jobs from every department or division is collected and assessed to produce a prioritized list of jobs that might benefit from a JHA. Part of the analysis should include a review of documentation related to the job such as accident history, incident reports, near hits, hazard reports, safety inspections, SOPs, and other available information. General observations of the workplace and jobs being performed are necessary to gain a better understanding of the types of activities and potential hazards in comparison to other jobs. Indications that a job is appropriate for a JHA include those with:
   - High injury or illness rate experience
   - High likelihood of risk
   - High severity of risk
   - History of human error or potential for human error that can result in severe consequences
   - Changes in the methods, material, or equipment
   - Multiple tasks or steps
   - Complex tasks
   - New jobs or existing jobs with newly added tasks.

   In addition to these indicators, jobs that make up a large portion of a workplace or have a large exposure level to the workforce should also be considered for analysis.

2. *Select a team.*

   Most hazard analyses and risk assessments benefit from a group of knowledgeable stakeholders. JHA's are no exception. A team facilitator, often a safety professional skilled in operational risk, hazard recognition, risk controls, and the JHA process, is needed to lead and guide the team. Other team members should include representatives from frontline management (supervisor or department manager), engineering, maintenance, quality, production, procurement, logistics, human resources, and most importantly, the operator or employee performing the job. The "operator" has first-hand knowledge and experience performing the job and knows what and how things can go wrong. The operator's unique understanding of their job helps ensure that the job's sequential tasks are properly laid out without omissions and that less obvious hazards are identified and addressed. Another benefit of having the affected stakeholder involved is that they are likely to have a sense of buy-in or ownership of the recommended control measures, helping ensure safer and more productive execution of the job.

3. *Prepare for the analysis.*

   Before conducting the job observations, a review of available documents related to the job and informal interviews with job operators is advised. Job-related documents such as written job instructions, safety procedures, job training, manufacturer specifications for equipment or tool, safety data sheets for chemicals, and other available documents are useful in the analysis. In addition, loss history and incident information including hazard reports, near-hit reports, accident investigations, and analysis reports and other related loss information related to the job can provide insight into the types of accident scenarios and the causal factors involved. Following a review of documents, informal interviews with operators should be performed to gather further information, verification, and insight regarding the job and its potential concerns. This is a key step that involves the employee in the safety process as well as enables the team to better understand the job and its sequence of steps, potential hazards, and existing controls. Questions should be raised that help identify specific job steps and activities that produce risk such as

- "What is the most critical part of the job?"
- "What part of the job is the most difficult to perform?"
- "Where can errors occur?"
- "What can go wrong?"
- "How can it go wrong?"

As a result, the team will have a better understanding of how the job is expected to be performed, the correct sequence of steps, as well as some insight in the associated hazards and existing controls.

4. *Observe the job.*

    The JHA involves three basic steps: (1) breaking down the job into a sequence of steps, (2) identifying the hazards and hazard scenarios, and (3) developing recommended control measures. To begin job observation, it is recommended that an experienced operator be selected for observation to ensure a more accurate accounting of the job. Reassure the operator that the purpose of the observation is to identify and define the correct sequence of job steps, related hazards, and necessary control measures and that it is not to evaluate the operator's performance. An adequate number of job cycles should be observed to provide a clear picture of the entire job and its individual steps. Digital video and photographs for later study are helpful in capturing details that are sometimes difficult to identify while in the operation due to surrounding activities and distractions normally encountered in the workplace. It is advised that the operator, as well as other workers who perform the same job, be involved in all phases of the analysis from reviewing the job steps and procedures to discussing potential hazards and recommended solutions.

5. *Break down job steps.*

    The job is broken down into a sequence of individual basic steps or tasks that are performed. Generally, a JHA contains 3–10 manageable steps to avoid making the breakdown of steps so broad that it does not include basic steps or so detailed that it becomes unnecessarily long. Each step should be described with action words (such as "remove part from bin," "place part in fixture," or "cut part into segments") providing enough information to communicate the activity without making it too complex. A review of the step descriptions with the operator(s) and team to ensure that descriptions are clear and that all steps are included and in the proper order should be performed. Once satisfied, the steps are documented in the first column of the JHA worksheet in sequence.

6. *Identify hazards for each step.*

    For each step, the team identifies associated hazards (existing and potential) from both the pre-analysis research and job observations. Hazard checklists and general questions are helpful in covering hazard classifications and types (Appendix 5.B provides a list of common hazards and descriptions). The team should develop potential hazard scenarios based on the information gather to determine what, how, and why certain hazards can occur. The hazard scenario would include the job step activities and resulting hazards. The following taken from the OSHA 3071 document provides a hazard scenario example:

'In the metal shop (environment), while clearing a snag (trigger), a worker's hand (exposure) comes into contact with a rotating pulley. It pulls his hand into the machine and severs his fingers (consequences) quickly.'

Development of creditable hazard scenarios associated with specific tasks is necessary to identify how and why specific hazards might occur, potential causal factors, and possible control measures needed. Hazard descriptions are listed in column 2 of the JHA document.

7. *Identify control measures for each step.*

Using the hierarchy of controls, the team will formulate necessary control measures to eliminate or reduce each identified hazard to an acceptable level. The team-based solutions may lead to redefining the method, changes in specific tasks, elimination of some tasks, redesign of workstation, or other changes. For hazards that are not eliminated, control measures to reduce the hazard to an acceptable level are developed and described in column 3 of the JHA document. Examples of controls might include safeguards, presence sensing devices, ergonomic tools and assists, specific equipment, job setup, lighting, ventilation, protective clothing and equipment, training in job procedures, and other engineering and administrative controls.

5.8.4 Fundamental Risk Assessment

A formal assessment of operational risk is defined as the systemic identification, analysis, and evaluation of risk, which is used in operational risk management efforts. For instance, a formal risk assessment should be considered before construction of new facilities or major repair projects; before new equipment is operated or when there has been a significant change in procedures; during operation and maintaining tasks and jobs that have an unknown risk level and high exposure level; and following incidents, especially those that have a significant impact or potential for impact on the operation. Taking more time to estimate potential harm and likelihood of occurrence would be appropriate at such times and could be accomplished by a small team planning a project or performing a walk-through. Such risk assessments are considered "formal" since they involve a formalized or structured assessment process that documents the resulting hazard analysis and risk assessment information used in making risk management decisions.

Generally, a task or job risk assessment follows a larger assessment of the workplace that has identified hazards and risks and prioritized tasks, jobs, and operations that require a more "job-specific" assessment. Fundamental risk assessments are used to determine a risk rating or score to express the risk level associated with a particular task or job. Individual risk ratings are used to judge whether the associated risks are deemed ALARP and therefore acceptable to an organization. It is important to keep the purpose of the assessment in mind, which is "controlling risk to an acceptable level." Stakeholders should avoid placing too much emphasis on the "precision" of risk ratings and focus on developing better risk controls to reduce risk levels to ALARP (Whiting, 2013).

When determining risk ratings, there are three methods used in analyzing, assessing, and scoring risks known as qualitative, semiquantitative, and quantitative previously discussed in Chapter 3. As described by ANSI Z690.3, the type of method selected is dependent upon the particular application, degree of detail required, availability of reliable data, and needs of an organization.

Qualitative risk assessment methods define the severity, likelihood, and level of risk by levels of significance such as "high," "medium," and "low." These qualitative or subjective methods of evaluation use descriptive definitions for risk categories, ranges, or ratings of severity and likelihood based on judgments, informed opinions, and beliefs. Their ease of use and application make qualitative methods the most common method used in fundamental assessments.

Semiquantitative risk assessment methods use numerical scoring systems for severity and likelihood and combine the two to produce a level of risk using a formula. Rating scores or scales may have a linear, logarithmic, or other numerical relationship. Although this method uses numerical values, the resulting scores are largely based on qualitative definitions to assist estimation, rather than quantifiable or empirical data.

Quantitative risk assessment methods assign fixed numerical values to both the severity and likelihood of an outcome using quantitative values such as statistical or historical data. It is important to note that it is objective only if the probabilities and severity outcomes are objective. True quantitative analysis is generally limited to more complex systems where there are sufficient data and information of the systems analyzed (Main, 2004).

5.8.5 Job Risk Assessment

Similar in process to the JHA, a job risk assessment follows the steps described previously and includes an additional component: risk estimation. When determining risk, the hazards identified in the job risk assessment are analyzed and evaluated to estimate both severity and likelihood of risk. Risk estimation and ranking require the organization/stakeholders to establish a set of definitions for risk severity and risk likelihood according to the organizations acceptable risk level or ALARP. The resulting risk levels are usually comprised in a risk matrix and risk priority action such as shown in Tables 5.2 and 5.3. Some organizations use numbers for likelihood and severity and multiply severity × likelihood to achieve a risk score or ranking. Others use only verbal descriptors.

The addition of the risk estimation provides a measurement component to the process allowing baselines and benchmarks to be established to aid decision making in the management of operational risk. Figure 5.6 provides an example of a job risk assessment and how it may be completed.

Generally, a job risk assessment includes a brief description of job steps in sequence, a description of hazards and their effect in each step, recommended control measures, and job risk ratings. Some risk assessments include both a precontrol risk rating and postcontrol rating as illustrated in Figure 5.6. Defining and verifying the potential credible hazard scenarios and how these scenarios can occur is an important part of an effective risk assessment. When defining hazard scenarios, adequate time and input from experienced stakeholders are necessary to properly develop and describe the scenarios.

TABLE 5.2 Risk Matrix Example

Risk Matrix		Likelihood			
		Very Unlikely (L1) Little or No Chance of Occurrence	Unlikely (L2) Could Occur but Unlikely	Likely (L3) Could Occur and is Likely	Very Likely (L4) Could Occur and is Very Likely
Severity	*Danger (S4)* Fatality, permanent injury/illness	Moderate risk 4	High risk 8	Very high risk 12	Very high risk 16
	Warning (S3) Long-term injury/illness	Moderate risk 3	Substantial risk 6	High risk 9	Very high risk 12
	Caution (S2) Medical attention	Low risk 2	Moderate risk 4	Substantial risk 6	High risk 8
	Notice (S1) First Aid	Very low risk 1	Low risk 2	Moderate risk 3	Moderate risk 4

FORMAL METHODS 111

TABLE 5.3 Risk Priority Levels Example

Risk Priority Level	Action and Response
Very high risk 12–16	• Immediate action is required before operation can begin
High risk 9–11	• Immediately report risk exposure at this level to management
Substantial risk 6–8	• Take action as soon as possible to prevent harm
Moderate risk 4–5	• Report risks to management as soon as possible during the shift • Ensure the ongoing effectiveness of existing risk controls
Low risk 2–3	• Take action when necessary and ensure risks remain low by verifying the continued effectiveness of existing controls
Very low risk 1	• Record risks and monitor for changes and control risks as needed

Job Risk Assessment

Dept:		Job:		Assessors:				Date:			
#	Job Steps	Hazards		Initial Risk				Controls	Residual Risk		
		Hazard and Effect	At Risk	Severity	Likelihood	Risk Rating			Severity	Likelihood	Risk Rating
	Basic description of tasks recorded in sequence	Description and effect of existing and potential hazards based on observations and experience	Population or groups exposed	S Code	L Code	Priority Risk Code		Describe necessary controls for each hazard using the Hierarchy of Controls method	S Code	L Code	Priority Risk Code
1	Worker reaches into bin next to machine, grasps with one hand a 15-lb casting, and carries to grinding wheel 10ft away every 2 min	(1) MSD to back, and hands from bending into bin and lifting	Metal workers in metal shop	3	3	9		(1) Open-side, tilted, and self-adjusting bins to reduce reach	2	1	2
		(2) Lacerations to hand from sharp burrs/edges of casting		2	3	6		(2) Cut-resistant gloves and sleeves	1	2	2
		(3) Tripping or slipping on items while carrying parts		3	2	6		(3) 5S system, hose reels, preinspection, and housekeeping	2	1	2
2	Worker holds piece with both hands at grinder to deburr edges for ~1 min	(4) MSD to hands from static muscle loading	Same	3	3	9		(4) Use fixture to hold piece while grinding	2	2	4
		(5) Abrasions from grinder		2	3	6		(5) Use fixture to hold piece while grinding	2	2	4
		(6) Eye and face injury from flying particles from grinding		3	4	12		(6) Use ANSI Z87 eye protection and face shield	2	2	4
3	Place piece in finished bin adjacent to grinder	(7) MSD to back twisting and tossing pieces	Same	3	3	9		(7) Place bin in position where twisting is not required	2	2	4

Figure 5.6 Job Risk Assessment Example

Simply stating that something is "possible" is not sufficient. A credible scenario for how it could occur is needed and then the scenario's likelihood can be estimated. Similarly, risk ratings for severity and likelihood should be well defined and agreed upon by stakeholders to achieve a consistent and credible risk scoring of jobs.

5.9 CONCLUSION

To summarize, fundamental hazard analysis and risk assessment methods are considered the foundation for all operational risk assessment methods. Their results often indicate the need for further more detailed and quantitative assessment techniques. Their strength of effectiveness is based on their simplicity and ease of use, acting as a quick filter for the numerous risks that need to be assessed.

As the key component of the risk management process, the effective application of fundamental hazard analysis and risk assessment methods is an essential skill. A firm understanding of these task and job level methods opens up the door to more advanced and specialized techniques of operational risk assessment. Proficiency in methods such as pretask, task, and JHA and risk assessment has become an expected function within an ORMS. Safety professionals who possess this skill and who can facilitate the process will be of greater value to an organization and more likely to advance in their profession.

REVIEW QUESTIONS

Fundamental hazard analyses and risk assessments are the commonly used and relied-upon techniques in the workplace. The following review questions are presented to reinforce some of the salient points regarding these tools and their application.

1. When selecting a method of hazard analysis and risk assessment, what factors and elements should the stakeholder consider?

2. Define the nature and use of a risk assessment and explain the similarities and difference between a hazard analysis and a risk assessment.

3. Provide three examples of hazards in a work environment that may not pose a significant risk. Using the same examples, describe how these hazards could present significant risk under different circumstances.

4. How are job hazard analyses and job risk assessments similar, and how are they different? What are the limitations and cautions of using job hazard analyses rather than risk assessments?

5. When is it most appropriate to use a formal hazard analysis and risk assessment technique?

6. Why are informal methods used, and where are they most beneficial?

7. When assessing risk, what elements must the assessor take into account? How is risk measured?

REFERENCES

ANSI/ASSE Z590.3-2011. *Prevention through Design: Guidelines for Addressing Occupational Hazards and Risks in Design and redesign Processes.* Des Plaines, IL: American Society of Safety Engineers, 2011.

ANSI/ASSE Z690.1-2011. *American National Standard – Vocabulary for Risk Management.* Des Plaines, IL: American Society of Safety Engineers, 2011.

ANSI/ASSE Z690.2-2011. *American National Standard – Risk Management Principles and Guidelines.* Des Plaines, IL: American Society of Safety Engineers, 2011.

ANSI/ASSE Z690.3-2011. *American National Standard – Risk Assessment Techniques.* Des Plaines, IL: The American Society of Safety Engineers, 2011.

ANSI/ASSE/AIHA Z10-2012. *American National Standard – Occupational Health and Safety Management Systems.* Fairfax, VA: American Industrial Hygiene Association, 2012.

BS OHSAS 18001:2007. *Occupational Health and Safety Management Systems – Requirements.* London, UK: British Standards Institution (BSI), 2007.

Hudson, E., D. Smith. *Field Level Risk Assessment: Manager's Handbook.* Alberta, Canada: Construction Owners Association of Alberta, 1998. Available at: www.coaa.ab.ca/portals/Library/Managers%20Handbook2_%20(1).pdf (accessed December 29, 2015).

ILO-OSH 2001. *Guidelines on Occupational Safety and Health Management Systems.* Geneva, Switzerland. International Labour Office, 2001.

ISO 14001:2004(E). *Environmental Management Systems – Requirements with Guidance for Use.* Geneva, Switzerland. International Organization for Standardization, 2004.

Lack, R., *The Dictionary of Terms Used in the Safety Professional*, 4th ed. Des Plaines, IL: American Society of Safety Engineers, 2001.

Main, B. W., 2004. "Risk assessment: a review of the fundamental principles." *Professional Safety* 49(12), 37–47.

Manuele, F. A. *Advanced Safety Management: Focusing on Z10 and Serious Injury Prevention.* Hoboken, NJ: Wiley, 2008.

Manuele, F. A. *On the Practice of Safety* 4th ed. Hoboken, NJ: Wiley, 2012.

MSHA. *OT 10-08-2. SLAM Risks: Coal and Metal/Nonmetal Fatalities.* January–June 2008. Washington, DC: MSHA, National Mine Health and Safety Academy, 2008.

MSHA. SLAM Risks the Smart Way, 2004. Available at http://www.msha.gov/webcasts/MNM2004SLAM/MNM2004SLAMWebcastscript.pdf (accessed December 29, 2015).

National Safety Council, *Accident Prevention Manual for Business & Industry: Engineering & Technology*, 10th ed. Itasca, IL: National Safety Council, 1992.

OSHA. *Publication 3071-2002. Job Hazard Analysis.* Washington, DC: Occupational Safety and Health Administration. 2002. Available at: https://www.osha.gov/Publications/osha3071.pdf?utm_source=rss&utm_medium=rss&utm_campaign=job-hazard-analysis-13 (accessed December 29, 2015).

OSHA. Case Study: Texas Operations Contractor Alliance for Safety at Dow Facility in Freeport, Texas, 2008. Available at: https://www.osha.gov/dcsp/success_stories/compliance_assistance/tocas_case_study.html (accessed December 29, 2015).

Whiting, J. F. Effective Risk Assessment in TA, JHA, JSA, JSEA, WMS, TAKE 5, and Incident Investigation, 2013. ASSE Professional Development Conference, June, Las Vegas, NV.

APPENDIX 5.A

ANSI Z690.3-2011 Risk Assessment Techniques	ANSI Z590.3-2011, Prevention through Design
Annex B – Risk Assessment Techniques (Informative)	Addendum G – Comments on Selected Hazard Analysis and Risk Assessment Techniques (Informative)
1. Brainstorming 2. Structured/semistructured interviews 3. Delphi technique 4. Checklists 5. Preliminary hazard analysis (PHA) 6. Hazard and operability (HAZOP) Study 7. Hazard analysis and critical control points (HACCP) 8. Toxicity assessment 9. Structured "what-if" technique (SWIFT) 10. Scenario analysis 11. Business impact analysis (BIA) 12. Root cause analysis (RCA) 13. Failure modes and effects analysis (FMEA) and failure modes and effects and criticality analysis (FMECA) 14. Fault tree analysis (FTA) 15. Event Tree Analysis (ETA) 16. Cause–Consequence Analysis 17. Cause-and-effect analysis 18. Layers of protection analysis (LOPA) 19. Decision tree analysis 20. Human reliability assessment (HRA) 21. Bow-Tie Analysis 22. Reliability-Centered Maintenance 23. Sneak Analysis (SA) and Sneak Circuit Analysis (SCI) 24. Markov analysis 25. Monte Carlo simulation 26. Bayesian Statistics and Bayes Nets 27. FN curves 28. Risk indices 29. Consequence/probability matrix 30. Cost/benefit analysis (CBA) 31. Multicriteria decision analysis (MCDA)	1. Preliminary hazard analysis (PHA) 2. What-if analysis 3. Checklist analysis 4. What-if/checklist analysis 5. Hazard and operability (HAZOP) analysis 6. Failure mode and effects analysis (FMEA) 7. Fault tree analysis (FTA) 8. Management oversight and risk tree (MORT)

APPENDIX 5.B

COMMON HAZARDS AND DESCRIPTIONS

Hazard Type	Hazard Description
Biological	Also called biohazards, biological hazards are produced from living organisms such as humans (such as blood-borne pathogens and medical waste), wild and domestic animals (carriers of disease, animal feces, etc.), fauna (insects that carry disease such as ticks or mosquitos), flora (plants such as poison ivy), bacteria, viruses, and fungus. These sources can cause a variety of health effects ranging from skin irritation and allergies to infections (e.g., tuberculosis, AIDS), cancer, and other serious illnesses
Chemical – toxic	Chemicals that present an exposure to people by absorption through the skin, inhalation, ingestion, or injection into the bloodstream that can cause illness, disease, or death. There can be an acute (immediate) effect or a chronic (medium- to long-term) effect from the accumulation of chemicals or substances in or on the body. The amount or dose of the chemical exposure determines the hazardous effect. Safety data sheets (SDSs) provide information regarding the toxic effects and toxic levels along with other chemical hazard information
Chemical – flammable	Chemicals with a lower flash point and boiling point present a fire hazard when an ignition source and oxygen are present. Generally, flammable liquids can and will ignite and burn easily at normal working temperatures
Chemical – corrosive	Chemicals with a corrosive property such as acids and bases that causes damage to the skin, metal, or other materials with contact or exposure
Collapse	Collapse of material such as soil (trenching or excavation), grain (silo or confined spaces), or other loose material causing entrapment, crushing forces, or struck-by hazards
Commercial pressure	Real or perceived or self-motivated pressure and demands for excessive productivity to complete activities at intolerable rates or amounts. This type of hazard is often a root cause of the other hazards listed
Explosion	Explosion hazards result from a sudden and violent release of a large amount of energy due to a chemical reaction (reactive chemicals in a confined space) or significant pressure difference (such as rupture of a compressed gas cylinder or boiler). Very finely divided material such as grain, metal, plastic, rubber, fiber, coal, or other combustible dust presents a fire or explosion hazard when dispersed and ignited in air
Electrical – shock	Passage of electrical current through the body due to contact with exposed conductors or a device that is not properly grounded (such as when equipment or tools come into contact with power lines). A relatively small current from a common house supply of 60 Hz alternating current can stop the heart. Also, electrical shock can cause secondary hazards such as falls when working at heights
Electrical – fire	Electrical overloading, overheating, or arcing can create an ignition source for fire, arc flash burns, and arc blast

Hazard Type	Hazard Description
Electrical – static discharge	Static electrical energy is created by friction of materials such as dispensing of liquids and movement of clothing, rotor blades, wheels, and other objects. This creates an excess or deficiency of electrons on the surface of materials that can discharges a high-voltage/low-current spark to the ground, resulting in an ignition sources or damage to electronics or the body's nervous system
Electrical – loss of power	Safety-critical equipment failure (such as life support in hospitals or supplied breathing air to workers in a confined space) resulting from a loss of power
Ergonomics – musculoskeletal	Damage to soft tissues such as muscle, nerve, and blood vessels, resulting from prolonged exposure to repetitive motion, force, awkward posture, static muscle loading, compression, and extreme environmental conditions
Ergonomics – human factors	A system design, procedure, equipment, or product that can cause misinterpretation, human error, omission, or false sense of security resulting in harm. Examples of human factor related hazards include poorly designed controls and displays (unable to easily distinguish different controls), equipment or products that do not work the way a person expects (counterintuitive), things that are hard to see (too small, obstructed, lack of contrast), poor warning, or instructions
Extreme temperature	Temperatures that result in heat stress or exhaustion (excessive heat) or metabolic slowdown and hypothermia (excessive cold)
Fall	Conditions that can result in falls from heights (ladders, man lifts, exposed ledges) or same level surfaces (such as slippery floors, poor housekeeping, uneven walking surfaces, floor opening, etc.)
Fatigue	Physiological and psychological impairment due to inadequate rest or sleep
Incompetent	Inability to complete an activity due to lack of knowledge and/or skill leading to other hazards
Lack of supervision/support	Inadequate supervision, support, or resources leading to the development of other hazards
Mechanical	Exposure to machinery movement or point-of-operation such as rotating, reciprocating, chipping, vibrating, cutting/shearing, punching, or crushing actions, causing physical damage to exposed body part. Injury can be caused by flying objects, crushing, caught-between, sharp edges, struck-by, puncture, twisting, or shearing
Noise	Exposure to excessive noise levels can result in hearing loss and interfere with verbal communication.

Hazard Type	Hazard Description
Psychosocial	Occupational stress caused by employees, management, the public, or conditions as work such as violence, conflict or aggression, harassment, excessive work pace or production demands, emotional or cognitive demands, poorly defined work roles, lack of job control, poorly managed change, and inadequate reward or recognition
Radiation – ionizing	Alpha, beta, neutron (particles), and gamma/X-rays that cause tissue damage by ionization of cellular components. Examples include naturally occurring radioactive materials, such as radon in mining; industrial and medical radioisotopes, such as tracer elements; high-voltage devices such as X-ray machines, radar generators, VDTs and TVs; and nuclear reactors
Radiation – nonionizing	Ultraviolet (UV), visible light, infrared (IR), radio-frequency (RF) and microwave (MW), and extremely low-frequency radiation (ELF) that cause injury to tissue by thermal or photochemical means. Sources of UV radiation include the sun, black lights, welding arcs, and UV lasers. Sources of IR radiation include furnaces, heat lamps, and IR lasers. Sources of RF and MW radiation include radio emitters and cell phones. Common sources of intense exposure include ELF induction furnaces and high-voltage power lines. Sources of electromagnetic include intense electric and magnetic fields created by MRI and high current electrorefining
Struck by	Accelerated mass that strikes the body, causing injury or death such as falling objects, vehicles, and projectiles
Struck against	Injury resulting from trauma or sudden force coming into contact with an object or surface initiated by the person such as a hammer striking the hand accidentally
Vibration	Damage to soft tissues resulting from frequent, long duration exposure to vibration such as power tools, heavy equipment cabs, and machinery
Visibility	Lack of adequate lighting or obstructed vision that results in error, omission, or failure to recognize existing hazards
Weather	Extreme weather conditions such as rain, snow, wind, or ice

APPENDIX 5.C
PERSONAL PROTECTIVE EQUIPMENT HAZARD ASSESSMENT FORM EXAMPLE

PPE Hazard Assessment From

Department:	Job Function/Activities:
WorkLocation(s):	

1. Hazards Present (check all that apply) (check "Other" and write "none" if no apparent hazards exist)	2. Describe source of Hazards (i.e., portable grinder, arcs from welding, work on steam lines, etc.)	Personal Protective Equipment to Consider (complete appropriate boxes with the specific PPE required e.g., splash goggles, face shields, nitrile gloves, etc.)				
		Eye	Hand/Arm	Head	Body/Legs	Foot
☐ Impact						
☐ Cuts/Penetration						
☐ Pinch/Crush/Roll Over						
☐ Thermal(Hot/Cold)						
☐ Light(optical)Radiation						
☐ Chemical						
☐ Biological						
☐ Dust						
☐ Electrical						
☐ Other						

Assessment completed by: _____

Signature: _____ Date: _____

APPENDIX

APPENDIX 5.D
JOB HAZARD ANALYSIS FORM EXAMPLE

Job Hazard Analysis

Job:
Job Description:
Analyzed by:
Analysis Date:
Approved by:

Task	Hazards	Controls

6

WHAT-IF HAZARD ANALYSIS

BRUCE K. LYON
Risk Control Services, Hays Companies, Kansas City, MO, USA

6.1 OBJECTIVES

- Introduce What-If analysis
- Review process hazard analysis methods
- Examine the use of techniques and their strengths and limitations
- Provide guidance on the use of What-If analysis techniques

6.2 INTRODUCTION

What if … ? What if indeed. Those that have experienced an incident involving serious injuries or fatalities might find themselves asking that very question. However, the time to ask "What-If" is up front during planning, development, and operational stages before incidents occur as part of an operational risk management system. Using methods such as What-If is a relatively easy and cost-effective way of identifying, analyzing, and controlling hazards and their risk to an acceptable level.

This chapter is designed to provide a primer in the concept, application, and use of a What-If hazard analysis and its variants such as What-If/checklist analysis and structured What-If technique (SWIFT) analysis, in general industry settings. Please note that there are numerous resources and texts that provide more in-depth coverage of this and other process hazard analysis (PHA) methodologies, specifically in chemical and petroleum industries, some of which are in response to regulatory requirements. In addition to the reference section of this chapter, an Internet search using the term "process hazard analysis" or "What-If hazard analysis" will lead to a number of possible resources.

6.3 OVERVIEW AND BACKGROUND

What-If hazard analysis is a well-established and widely used qualitative method for identifying and analyzing hazards, hazard scenarios, and existing and needed controls. It was originally developed in

---

Risk Assessment: A Practical Guide to Assessing Operational Risks, First Edition.
Edited by Georgi Popov, Bruce K. Lyon, and Bruce Hollcroft.
© 2016 John Wiley & Sons, Inc. Published 2016 by John Wiley & Sons, Inc.

the 1960s by the British chemical industry as an easier and less costly alternative to the hazard and operability (HAZOP) study (Nolan, 1994).

Although originally developed for chemical and petrochemical process hazard studies, the *What-If hazard analysis* and its variations have become widely used in many other industries including energy, manufacturing, high tech, food processing, transportation, and healthcare to mention a few. The method can be applied to a system, process, or operation or at a more specific focus such as a piece of equipment, procedure, or activity. Examples of where What-If might be applied include the following:

- Operations that contain hazardous chemical processes
- Operations with large refrigeration and chiller systems containing ammonia such as meat packing, food processing, and storage
- Nonroutine activities such as equipment installations, repair, or decommission
- "Tabletop drills" to develop emergency scenarios and necessary measures for preparedness, disaster recovery, and business continuity
- Design safety reviews of new facilities, systems, and equipment
- In operations where "management of change" is considered
- Analysis prior to selection and procurement of new technology, equipment, or materials.

As discussed in previous chapters, the act of identifying and analyzing hazards is fundamental to safety and operational risk management. Johnson (1980) expressed this in his statement: "Hazard identification is the most important safety process in that, if it fails, all other processes are likely to be ineffective." In other words, if an existing hazard is not identified, it goes unmanaged and uncontrolled, placing people and assets at risk.

Hazards and hazard exposure scenarios can be unique, diverse, and oftentimes complex. Therefore, it is important to select an appropriate method of hazard analysis and risk assessment for the situation. The American National Standards Institute's *ANSI/ASSE Z690.3-2011, Risk Assessment Techniques* standard (nationally adopted from IEC/ISO 31010:2009) lists 31 different techniques including *brainstorming*, *checklist analysis*, and the SWIFT discussed in this chapter. Further, the *ANSI/ASSE Z590.3-2011, Prevention through Design* standard lists eight techniques including *What-If analysis*, *checklist analysis*, and *What-If/checklist analysis*. Regarding the variety of methods, the ANSI Z590.3 Prevention through Design standard goes on to say the following:

> "Note 1: Over the past forty years, a large and unwieldy number of hazard analysis and risk assessment techniques have been developed. Descriptions of eight selected techniques are presented in Addendum G. As a practical matter, having knowledge of three risk assessment concepts will be sufficient to address most, but not all, risk situations. They are: Preliminary Hazard Analysis and Risk Assessment; the What-If/Checklist Analysis Methods; and Failure Mode and Effects Analysis."

Even though some techniques have become somewhat standard for certain applications or industries, it is advisable to consider the available information and expertise and select the most effective and efficient technique(s) to accomplish the objectives of the assessment.

6.4 PROCESS HAZARD ANALYSIS

A "process" can be defined as a series of actions or steps taken to produce something or achieve a particular result. In industrial settings, there are numerous processes that often include hazards and risks. In order to understand process-related hazards and their causes, a PHA is performed.

A PHA is a set of organized and systematic assessments of the potential hazards associated with an industrial process. It provides information intended to assist stakeholders in making decisions for improving safety and reducing operational risk associated with a process. The American Institute of Chemical Engineers' (AIChE, 2015) Center for Chemical Process Safety provides the following definition of a PHA:

An organized effort to identify and evaluate hazards associated with chemical processes and operations to enable their control. This review normally involves the use of qualitative techniques to identify and assess the significance of hazards. Conclusions and appropriate recommendations are developed. Occasionally, quantitative methods are used to help prioritized risk reduction.

A PHA is directed toward analyzing potential causes and consequences of fires, explosions, and releases of toxic or flammable chemicals and focuses on equipment, instrumentation, utilities, human actions, and external factors that may impact the process. In many cases, an additional benefit of conducting such an analysis is a more thorough understanding of the industrial process, leading to opportunities for improving process efficiency and cost reduction.

6.5 MANDATED ASSESSMENTS

Hazard analyses and risk assessments are more common in Europe, Australia, New Zealand, Canada, and other parts of the world. In the United Kingdom, risk assessments have been legally required of businesses since 1999 by the Health and Safety Executive. However, there are very few hazard analyses and risk assessments required by law in the United States. Two exceptions include the Occupational Safety and Health Administration (OSHA) Process Safety Management (PSM) of Highly Hazardous Chemicals and the Environmental Protection Agency (EPA) Risk Management Plan, both of which require PHAs. The following are brief summaries of these two standards:

- *OSHA 29 CFR 1910.119 PSM of Highly Hazardous Chemicals* standard, established in 1992, requires PHAs for regulated industrial processes containing 10,000 lb or more of a hazardous chemical for the purpose of protecting the employees working in and around such processes.
- *EPA 40 CFR PART 68 Chemical Accident Prevention Provisions, Risk Management Plan (RMP) Rule* issued in 1994 as a result of the Clean Air Act Amendments of 1990 mirrors the OSHA PSM requirements for PHAs in regulated facilities for the purpose of protecting the public and the environment from undesired consequences of explosions or releases.

In the PSM standard, OSHA defines process as "any activity involving a highly hazardous chemical including any use, storage, manufacturing, handling, or the on-site movement of such chemicals, or combination of these activities. For purposes of this definition, any group of vessels which are interconnected and separate vessels which are located such that a highly hazardous chemical could be involved in a potential release shall be considered a single process."

Specifically, OSHA's PSM standard addresses mandated PHAs in 1910.119 (e)(1) stating that "an initial process hazard analysis (hazard evaluation)" of covered processes be conducted by the operation. The PHA "shall be appropriate to the complexity of the process and shall identify, evaluate, and control the hazards involved in the process." Selection of the analysis method should be based on these considerations. OSHA also requires the operation to "determine and document the priority order for conducting process hazard analyses based on a rationale which includes such considerations as extent of the process hazards, number of potentially affected employees, age of the process, and operating history of the process." In essence, OSHA requires that PHAs be prioritized and performed in accordance to their *risk level*. The OSHA PSM and EPA RMP standards require PHAs to include the following:

- The hazards of the process
- The identification of any previous incident, which had a likely potential for catastrophic consequences in the workplace
- Engineering and administrative controls applicable to the hazards and their interrelationships such as appropriate application of detection methodologies to provide early warning of releases (acceptable detection methods might include process monitoring and control instrumentation with alarms and detection hardware such as hydrocarbon sensors)
- Consequences of failure of engineering and administrative controls
- Facility siting
- Human factors
- A qualitative evaluation of a range of the possible safety and health effects of failure of controls on employees in the workplace.

Other notable requirements related to performing PHAs in the OSHA's PSM 1910.119 standard include:

- "Performed by a team with expertise in engineering and process operations, and the team shall include at least one employee who has experience and knowledge specific to the process being evaluated."
- "One member of the team must be knowledgeable in the specific process hazard analysis methodology being used."
- Establish a system to promptly address the team's findings and recommendations; assure that the recommendations are resolved in a timely manner and that the resolution is documented.

What-If hazard analysis is one of several PHA methodologies referred to in the OSHA PSM standard and EPA RMP Rule as an acceptable method. The OSHA publication 3132, *Process Safety*

TABLE 6.1 Process Hazard Analysis Methodologies Listed in OSHA 1910.119(e)(2)

Process Hazard Analysis Method	Description
What-If	Uses a multiskilled team to create and answer a series of "What-If" type questions. This method has a relatively loose structure and is only as effective as the quality of the questions asked and the answers given
Checklist	Uses established codes, standards, and well-understood hazardous operations as a checklist against which to compare a process. A good checklist is dependent on the experience level and knowledge of those who develop it
What-If/checklist	A team-based, structured analysis that combines the creative, brainstorming aspects of the What-If with the systematic approach of the checklist. The combination of techniques can compensate for the weaknesses of each
Hazard and operability (HAZOP) study	A team-based, structured, systematic review of a system or product that identifies risks through the use of "guide words," which question how the design can fail due to certain limitations and deviations of the operation
Failure mode and effects analysis (FMEA)	Technique used to identify the ways systems and their components can fail and the resulting effect
Fault tree analysis	Technique used for identifying and analyzing factors that can contribute to a specified undesired event. Causal factors are deductively identified, organized in logical manner, and represented pictorially in a tree diagram

Management, dated 2000, states that "the process hazard analysis is a thorough, orderly, systematic approach for identifying, evaluating, and controlling the hazards of processes involving highly hazardous chemicals. The employer must perform an initial process hazard analysis (hazard evaluation) on all processes covered by this standard. The process hazard analysis methodology selected must be appropriate to the complexity of the process and must identify, evaluate, and control the hazards involved in the process."

The methods in *OSHA 1910.119(e)(2)* considered by OSHA as appropriate to determine and evaluate process hazards are listed and briefly described in Table 6.1.

In recent years, a number of consensus standards have introduced requirements for making risk assessments. For example, the National Fire Protection Association (NFPA) is beginning to incorporate risk assessment requirements in certain NFPA Codes and Standards, such as the 2013 edition of NFPA 654, *Standard for the Prevention of Fire and Dust Explosions from the Manufacturing, Processing, and Handling of Combustible Particulate Solids.* This trend is likely to continue as consensus standards are updated and introduced.

6.6 WHAT-IF ANALYSIS AND RELATED METHODS

The primary objectives of the What-If methodology are to identify and analyze the following: (1) major hazards and hazard exposure scenarios in a system; (2) causes, deviations, and weaknesses that can lead to major hazards; (3) control measures in the system; (4) and needed controls to achieve an acceptable risk level. It is important to remember that control measures should be selected in accordance with the hierarchy of controls.

Techniques associated with What-If methodology include (1) *brainstorming*, (2) *checklist analysis*, (3) *What-If hazard analysis*, (4) *What-If/checklist*, (5) *SWIFT, and* (6) *HAZOP study*. A description of each method, along with strengths and weaknesses, requirements, and process steps, are provided:

6.6.1 Brainstorming – Structured and Unstructured

Brainstorming is the first of 31 techniques listed in Annex B of ISO 31010/ANSI Z690.3 Risk Assessment Techniques standard. It is a common method used in a variety of situations in almost all industries, primarily as a supporting method to other analysis and assessment methods. Brainstorming is considered a relatively quick and easy means of collecting ideas surrounding a particular concern for further analysis. Sessions can be performed one-on-one or by a qualified, knowledgeable group of stakeholders. Its purpose is to stimulate and generate a free-flowing dialogue to identify potential failure modes, hazards, risks, and possible controls.

Brainstorming is conducted in a structured or unstructured fashion. In a structured session, each person is required to offer an idea as their turn arises. An "unstructured" brainstorming approach relies on spontaneity allowing the group to provide ideas as they come to mind, which can create a more relaxed atmosphere. However, this approach can allow more vocal members to dominate the session, causing other viewpoints and ideas to be overshadowed. In either case, to be effective, brainstorming sessions must ensure an environment free of judgment of ideas/items – no criticism or favoritism; encourage participation and creative thinking; and seek quantity over quality at this stage. Brainstorming is often the first of a sequence of techniques used.

Advantages:

- Ease and flexibility of use
- Promotes spontaneous, creative thinking
- Engages stakeholder participation.

Disadvantages:
- Dependent upon participants' skill level
- Difficult to assure comprehensiveness
- Group dynamics and personalities can affect or inhibit creative thinking
- Limited to identifying and collecting concerns or issues.

Requirements:
- Experienced facilitator
- Scribe
- Whiteboard, flip chart, or computer projector
- Knowledgeable team members that are freethinkers committed to finding a solution.

Process Steps:
1. *Define purpose and scope.* In a workshop setting, the facilitator first explains the purpose, scope, and ground rules of the brainstorming effort. The ground rules of the session typically include:
   (a) The purpose is to generate as many ideas as possible and stimulate new thoughts in a quick and energetic fashion.
   (b) No discussion, comments, or judgments should be given by other team members that may inhibit individuals while generating ideas. All ideas have value.
   (c) Each member is encouraged to participate and contribute as their turn arises.
2. *Pose problem statement.* The facilitator begins the brainstorming session by presenting a clear problem statement, question, or train of thought to the group. No discussion is allowed at this time.
3. *Generate ideas.* Each team member takes their turn to make suggestions and ideas related to the question/problem statement. This can be done randomly in an unstructured session or in a structured order.
4. *Record ideas.* A scribe is assigned to record each idea on a laptop projector, whiteboard, or flip chart with the display visible to all participants. It is important that team members are able to see the list of ideas to stimulate additional thoughts and ideas.
5. *Complete list.* Keeping the pace going, the facilitator will continue the process until the team exhausts their ideas.
6. *Finalize list.* The team then reviews the list of ideas, eliminates any duplication, and reaches a consensus on the final list. At this point, the brainstorming session is complete.

The brainstorming steps are presented in Figure 6.1.

6.6.2 Checklist Analysis

One of the most commonly used hazard identification and analysis methods is the checklist. Checklists typically consist of specific items or "yes/no" questions derived from published standards, codes, and industry practices for a specific application. Checklist analysis is used by individuals or teams to

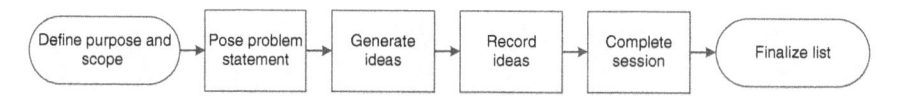

Figure 6.1 Brainstorming Steps

identify deviations and resulting hazards in a process or system. This method is relatively easy to use and cost effective; however, the quality of the analysis depends upon the quality and content of the checklist. For a checklist to be effective, it must target specific concerns, standards, or practices of the process or system being analyzed. Checklists should be selected and/or developed by someone experienced in the operation, potential deviations, types of hazards, and controls. If pertinent items or questions are omitted from the checklist, the analysis may miss existing hazards or hazard scenarios.

Advantages:

- Ease of use
- Limited training or preparation required
- Quick and cost effective.

Disadvantages:

- Quality of analysis depends on the quality of the checklist
- Omissions on checklist can result in missed potential hazards
- May not be suited for more complex situations.

Requirements:

- Specific and complete checklist
- Knowledgeable individual to perform checklist.

Process Steps:

1. *Define purpose and scope.* The stakeholders define the purpose and scope of the analysis.
2. *Select/construct checklist.* Experienced stakeholders examine the subject to be analyzed and select an existing checklist if available or construct a specific list of items or questions to address potential hazards.
3. *Perform checklist analysis.* Each team member takes their turn to make suggestions and ideas related to the question/problem statement. This can be done randomly in an unstructured session or in a structured order.
4. *List hazards and needed controls.* As a result of the checklist review, a scribe is assigned to record each idea on a laptop projector, whiteboard, or flip chart with the display visible to all participants. It is important that team members are able to see the list of ideas to stimulate additional thoughts and ideas. Figure 6.2 provides a flowchart of the process steps.

6.6.3 What-If Hazard Analysis

The *What-If hazard analysis* is a team-based, brainstorming process used to identify and analyze hazards of a system or process. Team members knowledgeable of the process discuss aspects in a

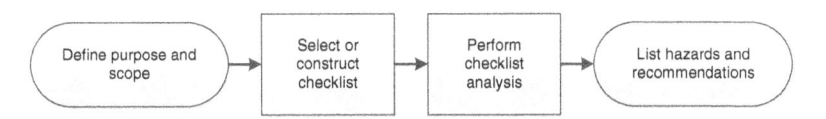

Figure 6.2 Checklist Analysis Steps

random, creative fashion asking "What-If" questions to identify any weaknesses, deviations, or hazards. As the group brainstorms, hazard scenarios are formed and potential causes, existing controls, and needed controls are identified. A spreadsheet is generated listing the tasks or elements and posed "What-If" questions along with resulting consequences, existing safeguards, and recommended additional controls. For this method to be successful, a skilled and knowledgeable leader able to facilitate the analysis and keep it on track is required. A recorder or scribe is needed to collect and document the findings. This method is one of the least structured types of PHA and is relatively easy to perform, which has advantages and disadvantages.

Advantages:

- Ease and flexibility of use
- Limited training in method required
- Promotes creative thinking
- Quick and cost effective
- Useful if relevant checklists or guidelines are not available.

Disadvantages:

- Quality of analysis depends on the quality of the facilitator's skills and team members' knowledge
- Unstructured approach can miss some potential hazards and causes
- Difficult to audit for completeness.

Requirements:

- Experienced facilitator
- Scribe
- Spreadsheet (hardcopy or Excel); whiteboard, flip chart, or computer projector
- Knowledgeable team members that are freethinkers committed to finding a solution
- Data and information on the activity or system.

Process Steps:

1. *Define purpose and scope.* A clear purpose statement that is specific and measureable in terms of its intended goal is first needed. The scope should define the activity or system, as well as boundaries of the analysis, and level of detail desired.
2. *Assemble team.* Based on the purpose and scope, the team should be selected to include experienced, knowledgeable individuals. The team should have representatives from engineering and design, production and operations, maintenance, and safety, health, and environmental. Knowledge of design standards, regulatory codes, operational error potential, incident history, maintenance needs, and other practical experience is required. A knowledgeable facilitator and a scribe are also needed.
3. *Communicate objectives and requirements to team.* The facilitator should clearly define the purpose, scope and objective of the analysis, and the role of team members. The boundaries and ground rules for the analysis should be clearly communicated. It is also important to identify the boundaries of acceptable consequences to determine when additional control measures are required.

4. *Gather and review information.* To support the analysis process, general information surrounding the system, the history, and manufacturer's specifications and instructions is gathered by the team facilitator/leader and provided to the team for review. This should include observation of the activity or system if in place. Essential reference materials include piping and instrumentation diagrams (P&IDs), schematics, drawings, instruction manuals, maintenance and service guidelines, component specifications, and safeguarding elements.
5. *Break down into tasks/elements.* Using information gathered, the activity or system is subdivided into sequential tasks or elements for individual analysis.
6. *Generate "What-If" questions for each task/element.* For each task/element, the team begins to think of questions that begin with "What-If" to identify possible hazards and hazard scenarios. For each step, this questioning process is applied dealing with procedural upsets, operator errors, equipment failures, and software errors. This process can be performed by an unstructured or structured brainstorming method. As team members pose specific "What-If" questions, the scribe records each question on a flip chart or laptop projection. This process spurs additional "What-If" questions by team member, which are recorded by the scribe. The facilitator completes and refines the list of "What-If" questions for the analysis.
7. *Respond to each "What-If" question.* For each "What-If" question, the team discusses and determines the cause(s), consequences, and existing safeguards or controls. In some versions, likelihood of occurrence is included in the spreadsheet analysis; however, most methods do not rank or quantify risk levels.
8. *Determine need for additional controls.* As the hazards, causes, and existing controls are identified, the need for additional controls is determined. Possible options for risk reduction are identified by team members and listed by the scribe. The team then selects the most effective and feasible control option(s) for recommendation to management.
9. *Report results to management.* Following the analysis, the facilitator and scribe finalize the spreadsheet. Each task/element with specific "What-If" questions, answers, consequences, controls, and recommendations are provided in a spreadsheet (as shown in Figure 6.3). The analysis spreadsheet is often used as an "action plan" for assigning responsibilities and completing recommended actions.

The process steps for What-If hazard analysis are shown in Figure 6.4:

What-If Hazard Analysis					
Facility/Operation/Process					
Date:	Team:					
ID #	What-If...	Causes	Consequences	Controls	Recommendations

Figure 6.3 What-If Hazard Analysis Form Example

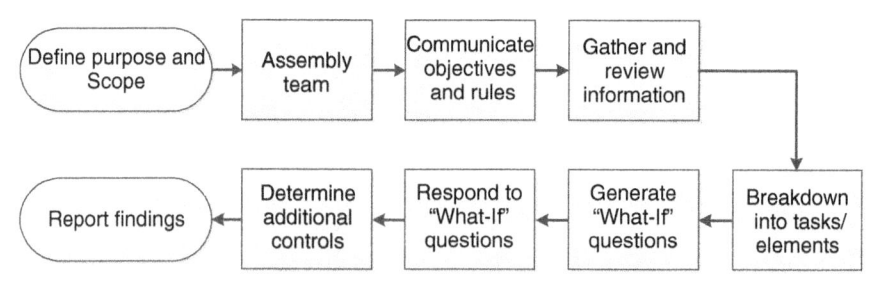

Figure 6.4 What-If Hazard Analysis Process Steps

6.6.4 What-If/Checklist

As indicated by its name, What-If/checklist is a hybrid of the checklist and What-If hazard analysis methods. It is conducted by a facilitator with a team and combines the creative, brainstorming aspects of the What-If with the more structured, systematic approach of the checklist. The combination of techniques can compensate for the weaknesses of each. The method uses a set of prewritten questions developed by knowledgeable, experienced, and qualified personnel to stimulate discussion and evaluate the potential hazards posed by a process. The facilitator and review team are selected to represent a wide range of disciplines such as production, maintenance, engineering, environmental, health, and safety. The team is provided with basic information on hazards of the process, procedures, equipment, instrumentation control, and materials. Information on any previous incidences or hazard reviews of the process is also provided. A tour of the process is performed to gain understanding, make observations, and examine the process from beginning to end. Then the team meets to collectively generate a list of "What-If" questions regarding the hazards and safety of the operation. When the list of questions is complete, the team systematically goes through a prepared checklist to stimulate additional questions. The team works to achieve a consensus on each issue. From these answers, recommendations are developed specifying additional action or study. The recommendations, along with the list of questions and answers, become the key elements of the hazard analysis report.

Advantages:
- Ease and flexibility of use
- Limited training needed
- More structured approach
- Creative thinking encouraged.

Disadvantages:
- Reliance upon checklists can inhibit creative thinking or limit identification of other risks
- Observation-based analysis may miss hazards not easily seen
- Relies upon quality of leader and team.

Requirements:
- Experienced facilitator
- Scribe
- Spreadsheet (hardcopy or Excel); whiteboard, flip chart, or projector
- Knowledgeable team members that are freethinkers committed to finding a solution
- Data and information on the activity or system
- Relevant checklists.

WHAT-IF ANALYSIS AND RELATED METHODS   131

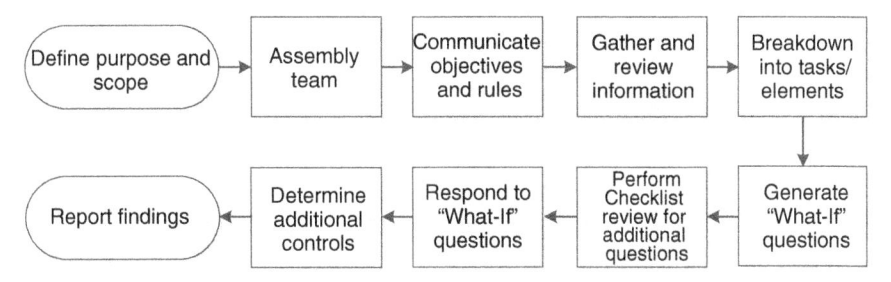

Figure 6.5 What-If/Checklist Process Steps

Process Steps:

The process steps (depicted in Figure 6.5) are similar to the What-If hazard analysis with the additional step of a systemic review of checklist items to each task or element being analyzed. The process steps are as follows:

1. *Define purpose and scope.*
2. *Assemble team.*
3. *Communicate objectives and requirements to team.*
4. *Gather and review information.*
5. *Break down into tasks/elements.*
6. *Generate "What-If" questions for each task/element.*
7. *Perform checklist review.* Upon completion of subdividing tasks or elements, and generating "What-If" questions through brainstorming, checklists relevant to the element/task are reviewed to fill in gaps or cover items that may have been missed. For instance, a checklist of "operator error" questions may be applied to a particular task to ensure certain human factor questions relevant to the activity are covered.
8. *Respond to each "What-If" question.*
9. *Determine need for additional controls.*
10. *Report results to management.*

There are many sources for existing checklists including some of the references in this chapter, the OSHA website or other sources.

6.6.5 Structured What-If Technique (SWIFT)

SWIFT is a more systematic method that relies on an experienced facilitator and team within a workshop to identify risks. Similar to the *What-If/checklist*, the key difference is the development of specific "What-If" questions by the facilitator prior to the workshop. This typically involves interviews of key stakeholders, thorough document reviews, and plans and diagram studies by the facilitator. As a result, the context of the study and key elements for analysis are formed. The team is selected to represent necessary disciplines including production, engineering, maintenance, human resources, environmental, health, and safety. Field observations of the process are conducted by the team. The workshop is then conducted with the leader posing specific "What-If" questions to the team in combination with prompts to investigate how a system, piece of equipment, organization, or procedure will be affected by deviations from normal operations and behavior. It is commonly applied at a more systems level providing less detail than *HAZOP* but great detail than other What-If methods. *SWIFT* can be used on a stand-alone basis or as part of a staged approach to make more efficient use of bottom-up methods like *failure mode and effects analysis (FMEA)*.

Advantages:

- Widely applicable
- Relatively easy and flexible
- Limited training required of team
- Provides greater detail than other What-If methods
- Targets specific concerns developed by facilitator.

Disadvantages:

- Dependent upon the skills of the facilitator
- Time required to gather data and perform analysis.

Requirements:

- Experienced facilitator
- Scribe
- Spreadsheet (hardcopy or Excel); whiteboard, flip chart, or computer projector
- Knowledgeable team members that are freethinkers committed to finding a solution
- Relevant data and information on the activity or system such as P&IDs, operating procedures, and accident database
- Interviews with key personnel
- Observation of the process or activity.

Process Steps:

The process steps shown in Figure 6.6 are similar to the *What-If/checklist* method; however, the key difference is the development of specific questions and prompt words by the facilitator prior to the workshop. The steps include the following:

1. *Define purpose and scope.*
2. *Assemble team.*
3. *Communicate objectives and requirements to team.*
4. *Gather specific information.* In addition to general information, the facilitator collects specific documentation related to the process and associated elements; conducts interviews with key individuals such as engineering, design, maintenance, and production; and observes the process.
5. *Formulate list of prompts/questions.* The facilitator reviews and analyzes the information collected to identify potential "What-If" questions for the team to consider.
6. *Break down the activity/system into tasks/elements.*
7. *Generate "What-If" questions for each task/element.* In addition to the typical brainstorming process, the facilitator uses the list of prompts and questions to create discussion and analysis of specific issues and concerns.
8. *Perform checklist review.* Specific checklists developed by the facilitator are used to complete the list of "What-If" questions.
9. *Respond to each "What-If" question.*

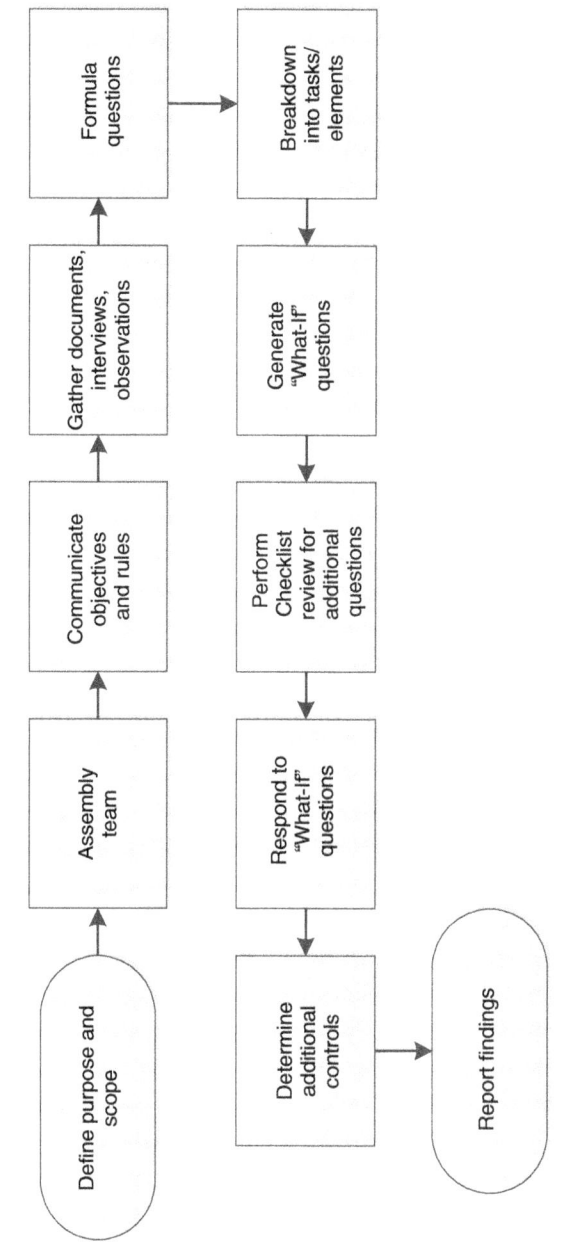

Figure 6.6 SWIFT Analysis Process Steps

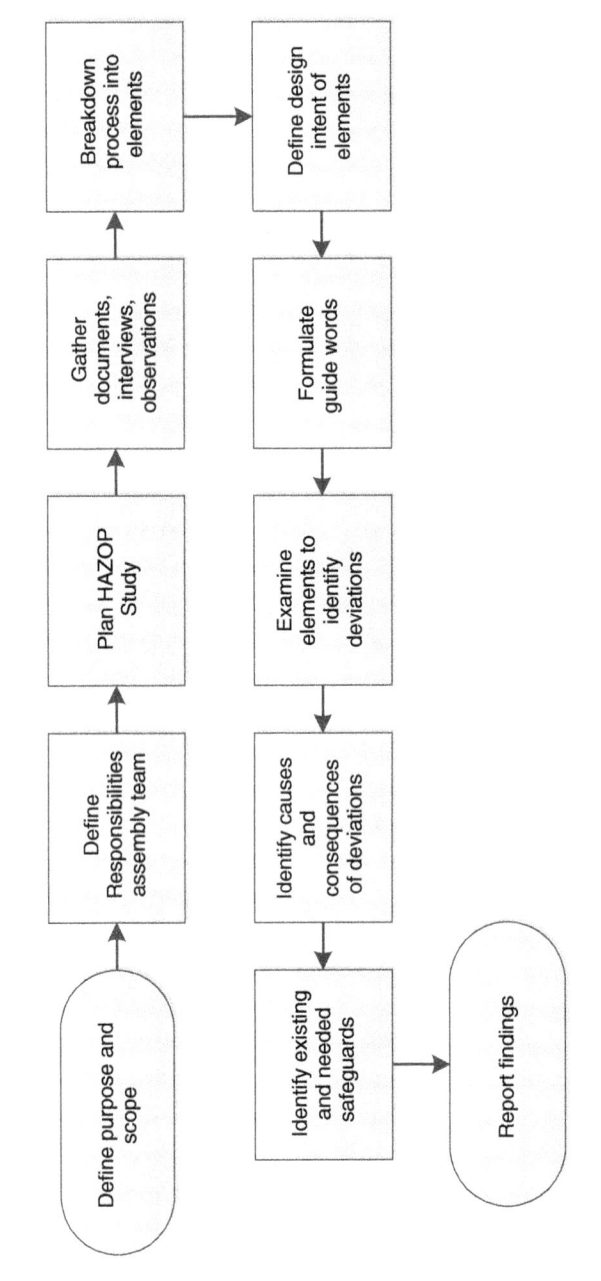

Figure 6.7 *HAZOP Study Process Steps*

WHAT-IF ANALYSIS AND RELATED METHODS

10. *Determine need for additional controls.*
11. *Report results to management.*

6.6.6 Hazard and Operability (HAZOP) Study

HAZOP is a more involved qualitative method used to identify both hazards and operability problems using "guide words" to prompt team members in identifying deviations that can lead to the failures. Similar to an FMEA, *HAZOP* identifies failure modes of a process, their causes, and resulting consequences. However, rather than starting with failures, HAZOP uses guide words to identify deviations from intended operations and then traces back to the possible causes and failure modes. It requires an experienced facilitator and a multidisciplinary team that typically meet multiple times to complete the analysis. Data is collected and the process is divided into individual elements. The intended design or performance for each element is clearly defined by the team. *HAZOP* guide words that express a specific type of deviation from the design intent (i.e., too much; too little; no; reversed; before; after) to be used during the analysis are established and agreed upon by the team. The team systematically reviews each element of the process using guide words to identify deviations, causes, consequences, existing safeguards, and recommended actions. The analysis is recorded in a spreadsheet similar to a What-If method.

Advantages:
- Useful in confronting more complex systems and hazards
- Provides greater detail than What-If methods
- Identifies operability problems as well as hazards
- Systematic and comprehensive methodology.

Disadvantages:
- Can be time consuming and expensive
- Often requires a high level of documentation and procedure specifications
- Relies on the expertise of the leader and team
- Typically does not include risk ranking or prioritization
- Does not assess effectiveness of existing or proposed controls.

Requirements:
- Experienced facilitator
- Scribe
- Software (Excel or other); whiteboard, flip chart, or computer projector
- Knowledgeable multidisciplinary team
- Relevant data and information on the process and its elements including diagrams, operating procedures, specifications, limits, P&IDs, and accident database
- Defined design intent of each element
- Interviews with key personnel
- Observation of the process or activity.

Process Steps:
The process steps shown in Figure 6.7 are adapted from the BSI IEC 61882:2001 Hazard and operability studies (*HAZOP* studies) – Application guide. The steps include the following:

1. *Define purpose and scope.* As in other risk assessment methods, a clear and specific purpose and scope is required. This includes defining boundaries, key interfaces, limitations, and assumptions of the study to be performed.
2. *Define responsibilities and assemble team.* A leader and a cross-functional team of experienced and knowledgeable members are formed. Specific responsibilities, along with the study's purpose and scope, are communicated to the team.
3. *Plan the study.* In preparation of the study, the leader and team members develop a schedule and determine logistics for conducting the study.
4. *Gather data.* The leader and team collect data relevant to the study including operating specification and limits, the design intent of system elements; conduct interviews with key individuals; and observe the process.
5. *Break down the process into elements.* The team breaks the system down into smaller components or specific steps to identify elements to be examined in the study.
6. *Define design intent for each element.* For each element or step, the team defines the intended functions and parameters under normal operating conditions used in the study.
7. *Formulate guide words.* During the planning stage, the leader and team determine guide words to be used in the study.
8. *Identify deviations using guide words on each element.* In addition to the typical brainstorming process, the facilitator uses the list of guide word prompts and questions to create discussion and analysis of specific issues and concerns. Through this process, the team identifies and documents potential deviations from the intended design.
9. *Identify causes and consequences for each deviation.* For each deviation, the team determines its causes and resulting consequences, which are recorded in their appropriate columns of the *HAZOP* spreadsheet as shown in Figure 6.8.
10. *Identify existing safeguard and recommended controls for each deviation.* For each deviation, the team determines current safeguards and additional controls needed. Each is recorded in the *HAZOP* spreadsheet in their respective columns.
11. *Report results to management.* The study findings are agreed upon by the team and documented in the *HAZOP* spreadsheet. Recommended corrective action is assigned to the responsible parties, and a final report is provided to management.

#	Guide Word	Element	Deviation	Causes	Consequences	Safeguards	Recommended Actions
1	Loss	Chemical mixer	Loss of agitation	(1) Agitator motor fails (2) Electrical utility lost/power outage (3) Agitator mechanical linkage fails (4) Operator fails to activate agitator	Unreacted HHC in the reactor carried over to storage tank and released to enclosed work area	HHC detector and alarm	(1) Add alarm shutdown of system for loss of agitator (2) Ensure adequate ventilation for enclosed work area (3) Update procedure

Figure 6.8 HAZOP Spreadsheet Example

6.7 RISK SCORING AND RANKING

A shortcoming in traditional hazard analysis methods is that they do not include the extra step of estimating risk (Main, 2012). Safety professionals are advised to include the risk assessment step in such methods to provide their organizations adequate decision-making information. Therefore, the following guidance is provided.

Following a What-If analysis, an analysis should be made of each hazard's likelihood of exposure scenario and its severity of consequences to estimate and rank risks for risk reduction. The scoring and ranking method selected is determined largely by the end purpose of the assessment. The primary reason for risk scoring and ranking is to identify and prioritize tasks and hazards that require further risk reduction to an acceptable level. These risk scores and rankings are used as a means of comparing risks internally against acceptability criteria to effectively direct resources to the greatest needs.

Some variations do incorporate risk scoring and risk classification in the analysis. For instance, an example is provided by OSHA in their Letter of Interpretation dated February 1, 2005 (OSHA, 2005). In the interpretation letter, the "Example Application of 1910.119(c)(3)(vii)" shows additional columns for consequence (c), likelihood (l), and risk class (r) as seen in Figure 6.9. The example identifies one hazard as well as its corresponding engineering and administrative controls, safeguards, recommendation/actions, and a quantitative description of consequence, likelihood, and the risk class ranking for the identified hazard. A description of the process, as well as tables for consequence (Table 6.2), likelihood (Table 6.3), risk matrix (Table 6.4), and priority legend (Table 6.5), is provided from the example as follows:

First, a qualitative description of consequence and likelihood/frequency of the hazard based on a failure of engineering and/or administrative controls is established. Table 6.2 is the Consequence Table. It is a qualitative description of the possible degrees of severity of consequences related to the identified hazard and its associated failure of controls. These consequences range from 1 to 4, with 4 being the most severe Consequence Class. Table 6.3 is the Likelihood Table; it is a qualitative

What If...	Consequences/ Hazard	Safeguards	C	L	R	Recommendations
Emergency shutdown valve 23 (ESD-23) fails to close when needed? (This can occur due to extremely cold weather, reliability due to inspection/testing/maintenance, or design problems)	Release of highly flammable materials in the operating area. Potential for fire/explosion with employee injuries/fatalities	1. Specific inspection/testing/ maintenance program for ESDs 2. Valve actuator sizing 3. ESD-23 is fail closed design	4	2	B	1. Due to cold weather modify MI procedures to increase ESD valve testing to 1/2 weeks. 2. Inspection records for ESD-23 not in file, follow-up to assure ESD-23 is inspected as required by MI procedures 3. No equipment data sheet was found for actuator for ESD-23, follow-up with engineering to assure design is correct 4. Consider oversizing valve actuator

Example Worksheet Excerpt from What-If/Checklist PHA Methodology
C = Consequence Class, L = Likelihood Class, R = Risk Class

Figure 6.9 OSHA Example of What-If/Checklist PHA Methodology

TABLE 6.2 Consequence Table Example

Consequence Class	Qualitative Employee Safety Consequence Criteria
1	No employee injuries
2	One lost time injury or illness
3	Multiple lost time injuries or illnesses
4	Multiple lost time injuries or illnesses w/one or more facilities

TABLE 6.3 Likelihood Table Example

Likelihood Class	Qualitative Likelihood Criteria
1	Not expected to occur during the lifetime of the process. Examples – Simultaneous failures of two or more independent instrument or mechanical systems
2	Expected to occur only a few times during the life of the process. Examples – Rupture of product piping, trained employees w/procedures injured during LOTO operation
3	Expected to occur several times during the life of the process. Examples – Hose rupture, pipe leaks, pump seal failure
4	Expected to occur yearly. Examples – nstrument component failures, valve failure, human error, hose leaks

description of the range of likelihood (probability or frequency) that an identified engineering or administrative control might fail. The likelihood ranges from 1 to 4, with 4 being the most likely to fail. Using the Consequence and Likelihood Class numbers, a Risk Priority Matrix (Table 6.4) can be constructed. The Risk Priority Matrix is used to identify the Risk Class. Once the Risk Class (e.g., C) is determined from the Risk Priority Matrix, the Risk Class can be correlated to the Risk Priority Legend (Table 6.5), which prioritizes the hazard as identified by the PHA team. In this case, the PHA team enters the chosen Consequence Class, the estimated Likelihood Class, and the resulting Risk Class on the PHA worksheets.

TABLE 6.4 Risk Priority Matrix Example

Consequence \ Likelihood	1	2	3	4
4	C	B	A	A
3	C	B	B	A
2	D	C	B	B
1	D	D	C	C

TABLE 6.5 Risk Priority Legend Example

	Risk Class	Explanation of Risk
Priority →	A	Risk intolerable – needs to be mitigated within 2 weeks to at least a Class C; if that cannot be accomplished, process needs to be shut down
	B	Risk undesirable – needs to be mitigated within 6 months to at least a Class C
	C	Risk tolerable with controls (engineering and administrative)
	D	Risk acceptable – no further action required

Authors' editorial notes: Regarding Table 6.5 from the OSHA example, several comments are made:

1) *Arbitrary time scales of "within 2 weeks" in Class A Explanation and "within 6 months" in Class B should be questioned if the risk is really "unacceptable" according to the organization's acceptability criteria.*
2) *The wording for Class B and Class C should be "Risk level is acceptable only if it can be continually shown that it is being managed/mitigated to ALARP. A Risk Review Frequency needs to be set on the basis of the estimated frequency of any changes of significant risk factors.*
3) *The wording for Class D should not indicate negligible nor no management attention required or equivalent. A low risk can become high as fast as a significant risk factor can change. Suggest wording - "Risk level is acceptable but needs a Risk Review Frequency that can be set lower than Class C but no risk is able to be ignored or neglected."*

6.8 APPLICATION OF "WHAT-IF"

The flexibility of the What-If analysis approach can be applied to nearly any operation, process, or activity, either existing or planned. It can be applied to routine and nonroutine activities, maintenance and service work, installations, and setup activities, among others. From a design review standpoint, this method can be used to identify single failures and obvious hazards of proposed changes or new designs.

In some industries, such as the semiconductor industry, the What-If method is useful in determining compliance with industry safety requirements (as indicated by International SEMATECH, 1999). As mentioned previously, it is one of the identified PHA methods in the OSHA PSM regulation making it suitable for operations that utilize hazardous chemical processes.

In addition, the technique is sometimes used in emergency planning and preparedness tabletop exercises. Brainstorming and What-If methods are used to identify potential emergency scenarios, analyze and evaluate existing emergency response procedures, and further develop contingency plans.

The success of a What-If analysis is dependent upon several key factors listed as follows:

- An experienced and skilled facilitator
- A qualified team of production, engineering, maintenance, management, and operators along with safety, health, and environmental professionals (generally 4–8 people)
- A clearly defined scope with set boundaries
- A thorough document review of available:
- Research on similar situations and potential scenarios; loss history; regulatory requirements; company guidelines, procedures, and protocols; training materials; etc.
  - Manufacturer specifications
  - Operating instructions

- Trouble shooting guides
- Pre-start-up checklists and inspections
- Job hazard analyses
- Schematics and diagrams
- Preventive maintenance (PM) schedule
- Warnings
- Chemical information and hazard ratings (safety data sheets).

Case Study Example: Structured What-If Analysis of a Vapor Combustion System

A rail tank car cleaning operation responsible for purging and cleaning tank car with various chemicals was planning to install a new Vapor Combustion System (VCS). The VCS is designed to destroy volatile organic compounds and hazardous air pollutants in order to meet EPA Clean Air Act air emissions regulations. The operation decided to perform a PHA using the *SWIFT* method.

The SWIFT analysis was used to identify (1) potential hazard scenario events; (2) causes and contributing factors leading to identified hazard scenarios; (3) existing safeguards to prevent identified hazard scenarios; and (4) needed additional safeguards and/or redundancies' to prevent identified hazard scenarios from occurring. Figure 6.10 is a process flowchart illustrating the sequence of steps in the process.

Document review: A review of documents and information from the process, equipment, and components as well as affected operations was performed in advance of the SWIFT analysis workshop. The purpose of the document and data review was to identify potential hazard scenario events associated with the process, potential causes and contributing factors, and necessary safeguards. Documents included the following information:

- Schematics, plans, and diagrams of system and surrounding area
- Vapor combustor manufacturer specifications, limitations, and maintenance requirements
- Steam condensate system information
- Rail car hookup information
- Safeguards information, manufacturers' specifications, function applications and limitations, maintenance requirements (antiflashback burners, detonation arrestors, shutoff valves, flame scanners, relief valves, rupture disks, thermocouples, pressure sensors, bonding and grounding, etc.)
- PLC control panel specifications, limitations, and maintenance requirements
- Mechanical integrity procedures – system/equipment PM schedule, inspection, testing, maintaining, repair procedures
- Job safety analyses, operator work instructions, training instructions, and related safety procedures

The document review was performed prior to the SWIFT analysis workshop by the facilitator. A list of 45 potential hazard scenarios was developed for use in the workshop brainstorming and analysis process.

On-site review: The SWIFT team members visited the site with engineering and technical staff to physically review the entire system including the rail car hookup equipment and platform, piping systems, control panel, and vapor combustion pad. Surrounding areas of the system that interact with the VCS were also reviewed. Each step of the process was observed with a description of the tasks provided by the staff to help establish potential hazard scenario events, causes, and safeguarding.

What-If analysis flow chart

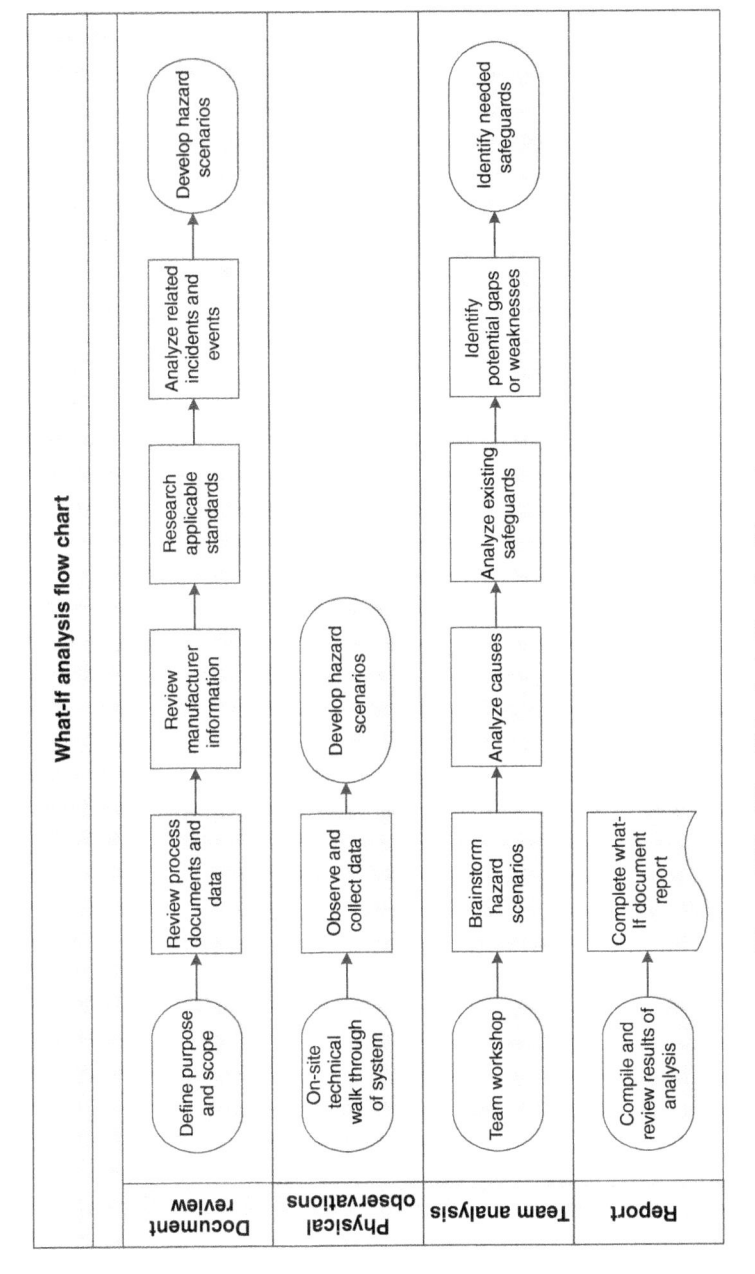

Document review: Define purpose and scope → Review process documents and data → Review manufacturer information → Research applicable standards → Analyze related incidents and events → Develop hazard scenarios

Physical observations: On-site technical walk through of system → Observe and collect data → Develop hazard scenarios

Team analysis: Team workshop → Brainstorm hazard scenarios → Analyze causes → Analyze existing safeguards → Identify potential gaps or weaknesses → Identify needed safeguards

Report: Compile and review results of analysis → Complete what-If document report

Figure 6.10 What-If Analysis Flowchart Example

Structured what-If technique analysis

Facility/operation/process: rail tank car cleaning : vapor combustion system

Date: June 12–17, 2012

Team: Bruce Lyon, facilitator; Deane H., fire protection specialist; Tom G., Engineering; Jay P., safety and health; Charles T., environmental; Don B., maintenance; Kevin S., production/tank car cleaning

ID	Steps/tasks
A.	Prestart-up and flare purging
B.	Combustor start-up
C.	Combustor shutdown
D.	Degas and steam cars to combustor
E.	Disconnecting car
F.	Depressure cars less than 25 psi vapor pressure
G.	Flare purging
H.	Combustor operation
I.	Maintenance

Figure 6.11 SWIFT Analysis Steps/Tasks

Structured what-If technique analysis

Facility/operation/process: rail tank car cleaning : vapor combustion system

Date: June 12–17, 2012

Team: Bruce Lyon, facilitator; Deane H., fire protection specialist; Tom G., Engineering; Jay P., safety and health; Charles T., environmental; Don B., maintenance; Kevin S., production/tank car cleaning

A. Prestart-up and Flare Purging

ID #	What If...	Causes	Consequences	Controls	Recommendations
A.1.	Insufficient purging of flare system	Inadequate amount of purge gas used at least ten system volumes) to drop O_2 level below 8%	Fire or explosion	Operator training in purging procedure	A.1.1 Automatic timing system for purge (Options: Gauge to show adequate volume of purge gas used; Oxygen testing of flare system after purge to verify) A.1.2 Purge point as close to relief valves as possible
A.2	Steam is used to purge the flare system	Human error – steam used to purge system	Fire or explosion – steam condenses in piping without displacing air	Operator training in purging procedure	A.2.1 Physical interlock to prevent steam from being used in purge A.2.2 Warning signage instructions
A.3	Igniting pilots before air is removed from system	Human error or omission – Inadequate purge; lack of purge	Fire or explosion	Operator training in purging procedure	A.3.1 Procedure to verify purging is complete before ignition

Figure 6.12 SWIFT Analysis: A. Prestart-Up and Flare Purging Section

Discussions about operational and equipment design perimeters, limitations, restrictions, deviations, human error potential, and consequences were held with technical staff.

SWIFT analysis: The team assembled in the conference room to perform a structured "What-If" analysis of the VCS. The analysis was facilitated by the leader and documented by the scribe. The prepared list of 45 potential hazard scenarios was presented to the team. An additional 12 scenarios were generated by team member in the brainstorming process and added to the list. A whiteboard was used to visually lay out the matrix of scenarios and their associated causes and safeguards. Each scenario was discussed with the team brainstorming to capture potential situations and contributing factors that could lead to specific hazard scenarios; the existing safeguards, redundancy systems; and any additional safeguarding required.

Structured what-If technique analysis						
Facility/operation/process: rail tank car cleaning : vapor combustion system						
Date: June 12–17, 2012		Team: Bruce Lyon, facilitator; Deane H., fire protection specialist; Tom G., Engineering; Jay P., safety and health; Charles T., environmental; Don B., maintenance; Kevin S., production/tank car cleaning				
B. Combustor start-up						
ID #	What If…		Causes	Consequences	Controls	Recommendations
B.1	Waste gas valve on degas rack is left open during combustor start-up operation		Human error or omission – waste gas valve is not closed or completely closed	Fire or explosion; damage to combustor	Operator training in VCS Instruction manual and VCS start-up JSA procedure	B.1.1 Gas valve open and closed positions marked and labeled B.1.2 Alarm if gas valve is not closed completely during start-up
B.2	Steam condensate is not drained from evaporator tank		Human error or omission – tank not completely drained before combustor start-up	Pollution – emissions – incomplete combustion	Operator training in VCS Instruction manual and VCS start-up JSA procedure	B.2.1 Bottom valve on evaporation tank with quarter-turn markings labeled
B.3	Igniting pilots before air is removed from system		Human error or omission – Inadequate purge; lack of purge	Fire or explosion	Operator training in purging procedure	B.3.1 Procedure to verify purging is complete before ignition

Figure 6.13 SWIFT Analysis: B. Combustor Start-Up Section

Report: Following the team analysis, the facilitator with the team's input compiled the data and information from the SWIFT analysis into a PHA report. The report highlighted needed safeguards, redundancies, additional training, and further engineering analysis to ensure safe operation of the VCS. The report was delivered to management with suggested action plan and priorities. The SWIFT spreadsheet examples are presented in Figures 6.11–6.13.

6.9 CONCLUSION

What-If hazard analysis is a relatively simple and flexible method of identifying and analyzing hazards in a process, activity, or system. It can be applied to a wide range of circumstances in almost all industries. As one of the PHA methods listed in the OSHA PSM standard, the What-If method has become a commonly used technique, both in regulated and nonregulated operations.

Safety professionals should take note of the fact that brainstorming and What-If methods rely heavily on the experience and skill of the facilitator, which determines the effectiveness of the analysis. Safety professionals that can successfully lead a team in this methodology will be of greater value to organizations and better positioned to manage operational risks.

REVIEW QUESTIONS

1. Describe the process of the What-If method. List several variations of this method.

2. List five examples of when What-If analysis can be applied.

3. Explain what a PHA entails, and list two PHAs that are mandated in the United States.

4. Describe the brainstorming process and why it is used.

5. Explain the basic differences between a HAZOP and a What-If method.

6. What element is missing in traditional What-If methods that should be included by the safety professional to aid decision makers?

REFERENCES

American Industrial Chemical Engineers (AIChE). 2015. Center for Chemical Process Safety, https://www.aiche.org/ccps (accessed December 30, 2015)

ANSI/ASSE Z590.3-2011. *Prevention Through Design: Guidelines for Addressing Occupational Hazards and Risks in Design and Redesign Processes.* Des Plaines, IL: American Society of Safety Engineers, 2011.

ANSI/ASSE Z690.3-2011. *American National Standard – Risk Assessment Techniques.* Des Plaines, IL: The American Society of Safety Engineers, 2011.

BSI. BS IEC 61882:2001. *Hazard and Operability Studies (HAZOP Studies) – Application Guide.* London, UK: British Standards Institution, 2001.

International SEMATECH. *Hazards Analysis Guide: A Reference Manual for Analyzing Safety Hazards on Semiconductor Manufacturing Equipment.* Austin, TX: SEMATECH Technology Transfer, 1999.

Johnson, W. G. *MORT Safety Assurance Systems.* New York, NY: Marcel Dekker, 1980.

Main, B. W. *Risk Assessment: Challenges and Opportunities.* Ann Harbor, MI: Design Safety Engineering, Inc., 2012.

Nolan, D. P., *Application of HAZOP and What-If Safety Reviews to the Petroleum, Petrochemical and Chemical Industries.* Park Ridge, NJ: Noyes Publications, 1994. Available at: http://nigc.ir/portal/File/ShowFile.aspx?ID=3e288b62-52b1-4e4f-8841-ac012a69f27d

OSHA. Standards Interpretation, Letter of Interpretation to Mr. Roygene Harmon, Appendix Example Application of A1910.119 (c)(3)(vii), February 1, 2005. Available at: https://www.osha.gov/OshDoc/Interp_pdf/I20050201A_appendix.pdf (accessed December 30, 2015)

OSHA. 1992. 1910.119 Process Safety Management of Highly Hazardous Chemical, https://www.osha.gov/pls/oshaweb/owadisp.show_document?p_table=STANDARDS&p_id=9763 (accessed December 30, 2015)

7

PRELIMINARY HAZARD ANALYSIS

GEORGI POPOV
School of Environmental, Physical & Applied Sciences, University of Central Missouri, Warrensburg, MO, USA

BRUCE K. LYON
Risk Control Services, Hays Companies, Kansas City, MO, USA

7.1 OBJECTIVES

- Introduce preliminary hazard analysis
- Review process
- Practical application
- Examples
- Practice exercises/questions

7.2 INTRODUCTION

As the name indicates, a preliminary hazard analysis (PHA) is a "preliminary" or initial analysis of a system design, facility, or process that is used in many industries and applications. PHA is used by safety professionals to identify hazards and necessary control measures and allow for risk levels to be prioritized for further risk assessment and management. It is one of the eight risk assessment techniques listed in the American National Standards Institute (ANSI), *ANSI/ASSE Z590.3-2011, Prevention through Design Guidelines for Addressing Occupational Hazards and Risks in Design and Redesign Processes*. The ANSI Z590.3 standard makes note that PHA, along with failure mode and effects analysis (FMEA), and What-If methods are sufficient to address most risk situations. This chapter will offer a careful review of the PHA process and evaluation of prevention measures.

PHA is a systematic approach originally developed in the 1960s by the US Army and published in the MIL-STD-882 standard as a method to identify hazards, assess the initial risks, and identify potential mitigation measures early in the design stage. It is referred to as a "preliminary" analysis

---

Risk Assessment: A Practical Guide to Assessing Operational Risks, First Edition.
Edited by Georgi Popov, Bruce K. Lyon, and Bruce Hollcroft.
© 2016 John Wiley & Sons, Inc. Published 2016 by John Wiley & Sons, Inc.

since it is usually followed by more refined hazard analysis and risk assessment studies in more complex systems. Variants of PHA have been developed including hazard identification (HAZID) and rapid risk ranking (RRR) methods according to Rausand.

In the *ISO 31010/ANSI/ASSE Z690.3-2011, Risk Assessment Techniques, Appendix B*, the standard defines a PHA as "a simple, inductive method of analysis whose objective is to identify the hazards and hazardous situations and events that can cause harm for a given activity, facility or system." It should be noted that PHA also stands for *process hazard analysis* in the Occupational Safety and Health Administration (OSHA) Process Safety Management of Highly Hazardous Chemicals (29 CFR 1910.119) and the Environmental Protection Agency (EPA) Risk Management Program for Chemical Accidental Release Prevention regulations.

Clemens states that a PHA "produces a hazard-by-hazard inventory of system hazards and an assessment of the risk of each of them. A PHA is also a screening or prioritizing operation. It helps separate hazards that pose obviously low, acceptable risk from the intolerable ones for which countermeasures must be developed." A limitation of PHA is indicated by Clemens in his following statement: "A PHA does not readily recognize calamities that can be brought about by co-existing faults/failures at scattered points in a system." Since hazards are identified and analyzed individually, the potential for synergistic effects from combined hazards can be missed. For example, a PHA may not recognize a combined exposure such as cold temperatures and vibration that can cause increased risk of soft tissue damage to hands, arms, feet, or other exposed areas.

The scope of PHA should consider worst-credible hazards that can result from the system and its function. The following elements to include in a PHA are adapted from the MIL-STD-882E standard:

- System components
- Energy sources
- Hazardous materials
- Material compatibilities
- Safety-related interfaces between system elements including software
- Interface considerations to other systems
- Environmental factors and constraints affecting the system
- Procedures for system's life-cycle modes including operating, test, maintenance, built-in test, diagnostics, emergencies, explosive ordnance render safe and emergency disposal
- Health hazards
- Environmental impacts
- Human factors engineering and human error analysis of operator functions, tasks, and requirements
- Inadvertent activation
- Life support requirements and safety implications in manned systems, including crash safety, egress, rescue, survival, and salvage
- Event-unique hazards
- Facilities, equipment, and training
- Safety-related equipment, safeguards, and alternate controls
- Malfunctions to system.

While *ISO 31010/ANSI Z690.3-2011, Risk Assessment Techniques*, describes PHA as a qualitative method, others define it as a semiquantitative assessment tool. Manuele in his book *Innovation in Safety Management* issues a word of caution against placing too much faith in numerical scores or values that are based on subjective judgments. Very few so-called "quantitative" scores are truly based on quantitative data. One advantage of using the numerical scores whether based on qualitative or quantitative data is the ease of recognizing and comparing risk levels within a risk matrix or profile.

7.3 PRELIMINARY HAZARD LIST

As described, a PHA is considered a fundamental system safety method of identifying hazards, which is best conducted early in the design process. Prior to a PHA, a preliminary hazard list (PHL) is commonly used to identify and compile a list of potential significant hazards associated with a system's design. The purpose of a PHL is to initially identify the most evident or worst-credible hazards that could occur in the system being designed. Such hazards may be inherent to the design or created by potential energy release in the system. A PHL is only a list of the hazards; however it can be the basis for an analysis that becomes a PHA or other risk assessments.

A PHL is normally developed by collecting information from available sources such as historical loss data and similar systems. Specifically, information on the system's specification and requirements is collected including potential energy sources and controls, potential hazardous materials and their containment, general and specific checklists, lessons learned from similar systems, incident reports and analyses, and interviews and discussions with system users or other knowledgeable parties.

Upon collection of the information, a team of qualified members reviews the information and conducts brainstorming to complete the list of potential significant hazards, a brief description of the hazard, and its causal factor(s).

7.4 PHAs AND THEIR APPLICATION

The PHA method was designed to be used as an exploratory or initial analysis early in the design stage when little information is available on design details or operating procedures. Early in a system's conceptual design and development, PHA is used to identify potential hazards and necessary design specifications to avoid, eliminate, and reduce identified hazards. Specific control measures and design specifications identified through a PHA can then be built into the system's design.

Taking the time to perform a PHA early on may actually speed up the design process and avoid costly mistakes. Any identified hazards that cannot be avoided or eliminated in the project design phase must be controlled so that the risk is reduced to an acceptable level. The hierarchy of controls model should be the basis of all risk control selections as required by the ANSI/ASSE Z590.3, Prevention through Design (PtD).

PHAs often lead to the need for more refined analyses and assessments such as FMEA; failure mode, effects, and criticality analyses (FMECA); and fault tree analyses (FTA). These methods are commonly used to further identify, evaluate, and avoid hazards in more complex or safety sensitive designs. As indicated by ISO 31010/ANSI Z690.3, PHA should be updated as necessary during phases of design, construction, and testing to detect any new hazards and controls needed.

While primarily used early in the design phase, a PHA may be performed at any point in a system's life cycle. PHAs are used in by many industries to examine existing systems, prioritize risk levels, and select those systems requiring further study. The use of a single PHA may also be appropriate for simple, less complex systems or when financial limitations will prevent more extensive techniques from being used.

7.5 THE CONTROL OF HAZARDOUS ENERGY

PHAs often include a basic review of potential energy or hazardous materials and their potential uncontrolled release (Rausand). Haddon's energy release theory developed by Dr William Haddon Jr in the 1970s provides a foundation for this review process. Haddon's "energy release theory" based on a system safety approach establishes a relationship between the causation and risk control method selected. Haddon's model should be considered when conducting a PHA or design safety review due to the fact that engineers understand systems thinking and can be related to the energy control strategies.

Haddon's "energy release theory" includes sequential control strategies listed in a hierarchy of control fashion that should be considered early in the design. Haddon's strategies are listed in the following in an abbreviated form:

1. *Prevent stored energy.* Prevent the marshaling of the form of energy in the first place, such as preventing the generation of thermal, kinetic, or electrical energy or ionizing radiation that can be potentially released.
2. *Reduce stored energy.* Reduce the amount of energy marshaled by its amount and concentration, such as limiting the amount of chemicals stored, reducing the size of materials handled, or reducing the speed of vehicles.
3. *Prevent energy release.* Prevent the release of the energy by incorporating physical containment.
4. *Reduce rate of release.* Modify the rate or spatial distribution of release of energy from its source such as reducing compressed air pressure to 30 pounds per square inch (psi) or reducing the slope of warehouse ramps for forklifts.
5. *Separate energy release from humans and assets by space or time.* Separate, in space or time, the energy being released from that which is susceptible to harm or damage. This strategy eliminates the intersection (exposure) of energy and humans or assets. Examples include increasing the distance between the point of operation of a punch press and the operator or scheduling human interaction with machine when its functions are neutralized.
6. *Separate energy release from humans and assets by physical barriers.* Separate by interposition of a material "barrier" such as the use of insulation on electrical lines, machine guards, or welding curtains.
7. *Modify contact surfaces.* Modify appropriately the contact surface, subsurface, or basic structure, as in eliminating, rounding, and softening corners edges and points with which people can come in contact.
8. *Strengthen susceptible structures.* Strengthen the structure, living or nonliving, that might otherwise be damaged by the energy transfer such as the reinforcement of storage racks exposed to forklift damage.
9. *Increase detectability and prevention of harm.* Move rapidly in detection and evaluation of damage that has occurred or is occurring, and counter its continuation or extension. Examples include fire alarms and sprinkler systems, proximity limit switches, or presence sensing devices.
10. *Prevent further damage.* After the emergency period following the damage energy exchange, stabilize the process. Examples include disaster recovery plans and emergency action and evacuation plans.

Haddon's control strategies validate the thinking that when appropriate energy controls are incorporated into the design, potential energy release is avoided, eliminated, or effectively controlled. Safety professionals using PHA or other risk assessment methods should pay close attention to the potential for hidden energies in products and systems.

7.6 FUNDAMENTAL SYSTEM SAFETY TENETS

PHA is a system safety analysis method. Therefore, it is appropriate to include a list of fundamental principles that apply to the system safety approach. The following listed tenets are taken from Richard Stephans' book *System Safety for the 21st Century: The Updated and Revised Edition of System Safety 2000*. These principles are consistent with those found in many safety and risk management texts as well as related standards. They are considered important to the safety professional and worth repeating:

1. Systematically identify, evaluate, and control hazards in order to prevent (or mitigate) accidents. [*As in the Practice of Hazard Analysis and Risk Assessment*]
2. Apply a precedence of controls to hazards starting with their elimination, designing to preclude hazards, and finally administrative controls. Administrative controls include signs, warnings, procedures, and trainings. (Of lowest precedence are those controls that rely on people.) [*The Hierarchy of Risk Controls*]
3. Perform proactively rather than reacting to events. This starts with a program plan.
4. Design and build safety into a system rather than modifying the system later in the acquisition process when any changes are increasingly more expensive. [*PtD*]
5. Develop and provide safety-related design guidance and give it to the designers as the program is initiated. [*Design Safety Reviews*; *Preliminary Hazard Analysis*]
6. Use appropriate evaluation/analysis techniques from the tabulated variety available. [*Hazard Analysis, Risk Assessment methods as described in this text, ANSI Z560.3, and ISO 31010/ANSI Z690.3*]
7. Rely on factual information, engineering, and science to form the basis of conclusions and recommendations.
8. Quantify risk by multiplying the ranking of undesired consequences of an event by the probability of occurrence. There are variations to this "equation."
9. Design, when allowed, to minimize or eliminate single-point failures that have an undesired consequence. Make at least 2-fault tolerant that is tolerant of multiple faults or system breakdown that would have adverse safety consequence. [*Redundancies in Controls, Layers of Protection*]
10. Identify, evaluate, and control hazards throughout the system's life and during the various operational phases for normal and abnormal environments. [*PtD*]
11. After application of controls to mitigate a hazard(s), the management must recognize and accept the residual risk. [*Acceptable Risk Level*]
12. Recognize the quality assurance interface: (1) decrease risk by using materials that are properly specified and possess adequate quality assurance and (2) implement to continually improve the system.
13. Tabulate and disseminate lessons learned and incorporate those lessons for future safety enhancement.
14. Apply system safety to systems to include processes, products, facilities, and services.
15. Recognize that near-miss [*undesired incidents that could cause harm*] conditions, if not corrected, most likely develop into accidents [*incidents resulting in harm*].

In MIL-STD-882E, the standard identifies a sequence of risk assessment steps used in system safety displayed in Figure 7.1. These essential steps can be found in other standards referring to the risk assessment process as well.

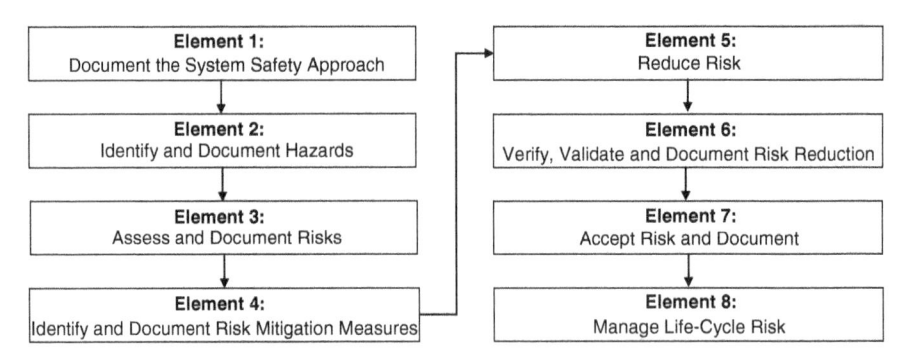

```
┌─────────────────────────────────┐         ┌─────────────────────────────────┐
│         Element 1:              │────────▶│         Element 5:              │
│ Document the System Safety      │         │         Reduce Risk             │
│          Approach               │         │                                 │
└─────────────────────────────────┘         └─────────────────────────────────┘
              │                                             │
              ▼                                             ▼
┌─────────────────────────────────┐         ┌─────────────────────────────────┐
│         Element 2:              │         │         Element 6:              │
│ Identify and Document Hazards   │         │ Verify, Validate and Document   │
│                                 │         │         Risk Reduction          │
└─────────────────────────────────┘         └─────────────────────────────────┘
              │                                             │
              ▼                                             ▼
┌─────────────────────────────────┐         ┌─────────────────────────────────┐
│         Element 3:              │         │         Element 7:              │
│  Assess and Document Risks      │         │    Accept Risk and Document     │
└─────────────────────────────────┘         └─────────────────────────────────┘
              │                                             │
              ▼                                             ▼
┌─────────────────────────────────┐         ┌─────────────────────────────────┐
│         Element 4:              │         │         Element 8:              │
│ Identify and Document Risk      │────────▶│    Manage Life-Cycle Risk       │
│       Mitigation Measures       │         │                                 │
└─────────────────────────────────┘         └─────────────────────────────────┘
```

Figure 7.1 System Safety Process Sequence. *Source*: Adapted from MIL-STD-882E (2012)

7.7 CONDUCTING A PHA

PHA is essential to the preventive and proactive aspect of a safety management system. The primary purpose of PHA is to identify and describe significant hazards that might arise from defects and unsafe conditions in the design and operation of a system or subsystem. The process steps for conducting a PHA are similar to other hazard analyses and risk assessments. Figure 7.2 outlined in ANSI/ASSE Z590.3 provides the process steps used:

Most, if not all, steps of the ANSI Z590.3 Risk Assessment Process can be applied in developing a PHA methodology. The PHA procedure is described in the following steps:

1. **Select an asset**. Identify systems, products, or assets of value to be analyzed.
2. **Select a matrix**. Select the risk assessment matrix to be used. The matrix should incorporate the organization's acceptable risk level. Definitions for severity, probability, and risk assessment codes (RAC) can be found in MIL-STD-882E or other resources. [*More is explained later in Scoring Systems.*]
3. **Establish context**. The PHA's purpose and scope are defined with its objectives and limitations. The scope should include a clear definition of the system to be assessed, including physical boundaries, operating phases, etc.
4. **Establish PHA team**. A team is recommended over a single individual when performing a PHA. The PHA team should consist of an experienced facilitator to lead the team, a scribe to document the analysis, and several team members with the necessary knowledge and experience in the system and associated hazards.
5. **Identify hazards**. Identify the system's potential hazards and their targets, including hazardous events or activities. A team approach using brainstorming to identify hazards is recommended. Resources that assist in identifying hazards may include checklists, similar designs/studies, codes and standards, historical loss data, interviews with system users, and a review of energy sources and their potential release.
6. **Assess severity**. For each identified hazard, the worst-credible case severity resulting from the hazard is assessed and scored according to the risk matrix.
7. **Assess probability**. For each identified hazard, the probability or likelihood of the hazard occurring is determined and scored according to the risk matrix.
8. **Assess risk**. For each hazard, the identified risk levels for severity and probability are entered into the risk assessment matrix.
9. **Determine acceptability**. For risks that are categorized as unacceptable, further action is required to reduce risk.

CONDUCTING A PHA 151

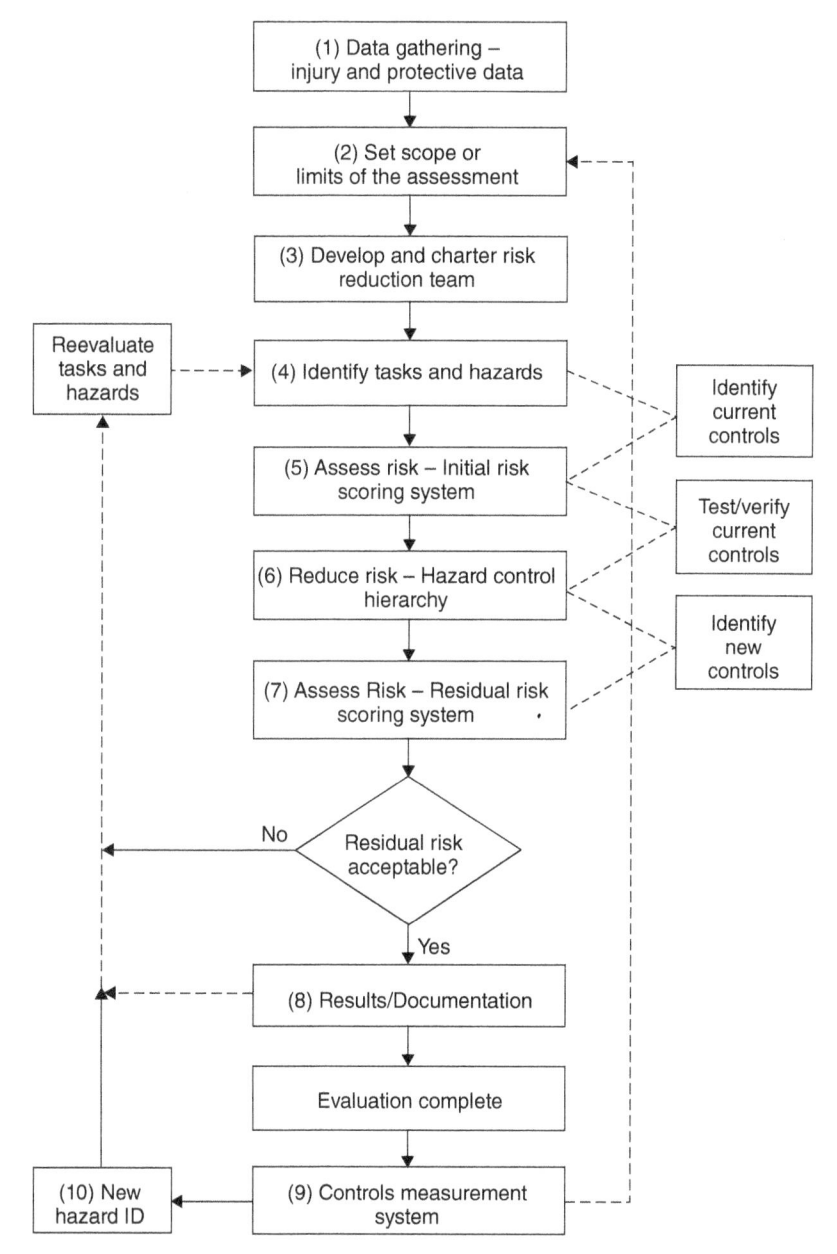

Figure 7.2 Risk Assessment Process. *Source*: Reprinted with Permission from ANSI/ASSE Z590.3-2011 (Courtesy of the American Society of Safety Engineers)

10. **Select controls**. For risks that are unacceptable, the hierarchy of controls is used to select the most effective controls feasible to be incorporated into the design.
11. **Reevaluate risk**. With control measures in place, determine whether new hazards have been introduced or if existing hazards have been increased that require additional control measures.
12. **Determine acceptability**. After all control measures have been implemented, the risk is reevaluated and scored. For risks that remain unacceptable, a decision is made to either abandon the project or continue evaluating new control measures that will reduce risk to an acceptable level.

7.8 SCORING SYSTEMS

One of the first steps in conducting a PHA or any other risk assessment is selecting the appropriate risk assessment matrix to be used. A risk assessment matrix is defined and explained by ANSI Z590.3 as follows:

> *A risk assessment matrix provides a method to categorize combinations of probability of occurrence and severity of harm, thus establishing risk levels. A matrix helps in communicating with decision makers on risk reduction actions to be taken. Also, risk assessment matrices assist in comparing and prioritizing risks, and in effectively allocating mitigation resources.*

Guidance from the ANSI Z590.3 PtD standards goes on to says that an organization "shall" create and agree upon a risk assessment matrix or other validated process that is suitable to the hazards and risks being assessed. The selected risk assessment matrix or validated method is used to determine and document risk. Stakeholders involved in risk assessment should understand that the definitions of the terms used for incident probability and severity and for risk levels vary greatly in various risk assessment matrices available.

In order to establish initial scoring system, a well-established system such as those outlined in ANSI Z590.3 can be used. Addendum F of ANSI Z590.3 offers several examples of risk assessment matrices and definitions of terms. An example is also provided in MIL-STD-882E, the Department of Defense Standard Practice for System Safety, shown in Table 7.1.

TABLE 7.1   Risk Assessment Matrix

Severity / Probability	Catastrophic (1)	Critical (2)	Marginal (3)	Negligible (4)
Frequent (A)	High	High	Serious	Medium
Probable (B)	High	High	Serious	Medium
Occasional (C)	High	Serious	Medium	Low
Remote (D)	Serious	Medium	Medium	Low
Improbable (F)	Medium	Medium	Medium	Low
Eliminated (E)	Eliminated			

Source: Adapted from MIL-STD-882E (2012).

The use of a composite of matrices that include numerical values for probability and severity levels can be used. For practical reasons, the combinations are expressed as numerical risk scorings for those who prefer to deal with numbers rather than qualitative indicators. More detailed FMEA and Bow-Tie risk assessment methodologies described in this book also utilize numerical risk scorings.

It should be noted that the numbers in Table 7.2 were judgmentally determined and are qualitative in nature. Table 7.3 provides definitions for each of the risk levels for severity and probability.

For each level of severity and probability, a clear description of each level is required. Table 7.3 provides descriptions adopted from the ANSI Z590.3 PtD standard used in the practice example in this chapter. For this text, the following risk ranking criteria from the PtD standard is used:

1. Very high risk: 15 or greater
2. High risk: 10–14
3. Moderate risk: 6–9
4. Low risk: Under 1–5.

PRACTICAL APPLICATION 153

TABLE 7.2 Example Risk Assessment Matrix: Numerical Grading and Scoring

Severity ranking: 5
Probability ranking: 5

Occurrence Probabilities and Values

RA Matrix	Probability Level/Exposure Code				
Severity level	1	2	3	4	5
5	5	10	15	20	25
4	4	8	12	16	20
3	3	6	9	12	15
2	2	4	6	8	10
1	1	2	3	4	5

TABLE 7.3 Severity and Probability Descriptions

	Incident or Exposure Severity Descriptions	
5	Catastrophic	One or more fatalities, total system loss, chemical release with lasting environmental or public health impact
4	Critical	Disabling injury or illness, major property damage and business downtime, chemical release with temporary environmental or public health impact
3	Marginal	Medical treatment or restricted work, minor subsystem loss or damage, chemical release triggering external reporting requirements
2	Negligible	First aid or minor medical treatment or minor medical treatment only, non-serious equipment or facility damage, chemical release requiring routine cleanup without reporting
1	Insignificant	Inconsequential with respect to injuries or illnesses, system loss or downtime, or environmental chemical release
	Incident or Exposure Probability Descriptions	
5	Frequent	Likely to occur repeatedly
4	Likely	Probably will occur several times
3	Occasional	Could occur intermittently
2	Seldom	Could occur, but hardly ever
1	Unlikely	Improbable, may assume incident or exposure will not occur

7.9 PRACTICAL APPLICATION

As previously described, PHA is used to identify hazards, assess the initial risks, and identify potential mitigation measures. Its methodology is well described in MIL-STD-882E and can be applied to both military and civilian projects and products. The following steps outline the basic application of a PHA:

(a) *Develop list of hazards.* In order to identify the hazards associated with the system to be analyzed, a PHL should be developed as indicated earlier in this chapter. Other options include transferring the identified hazards from existing job hazard analysis, job safety analysis, task analysis, job risk assessment, or "What-If/checklist analysis."

(b) *Select/develop PHA worksheet.* Those involved in performing PHA are advised to select or develop PHA models that meet the specific needs of the system, analysis, and organization. PHA worksheets typically include columns for (1) hazard description or scenario, (2) task or process description, (3) exposed asset(s), (4) probability of hazard event, (5) severity of hazard, (6) existing controls, (7) a RAC, and (8) remedial action needed. A process map or a flowchart such as the one presented in Figure 7.3 can be used to help guide the execution of a PHA.

(c) *Complete worksheet.* The worksheet requires the PHA team to identify the process step, hazards, hazard category (chemical, physical, biological, etc.), system effects/employee exposure,

QAP BC

Hazards Associated With the Problem

Potential Effect # and a short name	Hazard	Potential Effects	Process	Business Unit/Department	Description of Current Controls	ANSI/ASSE Z590.3-2011: Prevention Through Design, Hierarchy of Controls
1 EXP	Chemical/Ph	Explosion				
2 Health	Chemical	Health/Benzene exposure				
3 Env	Chemical	Environmental				

High-Level Flowchart
Step 1 — Step 2 — Step 3 — Process complete

Figure 7.3 Simple PHA Worksheet and Process Flowchart

PRACTICAL APPLICATION 155

PRELIMINARY HAZARD ANALYSIS WITH TRACKING LOG

Date:

Project/Process	
Prepared by:	

Methods Used

Process	Hazard or Hazardous Event	Hazard Category	System Effects – Employees Exposure	Current Controls

Figure 7.4 PHA Worksheet Example

and current controls as identified in the example in "potential effects". In filling out the worksheet, the PHA team will determine if current controls exist or if additional controls are needed. The goal is to build a logical system from hazard identification to categorizing the hazards, systems effects, and current controls. Figure 7.4 provides another example of a PHA worksheet.

(d) *Document risk levels.* For each hazard, the PHA team evaluates risk levels for severity and probability and enters the corresponding scores for each in the worksheet. This typically takes the form of a simple severity (S) × probability (P) risk factor or a risk priority number (RPN) if a third factor such as frequency (F) is included as used in FMEA. Some of the risk assessment tools discussed in the book require simultaneous use or deriving quantitative data from other risk assessment techniques. Figure 7.5 is an example of initial PHA including a RAC or RPN column. The PHA is qualitative in nature; however, PHAs often lead to the need for more in-depth quantitative risk assessment.

(e) *Rank risks.* Using the risk factor multiples, RPNs, or RACs determined in the PHA, hazards and risks can be prioritized for proper risk management. When PHA is used for prioritizing multiple systems or assets, the risk scores are used to rank their importance in the application of controls.

The example in Figure 7.5 is a part of a more advanced risk assessment model that includes business objectives for the organization. In addition, the column "Recommended Actions" allows for actions based on the ANSI Z590.3 PtD hierarchy of controls that are aimed at reducing the probability and/or severity of an incident resulting from the hazard.

PRELIMINARY HAZARD ANALYSIS WITH TRACKING LOG

Project/Process	
Prepared by:	

Date:

Main

Methods Used	

Hazardous Event	System Effects	RAM or RPN	Business sustainability	Excellence in EHS	Retain talented employees	Increase profits	Respect for the individual	Increase market share	Total	Comments	Recommended Actions	Controlled RAC or RPN	Standards	Action	Ref. Document	Signature
Back injuries	16 RC affected	36	23	17	13	11	17	5	122							
Shoulder, arms, neck injury	16 RC affected	24	23	17	13	11	17	5	110							
Fall/slip injuries	16 RC affected	12	23	17	13	11	17	5	98							

Figure 7.5 Advanced Risk Assessment Model

7.10 SUMMARY

PHA relies on a combination of team brainstorming, professional judgment, and qualitative or semi-quantitative methods to assess the significance of hazards and assign a ranking to each task or process. Utilizing PHA may help in prioritizing recommendations for reducing risks. The method is flexible enough and it may be applicable to any activity or system. PHA is usually used as a high-level analysis early in the design phase of the project. However, it may be used to evaluate hazards or tasks of existing processes or systems.

The PHA is a simple and efficient method for many applications. However, PHA does have some limitation that should be kept in mind when selecting the appropriate method. Some of these are listed as follows:

- PHA provides a hazard-by-hazard inventory that does not easily recognize synergistic effects from combined hazards and provides only preliminary information.
- PHA is not comprehensive and does not typically provide detailed information on risks or control measures.
- PHA usually does not require listing of barriers.
- PHA does not address consequences of the listed hazards and generally requires additional follow-up analyses or utilization of more complex risk assessment techniques.

The quality of the evaluation depends mainly on the quality and availability of documentation, the expertise of the safety professional with respect to the various analytical techniques, and the capability of the review team.

REVIEW QUESTIONS

1. State the primary purpose of PHA.
2. Describe how PHA is used by safety professionals to identify hazards and necessary control measures.
3. Explain why taking the time to perform a PHA early on may actually speed up the design process and avoid costly mistakes.
4. Summarize the fundamental principles that apply to the system safety approach.
5. Identify your favorite initial scoring system outlined in ANSI Z590.3.
6. Briefly explain the advantages of PHA.
7. Name three of the difficulties associated with using PHA without any other risk assessment methods.

REFERENCES

ANSI/ASSE Z590.3-2011. *Prevention Through Design: Guidelines for Addressing Occupational Hazards and Risks in Design and Redesign Processes*. Des Plaines, IL: American Society of Safety Engineers, 2011.
ANSI/ASSE Z690.3-2011. *American National Standard – Risk Assessment Techniques*. Des Plains, IL: The American Society of Safety Engineers, 2011.
MIL-STD-882E. *Standard Practice for System Safety*. Washington, DC: Department of Defense, 2012.

PROCESS CURRENT STATE

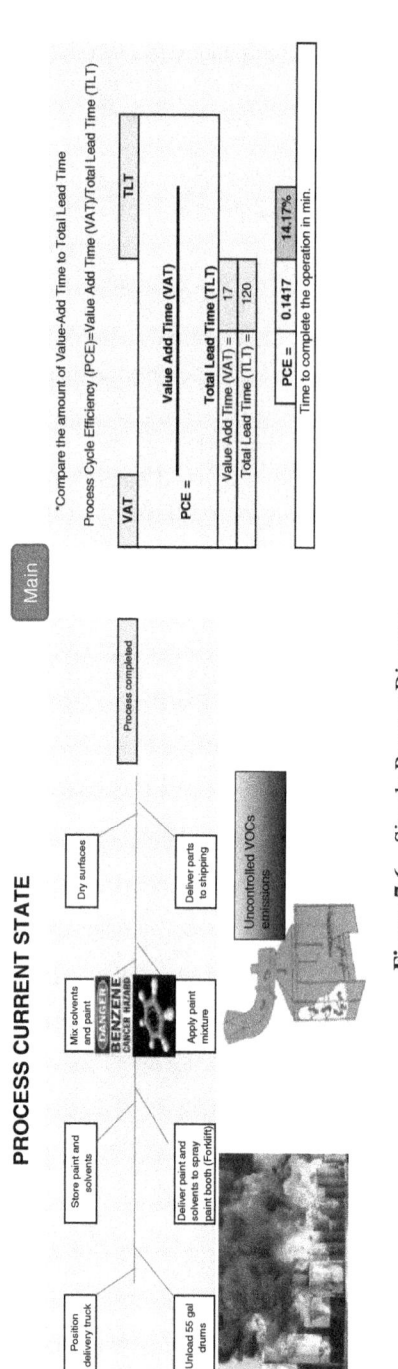

Position delivery truck	Store paint and solvents	Mix solvents and paint	Dry surfaces	Process completed
Unload 55 gal drums	Deliver paint and solvents to spray paint booth (Forklift)	Apply paint mixture	Deliver parts to shipping	

Uncontrolled VOCs emissions

Main

*Compare the amount of Value-Add Time to Total Lead Time
Process Cycle Efficiency (PCE)=Value Add Time (VAT)/Total Lead Time (TLT)

VAT		TLT
PCE =	Value Add Time (VAT) / Total Lead Time (TLT)	

Value Add Time (VAT) =	17
Total Lead Time (TLT) =	120

PCE =	0.1417	14.17%

Time to complete the operation in min.

Figure 7.6 Simple Process Diagram

Potential Effect # and a Short Name	Hazard	Potential Effects	Process	Business Unit/ Department	Description of Current Controls	ANSI/ASSE Z590.3-2011: Prevention Through Design Hierarchy of Controls
1 EXP	Chemical/ph	Explosion	Housekeeping and flammable liquid storage	Spray paint department	None	None
2 Health	Chemical	Health/Benzene exposure	Oil based flammable liquid application in the spray booth	Spray paint department	SOPs and PPE	Admin and PPE
3 Env	Chemical	Environmental	Oil based flammable liquid application in the spray booth	Spray paint department	None	None

Figure 7.7 A Simplified Version of the PHA

Potential Effect # and a Short Name	Hazard	Potential Effects	Process	RF	Business Unit/ Department	Description of Current Controls	ANSI/ASSE Z590.3-2011: Prevention Through Design Hierarchy of Controls
1 EXP	Chemical/ph	Explosion	Housekeeping and flammable liquid storage	15	Spray paint department	None	None
2 Health	Chemical	Health/Benzene exposure	Oil based flammable liquid application in the spray booth	12	Spray paint department	SOPs and PPE	Admin and PPE
3 Env	Chemical	Environmental	Oil based flammable liquid application in the spray booth	16	Spray paint department	None	None

Figure 7.8 Simplified PHA Version of the Spray Paint Operation with Risk Factor

	Hazard#	1 EXP	2 Health	3 Env
Severity Ranking:		5	4	4
Probability Ranking:		3	3	4
Total		15	12	16

Figure 7.9 Suggested Risk Level for Spray Paint Operation

Practical Example

XYZ Manufacturing is a company that produces high-quality aluminum parts for the automotive industry. The company stores oil-based paint in 55 gal drums in the storage room. Flammable solvents are also stored in the storage room. The solvents contain up to 5% of benzene.

The paint and the solvents are moved with a forklift truck to the conventional spray booth. Conventional air-spray guns are the standard spray equipment used to apply coatings in the automotive refinishing industry. The employees are using this type of spray gun. A low volume between 2 and 10 cubic feet per meter (cfm) of air is pressurized and forced through a nozzle; the paint is atomized in the air at the nozzle throat. The spray guns are operated with air pressures of 30–90 psi at a fluid pressure of 10–20 psi. The air is supplied by air compressors during spraying operations. There are two basic types of conventional spray guns: siphon feed and gravity feed. The company is using siphon-feed guns, where the paint cup is attached below the spray gun, and the rapid flow of air through the gun creates a vacuum that siphons the paint out of the cup.

The coating operators are using only N95 respirators and safety glasses. However, the glasses are "coated" with paint after just 10 min of use. The spray paint operators are complaining that they can't see very well the parts, and the quality control (QC) manager determined that only 70% of the parts are coated properly. The following diagram (Figure 7.6) presents the sequence of the process.

A simplified version of the PHA is presented in Figure 7.7.

Students' task: assign risk factor (severity × probability) to the potential effects from the hazards. Our suggested risk level is presented in Figures 7.8 and 7.9.

8

FAILURE MODE AND EFFECTS ANALYSIS

GEORGI POPOV
School of Environmental, Physical & Applied Sciences, University of Central Missouri, Warrensburg, MO, USA

BRUCE K. LYON
Risk Control Services, Hays Companies, Kansas City, MO, USA

8.1 OBJECTIVES

- Introduction, overview, and background
- Purpose and use
- Practical application
- Examples
- Practice exercises/questions

8.2 INTRODUCTION

Failure mode and effects analysis (FMEA) is one of the most commonly used techniques for hazard analysis and risk assessment. As its name implies, FMEA is used to identify and analyze the ways in which system components can fail to fulfill their designed intent and the resulting effects to the system. To state plainly, FMEA focuses on failures and their effects to understand *how* each failure can be prevented and their effects reduced. It is generally considered a qualitative or semiquantitative method that lists systematically the failure modes and their effects, existing safeguards, and any additional controls that are needed to reduce risk to an acceptable level.

Traditionally, the FMEA method has been used as a form of "reliability" analysis of systems, subsystems, processes, and hardware. However, FMEA can also be used to analyze jobs and individual tasks and to identify potential deviations from desired performance criteria that can cause failures (exposure to hazards), how these failures can occur, and their resulting effect on the workers, as well as the work environment. This chapter will address the practical use of the FMEA technique in identifying failure modes and their effects, existing control measures, and ranking systems according

---

Risk Assessment: A Practical Guide to Assessing Operational Risks, First Edition.
Edited by Georgi Popov, Bruce K. Lyon, and Bruce Hollcroft.
© 2016 John Wiley & Sons, Inc. Published 2016 by John Wiley & Sons, Inc.

to their importance or criticality. The text will provide qualitative, semi-quantitative, and quantitative information for additional analytical techniques such as Bow-Tie analysis.

FMEA, one of the first failure analysis methods, was developed by the US Department of Defense (DoD) in 1949. It was originally published in the Military Procedure *MIL-P-1629 – Procedures for performing a failure mode effect and critical analysis*, and its objective was to classify possible failures related to personnel and equipment. Later, the methodology was used by National Aeronautics and Space Administration (NASA). During the 1960s, FMEA was successfully used by the nuclear industry and space exploration programs.

The US automotive industry adopted FMEA methodology in the late 1970s. Numerous standards and publications addressing FMEA were issued since the initial MIL-P-1629 standard was published. The US DoD updated the standard during the 1970s and 1980s. The FMEA methodology is now widely used by various industries including manufacturing, automotive, semiconductors, food processing, and healthcare. In some applications, a critical items list (CIL) is produced from an FMEA.

A similar method that incorporates an additional step of performing a formal criticality calculation is called failure mode effects and criticality analysis (FMECA). It was developed by NASA to improve and verify the reliability of space program hardware (Carlson, 2012). FMECA requires objective data to support its criticality analysis and calculation, as well as more detailed risk ranking information and is not specifically covered in this text.

FMEA is included in many US and international standards. For example, the International Standards for Organization *(ISO) 31010:2009* and American National Standard Institute *ANSI/ASSE Z690.3-2011 Risk Assessment Techniques* standard suggests using FMEA as one of the risk assessment techniques. In addition, FMEA is one of the eight risk assessment techniques listed in *ANSI/ASSE Z590.3-2011, Prevention through Design Guidelines for Addressing Occupational Hazards and Risks in Design and Redesign Processes*. The ANSI Z590.3 standard states that FMEA, along with preliminary hazard analysis (PHA), and What-If methods are sufficient to address most risk situations.

8.3 PURPOSE AND USE

An FMEA is performed to review the defined system's components individually to identify failure modes and causes and effects of such failures on the system. In other words, its purpose is to identify the ways in which systems, components, or processes may fail and the effect the failure may have on the system and users. Frequently used as a first step of a system reliability study or product development, FMEA is also used in many different applications. According to the ISO 31010/ANSI Z690.3, *Risk Assessment Techniques*, there are different FMEA applications:

- Design or product FMEA – used for components or product design
- System FMEA – used for an entire system
- Process FMEA – used for manufacturing, assembly, or other process
- Service FMEA – used for installation or service of equipment during operation
- Software FMEA – used for software systems and controls.

Other FMEA types exist and are described by Carlson in his book, *Effective FMEAs – Achieving Safe, Reliable, and Economical Products and Processes Using Failure Mode and Effects Analysis (Carlson, 2012)*. Several of interest to the safety profession include (1) human factors FMEA, used for interactions between users and equipment; (2) concept FMEA, a condense version used for analyzing alternative concepts; (3) hazard analysis, used for systems throughout their life cycle to analyze safety-related risks; (4) failure mode effects and diagnostic analysis (FMEDA), used as an extension

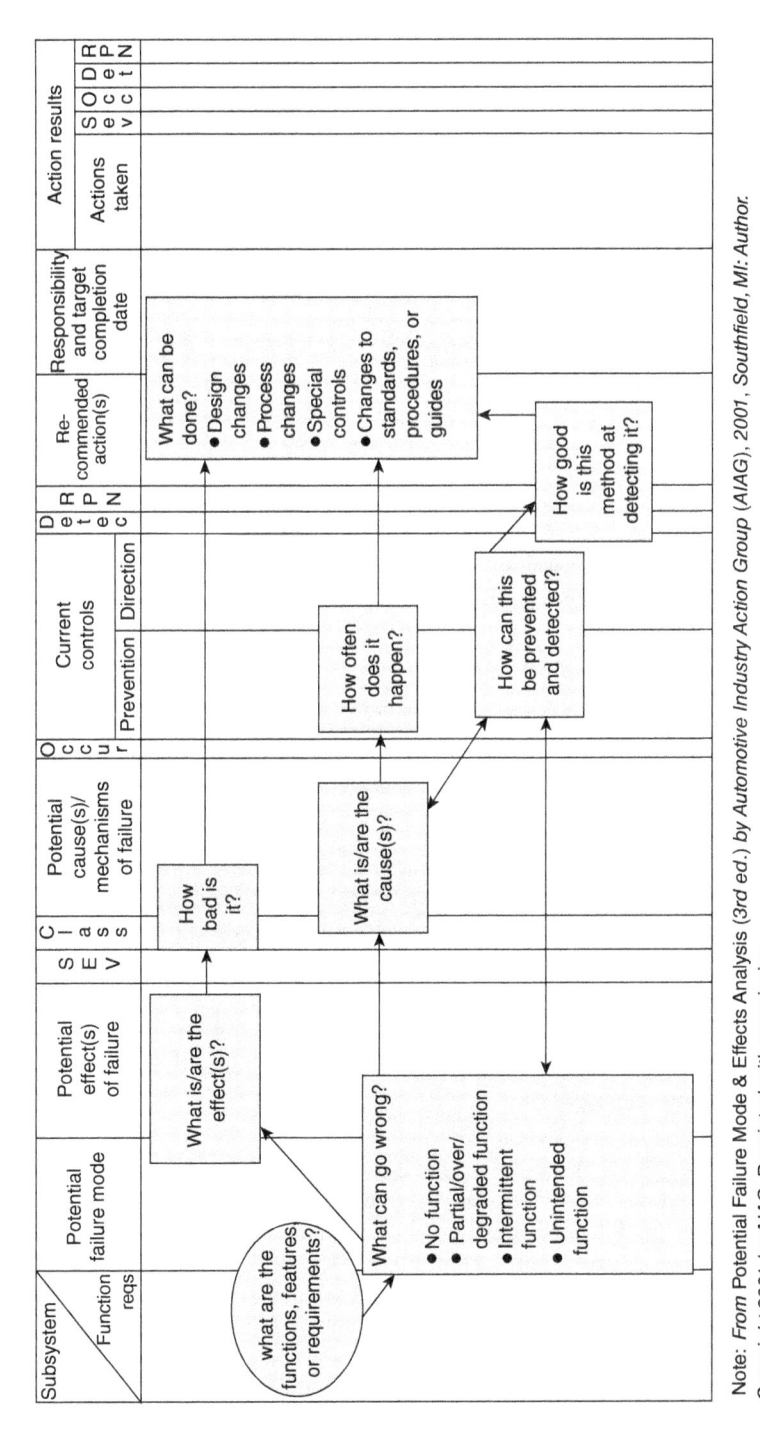

Figure 8.1 Potential Failure Mode and Effects Analysis Sequence (Informative). *Source:* Taken from ANSI/ASSE Z590.3-2011 (Courtesy of the American Society of Safety Engineers)

of FMEA to systematically diagnose failures and effects; and (5) FMECA, used as an extension of FMEA so that each fault mode identified is ranked according to its importance or criticality.

As with any tool, there are certain limitations that should be recognized with the use of FMEA. Some of these include the following:

- FMEA is only used to identify single failure modes or hazards and does not address synergetic effects from multiple hazards.
- It may be time consuming and costly.
- It may be difficult to use for complex multilayered systems.
- Generally, it does not address consequences of the listed hazards. Therefore, FMEA usually requires additional follow-up analyses or utilization of more complex risk assessment techniques.

Traditionally, FMEA is designed to identify all potential failure modes of the parts of the system and subsystems or various steps of the process. In addition, FMEA seeks to identify the mechanisms of the failure and how to avoid the failures and/or mitigate the effects of the failures on the system or the process. For each part, system/subsystem, or process, the failure modes and their potential effects on the rest of the system are recorded in a specific FMEA worksheet.

As indicated by ISO 31010/ANSI Z690.3, FMEA can be used in a number of ways to

- assist in selecting design alternatives with high dependability;
- ensure that all failure modes of systems and processes and their effects on operational success have been considered;
- identify human error modes and effects;
- provide a basis for planning, testing, and maintenance of physical systems;
- improve the design of procedures and processes;
- provide qualitative, semi-quantitative or quantitative information for analysis techniques such as fault tree analysis.

There are many different variations of FMEA formats and worksheets. An example of an FMEA worksheet reprinted with permission from ANSI/ASSE Z590.3 is presented in Figure 8.1.

The FMEA process presented in the previous example requires entry of probability, severity, and detection codes or rankings. It should be noted that there are various severity, probability, and detection ranking scales described in existing standards and texts. Further discussion on these is included in this chapter as well as Chapter 4.

8.4 DEFINING FAILURE MODES

A clear meaning of a "failure mode" is necessary for those using FMEA and other failure analysis techniques. Carlson (2012) defines failure mode as "the manner in which the item or operation potentially fails to meet or deliver the intended function and associated requirements." For each FMEA effort, the scope should include a clear definition for failure modes in context of the system to be analyzed. Depending on the definition established, examples of failures may include failure to perform within defined limits, inadequate or poor performance, intermittent or inconsistent performance, and unintended or undesired functions.

ANSI Z590.3 defines failure mode as "what is observed to fail or to perform incorrectly." The standard further describes failure mode considerations as follows:

The possible failure modes that could result in hazardous situations shall be considered, including the reasonably foreseeable uses and misuses of facilities, materials, and equipment.

Credible circumstances that could arise that would result in the occurrence of an undesirable incident or exposure shall be identified. Determine how and under what circumstances this situation could be harmful.

For each function, all potential failure modes are identified. Each failure mode should be considered independently from other failure modes and sufficiently described to allow the FMEA team to determine its cause(s).

8.5 RISK DESCRIPTION CONSIDERATIONS

The FMEA technique is designed to identify failures and their resulting risk exposures, making the risk descriptors and scoring system vital to an effective analysis. Many risk level scoring models are available ranging from basic two-factor systems to more complex risk factor systems including four or more variables. These risk descriptors often used in FMEA and other risk assessments include the following:

- Severity of consequence (S)
- Occurrence probability (O)
- Frequency of exposure (E)
- Detection of failure (D)
- Prevention effectiveness (PE)
- Risk priority number (RPN).

The ANSI Z590.3 *Prevention through Design* standard presents a two-factor scoring system with risk codes for the severity of consequences and occurrence probability. Severity of consequence rating (S) is described as an estimate of the magnitude of harm or damage that could reasonably result from a hazard-related incident or exposure. The occurrence probability (O) rating is based on the likelihood of the potential failure or hazardous event occurring. The ANSI Z590.3 standard defines probability as "an estimate of the likelihood of an incident or exposure occurring that could result in harm or damage for a selected unit of time, events, population, items, or activity being considered." In some FMEA forms, occurrence and probability are used interchangeably.

In many FMEA models, a third factor, "detection of failure" (D), is used in the risk level scoring system. The detection rating is based on an estimate of how easily the potential failure could be detected prior to its occurrence. In other applications, where hazard control measures are analyzed, the use of "prevention effectiveness" (PE) is used in place of detection. Prevention effectiveness is an estimate of a control measure's efficacy in controlling the failure and its effect and is determined according to the hierarchy of control model found in ANSI Z590.3 and other safety standards. The prevention effectiveness (PE) code may be preferred since it aligns closely with the "prevention through design" concept found in the ANSI Z590.3 standard.

As described by ANSI Z560.3, the RPN is a semiquantitative measure of criticality obtained by multiplying numbers from rating scales (usually between 1 and 10) for consequence of failure, likelihood of failure, and ability to detect the problem. (A failure is given a higher priority if it is difficult to detect.)

There are other models suggesting a four-variable scoring system including severity, probability, frequency of exposure, and detection. The use of three- and four-risk factor systems should be carefully examined. As indicated by Manuele (2014), a four-factor risk scoring system can be problematic. Manuele presents a hypothetical scenario of a fatality that is obviously an unacceptable risk; however, when applying a four-factor risk score, it is rated as acceptable. This occurs due to a dilution of severity by the other three factors through the mathematical scoring, giving each

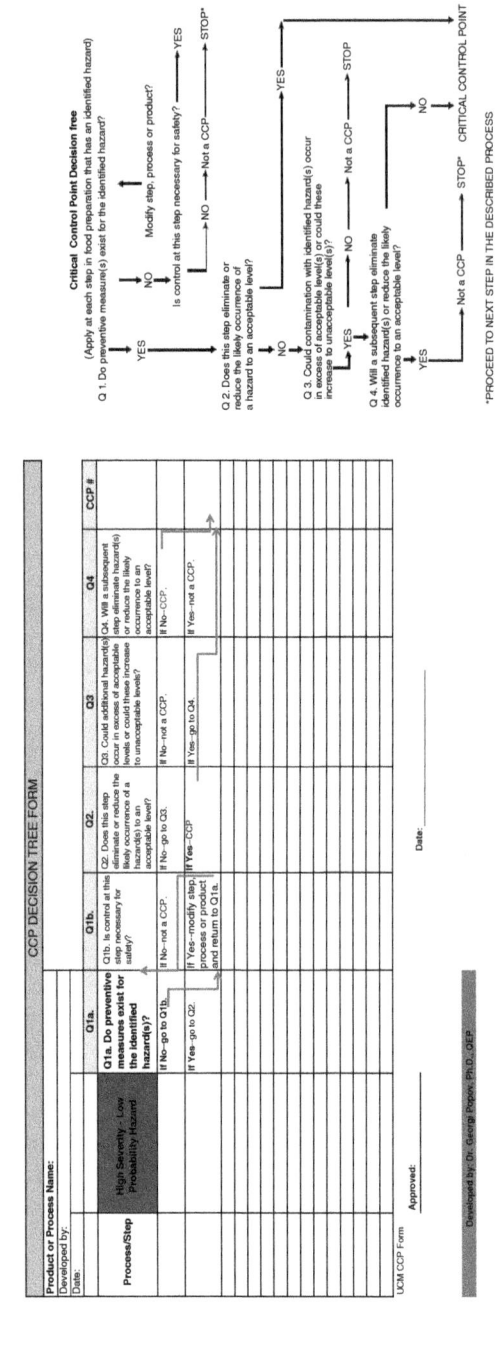

Figure 8.2 Modified CCP Decision-Making Model

RISK DESCRIPTION CONSIDERATIONS

risk factor a weighting of 25%. In a three-factor risk scoring system of severity, probability, and frequency of exposure, the severity factor is also "discounted" in the final score. With three factors in the equation, each has a weighting of 33% of the final risk score as shown in the following:

$$\text{Severity} \times \text{Probability} \times \text{Frequency of exposure} = \text{risk}$$

To more accurately score risk levels, Manuele suggests that severity receive proper weighting to reflect the impact severity has on incident outcomes as shown in the following:

$$\text{Severity} \times (\text{Probability} + \text{Frequency of exposure}) = \text{risk}$$

Hazards that present high-severity and low-probability risks should be reviewed very carefully. Traditional measures of safety performance have focused on the number of accidents/incidents and incident rates such as the total case rate (TCR); the days away, restricted, and transfer (DART) case rate; and the days away from work (DAFWII) case rate. Special attention should be given to potential low probability–high severity of consequences cases, such as fatalities, severe injuries, property damage, or environmental impacts.

To systematically reduce the potential for *high-severity/low-probability events*, it is necessary to concentrate on preventive measures and specifically the strength of the risk control measures or barriers. Risk professionals can essentially "borrow" ideas and methodologies from other industries such as the food processing industry (see Chapter 15, Food Processing Risk Assessment) and apply them to their industry (US FDA, 1997). For example, a critical control point (CCP) food safety methodology can be used to assess high-severity/low-probability risks in nonfood related exposures. An example of a modified CCP decision-making model is presented in Figure 8.2.

It may be suggested that the CCP decision tree method or similar methodology be used to evaluate all high-severity hazards. For each procedure/process that may cause or have potential severe consequences, the first consideration is (Q1) whether any controls exist for the identified severe hazard. If controls exist for the severe hazard, an evaluation of the control's effectiveness is performed (Q2). This involves determination whether the control(s) will avoid/eliminate the hazard or reduce the risk of occurrence to an acceptable level. If the risk is reduced to an acceptable level, the step is considered a CCP and included in the health and safety plan. If the control does not reduce risk of occurrence sufficiently, then the severity of the hazards is evaluated. If no safety/health/environmental threat exists from the identified hazard, it is not a CCP and the process stops. However, if the procedure/process presents a severe hazard, then the subsequent step is considered. If no efficient subsequent step exists, the evaluated step is considered a CCP and included in FMEA with a notation or color coding. Figure 8.3 illustrates the CCP decision tree process steps taken.

When developing an FMEA risk level scoring system, a rating scale with numerical grading as provided in Table 8.1 reprinted with permission from ANSI/ASSE Z590.3 standard should be considered. Combining severity and probability values produces a simple yet effective, risk level score method.

Author's Cautionary Note: Some publications and previous military standards suggest the criteria used for the assignment of the severity, detection, and occurrence ratings be based on a 10-point scale, rather than five-point scale for better discrimination of the relative risk. When a 10-point scale is used, the RPN calculation ($RPN = S \times O \times D$) will be a number between 0 and 1000. The 10-point system creates a large range and may not fit all applications. A 5-point scale is likely more appropriate for most applications.

Various organizations may use different rating scales based on customer or project requirements; however, it is important that consistent evaluation rating criteria be applied. Rating criteria should be standardized so that a lower RPN value indicates a lower risk level. The following

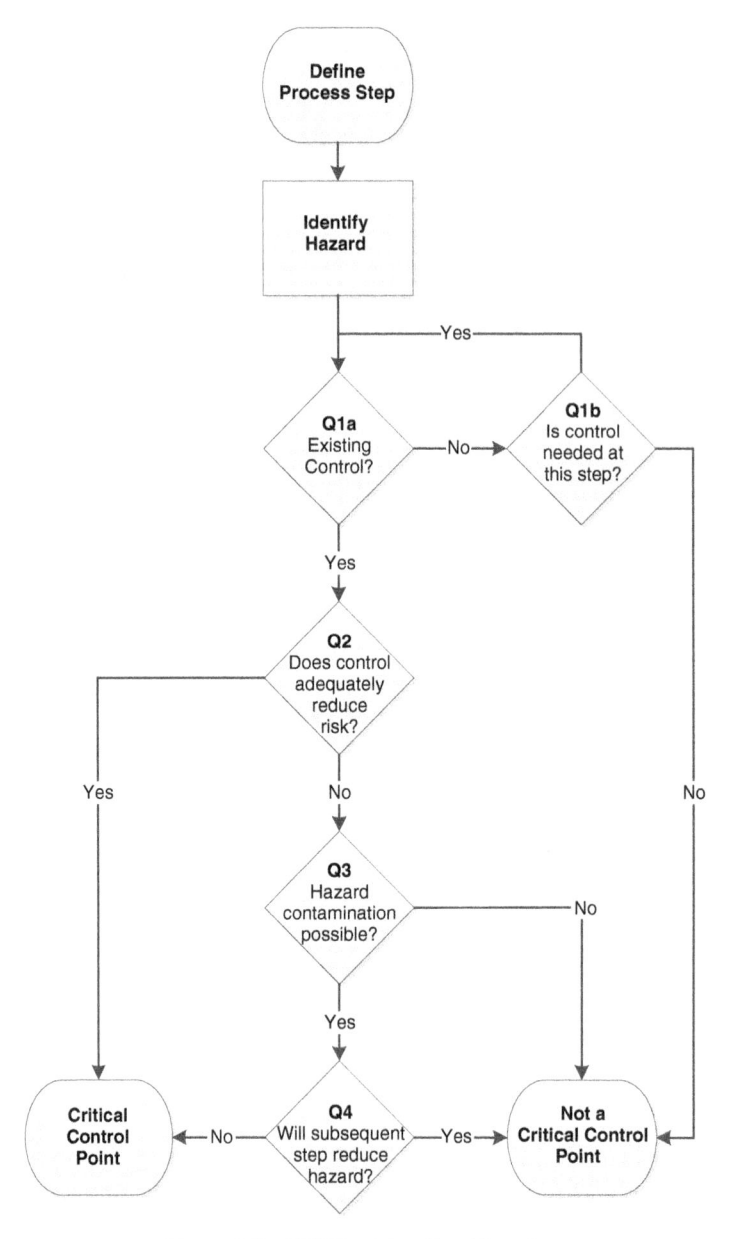

Figure 8.3 CCP Decision Tree Flowchart

risk criteria in Table 8.1 using severity of consequence and occurrence probability factors (Risk level = severity × probability) are very high risk (15 or greater), high risk (10–14), moderate risk (6–9), and low risk (1–5). Risk score action for two variables and 1–5 scoring matrix (5 × 5 model) reprinted with permission from ANSI/ASSE Z590.3 is presented in Table 8.2.

As previously mentioned, FMEA typically adds a third variable such as prevention effectiveness. Using a 5-point risk assessment scale for three variables will result in a RPN between 1 and 125. This calculation is displayed as follows:

$$\text{Risk priority number} = \text{Severity} \times \text{Probability} \times \text{Prevention effectiveness}$$

RISK DESCRIPTION CONSIDERATIONS

TABLE 8.1 Example Five-Point Risk Assessment Matrix

Severity Levels and Values	Occurrence Probabilities and Values				
	Unlikely (1)	Seldom (2)	Occasional (3)	Likely (4)	Frequent (5)
Catastrophic (5)	5	10	15	20	25
Critical (4)	4	8	12	16	20
Marginal (3)	3	6	9	12	15
Negligible (2)	2	4	6	8	10
Insignificant (1)	1	2	3	4	5
Very high risk: 15 or greater High risk: 10–14 Moderate risk: 6–9 Low risk: 1–5					

Source: Taken from ANSI/ASSE Z590.3-2011 (Courtesy of the American Society of Safety Engineers)

TABLE 8.2 Risk Scoring Tables for 5 × 5 Matrix

Risk Level 5 × 5	Min	Max
Very high risk	15	25
High risk	10	14
Moderate risk	6	9
Low risk	1	5

Category	Risk Score Action
Very high risk	15 or greater – operation not permissible. Immediate action necessary
High risk	10 to 14 – remedial actions to be given high priority
Moderate risk	6 to 9 – remedial action to be taken at appropriate time
Low risk	1 to 5 – remedial action discretionary

Source: Taken from ANSI/ASSE Z590.3-2011 (Courtesy of the American Society of Safety Engineers)

Suggested scoring methodology	
Incident or exposure severity descriptions	
Catastrophic (5)	One or more fatalities, total system loss, chemical release with lasting environmental or public health impact.
Critical (4)	Disabling injury or illness, major property damage and business downtime, chemical release with temporary environmental or public health impact.
Marginal (3)	Medical treatment or restricted work, minor subsystem loss or damage, chemical release triggering external reporting requirements.
Negligible (2)	First aid or minor medical treatment or minor medical treatment only, nonserious equipment or facility damage, chemical release requiring routine cleanup without reporting.
Insignificant (1)	Inconsequential with respect to injuries or illnesses, system loss or downtime, or environmental chemical release.
Incident or exposure probability descriptions	
Frequent (5)	Likely to occur repeatedly.
Likely (4)	Probably will occur several times.
Occasional (3)	Could occur intermittently.
Seldom (2)	Could occur, but hardly ever.
Unlikely (1)	Improbable, may assume incident or exposure will not occur.
Prevention effectiveness	
Poor (5)	None
Low (4)	PPE
Moderate (3)	Warning, Administrative Controls
Better (2)	Engineering Controls
Best (1)	Avoid, Eliminate, Substitute

Figure 8.4 Example of Severity, Probability, and Prevention Effectiveness Rating Scales

TABLE 8.3 Risk Scoring Tables for 5 × 5 × 5 Model

Risk Level 5 × 5 × 5	Min	Max
Very high risk	75	125
High risk	45	74
Moderate risk	25	44
Low risk	1	24

Category	Risk Score Action
Very high risk	75 or greater – operation not permissible. Immediate action necessary
High risk	45 to 74 – remedial actions to be given high priority
Moderate risk	25 to 44 – remedial action to be taken at appropriate time
Low risk	1 to 24 – remedial action discretionary

As indicated in Figure 8.4, rating scales and descriptions for severity, probability, and prevention effectiveness rank from 1 to 5. The higher the number, the higher the risk exposure.

A simple mathematical extrapolation can be used to derive risk score action for three variables and 1–5 scoring matrix (5 × 5 × 5) as provide in Table 8.3.

8.6 FMEA PROCESS STEPS

FMEA methodology is well described in military and civilian standards and can be applied to both military and civilian projects and products. As the name implies, FMEA takes a system, breaks it down into individual components, and then systematically looks at the different ways each component could fail and the effects of each failure on the system.

In the SEMATECH (1992) document "Failure Mode and Effects Analysis (FMEA): A Guide for Continuous Improvement for the Semiconductor Equipment Industry," a good example of an FMEA process is provided. The document contains guidelines in the use of FMEA at the design stage for semiconductor manufacturing equipment to ensure its reliability. The basic process steps listed in the document include (1) FMEA prerequisites, (2) functional block diagram, (3) failure mode analysis and preparation of work sheets, (4) team review, and (5) corrective action. The flowchart in Figure 8.5 provides an example of the process steps in an FMEA taken from "Failure Mode and Effects Analysis (FMEA): A Guide for Continuous Improvement for the Semiconductor Equipment Industry," SEMATECH Technology Transfer #92020963B-ENG. Reprinted with permission.

The following steps summarize the FMEA process. It should be noted that an FMEA worksheet can be completed *by column*, entering in sequence all functions and failure modes, followed by all effects, or *by row*, completing each row in sequence list one function, one failure mode, one effect, one cause, and one control.

1. *Establish context.* As in all risk assessment efforts, a specific purpose and scope with limitations for the FMEA is established. The FMEA context will include the purpose of the project, the systems or functions to be analyzed, the project's time frame, resources necessary, as well as the risk matrix, risk criteria, and definitions to be used.
2. *Select FMEA team.* An experienced and qualified team and a facilitator are selected to conduct the FMEA. The team should be multidisciplinary and include individuals from areas such as safety and risk management, business management, accounting, purchasing, engineering, production, quality assurance, human resources, employees, and others. As suggested in ANSI Z10 and OSHA's Process Safety Management standard, the team should also include employees who are involved in the operations as they are more familiar with the variability and limitations of the operation. This promotes a sense of ownership among those who must implement the plan. The risk assessment team may need assistance from outside experts

FMEA PROCESS STEPS

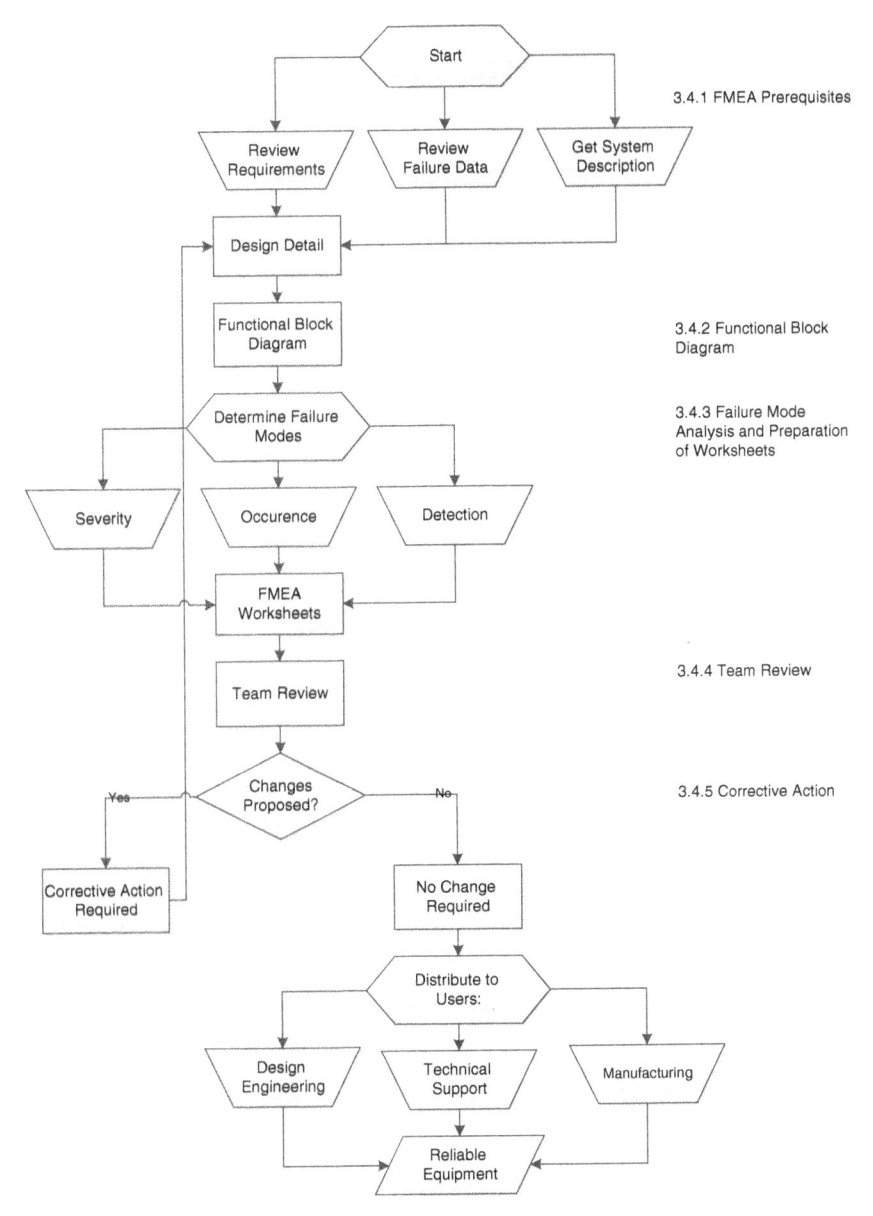

Figure 8.5 FMEA Process. *Source:* International SEMATECH (1992). Reprinted with permission

who are knowledgeable in the potential chemical, physical, biological, or ergonomics hazards associated with the product and the process. However, FMEA that is developed totally by external sources may be inaccurate, incomplete, and lacking in support at the local level. In addition, members of the team should be at least somewhat familiar with FMEA. The team leader or facilitator should be an experienced professional with system safety knowledge and various risk assessment techniques.

3. *Gather data.* The team collects related documents and reference resources including description of the system or similar systems, operating procedures, design specifications, diagrams and drawings, loss history, lessons learned, and other available data that may help identify

failure modes and effects. Preliminary hazard lists (PHL), PHA, or other hazard analyses previously performed should be gathered and reviewed.

4. *Enter process function.* After reviewing the data and identifying the primary functions, the process steps or functions to be analyzed are entered in sequential order in column #1 (see Figure 8.6 FMEA example). This may include a brief description of the parts of the system under study broken down into processes, levels, or tasks.
5. *Identify failure modes.* Beginning with the first function listed, all identified failure modes and their causes are listed in the Potential Failure Modes column #2. Assessing the current situation involves identifying potential failures or safety concerns associated with the function.
6. *Identify effects.* For each failure mode, all potential effects are listed in the Potential Effects of Failure column #3. This involves an in-depth review of potential hazards and risks that may result from the failure or exposure.
7. *Determine severity.* Using the selected risk criteria for severity of consequences, each failure's effect is rated for its estimated severity level and entered in the Severity column #4.
8. *Identify failure cause(s).* For each potential failure mode, causes or failure conditions that can lead to the failure are identified and listed in Potential Cause(s) of Failure column #5.
9. *Determine occurrence.* For each cause, the occurrence probability is assessed and recorded in the Occurrence column #6. Likelihood of the occurrence is determined using the established risk criteria for occurrence probability.
10. *Identify existing controls.* For each cause, the team identifies the existing controls and records them in the Existing Controls column #7.
11. *Determine prevention effectiveness.* The team evaluates the effectiveness of existing controls according to the hierarchy of controls and assigns a prevention effectiveness rating in column #8.
12. *Enter severity and occurrence rating.* The severity rating taken from column #4 and occurrence rating from column #6 are combined (or multiplied depending upon the risk matrix and criteria selected) producing a risk rating, which is entered in the Severity + Occurrence column #9.
13. *Calculate RPN.* The prevention effectiveness (PE) rating from column #8 and the risk rating from column #9 are combined (or multiplied) to produce an initial RPN, which is entered into column #10.
14. *Additional controls.* For RPNs that are above the acceptable risk level, additional controls are necessary. Using the hierarchy of controls model, appropriate controls are selected and entered in the Needed Action column #11. For each selected additional control measure, the appropriate party for its implement is identified and recorded in Responsible Party column #12.
15. *Verify results.* Following the implement of additional controls, a review of the process function is made to ensure that no additional hazards are created and that the control is effective as anticipated. These results are entered in the Results column #13.
16. *Reassess severity.* After additional controls are implemented and verified, the severity level is reassessed and entered in the Severity 2 column #14.
17. *Reassess occurrence.* After additional controls are implemented and verified, the occurrence level is reassessed and entered in the Occurrence 2 column #15.
18. *Reassess prevention effectiveness.* After additional controls are implemented and verified, the effectiveness of prevention is reassessed and entered in the PE 2 column #16.
19. *Recalculated RPN.* After additional controls are implemented and verified, the adjusted severity, occurrence, and prevention effectiveness levels are combined producing a second RPN entered in the RPN 2 column #17.

PRACTICAL APPLICATION 175

	Failure Mode and Effects Analysis																
Process:			Operation:							Prepared by:							
										Date:							
1	2	3	4	5	6	7	8	9	10	11	12	13	14	15	16	17
Process Function	Potential Failure Mode	Potential Effect(s) of Failure	Severity	Potential Cause(s) of Failure	Occurrence	Existing Controls	PE	Sev + Occ	RPN	Needed Action	Responsible Party	Results	Severity 2	Occurrence 2	PE 2	RPN 2

Figure 8.6 Failure Mode and Effects Analysis Example

8.7 PRACTICAL APPLICATION

Specifically, hazards are identified and recorded in "Process Operation, Function, or Purpose" column so that the risks arising from those hazards can be evaluated and determined if they are acceptable or not. Hazards include all aspects of technology, human factors, and activity that produce risk. Hazards can be physical, biological, chemical, mechanical, psychosocial, etc.; risks can be focused on the health and safety of the worker, property, or the environment.

The example in Figure 8.7 is an FMEA used to evaluate potential hazards during rebar operations performed in concrete construction. The NIOSH evaluated reinforcing ironworkers' (rodbusters) exposures to risk levels for developing low-back and hand disorders when tying together reinforcing steel bars (rebar) on a freeway bridge (National Institute for Occupational Safety and Health, 2009).

The three main potential effects of the hazards were low-back, shoulder/neck, and hand/wrist/finger injuries. The potential effects are recorded in the FMEA form. Severity ratings of potential injuries are properly entered in the form. Potential causes of exposure are also recorded and occurrence/probability rankings entered in the OCC column. Current controls are identified and PE ratings are entered in the PE column. In this case, training was rated 3 out of 5 on a PE scale. Risk level (S × O) is automatically calculated on the FMEA tool. RPN is also automatically calculated. In this case, low-back injuries have the highest RPN and it is advisable to implement potential preventive measures.

The next step begins by identifying the solution(s) to hazard(s) recognized in current state initial risk assessment. Consideration of PtD concepts is used to evaluate and select possible solutions for continued analysis. One possible solution, as recommended by NIOSH researchers is the use of power tool. Suggestions are entered on the right side of the extended FMEA form presented in Figure 8.8.

It should be noted that severity rating remains the same. Probability/occurrence was slightly reduced due to reduction of exposure (bent posture) time. PE remains unchanged for potential low-back and shoulder/neck injuries. However, PE was reduced from 3 to 2 for potential hand/wrist/finger injuries due to implementation of new control measures (new tool). The initial RPN should be recalculated after initial corrective and preventive actions are taken. This should be documented on the right side of the FMEA form.

Unfortunately, our highest RPN for low-back injuries was not reduced sufficiently enough. The researchers suggested a solution for rebar tying using a MAX USA RB392 power tool with adjustable

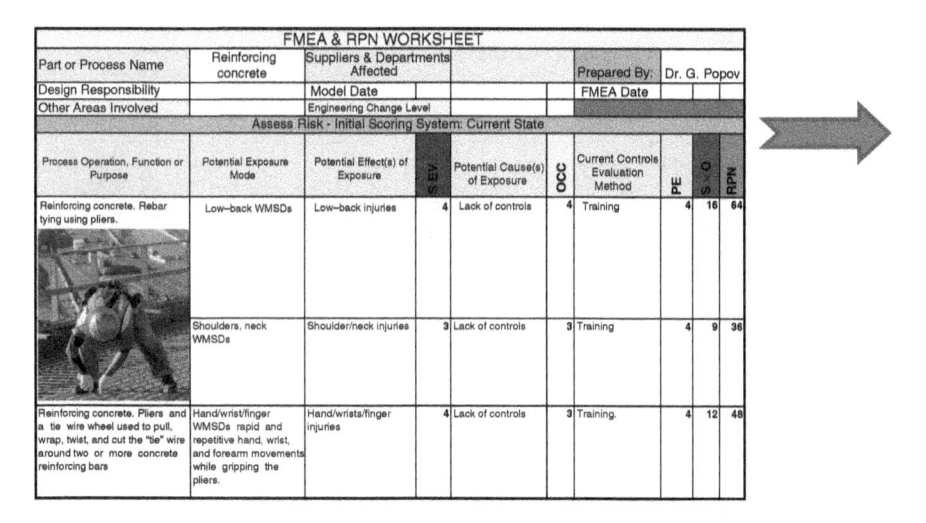

Figure 8.7 FMEA Hazard Analysis Example

extension. Therefore, we will recalculate RPN with the suggested new control measures (adjustable extension). Improved future state FMEA is presented in Figure 8.9.

It should be noted that probability and prevention effectiveness were reduced. Again, the original RPN should be recalculated after suggested corrective and preventive actions (adjustable extension) are taken. The future state RPN is documented on the right side of the FMEA form.

The practical example in the previous text demonstrated the use of process FMEA in the construction industry. FMEA could be utilized in a variety of different processes or design projects.

8.8 SUMMARY

FMEA is applicable to human, equipment, and system failure modes as well as software, hardware, or processes. It also presents failure modes, causes, and effects in systemic and easy-to-read format highlighting the highest RPNs.

FMEA relies on professional judgment and semiquantitative methods to assess the significance of hazards and assign a ranking to each task or process. Utilizing FMEA helps in prioritizing recommendations for reducing risks.

FMEA is flexible enough and it may be applicable to any process, new product design, or a system. FMEA may be used as a high-level analysis early in the design phase of the project or detailed risk assessment of low-level processes or systems.

SUMMARY

Figure 8.8 FMEA Current State Risk Example

178 FAILURE MODE AND EFFECTS ANALYSIS

Part or Process Name	Reinforcing concrete	Suppliers & Departments Affected		Prepared By					Back to RPN 1 Fig. 8.8				
Design/ Mfg Responsibility		Model Date		FMEA Date									
Other Areas Involved		Engineering Change Level											
	Assess Risk - Initial Scoring System: Current State								Assess Risk - Residual Risk Scoring System: Future State				
Process Operation, Function or Purpose	Potential Exposure Mode	Potential Effect(s) of Exposure	SEV	Potential Cause(s) of Exposure	OCC	Current Controls Evaluation Method	PE	S×O	RPN	Recommended Action(s)			SEV 2	OCC 2	PE 2	RPN 2
Reinforcing concrete. Rebar tying using pliers. 	Low–back WMSDs	Low–back injuries		Lack of controls		Training	4	16	64	Tools design changes. Adjustable extension 	Management	Reduced Exposure	4	1	2	8
	Shoulders, neck WMSDs	Shoulder/neck injuries	3	Lack of controls	3	Training	4	9	36			Reduced Exposure	3	1	2	6
Reinforcing concrete. Pliers and a tie wire wheel used to pull, wrap, twist, and cut the 'tie' wire around two or more concrete reinforcing bars	Hands/wrists/ fingers WMSDs. rapid and repetitive hand, wrist, and forearm movements while gripping the pliers.	Hands/wrists/ fingers injuries.	4	Lack of controls	3	Training.	4	12	48	MAX–USA RB-392 power tool 	Management	Reduced exposure/ probability	3	1	2	6

		Incident or Exposure Severity Descriptions
5	Catastrophic	One or more fatalities, total system loss, chemical release with lasting environmental or public health impact.
4	Critical	Disabling injury or illness, major property damage and business downtime, chemical release with temporary environmental or public health impact.
3	Marginal	Medical treatment or restricted work, minor subsystem loss or damage, chemical release triggering external reporting requirements.
2	Negligible	First aid or minor medical treatment or minor medical treatment only, non-serious equipment or facility damage, chemical release requiring routine cleanup without reporting.
1	Insignificant	Inconsequential with respect to injuries or illnesses, system loss or downtime, or environmental chemical release.
		Incident or Exposure Probability Descriptions
5	Frequent	Likely to occur repeatedly.
4	Likely	Probably will occur several times.
3	Occasional	Could occur intermittently.
2	Seldom	Could occur, but hardly ever.
1	Unlikely	Improbable, may assume incident or exposure will not occur.
		Prevention Effectiveness
5	Poor	None
4	Low	PPE
3	Moderate	Warning, administrative controls
2	Better	Engineering control
1	Best	Avoid, eliminate, or substitute

Developed by: Georgi Popov, Ph.D., QEP
SEPAS
University of Central Missouri
v. 1.0 / 11/7/11 ® 2011

Figure 8.9 FMEA Future State Risk Example

The quality of the analysis depends mainly on the knowledge of the team members, quality and availability of documentation, the expertise of the safety leader, and the management of the organization.

REVIEW QUESTIONS

1. Describe the purpose of FMEA.
2. Summarize the historical developments of FMEA methodology.
3. Identify your favorite uses of FMEA as a safety risk assessment method.
4. Explain how FMEA can be used for quality risk assessments.
5. Categorize the scoring systems used for FMEA risk assessments.
6. Differentiate between the two ways an FMEA can be completed (by column and by row).
7. Briefly explain the advantages of FMEA.

REFERENCES

ANSI/ASSE Z590.3-2011. *Prevention Through Design: Guidelines for Addressing Occupational Hazards and Risks in Design and Redesign Processes.* Des Plaines, IL: American Society of Safety Engineers, 2011.

ANSI/ASSE Z690.3-2011. *American National Standard – Risk Assessment Techniques.* Des Plaines, IL: The American Society of Safety Engineers, 2011.

Carlson, C. S., *Effective FMEAs – Achieving Safe, Reliable, and Economical Products and Processes Using Failure Mode and Effects Analysis.* Hoboken, NJ: John Wiley & Sons, Inc., 2012.

International Organization for Standardization. 2009. *ISO 31010 – Risk management – Risk assessment techniques.* Retrieved from: http://www.iso.org/iso/catalogue_detail?csnumber=51073 (accessed December 22, 2015)

International SEMATECH. *Failure Mode and Effects Analysis (FMEA): A Guide for Continuous Improvement for the Semiconductor Equipment Industry.* Technology Transfer #92020963B-ENG. Albany, NY: SEMATECH, 1992. Also available at http://www.sematech.org/docubase/document/0963beng.pdf. (accessed December 22, 2015)

Manuele, F. A., *Advanced Safety Management: Focusing on Z10 and Serious Injury Prevention,* 2nd ed. New York: John Wiley & Sons, Inc., 2014.

National Institute for Occupational Safety and Health. 2009. Reducing Work-Related Musculoskeletal Disorders among Rodbusters. Retrieved from: http://www.cdc.gov/niosh/docs/wp-solutions/2010-103/pdfs/2010-103.pdf (accessed December 22, 2015)

U.S. Food and Drug Administration. 1997. HACCP Principles & Application Guidelines. Retrieved from: http://www.fda.gov/Food/GuidanceRegulation/HACCP/ucm2006801.htm (accessed February 7, 2016)

Practical Example – Assignment #2 – FMEA

Read the following OSHA hazard alert.

"Worker Exposure to Silica during Hydraulic Fracturing" The alert is located at https://www.osha.gov/dts/hazardalerts/hydraulic_frac_hazard_alert.html.

OSHA publications are publically accessible. The hazard alert is also accessible as a pdf file.

For your convenience, the authors have prepared process 1 FMEA. Please see FMEA image in the following figure. Your assignment is to complete processes 2 and 3. An Excel FMEA form is prepared for this chapter as well. Use the same logic as the presented construction FMEA example. Select hazards 2 and 3 from the OSHA hazard alert. There are more than three hazards listed in the publication. Select two and complete FMEA columns. Rank severity (SEV), probability/occurrence (Prob/OCC), and PE on a 1–5 rating scale. Remember that these rankings are somewhat subjective and based on your experience. Once you enter the ranking, S × O and RPN will be automatically calculated for you. Next, identify "recommended actions," responsible party, action results, and residual SEV, OCC, and PE. Ideally, we can reduce all the RPN scores from the initially identified hazards. For your convenience, the authors left some of the digital images embedded in the Excel FMEA file. You can use some of them or delete them.

Example of FMEA – Process 1 completed.

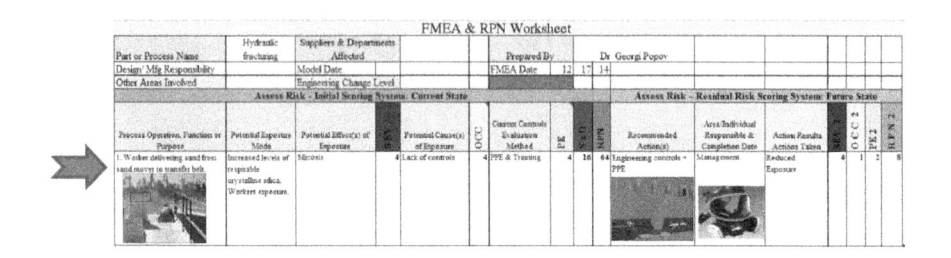

After you complete the student FMEA assignment, save it as Your Name_FMEA Assignment and submit to the instructor.

9

BOW-TIE RISK ASSESSMENT METHODOLOGY

GEORGI POPOV
School of Environmental, Physical and Applied Sciences, University of Central Missouri, Warrensburg, MO, USA

BRUCE K. LYON
Risk Control Services, Hays Companies, Kansas City, MO, USA

9.1 OBJECTIVES

- Introduction
- History and overview
- Review methodology
- Provide guidance on the use of Bow-Tie methodologies
- Examine the use of Bow-Tie methodologies, their strengths, and limitations

9.2 INTRODUCTION

Bow-Tie analysis is a relatively new and simple tool that is used to analyze and communicate risk pathways and controls in selected hazard scenarios. It is one of many "barrier-type" risk models available to assist in the identification and management of risk. The "Bow-Tie" name is derived from the shape of the diagram created in the analysis.

One of the benefits of a Bow-Tie diagram is that it gives an overview of multiple plausible scenarios and associated controls and consequences, in a single picture that would be difficult to explain otherwise. It helps stakeholder to identify, evaluate, and communicate existing controls for selected hazards in a given scenario. The International Organization for Standardization (ISO) 31010 nationally adopted by the American National Standards Institute *ANSI/ASSE Z690.3-2011 Risk Management Techniques* standard describes Bow-Tie analysis as "a simple diagrammatic way of describing and analyzing the pathways of a risk from causes to consequences" (ANSI/ASSE Z690.3-2011).

Risk Assessment: A Practical Guide to Assessing Operational Risks, First Edition.
Edited by Georgi Popov, Bruce K. Lyon, and Bruce Hollcroft.
© 2016 John Wiley & Sons, Inc. Published 2016 by John Wiley & Sons, Inc.

This will demonstrate the use of the Bow-Tie method and its applications in SH&E hazard analysis and interventions. Simple case studies and more innovative practical uses of the methodology as well as practical examples and benefits of this versatile tool will be presented.

9.3 HISTORY

The origins of Bow-Tie analysis are unclear; however, one of the earliest documented uses can be found in course notes from a lecture on Hazard Analysis (HAZAN) at the University of Queensland, Australia, in 1979 (Gill, 1979).

Following the 1988 disaster on the Piper Alpha Platform, the oil and gas industry was under great pressure to exert more efforts in systemic hazard analysis and risk assessment of their operations. As a result, the Bow-Tie methodology began to appear. The Royal Dutch/Shell Group was one of the first global companies to effectively integrate Bow-Tie analysis into its business practices. The Bow-Tie method became a company standard within the Royal Dutch/Shell Group for analyzing and managing risks and soon became adopted by others in the industry (HSSE in Shell, 2010). The Bow-Tie method has since spread outside of the oil and gas industry and is used by aviation, healthcare, military/defense, mining, chemical, maritime, and other industries.

Examples of Bow-Tie analysis have been published by a number governments and industry associations including the United Kingdom, France, Australia, New Zealand, the European aviation industry, and the US Federal Aviation Authority.

One of the benefits quickly realized with a Bow-Tie diagram was the fact that it provides a "big picture" view of a process or system to effectively explain risk exposures and controls. Its ability to provide a readily understandable visualization of the relationships between the hazards, the proactive and reactive controls, and the consequences for a hazardous event has made it an important risk assessment tool.

9.4 OVERVIEW

The Bow-Tie analysis methodology is used to communicate risk by providing a clear road map for risks and their controls. As a barrier-based approach to risk, the Bow-Tie method follows James Reason's "Swiss Cheese" Model of Defenses (Reason, 1997). In his model, he illustrates how barriers have weaknesses and gaps that under specific circumstances can align and allow a hazard exposure to occur, such as the holes in Swiss cheese. Should escalating factors cause the barriers weaken and align allowing a hole to form, the hazard can penetrate the defenses and cause harm.

The Bow-Tie method is unique in that it combines the use of a simplified fault tree analysis (left-hand side of Bow-Tie) to analyze causation of a hazardous scenario or event (the center knot) and a simplified event tree analysis of the resulting consequences (right-hand side). The analysis is on the barriers or controls between the causes and risk, and barriers or controls between the risk and the resulting consequences. Figure 9.1 provides a visual layout of the Bow-Tie analysis pathways.

A Bow-Tie analysis provides a clear understandable view of the existing "barriers" to prevent the causes of hazards and "reactive" measures to mitigate or reduce the severity of the consequences should the hazardous event occur. Such diagrams demonstrate the critical connection between hazards, preventive barriers, risks, mitigation controls, and business consequences. Furthermore, the preventive and mitigating measures can be linked to tasks, procedures, responsible individuals, and systems.

Bow-Tie analysis can be used to qualitatively, semiquantitatively, and quantitatively demonstrate controls, countermeasures, and risk reduction measures and can be applied in many industries, services, and business sectors. The methodology can be used as an important element of an operational

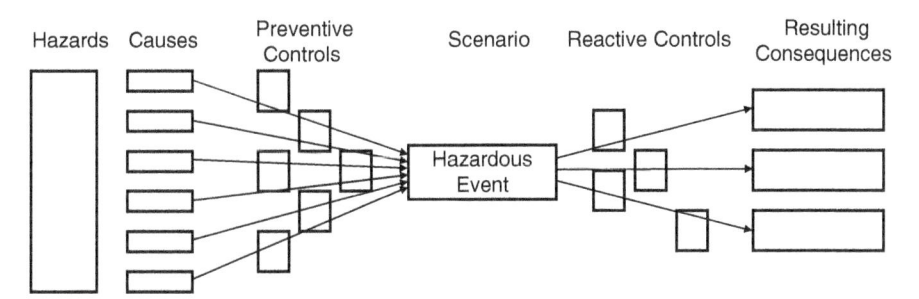

Figure 9.1 Bow-Tie Diagram Model

risk management system and included in the development of an effective business case for SH&E interventions. Some of the strengths of a Bow-Tie analysis are summarized as follows:

- The method visually illustrates the hazards, their causes and consequences, and the controls to minimize the risk.
- It is simple to understand and gives a clear graphic representation of the problem. The Bow-Tie diagram can be readily understood at all levels, from senior managers, engineers, and operations personnel to regulators, line supervisors, and trained workers.
- Bow-Tie diagrams present the big picture and can capture previous incidents as well as future state of hazards and consequences after SH&E interventions.
- The modified Bow-Tie diagrams can add another layer of complexity and provide better visualization of the severity and probability of the hazards and consequences.
- The modified Bow-Tie methodology can identify where resources should be focused for risk reduction.
- The method could potentially reduce the time and extent of SH&E hazard analysis and can lead to a potential reduction in unnecessary/less important barriers.
- The method can illustrate the hierarchy of controls that are currently in place (current state) and the need for additional "layers of protection."
- Bow-Tie diagram can visualize the links between the elements of the organization's management system to specific controls and provide a platform for continuous improvement and future SH&E interventions.
- Bow-Tie methodology provides a logical and structured approach to consider all aspects of the risk management and potential areas for risk reduction.
- The method can also be used to identify gaps and issues that are missed by other risk assessment techniques.
- Bow-Tie diagram can also be used for desirable consequences.

Bow-Tie analysis, for all its strengths, has certain limitations to consider when selecting the appropriate hazard analysis and risk assessment techniques. Some of these limitations are as follows:

- Execution of Bow-Tie methods can be time consuming.
- The analysis is generally limited to qualitative measures.
- A Bow-Tie diagram usually cannot depict where multiple hazards or exposures occur simultaneously; therefore, it is not effective in addressing potential synergetic effects of such hazards and exposures.

- The methodology can sometimes oversimplify complex situations, particularly where quantification is attempted.
- If the purpose is to model complex relationships between prevention controls, additional risk assessment methodologies are required.

9.5 BOW-TIE METHODOLOGY

As previously mentioned, the Bow-Tie method provides a big picture overview of a hazard scenario and its relationships between hazards and causes, barriers to prevent occurrence, and mitigating controls to reduce the impact should an event occur. A conventional Bow-Tie diagram for the Bhopal disaster is presented in Figure 9.2.

The beauty of a Bow-Tie analysis is its ability to provide a clear visual road map of how hazards are managed and risks reduced. Some important questions to consider when developing a Bow-Tie diagram include the below questions:

- What are the hazards and their causes?
- What occurs when control is lost?
- What are the potential consequences?
- How can the hazard event be avoided?
- How can the organization recover should the event occur?
- How can the likelihood and/or severity be limited?
- How might controls fail or effectiveness be reduced?
- How are control failures prevented?

The following are steps for conducting a conventional Bow-Tie analysis:

1. *Select hazard scenario* – A single hazard scenario or "top event" where control of a hazard is lost is selected for analysis and placed in the center knot of the Bow-Tie diagram. Selection is based on a significant and plausible worst-case scenario that is of concern to the organization. For analysis purposes, the selected top event should be well defined with a description of the event, systems and processes involved, where and when it occurs, and elements associated.

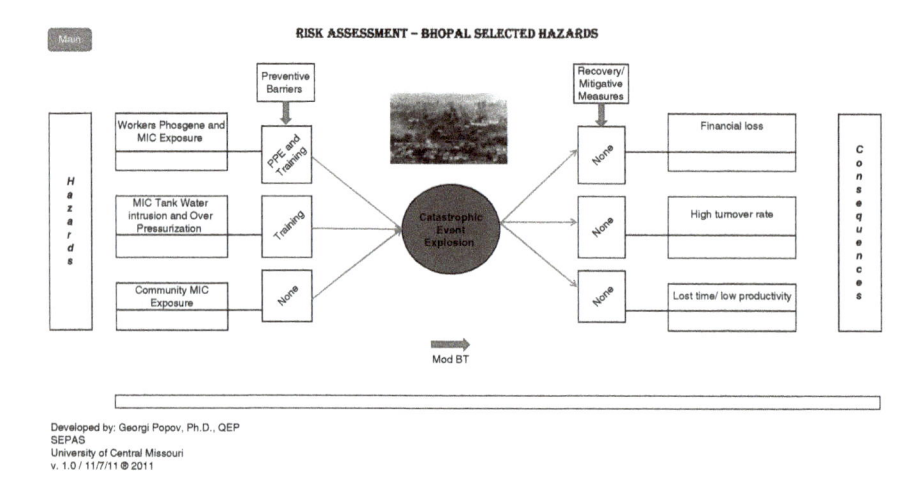

Figure 9.2 Bhopal Disaster: Conventional Bow-Tie Diagram

2. *Identify hazards* – Hazards that can lead to the selected hazard scenario are identified and listed in the far left-hand side of the Bow-Tie diagram. A "conventional" Bow-Tie analysis does not include an option to "quantify" the risks from the hazards and business risk or consequences. Therefore, it is important to identify significant and plausible hazards to be included in the analysis. There are a variety of hazard identification methods available, including preliminary hazard analysis (PHA), failure mode and effects analysis (FMEA), hazard and operability (HAZOP) study, what-if and checklist methods, and risk assessment matrix (RAM), among others. Similar tools and methodologies can be used to identify and document the hazards, to perform semiquantitative risk assessment, and to include results in the Bow-Tie diagram.

3. *Identify causes* – For each hazard, identify potential causes (also referred to as triggers or threats) that could lead to the hazard scenario. Trigger mechanisms can be singular or multiple and require experienced team members to perform a degree of research, review, and brainstorming to adequately identify pertinent causes. Causes are listed in the second column on the left-hand side of the diagram.

4. *Link causes to hazard scenario* – To connect hazard causes to the scenario, a line is drawn from each cause to the hazard scenario forming the left-hand side of the Bow-Tie.

5. *Identify escalating factors* – Conditions, factors, or failure modes that could escalate the hazard and/or reduce the effectiveness of existing controls are added to the hazard linkage in the diagram. Judgment is necessary in identifying only those potential failure modes that present a real weakness and potential for escalating the hazard.

6. *Identify preventive controls* – Existing preventive control measures designed to *prevent* the hazard from occurring are identified for each hazard and added to the diagram between the cause and the scenario. Preventive control barriers are considered "active" controls designed to prevent the exposure or release of a hazard, which leads to the hazard scenario. Examples of preventive controls might include a machine guard on a punch press to prevent contact with point of operation, a hotwork permit system to prevent fire from starting during welding, and concrete pylons around chemical storage tanks to prevent damage from forklifts.

7. *Identify consequences* – Potential consequences resulting from the scenario are identified and listed on the right-hand side of the Bow-Tie diagram. Lines are drawn to radiate out from the event scenario to each consequence. Potential consequences are defined as credible worst-case effects without any mitigation controls. A consequence should be defined to represent the particular harmful outcome that would be prevented by the set of safeguarding measures in place.

8. *Identify mitigating controls* – Existing mitigating controls for specific hazard exposures are identified and placed across the radial lines between the scenario and the consequence. These controls are considered "passive" in nature, protecting against or reacting to an exposed hazard to reduce severity of the consequence. Several examples of mitigating controls might include fire suppression/sprinkler systems to mitigate fire spread, secondary containment to contain a liquid chemical spill or release, and fall arrest system to reduce injury to the individual during a fall.

9. *Identify control influences* – Influences from management, engineering, operation, or maintenance activities on existing "active" and "passive" controls are identified and shown under the Bow-Tie with linkage to the respective control. For instance, machine interlock controls would be linked to engineering, energy isolation and lockout practices linked to maintenance, and operator training linked to operations or management activities.

A conventional Bow-Tie analysis is qualitative tool; however, some quantification is possible if pathways are independent, controls are reliable, and probability of a consequence is known. For pathways and controls that are not independent, or control effectiveness is unknown, quantification is less effective.

9.6 PRACTICAL APPLICATION

A Bow-Tie analysis primarily focuses on the control factors: preventive controls or barriers used to prevent the event from occurring and the mitigating controls to reduce the impact should the event occur. Oftentimes it is necessary to use multiple risk assessment tools to identify, analyze, evaluate, control, and communicate the risks. Bow-Tie diagrams are a useful component in such a process providing a clear visual of the pathways between hazards, controls, and consequences. In simple systems, a Bow-Tie analysis may be adequate on its own. In more complex events, several hazard analyses and risk assessment tools may be needed as shown in the following case studies.

9.6.1 Case Study #1: Spray Paint Operation

A spray paint operation in a manufacturing plant uses flammable liquid solvents in the paint thinning and cleaning process. The solvent is stored in 55 gal drums kept near the paint booth where it is periodically dispensed into smaller open containers. The solvent is 100% benzene with a flash point of −11 °C (12 °F) and a lower explosion level of 1.2% and an upper explosion level of 7.8%, making it highly volatile and flammable. Potential ignition sources from electrostatic charge accumulation can create a hazardous condition when handling this material. To minimize this hazard, bonding and grounding are necessary but may not by themselves be sufficient. Health hazards are also a significant concern with low permissible exposure limits including an American Conference of Governmental Industrial Hygienists (ACGIH) short-term exposure limit of 2.5 ppm, an ACGIH time-weighted average (TWA) of 0.5 ppm, and an OSHA TWA of 1 ppm.

The sequence of hazard analysis and risk assessment tools used in the spray paint booth case study is depicted in Figure 9.3.

Step 1. Identify hazards, causes, and controls – To initially identify hazards, their causes, and existing control strategies, a PHA (Chapter 7) is used. Brainstorming sessions, checklists, or other hazard identification techniques are used to identify specific hazards, their potential effects, the affected process steps and business units and existing controls. The following PHA in Figure 9.4 presents

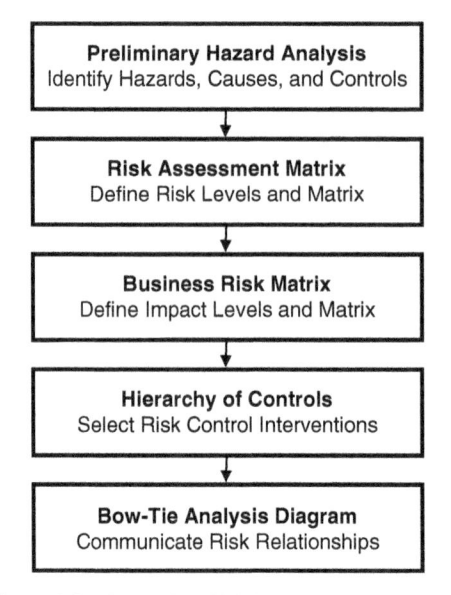

Figure 9.3 Example of Risk Assessment Sequence

PRACTICAL APPLICATION 187

Figure 9.4 Preliminary Hazard Analysis for Paint Booth Example

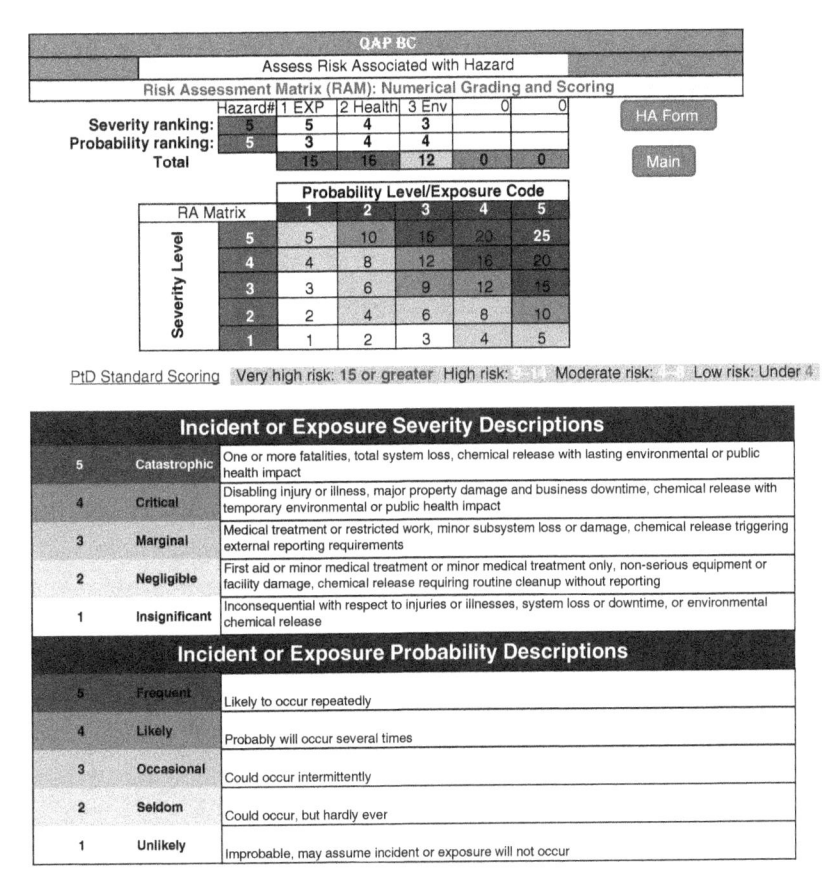

Figure 9.5 Risk Assessment Matrix Example. *Source:* Taken from ANSI/ASSE Z590.3-2011 (Courtesy of American Society of Safety Engineers)

the paint booth case study. An expanded version of the case study related to this chapter is available in Appendix 9.A.

Step 2. Define risk levels – To define operational risk factors and levels to be used in the assessment, a RAM with specific risk descriptions is selected. In Figure 9.5, a conventional two-factor RAM (severity level × probability of occurrence) from the ANSI/ASSE Z590.3-2011 PtD standard is presented. [Note: Other options exist including three or more risk factor systems, which produce a risk priority number (RPN) most often used in FMEA.] The following risk descriptions are adapted from ANSI Z590.3. 2011 Prevention through Design standard.

Severity of Consequence Level Descriptions

5. *Catastrophic*: One or more fatalities, total system loss, or chemical release with lasting environmental or public health impact
4. *Critical:* Disabling injury or illness, major property damage and business downtime, or chemical release with temporary environmental or public health impact
3. *Marginal:* Medical treatment or restricted work, minor subsystem loss or damage, or chemical release triggering external reporting requirements
2. *Negligible:* First aid or minor medical treatment only, nonserious equipment or facility damage, or chemical release requiring routine cleanup without reporting
1. *Insignificant:* Inconsequential with respect to injuries or illnesses, system loss or downtime, or environmental chemical release.

Probability of Occurrence Descriptions

1. *Unlikely:* Improbable, may assume incident or exposure will not occur
2. *Seldom:* Could occur, but hardly ever
3. *Occasional:* Could occur intermittently
4. *Likely:* Probably will occur several times
5. *Frequent:* Likely to occur repeatedly.

Step 3. Apply risk scores – Taking the three hazards listed in the PHA (Figure 9.4), the RAM severity and probability ratings are applied in Figure 9.5. The explosion hazard (1 EXP) RAM rating is determined as 15 or very high risk, the health hazard resulting from benzene exposure (2 Health) is rated at 16 or very high risk, and the environmental hazard (3 Env) is rated at 12 or high risk.

Step 4. Transfer risk factors to PHA – These risk factors (severity × probability) are transferred to the risk factor (RF) column in the PHA worksheet shown in Figure 9.6.

Step 5. Assess business risks – A business risk assessment matrix (BRAM) similar to the RAM is used to define business impact risk levels. However, the "severity" rating is replaced with "extent of impact" on business, and the "probability" rating is replaced with "likelihood of business losses" as shown in Figure 9.7.

Extent of Business Losses Impact Descriptions

1. Insignificant: Inconsequential with respect to business losses
2. Negligible: Minor business losses
3. Marginal: Business losses triggering external reporting requirements
4. Critical: Business downtime, significant business losses, or corporate image impact
5. Catastrophic: Unsustainable losses, total business loss, or inability to continue business operations.

Likelihood of Business Losses Descriptions

1. Unlikely: Improbable, may assume business loss will not occur
2. Seldom: Could occur, but hardly ever
3. Occasional: Could occur intermittently
4. Likely: Likely to occur several times
5. Frequent: Likely to occur repeatedly.

PRACTICAL APPLICATION 189

Figure 9.6 Preliminary Hazard Analysis with RF Column

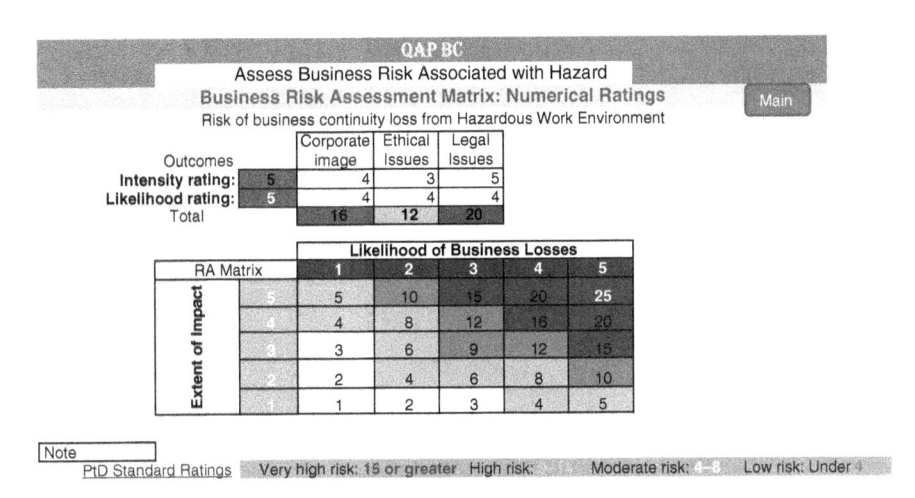

Figure 9.7 Business Risk Assessment Matrix

Step 6. Apply results to Bow-Tie diagram – As stated earlier, a Bow-Tie diagram is used to communicate the relationship between hazards, controls, and consequences. Once the hazards have been identified and prioritized using the simplified RAM (Figure 9.5), the results are transferred to a Bow-Tie analysis diagram. In a "conventional" Bow-Tie analysis, only "qualitative" risk descriptions are used. However, by incorporating semiquantitative risk factors for severity and probability as shown in Figure 9.8 example, a modified Bow-Tie can be used.

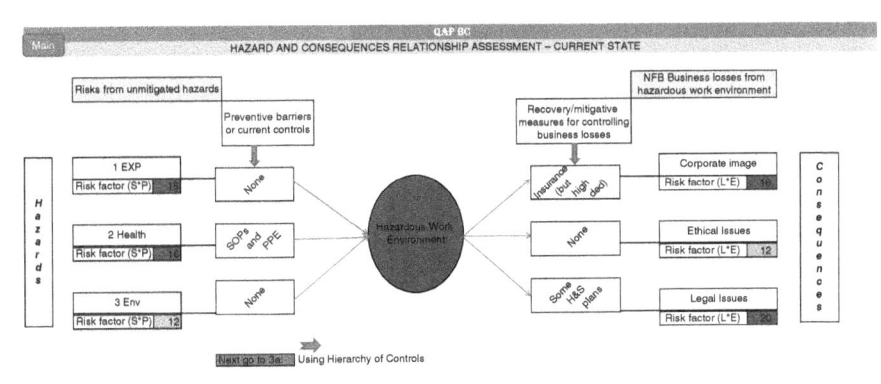

Figure 9.8 Modified Bow-Tie Diagram

Step 7. Select controls – A hierarchy of controls is used in evaluating and selecting the most effective risk control interventions for each hazard. The model presented in Figure 9.9 is based on the concepts described in the ANSI Z590.3 PtD standard.

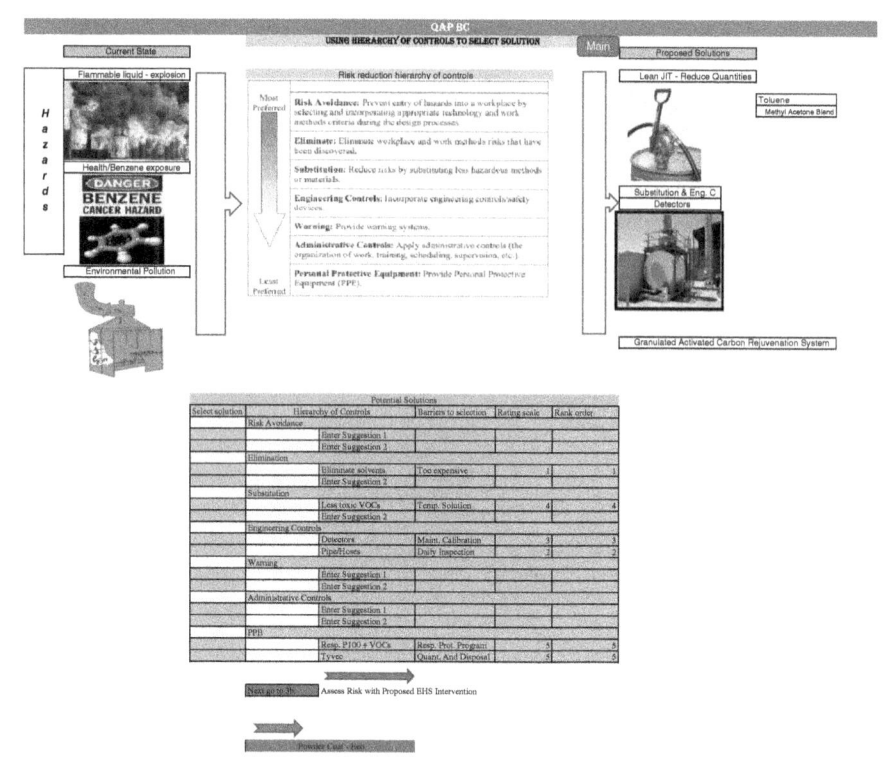

Figure 9.9 Hierarchy of Controls. *Source:* Taken from ANSI/ASSE Z590.3-2011 (Courtesy of American Society of Safety Engineers)

For this exercise, the proposed solutions will include minor modifications of the process:

- (1 EXP) Explosion hazard intervention – To reduce the risk of explosion hazards from the flammable liquids, a hose/piping system to ensure just-in-time (JIT) delivery of the flammable liquid is selected. This control measure is designed to deliver the exact amount of solvents from the storage room to the paint booth instead of storing the 55 gal drums of flammable solvents in the booth. By employing this measure, the quantity of flammable liquid present in the paint booth system is significantly reduced.
- (2 Health) Health hazard intervention – To address the health hazards from benzene solvent exposure, a less hazardous mixture is substituted. Less toxic chemical options that can be substituted include a blend of toluene and methyl acetone, or methyl acetate (MeOAc). MeOAc is a fast-evaporating, active solvent that can be used with a broad range of coating and ink resins.
- (3 Env) Environmental hazard intervention – To address the environmental pollution hazard, the company can install granulated activated carbon system to capture VOCs and hazardous air pollutants (HAPs).

Step 8. Apply controls – By addressing the hazards and applying prevention through design principles, the risks can be substantially reduced as indicated in the adjusted risk levels in Figure 9.10. The new risk-level estimates for the three top hazard events are entered in the RAM. The adjusted risk rankings found in the RAM are listed as follows:

PRACTICAL APPLICATION

- 1 EXP – Explosion hazard rating is reduced from 15 (very high) to 10 (high).
- 2 Health – Health hazard exposure is reduced from 16 (very high) to 8 (moderate).
- 3 Env – Environmental hazard is reduced from 12 (high) to 6 (moderate).

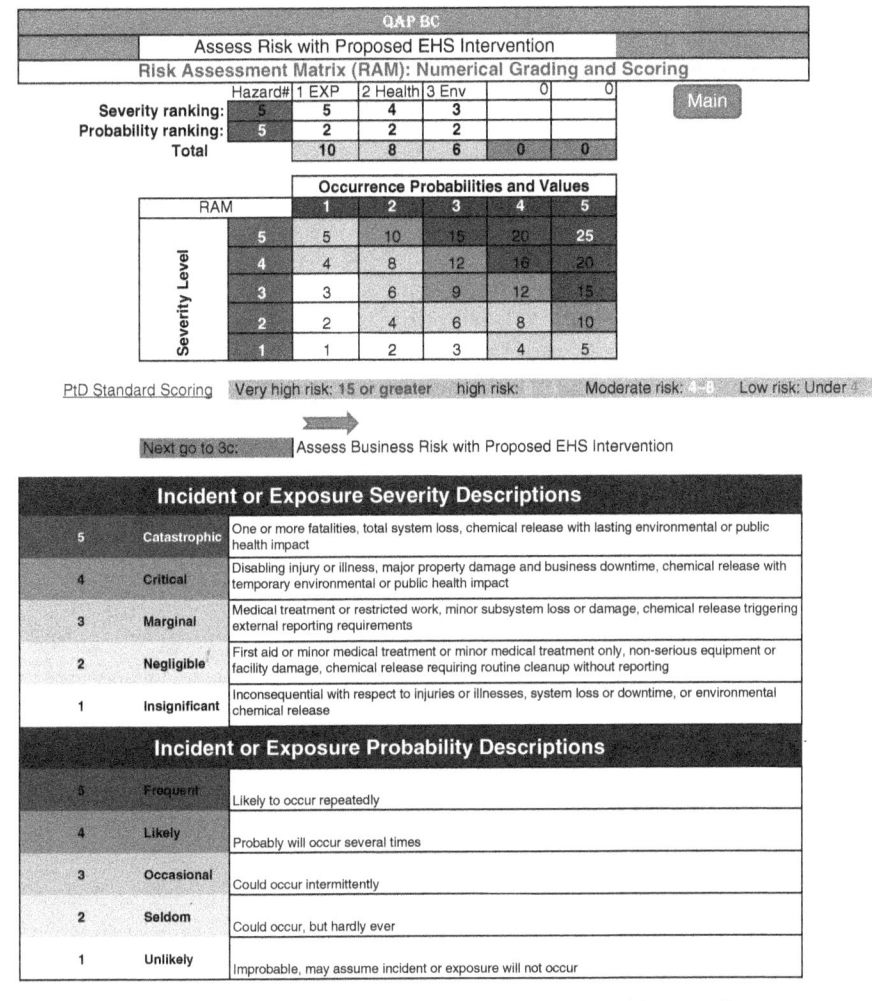

Figure 9.10 Risk Assessment Matrix with Proposed Interventions

After selecting the control measures and estimating the resulting risk levels, the new risk factors (RF) are transferred to the updated PHA shown in Figure 9.11. Notice that the RF column rankings reflect the new risk level after the proposed intervention.

Figure 9.11 Preliminary Hazard Analysis with Updated RF Scores

Step 9. Reassess business risks – Similarly, business risk impact levels for the organization's (1) corporate image, (2) ethical issues, and (3) legal issues are evaluated using an updated business RAM. Logically, the business risks will be reduced with the proposed interventions. An example of the three business continuity risk ratings after the proposed changes is presented in Figure 9.12.

Step 10. Transfer results to modified Bow-Tie – The hazards with proposed intervention have been reevaluated and prioritized with the simplified RAM. The results are then transferred to a new "modified" Bow-Tie diagram as shown in Figure 9.13 to better communicate the risk pathways. Overall, the proposed SH&E intervention will lead to risk reduction. However, minor modifications of the process may not be sufficient. The risk assessment team may suggest more drastic changes. For instance, one suggestion may be to completely overhaul the operation and introduce powder coating (substituting a less hazardous process). Powder coating does not require a solvent to keep the binder and filler parts in a liquid suspension form. Therefore, elimination of solvents will lead to substantial risk reduction. However, experienced safety professionals will suggest a new risk assessment be performed, since powder coating operations represent different hazards. [The readers and instructors are encouraged to develop risk assessment methodology for powder coating operations. An example is included in the interactive Bow-Tie Excel tool.] (An example is provided in the companion website)

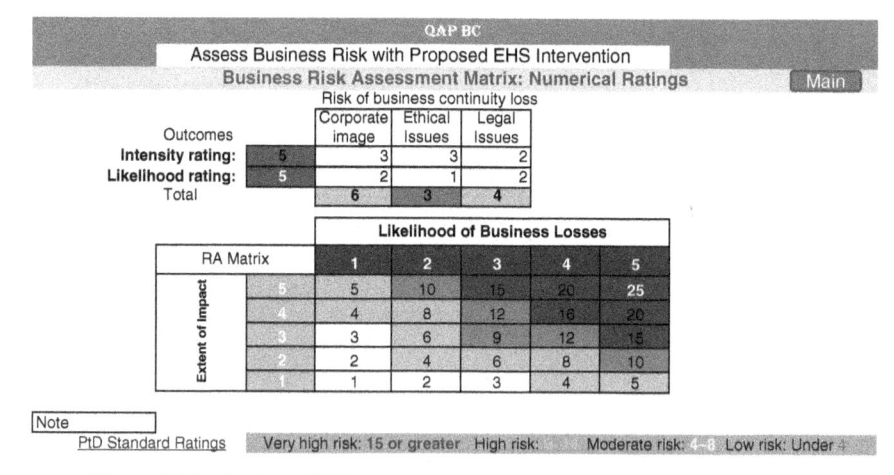

Figure 9.12 Business Risk Assessment Matrix after the SH&E Interventions

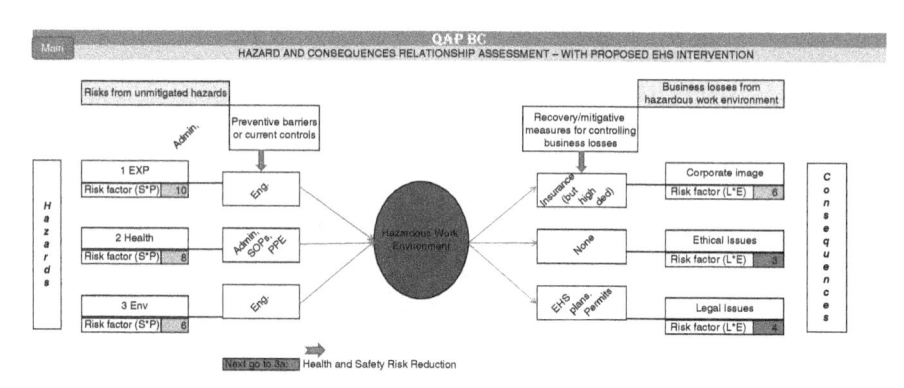

Figure 9.13 Modified Bow-Tie Diagram with Proposed SH&E Interventions

PRACTICAL APPLICATION 193

9.6.2 Case Study #2: Bhopal Disaster

The Bow-Tie method can be combined with other techniques to address more sophisticated risks. To demonstrate this concept, a FMEA and Bow-Tie are combined in the following Bhopal disaster case study. In this example, a FMEA rating scale with numerical grading is used for severity (S), probability occurrence (O), and prevention effectiveness (PE) as described in Chapter 8. Regarding PE scales, it should be recognized that higher effectiveness translates to lower risk and therefore a lower PE number. Likewise, less effectiveness of controls will result in higher risk and a higher PE number. Therefore, insufficient preventive measures will result in the higher PE scores, leading to higher RPNs and higher risk. For consistency, it is suggested that the same 1–5 rating scale be used as in the previous chapters. Examples of severity, probability, and prevention effectiveness rating scales are provided in Figure 9.14 and risk factor descriptions in Figure 9.15.

Severity:	Probability/Occurrence:	Prevention Effectiveness:
1 – 5 scale. Where:	1 – 5 scale. Where:	1 – 5 scale. Where:
1. Insignificant	1. Unlikely	1. Avoid, eliminate, substitute
2. Negligible	2. Seldom	2. Engineering control
3. Marginal	3. Occasional	3. Warning, administrative
4. Critical	4. Likely	4. PPE
5. Catastrophic	5. Frequent	5. None

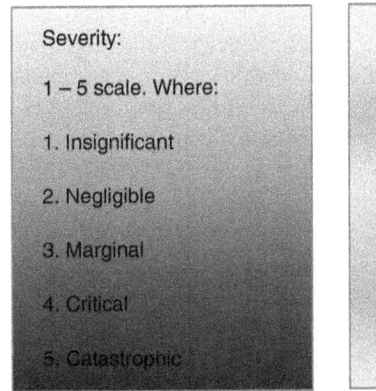

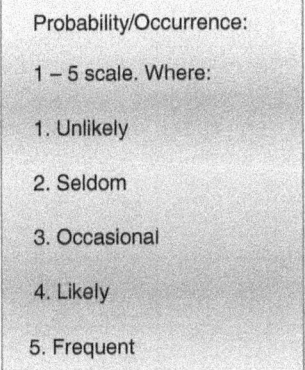

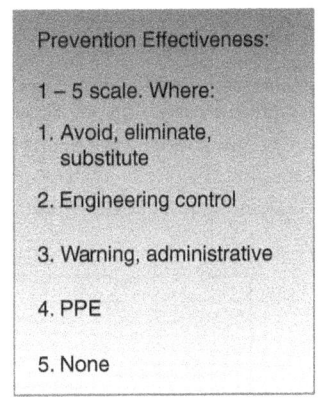

Figure 9.14 Risk Factor Rating Scales Example

Suggested scoring methodology	
Incident or Exposure Severity Descriptions	
Catastrophic (5)	One or more fatalities, total system loss, chemical release with lasting environmental or public health impact
Critical (4)	Disabling injury or illness, major property damage and business downtime, chemical release with temporary environmental or public health impact
Marginal (3)	Medical treatment or restricted work, minor subsystem loss or damage, chemical release triggering external reporting requirements
Negligible (2)	First aid or minor medical treatment or minor medical treatment only, nonserious equipment or facility damage, chemical release requiring routine cleanup without
Insignificant (1)	Inconsequential with respect to injuries or illnesses, system loss or downtime, or environmental chemical release
Incident or Exposure Probability Descriptions	
Frequent (5)	Likely to occur repeatedly
Likely (4)	Probably will occur several times
Occasional (3)	Could occur intermittently
Seldom (2)	Could occur, but hardly ever
Unlikely (1)	Improbable, may assume incident or exposure will not occur
Prevention Effectiveness	
Poor (5)	None
Low (4)	PPE
Moderate (3)	Warning, Administrative Controls
Better (2)	Engineering Controls
Best (1)	Avoid, Eliminate, Substitute

Figure 9.15 Risk Factor Descriptions Example

A combination of both the risk level (S×O) and RPN (S×O×PE) for the Bhopal disaster is presented here. Only three hazards were selected for evaluations, which are presented in Figure 9.16:

1. *Sevin (pesticide chemical) hazard* – The severity (S) resulting from worker exposure to Sevin is rated at 4 (critical), as the exposure may result in disabling injury or serious illness. Probability of occurrence (O) is rated at 4, since exposure will "probably occur several times." Prevention effectiveness (PE) is rated at 3 (moderate), since some administrative controls were available as well as personal protective equipment (PPE).
2. *Methyl isocyanate hazard* – Similarly, methyl isocyanate (MIC) storage severity is rated at 5 (catastrophic) due to explosion hazards, possibility of total system loss, or a chemical release with lasting environmental and public health impact. However, probability of occurrence is rated at 2 (seldom). Prevention effectiveness is rated at 3 (moderate), since warning signs were in place.
3. *MIC reaction/release hazard* – MIC reaction resulting a release of Sevin to the community is considered in this case. Severity is rated at 4 (critical) due to the distance to the populated area. Probability of occurrence is rated at 1 (unlikely). However, prevention effectiveness is rated at 5 (none), since no preventive measures for community exposures were considered (Ref: http://www.lakareformiljon.org/images/stories/dokument/2009/bhopal_gas_disaster.pdf) (Eckerman, 2001).

FMEA & RPN WORKSHEET									
Part or Process Name	Sevin Production	Suppliers and Deptartments Affected				Prepared By			
Design/Mfg Responsibility		Model Date				FMEA Date			
Other Areas Involved		Engineering Change Level							
Process Operation, Function, or Purpose	Potential Failure Mode	Potential Effect(s) of Failure	SEV	Potential Cause(s) of Failure	OCC	Current controls evaluation method	PE	S×O	RPN
Sevin	Worker exposure	Respiratory issues	4	Improper PPE	4	PPE and training	3	16	48
MIC storage	MIC leak	Explosion	5	Overpressurization	2	Clean lines use PPE	3	10	30
MIC reaction to produce sevin	Community MIC exposure	Health issues	4	MIC leak	1	Training	5	4	20

Figure 9.16 Bhopal Disaster FMEA and RPN Example

Once hazards have been identified and prioritized with the FMEA, the Bow-Tie method can be applied to further assess risks and provide a framework for demonstrating their consequences. Bow-Tie diagrams can be modified to include risk-level scores and RPN scores to provide a semiquantitative aspect to the risk communication tool. In addition, a modified layers of protection analysis (LOPA) can be added to the Bow-Tie diagram. Figure 9.17 provides an example of a modified Bow-Tie risk assessment for the Bhopal disaster example.

Notice that below the preventive barriers or controls, in the modified Bow-Tie analysis, there are added layers of protection that were either (1) existent but not operational or (2) nonexistent. For instance, the following controls were existent but not operational:

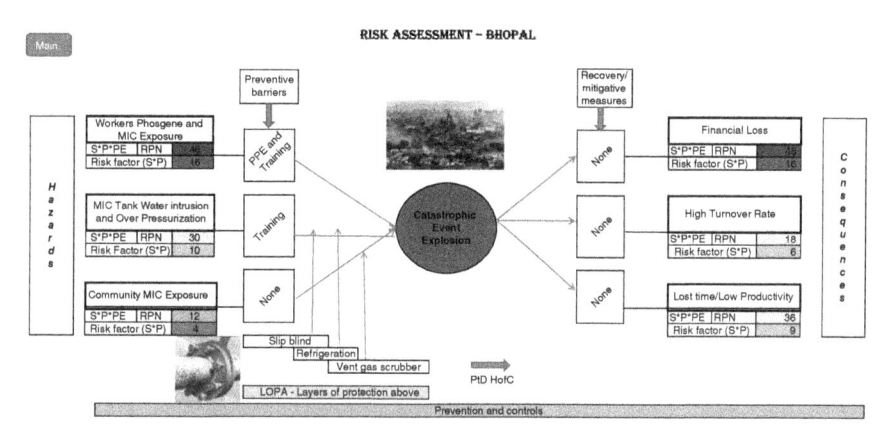

Figure 9.17 Modified Bow-Tie with RPN and LOPA Example

- A slip blind was to be inserted in order to prevent water from entering the storage tanks as required by the training procedure (administrative control). However, the slip blind was not inserted due to corrosion buildup (escalating factor), making it difficult to insert and to save time.
- The refrigeration system (engineering control) was not operational at the time of the accident, which indicates a lack of proper maintenance (escalating factor).
- The vent gas scrubber (engineering control) was switched off for repair and was too small by the design for the system.

In addition, no reactive or mitigating controls were identified to reduce the harm of the event, thus allowing the consequences to be catastrophic.

9.7 SUMMARY

The Bow-Tie analysis method enables visualization of the relationship between hazards and their causes, existing preventive measures, the event scenario, and mitigation measures to limit their consequences. To put simply, it provides a "30,000 ft." overview of the evaluated system and its controls.

The structure of the Bow-Tie approach forces an assessment of how effectively preventive barriers and controls are on preventing the hazard from occurring and how well prepared the organization is to recover should undesirable event occur. The method is relatively simple and frequently identifies gaps and issues that are missed by other techniques.

Bow-Tie methodologies are extremely versatile and are used for various applications in most all industries. However, it is highly recommended that a multidisciplinary team is formed to successfully apply the risk assessment process early in the design phase. Such multidisciplinary team will identify the potential hazards and potentially evaluate the proposed barriers based on the hierarchy of controls. As demonstrated in the Bhopal example, more than one barrier per hazard may be proposed. Furthermore, multiple mitigation measures may be proposed to limit undesirable event consequences.

REVIEW QUESTIONS

1. Summarize Bow-Tie analysis from a historical prospective.
2. Discuss the uses of Bow-Tie assessment methodology.
3. Explain the importance of the existing "barriers" to reduce the risk and minimize consequences.

4. Describe how Bow-Tie analysis can be used to qualitatively or semiquantitatively demonstrate controls, countermeasures, and risk reduction measures.

5. Compare Bow-Tie analysis, preliminary hazard assessment, and failure mode and effects analysis.

6. Identify possible members of the multidisciplinary risk assessment team.

7. Examine some of the advantages of Bow-Tie analysis compared to other risk assessment methods.

REFERENCES

ANSI/ASSE Z590.3-2011. *Prevention through Design: Guidelines for Addressing Occupational Hazards and Risks in Design and Redesign Processes*. Des Plaines, IL: American Society of Safety Engineers, 2011.

ANSI/ASSE Z690.3-2011. *American National Standard – Risk Assessment Techniques*. Des Plaines, IL: The American Society of Safety Engineers, 2011.

Eckerman, I. *Chemical Industry and Public Health Bhopal as an Example*. Available at: http://www.lakareformiljon.org/images/stories/dokument/2009/bhopal_gas_disaster.pdf (accessed February 7, 2016), 2001.

Gill, D. 1979. ICI place Hazan Course Notes, presented at The University of Queensland, Australia.

HSSE in Shell, 2010. Available at: http://www.shell.com/content/dam/shell/static/environment-society/downloads/safety/hsse-in-shell-lr.pdf (accessed February 7, 2016).

Reason, J. *Managing the Risks of Organizational Accidents*. Burlington, VT: Ashgate, 1997.

APPENDIX 9.A

QAP CORPORATION – ANNUAL REPORT

QAP Corporation case study is an expanded version of Chapter 7 spray paint practical example. Use the case study to follow Bow-Tie Excel risk assessment (RA) tool steps. Read the annual reports in the following report, and think about the hazards associated with oil-based spray painting. Also, think about the business consequences. The tool provides risk assessment of base case/current state and one comparison case.

BUSINESS MANAGER ANNUAL REPORT

As I reflect on the decades since the founding of QAP Corporation, I can say without hesitation that these last few years were one of the most challenging in our history. Faced with the current global economic downturn many of us have ever experienced and increased competition across our markets, our people delivered results that were as impressive considering the business conditions. We were informed by our WC insurance carrier that we are approaching EMR of 1. We had seven reportable injury and illnesses last year. Our IIR is significantly higher than our competitors.

Injury And Illness Incidence Rate Calculator and Comparison Tool

Year:	2012
Area:	Private Industry, All U.S.
Supersector:	Manufacturing
Industry:	Fabricated metal product manufacturing

Case Type	Your Establishment	Private Industry, All U.S.
Total	12.3	5.7
Days Away	3.5	1.5
Job Transfer/Restriction	5.3	1.3
DART	8.8	2.9

Direct and indirect costs related to the injuries and illnesses were estimated at $357,789. Obviously changes are necessary. We can't continue to operate that way.

Estimated Total Cost

The extent to which the employer pays the direct costs depends on the nature of the employer's workers' compensation insurance policy. The employer always pays the indirect costs.

Injury Type	Instances	Direct Cost	Indirect Cost	Total Cost	Additional Sale (Indirect)	Additional Sale (Total)	
Respiratory Disorders (gases, fumes, chemicals, etc.)	3	$ 92,136	$ 101,349	$ 193,485	$ 1,447,851	$ 2,764,071	Remove
Carpal Tunnel Syndrome	1	$ 29,647	$ 31,511	$ 60,158	$ 450,167	$ 859,400	Remove
Dermatitis	2	$ 18,588	$ 22,304	$ 40,892	$ 318,650	$ 584,170	Remove
Poisoning - Chemical (other than metals)	1	$ 30,121	$ 33,133	$ 63,254	$ 473,330	$ 903,628	Remove

Totals

Estimated Direct Costs:	$ 169,492
Estimated Indirect Costs:	$ 188,297
Combined Total (Direct and Indirect Costs):	$ 357,789
Sales To Cover Indirect Costs:	$ 2,689,998
Sales To Cover Total Costs:	$ 5,111,269

Unfortunately, we had a State Department of Health and Environment inspection that led to severe fines. In addition, they set strict deadlines to complete the recommended corrective measures. We were notified that if we continue to release uncontrolled VOC emissions, we'll have to close the operations next year.

Operating Status:	X	HPV Flag:	
Operating Status Description:	Notification for closure	State Registration Number:	
State County Compliance Source:	2910100013	Government Facility Code Description:	PRIVATELY OWNED/ OPERATED
Region Code:	7	Class Code:	B
Primary SIC Code:	3999	Class Code Description:	Potential uncontrolled EM 9
Primary SIC Description:	MANUFACTURING INDUSTRIES	Compliance Status:	Compliance Violation
NAICS Code:	325998	Compliance Status Description:	Notified of potential shut down
NAICS Code Description:	All Other miscellaneous chemical product and preparation Manufacturing	Date Plant Information Last Updated:	10/24/2013

Pollutant Data

Air Program Code	Pollutant Code/CAS Number	Pollutant/CAS Description	Attain Indicator	Attain Indicator Description	Pollutant Compliance Status	ES Pollutant Compliance Description	Pollutant Class Code	Pollutant Class Description
3	VOC	Volatile organic compounds	A	Nonattainment area	9	Compliance violations – notified of potential SHUT down	B	Potential uncontrolled EM

In addition, our turnover rate is unsustainable. Our employee turnover rate last year was 54.28%.

QAP BC				
Employee Turnover Rate				
Month	Beginning E	New Hires	VQ	End E
January	79	10	7	82
February	82	2	7	77
March	77	4	1	80
April	80	6	3	83
May	83	5	1	87
June	87	5	2	90
July	90	2	5	87
August	87	3	5	85
September	85	2	6	81
October	81	4	2	83
November	83	10	5	88
December	88	8	2	94
Mean # of employees=				84.75
VQ=				46
Employee turnover=				54.28%

where VQ is voluntary quit/voluntary leaving/separation. End E is end year employees.

Our quality control (QC) manager determined that only 72% of the parts are coated properly leading to significant losses. Overall labor efficiency is just above 54%.

As a result, our R&D team recommended changes in the process.

I'm sure we will deliver on our EHS and financial commitments and emerge stronger and well positioned for a sustainable growth.

Current Year

As you very well know, we were acquired by XYZ Corporation. XYZ Corporation is ISO 14001 and OHSAS 18001/ANSI/AIHA Z10 certified. We are tasked with achieving ISO 14001 and ANSI/AIHA

Z10 certification within three years. Our new management believes that "worker safety is a keystone habit – a habit that can set off a chain reaction. And by changing that, he could actually transform the company."

In addition, our parent company strengthened core businesses and invested in the launches of a number of recently approved innovative products. We also continue to play a role in helping to shape automotive industry policy around the world. The future of our core products is promising and exciting. We have formed a team of professionals, who are diligently working to address all EHS-related challenges.

Moving Ahead

Every difficult period brings with it a corresponding opportunity for growth. Despite a challenging year, we believe, we'll be stronger next year. We have outstanding new products, robust pipelines, and talented employees working in a streamlined organization with more resources for growth. Our EHS team is working with our parent company engineers, accountants, HR, supervisors, and **employees**, to develop even better solutions.

The global automotive market is expected to grow almost 5%/year over the next 5 years. Our EHS objectives and business strategies are aligning with many evolving trends in automotive and the plastics industries.

I believe that the brightest and most innovative automotive companies – with dedicated people who care about our **business**, **employee**, **community**, product safety, quality, and sustainable development – will thrive in this evolving and still-changing environment. QAP Corporation will be one of these companies.

Growth Priorities and Business Objectives

QAP Corporation has tremendous opportunities for growth: our employees, products, pipeline, and now global presence. Our unwavering operating model includes a commitment to being broadly based in automotive and the plastics industries, a decentralized management approach that keeps our people close to customers, managing for the long term, and a focus on people and values.

Within our new strategic framework, we galvanize our organization around high-level business objectives that reflect the changing global environment. These provide leaders with a common set of growth priorities:

- Our growth has always been based on sustainable business model, EHS excellence, and innovative quality products that serve customer needs in a meaningful way.
- A mix of internal and external sources to sustain a robust pipeline of new products that provides a competitive advantage.
- Maintain a global presence and continue to expand our presence in an appropriate way for diverse markets and customers. Our approach will be strategic, sustainable, effective, and cost efficient to address local and global needs.
- The hallmark of QAP Corporation is our talented employees. Make sure we have the right people and core values in place to help this company excel. Our ability to develop, challenge, motivate, and reward a diverse workforce is our cornerstone for sustained growth and increased market share.

- We are committed to EHS excellence, managing our products' life cycle in a social responsible manner; protecting our workers, customers, and the public; producing safer products; and maintaining sustainable operations.
- We use our values to build financial success, environmental excellence, stewardship, and social responsibility in order to deliver net long-term benefits to our shareholders, employees, customers, suppliers, and the communities in which we operate.
- We will continually strive to improve our operations and products to protect our environment and resources for future generations.

QAP CORPORATION WORKSHOP – NEW PROCESS DEVELOPMENT

Quality Auto Parts: Painting Operation

Description of Operation QAP Corporation is a company that produces high-quality aluminum and plastic parts for the automotive industry. The company stores oil-based paint in 55 gal drums in the storage room. Flammable solvents are also stored in the storage room. The solvents contain up to 5% of benzene. The paint and the solvent drums are moved with a forklift truck to the conventional spray booth (OSHA Forklift S&H Topic).

Source: OSHA Forklifts: https://www.osha.gov/SLTC/poweredindustrialtrucks/standards.html.

Conventional air spray guns are the standard spray equipment used to apply coatings in the automotive refinishing industry. The employees are using this type of spray gun.

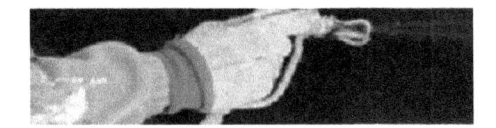

Source: OSHA Spray Operations: https://www.osha.gov/SLTC/sprayoperations/.

A low volume (2–10 cubic feet per meter (cfm)) of air is pressurized and forced through a nozzle; the paint is atomized in the air at the nozzle throat. The spray guns are operated with air pressures of 30–90 pounds per square inch (psi) at a fluid pressure of 10–20 psi. The air is supplied by air compressors during spraying operations. There are two basic types of conventional spray guns: siphon fed and gravity fed. The company is using siphon-fed guns, where the paint cup is attached below the spray gun, and the rapid flow of air through the gun creates a vacuum that siphons the paint out of the cup.

QAP CORPORATION WORKSHOP – NEW PROCESS DEVELOPMENT

The coating operators are using only N95 respirators and safety glasses. However, the glasses are "coated" with paint after just 10 min of use. The spray paint operators are complaining that they can't see very well the parts, and the QC manager determined that only 72% of the parts are coated properly. Overall labor efficiency is 54.08%.

Overall Labor Effectiveness (OLE) Tool					OLE = Availability × Performance × Quality			
Availability = Time operators are working productively / Time scheduled						Quality = QC parts / Total parts produced		
Example	Operators	Scheduled Time/min	Downtime	Total Sch. Time		Example	QCd Units	Started Units
	10	450	0	4500		See F13	4500	6250
	10	450	60	3900	0.8	Quality =	0.72	
				0.866666667		Quality % =	72.00%	
Availability % =				86.67%				

Performance = Actual output of the operators / the expected output						
Example	Unit/Parts/hr	Unit/Parts/min				
	60	1				
Example	Operators	Scheduled Time/min	Downtime	Available Time	Total Units	
	10	450	0	4500	4500	
					Actual Units	
	10	450	60	3900	3900	
				0.866666667	0.866667	
Performance % =				86.67%	86.67%	

OLE=	0.5408
OLE %=	54.08%

The following diagram presents the sequence of the process.

A modified fish-bone diagram of the process is provided for simplicity and visualization.

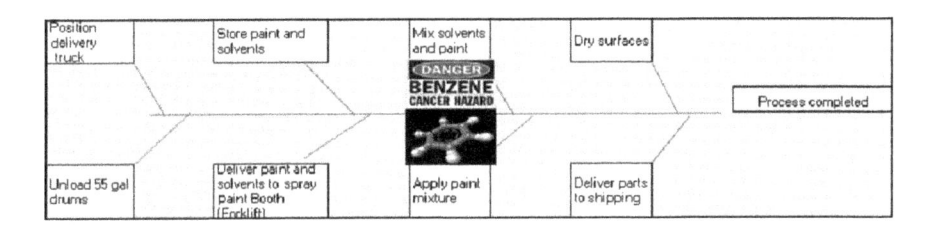

Suggested changes for next year: Our county was designated as a nonattainment area by EPA. Therefore, VOC control has a very high priority. The allowable solvent content in surface coating formulations used by spray painting operations will be progressively reduced by legislative pressure. In addition, our R&D team developed lower-VOC emission formulations. Furthermore, we are going to substitute benzene with toluene and methyl acetone blend and improve paint booth operators PPE. It is estimated that this new process will reduce our process cycle efficiency (PCE) by 5.0%. However, our parent company IH department assured us that we will significantly reduce the risk of exposure.

Other changes in the process: We are not going to use fork lift to deliver 55 gal drums to the painting booth anymore.

Source: OSHA Forklifts: https://www.osha.gov/SLTC/poweredindustrialtrucks/standards.html.

Instead, we'll be using a strictly controlled piping system to deliver just-in-time solvents and paint from the storage room. We will also install carbon adsorbers (granulated activated carbon rejuvenation system) to reduce VOC emissions.

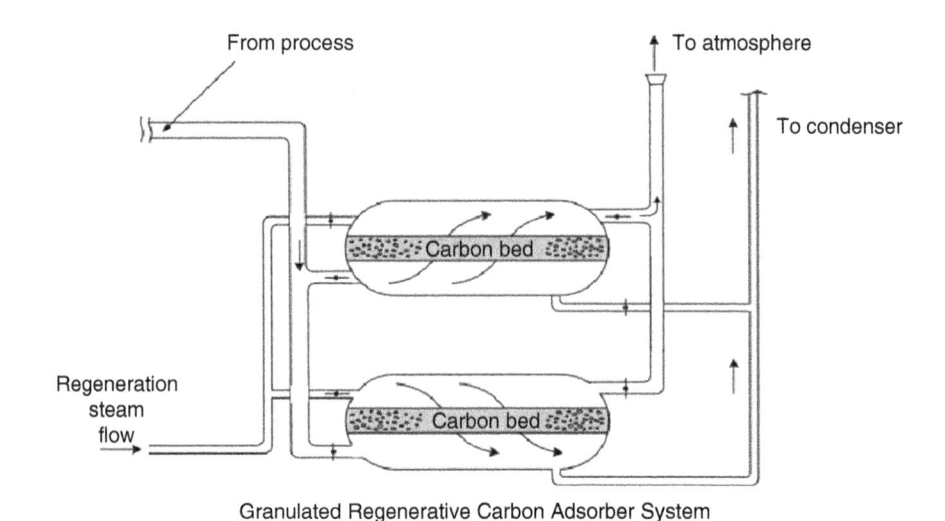

Granulated Regenerative Carbon Adsorber System

Source: EPA http://www.epa.gov/ttnchie1/mkb/documents/Regenerative%20Carbon%20Adsorber_5-5-05.pdf.

New process diagram.

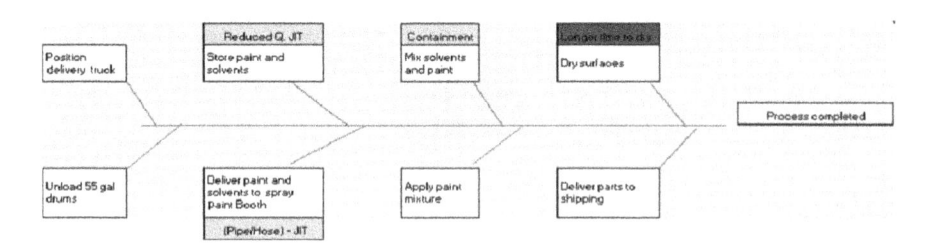

Risk Assessment Strategies and Nonfinancial Benefits

The current state risk assessment indicates that the workers were exposed to organic compounds, there is a risk of potential explosion due to solvents storage, and uncontrolled VOC emissions are of a serious concern for the community. The current controls do not adequately protect the workers during this high-risk operation.

We will apply risk assessment strategies, evaluate nonfinancial benefits (NFB), calculate risk reduction, and incorporate PtD principles.

Hazard Intervention

The company formed a project team to determine worker exposure control methods that also meet the requirements for EHS excellence, sustainable development, operability, cost containment, and worker risk minimization.

The project team interviewed operations management to develop a set of wants and needs. Two possible interventions were proposed. The project team then utilized DMAIC model to further develop their understanding of the project requirements. The project team summarized the new process for each of the two possible interventions, including the potential NFB impacts. Bow-Tie risk assessment tools were developed to visualize the processes, identify SH&E intervention opportunities, and calculate risk reduction and residual risk (**R2 and R2**).

Beyond Compliance: QAP Case Study Risk Assessment Tool

Bow-Tie risk assessment (RA) will enable the users to:

1. Implement the steps of the RA process
2. Identify and prioritize organizational hazards
3. Establish risk reduction (R2) plans that will align with the business goals of the organization
4. Influence the management to support R2 plans
5. Measure the effectiveness of the RA process and outcomes of the goals

Implement the steps of the RA process – Tools to address all three steps of the RA process were developed. The tools are based on PtD and ISO 31000 series standards.

Identify and prioritize organizational hazards – A modified PHA tool was developed to identify and prioritize the top three safety, health, and environmental (SH&E) hazards identified in the case study.

Establish risk reduction (R2) plans that will align with the business goals of the organization – Business objectives (BO) prioritization tool, develop SH&E intervention implementation plan, and later identify the impacts on BO.

Influence the management to support risk reduction (R2) plans – This is a critical area and SH&E professionals will have to learn how to influence the management. Based on our experience, the best way to influence the management is to develop R2 and business case for SH&E intervention. Numerous tools are available. However, there is no universal tool to collect cost associated with injuries and illnesses. Different tools are presented in this chapter and Chapter 19. They can be expanded or you can skip some of the tools/steps.

Measure the effectiveness of the RA process and outcomes of the goals – This is another very important requirement. To measure the effectiveness of the RA process, we have to look at the standards that were referenced in this book. ANSI Z 10 provides suggestions for the following categories: *R2, productivity, financial performance, quality,* and *other business objectives*. RA tools used in this project are derived from the PtD standard and Bow-Tie tool is described in greater details in ISO 31010. Some of the tools were modified to provide visualization. The R2 and R2 tool is simple percent risk reduction calculation and residual risk calculation. In order to address productivity, two different tools were used: overall labor effectiveness (OLE) and PCE. Both productivity tools are modified Lean tools. It is a well-known fact that SH&E professionals do not take enough credit for improved productivity and quality improvements. Therefore, OLE and PCE tools are included to address some quality issues. Financial performance will be addressed in Chapter 19 of the book. Cost/benefit analysis tool will be included in Chapter 19. In addition, the financial analysis tool was developed to calculate the financial benefits (FBs) of the SH&E intervention. Other business objectives may be addressed by capturing the NFB of the intervention.

"*Measure the effectiveness of the RA (ERA) process*" equation can be presented as

$$ERA = RR + FB + NFB + PCE + OLE + ETR$$

The RA tool was developed to help SH&E students and early career professionals diversify their knowledge and become important members of the decision-making team. The RA tool uses five steps to determine and illustrate the benefits and business value of SH&E projects, programs, and activities designed to eliminate, minimize, or mitigate the risks to reduce the (associated occupational injuries and illnesses) or SH&E risk in general. RA tools are based on PtD and ISO 31000 series standards. ANSI Z 10 requires linking SH&E objectives to financial considerations (p. 57). Provided in the standard is the blueprint for widespread benefits in H&S, as well as *productivity, financial performance, quality*, and other business objectives. In order to address the benefits of RA and SH&E interventions, additional tools that provide a blueprint for integration of ANSI Z10, ISO 31000, and PtD standards had to be developed. We encourage the readers to use a variety of RA techniques listed in ISO 31010 and modify them as needed. A system approach for integration of various methodologies can be presented as a final project.

The particular project addresses initial RA, suggested controls (according to the PtD hierarchy of controls), two options for SH&E interventions (including two additional RA), productivity (PCE), quality (OLE), FB and NFB of the proposed interventions, and contributions to the business objectives.

The following deliverables may be included in the project:

A list of hazards with definitions – See Tool 2a Identify hazards. Tool 2a is a modified PHA form that includes risk-level calculation. Risk level is transferred from Tool 2b.

Sample risk assessment matrix/methodology – See Tool 2b. RAM "is a composite of matrices that include numerical values for probability and severity levels and their combinations are expressed as risk scorings. It is presented here for people who prefer to deal with numbers rather than qualitative indicators."

Source: ANSI/ASSE Z590.3-2011: Prevention Through Design.

Risk levels – To define operational risk factors and levels to be used in the case study assessment, a RAM with specific risk descriptions was selected. A conventional two-factor RAM (severity level × probability of occurrence) from the ANSI Z590.3-2011 PtD standard is presented. [Note: Other options exist including three or more risk factor systems, which produce a risk priority number (RPN) most often used in failure mode and effects analysis (FMEA).] Two-dimensional RAM was selected for this case study. The following risk descriptions are adapted from ANSI Z590.3. 2011 Prevention through Design standard.

Severity of Consequence Level Descriptions

5. *Catastrophic*: One or more fatalities, total system loss, or chemical release with lasting environmental or public health impact.
4. *Critical:* Disabling injury or illness, major property damage and business downtime, or chemical release with temporary environmental or public health impact.
3. *Marginal:* Medical treatment or restricted work, minor subsystem loss or damage, or chemical release triggering external reporting requirements.
2. *Negligible:* First aid or minor medical treatment only, nonserious equipment or facility damage, or chemical release requiring routine cleanup without reporting.
1. *Insignificant:* Inconsequential with respect to injuries or illnesses, system loss or downtime, or environmental chemical release.

Probability of Occurrence Descriptions

1. *Unlikely:* Improbable, may assume incident or exposure will not occur.
2. *Seldom:* Could occur, but hardly ever.
3. *Occasional:* Could occur intermittently.
4. *Likely:* Probably will occur several times.
5. *Frequent:* Likely to occur repeatedly.

Apply risk scores – Taking the three hazards listed in the PHA (Tool 2a), use the RAM severity and probability ratings (Tool 2b) and the ratings are automatically transferred to PHA (Tool 2a). The explosion hazard (1 EXP) RAM rating is determined as 15 or very high risk (critical to safety risk), the health hazard resulting from benzene exposure (2 Health) is rated at 16 or very high risk, and the environmental hazard (3 Env) is rated at 12 or high risk.

Assess business risks – A business risk assessment matrix (BRAM) similar to the RAM is used to define business impact risk levels. However, the "severity" rating is replaced with "extent of impact" on business, and the "probability" rating is replaced with "likelihood of business losses" as shown in Tool 2c (assess current business risk). The ratings are provided as follows.

Extent of Business Losses Impact Descriptions

1. Insignificant: Inconsequential with respect to business losses.
2. Negligible: Minor business losses.
3. Marginal: Business losses triggering external reporting requirements.
4. Critical: Business downtime, significant business losses, or corporate image impact.
5. Catastrophic: Unsustainable losses, total business loss, or inability to continue business operations.

Likelihood of Business Losses Descriptions

1. Unlikely: Improbable, may assume business loss will not occur.
2. Seldom: Could occur, but hardly ever.
3. Occasional: Could occur intermittently.
4. Likely: Likely to occur several times.
5. Frequent: Likely to occur repeatedly.

Apply results to Bow-Tie diagram – Bow-Tie diagram is used to communicate the relationship between hazards, controls, and consequences. Once the hazards have been identified and prioritized using the simplified RAM (Tool 2b), the results are transferred to a Bow-Tie analysis diagram. In a "conventional" Bow-Tie analysis, only "qualitative" risk descriptions are used. However, by incorporating semiquantitative risk factors for severity and probability as shown in Tool 2d, a modified Bow-Tie can be used. The modified Bow-Tie risk assessment methodology includes severity and probability numerical ratings for SH&E hazards and extent and likelihood of business losses. Corporate image and ethical and legal issues were selected for the consequence analysis. Corporate image was selected based on 12/31/2014 ASSE's President Message, available at http://www.asse.org/risk-management-reputation-is-key/. Legal consequences were selected because it is critical to business, and it is one of the outcomes of ISO 31010 tool B11: Business Impact Analysis (BIA).

Completion of at least three risk assessments – Initial RA: Tools 2a, b, c, and d. Two SH&E intervention options: Tools CH1 and CH2, Tools 3b1, 3b2, 3c1, 3c2, 3d1, 3d2.

It is recommended that the project participants include:

Creating a risk assessment team – Tool 1c

Completing the initial risk determination – Step 2: Tools 2a–d

Documentation of the agreed-upon initial controls and determination of residual risk – Step 3 Tools 3a1, Process FS1, Tool CH1, Tools 3b1, 3c1, 3d1, and 3e1

One of the key messages of RA is that only avoidance/elimination and substitution can reduce severity. Therefore, substitution with less toxic chemicals was considered. Other control measures were added as well. A more advanced version of this RA tool includes Bow-Tie and layers of protection analysis (LOPA) integration. LOPA is another ISO 31010 tool used for risk assessment. However, the methodology described earlier is not included in this project to avoid overcomplication.

ALARP – In our case, acceptable/tolerable risk required significant reduction of the severity. Therefore, completely new EcoDryScrubber system had to be considered. The system eliminates the need for respirators and respiratory protection program. It eliminates the need for expensive activated carbon filters to control VOCs. The new system creates different hazards (combustible dust); however, they are so well controlled that the probability of explosion is negligible. Please see sampling data (EcoData blue button). Students and practitioners are encouraged to develop RA methodology for this option.

Consideration of additional controls and determination of revised residual risk – Step 3 Tools 3a2, Process FS2, Tool CH2, Tools 3b2, 3c2, 3d2, 3e2.

Note: Step 4 will be included in Chapter 19 to "*measure the effectiveness of the RA (ERA) process.*"

ISO 22301-2012 and ISO 31010 tool B11 Business Impact Analysis (BIA) provide a more detailed approach for safety and business objectives integration. For instance, BIA suggests that the RA team should provide an agreed understanding of the identification and criticality of key business processes. In addition, the inputs include financial and operational consequences of loss of critical process. In our case, State Department of Health and Environment inspection led to severe fines and they set strict deadlines to complete the recommended corrective measures. We were also notified that if we continue to release uncontrolled VOC emissions, we'll have to discontinue the operations next year. It is obvious that Bow-Tie analysis is a very good tool to present hazards, preventive barriers, mitigation measures, and consequences. In addition, the modified Bow-Tie diagram presents opportunities to include semiquantifiable risk level, consequences, and color coding.

Completion of a risk register – Step 5: Tool 5.

Note: If the user changes any of the values in the RA tool, they may change colors, financial measures, or nonfinancial values. The tool is intended to be interactive.

Assignment

For your convenience, the authors have prepared process 1 (current state) risk assessment Excel tool. We also developed comparison case for substitution with less toxic chemicals. Your assignment is to complete process 3. QAP Manufacturing decided to invest in powder coating system. DryScrubber system is completely integrated into the booth and fully automated. For more information, please visit http://www.durr-usa.com/fileadmin/user_upload/fas/02_psa/pdf_e/EcoDryScrubber_Brochure_LowResolution_EN.pdf.

Become familiar with applicable OSHA standards: "Spray finishing using flammable and combustible materials." Review 29CFR1910.107 standard: https://www.osha.gov/pls/oshaweb/owadisp.show_document?p_table=STANDARDS&p_id=9753 and http://www.durr.com/fileadmin/user_upload/fas/02_psa/pdf_e/EcoDryScrubber_e_01.pdf Video: https://www.youtube.com/watch?v=zdyz9ubJUTU.

Complete Excel Tool 3a2 and identify the control measures with DryScrubber system. Process FS2 (powder coating) is developed for your convenience. Next, complete Tool 3 CH2 hazard analysis with proposed controls. Calculate risk factor using Tool 3b2. Risk factor values will be transferred to Tool 3 CH2 automatically. Assess changes in business risk (NFB) using Tool 3b2. The results will be automatically transferred to the Bow-Tie diagram (Tool 3d2). Lastly, review risk reduction and residual risk comparison.

10

DESIGN SAFETY REVIEWS

BRUCE K. LYON

Risk Control Services, Hays Companies, Kansas City, MO, USA

10.1 OBJECTIVES

- Introduce design safety reviews
- Review challenges and obstacles
- Examine elements within design reviews
- Provide guidance on conducting design safety reviews

10.2 INTRODUCTION

Arguably the greatest missed opportunity for organizations to reduce risk is the failure to adequately identify, assess, and control risks during the design and redesign phase. At first glance, a safety design review seems like a fairly easy enough process that would be universally practiced within organizations and engineering circles. In reality, there are obstacles that must be overcome to be successful in putting this concept into practice.

Manuele stated that "over time, the level of safety achieved will relate directly to whether acceptable risk levels are achieved or not achieved in the design and redesign processes" (Manuele, 2014). In accordance with the *hierarchy of controls* premise, the most effective and economical way to avoid, eliminate, or control hazards is to address them "upstream" during design and redesign. Most will agree with this concept; however, in practice, few organizations take full advantage of incorporating safety into the preoperational phase. This presents a major opportunity for safety professionals equipped with the skills and desire to advise and guide organizations through the process of identifying hazards and reducing risks during design and redesign.

To begin, proper definitions related to design safety reviews are required. In the foreword of the American National Standards Institute (ANSI)/American Society of Safety Engineers (ASSE) ANSI/ASSE Z590.3-2011, *Prevention through Design: Guidelines for Addressing Occupational Hazards and Risks in Design and Redesign Processes* standard, it refers to the "stated intent of the ASSE,

---

Risk Assessment: A Practical Guide to Assessing Operational Risks, First Edition.
Edited by Georgi Popov, Bruce K. Lyon, and Bruce Hollcroft.
© 2016 John Wiley & Sons, Inc. Published 2016 by John Wiley & Sons, Inc.

in a Position Paper that was approved by the Board of Directors in 1994," which defines *designing for safety* as follows:

> *Designing for Safety (DFS) is a principle for design planning for new facilities, equipment, and operations (public and private) to conserve human and natural resources, and thereby protect people, property and the environment. DFS advocates systematic process to ensure state-of-the-art engineering and management principles are used and incorporated into the design of facilities and overall operations to assure safety and health of workers, as well as protection of the environment and compliance with current codes and standards.*

This definition remains valid and to the point. Other key definitions for terms used in this chapter are taken from ANSI Z590.3 and provided as follows:

Design. The process of converting an idea or market need into the detailed information from which a product, process, or technical system can be produced.

Design safety reviews. An important management process tool for integrating safety and health into the design process for new facilities, processes, or operations, and for changes in existing operations. Design safety reviews are most effective when performed at an early stage when design objectives are being discussed.

Hazard. The potential for harm. Hazards include all aspects of technology and activity that produce risk. Hazards include the characteristics of things (e.g., equipment, technology, processes, dusts, fibers, gases, materials, and chemicals) and the actions or inactions of people.

Hazard analysis. A process that commences with the identification of a hazard or hazards and proceeds into an estimate of the severity of harm or damage that could result if the potential of an incident or exposure occurs.

Life cycle. The phases of design, construction, operation, maintenance, and disposal for a facility, equipment, process, and material.

Prevention through design. Addressing occupational safety and health needs in the design and redesign process to prevent or minimize the work-related hazards and risks associated with the construction, manufacture, use, maintenance, retrofitting, and disposal of facilities, processes, materials, and equipment.

Redesign. A design activity that includes all retrofitting and altering activities affecting existing facilities, equipment, technologies, materials, and processes, and the work methods.

Risk assessment. A process that commences with hazard identification and analysis, through which the probable severity of harm or damage is established, followed by an estimate of the probability of the incident or exposure occurring, and concluding with a statement of risk.

Safety. Freedom from unacceptable risk.

Outside of ANSI Z590.3 *Prevention through Design* and ANSI Z10 *Occupational Health and Safety Management Systems* standards, there is relatively little guidance available in the area of conducting design-phase risk assessments. "Safety-through-design" pioneers such as Bruce Main, Fred Manuele, Wayne Christensen, and Paul Adams, among others, have provided notable groundwork in this area as well. This chapter provides a practical approach to establishing a method for anticipating, recognizing, eliminating, or minimizing operational hazards and risks before they are introduced into the workplace.

10.3 CHALLENGES AND OBSTACLES TO OVERCOME

The concept of addressing safety during design seems logical and desirable to most. However, in practice, it is far too rare for organizations to perform appropriate risk assessments in their design and redesign processes. More concerning is the fact that some organizations totally exclude the safety, health, and environmental (SH&E) aspect from design or redesign. This obviously leads to the likelihood of "embedding hazards" into products and systems that can only be removed or reduced through costly retrofits.

Not long ago, this author was asked by a large global organization to participate in a planning session for a new manufacturing facility to be built. The purpose of the planning session was to determine specific project tasks, resources, and time frames needed to successfully launch operations for the new facility within the 18-month target date. Involved in the session were members of management, production, maintenance, quality, engineering, human resources, and safety, health, and environmental staff. As the team worked through the planning steps, it became apparent to the author that a safety review of the new facility's design had not been discussed. When the author suggested such a risk assessment, there was hesitation from team members. Some did not see this as a feasible time for an assessment and indicated that a comprehensive safety analysis would be performed once the facility was fully operational. Others, including two safety representatives, stated that it would be difficult to identify hazards and assess risks without the physical structures, equipment, and employees in place. In addition, it was the group's general belief that their corporate design and engineering departments were addressing all necessary requirements including safety and code compliance issues in the design. As a result, no formal safety review of the design was scheduled. However, the author used this as an opportunity to work with local team members over the next 18 months identifying potential hazards and risk reduction measures as the facility was built, which proved to be beneficial.

As stated earlier, the lack of design safety reviews can lead to embedded problems in facilities, processes, equipment, workstations, and products. The following are examples observed by this author over the past several years:

- Emergency shower and eye wash stations placed directly in front of or near electrical panels. Specifically, these stations were located in the forklift recharging bays at each of the organization's similarly designed facilities.
- Lack of ventilation and local exhaust systems. In a metal fabrication operation, a quality department's destructive testing laboratory performs tests including chemical reactions, grinding, cutting, brazing, and welding, which produced air contaminants; however, these health hazards were not considered during the design phase.
- Lack of adequate emergency/tornado shelter space for planned occupancy in new facility. In addition, conveyor systems and equipment created obstacles for emergency evacuation routes in some areas of the facility.
- No containment for indoor tank storage of chemicals to prevent spillage from entering floor drains.
- Multiple blind corners and bottlenecks for forklift and pedestrian traffic.
- Elevated work platforms requiring fixed ladders and stairs and lifting and lowering of materials and equipment.
- Poor ergonomic workstation designs such as nonadjustable work surfaces or seating, excessively wide conveyors, excessive manual material handling requirements, poor placement of storage, high noise areas, and poor lighting.

Unfortunately, these types of "design problems" are not all that uncommon. Part of the reason for this disparity is that most SH&E practitioners spend most, it not all, of their time and efforts in the "operational phase." David Walline, the founding chair of ASSE's Risk Assessment Institute, estimates that only 10% of the traditional safety role is dedicated to the "preoperational" or design stage (Walline, 2014). To test this assumption, one of the authors performed a review of job descriptions for SH&E positions listed on the ASSE Job Board website (ASSE, 2014). The review revealed that a majority of the listed job responsibilities described duties such as SH&E program management, regulatory compliance, workplace audits and inspections, incident investigations, employee training, loss analyses, and other activities associated with normal operations. Less than 12% of the job postings reviewed made mention of a "preoperational activity" such as reviews of new systems/equipment, preplanning for construction or expansions, and process changes analysis. No mention of prevention through design, designing in safety, or design safety reviews was found in any of the job descriptions. This anecdotal evidence supports Walline's assumptions.

However, there is some reason for optimism that this is beginning to change. During the 2012 ASSE Professional Development Conference, two of the authors made a presentation on risk assessment to a large audience. During the presentation, the attendees were asked to indicate by a show of hands, if they actively perform risk assessments in the design phase. To the delight of the authors, over half of the attendees indicated favorably.

For those organizations that do not incorporate safety reviews or risk assessments during design, the question is "why." The answer may fall into one or more of the following categories of common barriers: tradition, training, turf, and time.

1. Tradition – An organization's culture and traditions are deeply embedded and difficult to change. Within those cultures and traditions are organizational structures and operating systems. Many organizations operate in highly compartmentalized departments (or silos) with a strict chain of command making it difficult for the information to flow "horizontally" (from department to department) or "vertically" (around a particular manager). Documented procedures often leave out the critically important step of communicating with other departments. These interdepartmental barriers can prevent open sharing and collaborating among departments in many cases. The following scenario may apply to such organizations:

    (a) *Engineering department* – Designers and engineers develop their designs according to design criteria that include project goals, cost, quality, and performance within their department as expected by the organization. Their primary goal is to produce a design that works. There is no requirement, nor is there any desire to seek peer review by other departments concerning the design.

    (b) *Procurement department* – Components and materials specified by the engineering department are provided to the procurement department. Typically, there are very few specifications for safety, health, environmental, or ergonomics requirements other than code compliance or regulatory requirements. The procurement department's goal is to acquire the necessary materials by the specified time, at the lowest cost possible (often within a specified budget). Again, there are no requirements or protocols for peer reviews of materials considered or selected by other departments.

    (c) *Supplier/contractor* – Once the design is complete, it is handed over to an outside supplier/contractor responsible for building and installing the unit. There is limited oversight by the facilities engineering/maintenance and/or production departments during the installation. The supplier/contractor's goal is to have the unit completed and installed according to the design specification, by the required time frame and within specified budget.

    (d) *Production department* – Once installed, the design is put into operation by the production department. Production operations may include pre-start-up inspections and procedures, programming, adjustments, oiling and lubricating, clearing jams, and resetting machines. The production department's primary goal is to achieve and maintain maximum production

CHALLENGES AND OBSTACLES TO OVERCOME

(as measured by number units produced), meeting quality standards, at the lowest cost possible. Again, there are no requirements for the production department to involve the SH&E department until problems arise.

(e) *Maintenance department* – The operation requires regular scheduled preventive maintenance and occasional breakdown repair service performed by the maintenance department. Maintenance's primary goal is to return the system to normal as quickly as possible to reduce downtime and prevent business interruption. Outside of the Occupational Safety and Health Administration's (OSHA) lockout/tagout requirements, the organization does not require involvement from the SH&E department until incidents occur.

(f) *Maintenance/decommission contractor* – At some point, the system/product completes its life span or usefulness. It is removed from service by maintenance or an outside contractor, making way for a new unit or technology to take its place. Again, the SH&E department has little involvement in decommissioning equipment accept to respond to incidents that may occur during the process.

(g) *Safety, health, and environmental department* – Throughout the life span of the design, exposure to hazards can cause harm to people, assets, and/or the environment. The SH&E department is responsible for identifying and controlling existing hazards, as well as responding to incidents that occur. For those systems/products that have not adequately addressed safety in the design, a greater amount of control measures are required to achieve the organization's acceptable level of risk. Often, the primary goals of the SH&E department are to meet compliance and reduce losses.

This scenario illustrates how individual department's goals can conflict and impede interdepartmental communication and cooperation if there are no formal mechanisms or specific requirements to do so. Whether an organization realizes it or not, their traditions and culture determine the level of importance placed on safety criteria throughout the operation.

2. Training – Formal education and training provided to engineers has typically not included occupational safety and health principles. Many engineers have little to no experience in anticipating or recognizing hazards and are not familiar with the concept of the hierarchy of controls. In regard to the National Institute for Occupational Safety and Health's (NIOSH) Prevention through Design (PtD) initiative, director John Howard, M.D., stated, "One important area of emphasis will be to examine ways to create a demand for graduates of business, architecture and engineering schools to have basic knowledge in occupational health and safety principles and concepts." However, this is not a one-way street. Safety professionals need to better understand their organization's cost drivers and internal protocols for engineering and design as well.

3. Turf – Design engineers are responsible for designing systems/products according to established design criteria, within set time and budget constraints. Their primary goal is to design things to work. Understandably, it is not in an engineer's nature to deviate from their formal education and training or established protocols. As a result, there is often reluctance to seek input from nonengineering departments. This presents an opportunity for safety professionals to prove their worth to engineers (and management) by facilitating preoperational risk assessments that enable the organization to create safer designs that are more cost effective. Financial measures such as cost/benefit analysis and return on investment (ROI) will aid in communicating the value of design safety reviews.

4. Time – The critical path from design concept to production is time limited. Engineers are on a tight schedule and expected to meet their deadlines. Lack of forethought or time allotted for safety reviews at the design phase is common. Risk assessments take time to conduct effectively, typically more time than can occur within a design review session. As pointed out by Bruce Main, risk assessments may need to be performed separately from the engineering design review and should be incorporated into the design process as early as possible.

To overcome barriers, organizations must establish a strong safety culture that requires acceptable risk levels be attained and maintained, and understand that incorporating safety into the design and redesign processes has an exceptional role in achieving that purpose (Manuele, 2008). Communication and cooperation among engineering and nonengineering departments must improve. Expectations and accountabilities for safety in design and redesign efforts must be well defined and communicated to all parties. And more training and education for engineers and designer in hazard recognition and control should be provided. Paul Adams (Adams, 1999) sums up these needs well in the book *Safety Through Design* with the following recommendations:

1. Safety needs to be addressed at the earliest possible point in a project, preferably at the scoping and specification phase.
2. Safety-focused events are needed to give proper attention to hazard identification and elimination.
3. Engineers need a model for understanding their role in preventing incidents.
4. Engineers need to be trained on the process for designing safe systems.

Safety practitioners, for the most part, have not done a good job engaging in the design process. Many reasons can be cited including job descriptions, daily work demands, lack of notification or invitation to participate in design reviews, position and status with an organization, and general lack of knowledge in the design and engineering process. Safety professionals that are more influential in the design process will have greater impact on safety and will increase their overall value.

10.4 STANDARDS REQUIRING DESIGN SAFETY

One of the first standards in the United States to include requirements for design safety reviews as well as risk assessments, procurement, and management of change is ANSI Z10 *Occupational Health and Safety Management Systems* originally released in 2005. The consensus standard was updated in 2012 and is considered a "state-of-the-art" blueprint for the development of an operational risk management system. ANSI Z10 provides guidance in the implementation of a continuous improvement system that requires management leadership and employee involvement, planning, implementation and operation, evaluation and corrective action, and management review.

Within ANSI Z10's implement and operation section, requirements for design reviews and management of change are specified. The standard states that "the organization shall establish a process to identify, and take appropriate steps to prevent or otherwise control hazards at the design and redesign stages, and for situations requiring Management of Change to reduce potential risks to an acceptable level." Such design reviews should be considered with the anticipated introduction of new technology, equipment, or facilities; new procedures and work practices; new raw materials; or new designs. The specific inclusion of design safety and risk assessment requirements in this management system standard indicates their importance.

A second significant US standard was released in 2011. The publication of ANSI Z590.3 *Prevention through Design: Guidelines for Addressing Occupational Hazards and Risks in Design and Redesign Processes* brought to light the need for design safety. The standard sets forth principles and methodologies in addressing hazards and risks in the design and redesign process through supplier relationships, design safety reviews, risk assessments, and the use of the hierarchy of control in reducing risk to an acceptable level. It was developed as a result of initial efforts by the *Advisory Committee of the Institute for Safety through Design* at the National Safety Council, followed by the NIOSH PtD initiative. According to NIOSH, one of the key elements of ANSI Z590.3 is that it provides guidance for "life-cycle" assessments and a design model that balances environmental and occupational safety

THE REVIEW OF DESIGNS 215

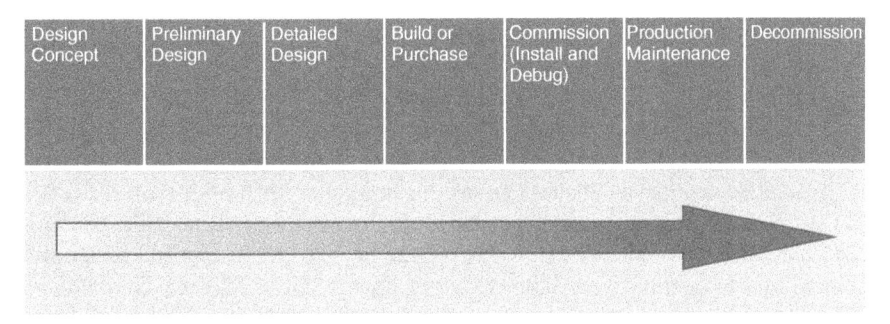

Figure 10.1 Design Concept through Decommissioning Process *Source:* Reprinted with Permission from ANSI/ASSE Z590.3-2011 (Courtesy of the American Society of Safety Engineers)

and health goals over the life span of a facility, process, or product. All facilities, equipment, and products have a defined life cycle in which the risks will change. This necessitates, as prescribed in ANSI Z590.3, that risk assessments and risk reduction be incorporated into each stage of a product or system's life cycle. Figure 10.1 reprinted with permission from ANSI/ASSE Z590.3-2011 shows the progression of life-cycle stages from design to decommission.

The importance of the PtD concept is immense. Similarly to the *hierarchy of controls* premise, the most effective control in reducing risk is to eliminate hazards through design. When hazards are eliminated before they are introduced into the workplace, risks derived from those hazards are also removed, helping the organization achieve and maintain acceptable risk levels. Even though all risks cannot be completely eliminated, the most "risk-effective" as well as cost-effective place for an organization to apply risk management efforts is in its design and redesign phase.

Other safety standards are beginning to incorporate PtD requirements. For example, ANSI B11.0-2015, *Safety of Machinery – General Requirements and Risk Assessments* standard lists design safety requirements and responsibilities for both the supplier and user of machines. Specifically, B11.0 requires the supplier to identify hazards and assess and reduce risks to an acceptable level during the design, construction and installation of the machine, as well as develop safety-related information for operation and maintenance of the machine. An important aspect of B11.0 is the requirement for "collaborative efforts" between the supplier and end user in the conceptual design and building of a machine, which is echoed in Z590.3's requirements for supplier relationships. Such efforts require meaningful and effective communication between the user and the supplier from beginning to end. Procurement requirements and machine safety specifications provided by the user should be communicated up front to the supplier for incorporation into the machine's design. Any residual risks identified by the supplier's design safety review and risk assessments are to be communicated to the user so that appropriate risk reduction measures are applied to the machine. B11.0 states that any risk transferred between the supplier and user must be documented and communicated. Collaborative efforts between suppliers and users should be interactive and ongoing to the degree necessary.

10.5 THE REVIEW OF DESIGNS

As described by Main, a *design review* is typically a formal evaluation of a design to ensure that the design meets specified criteria. Safety is only one element to consider in a design review, and the complexity and criticality of the design will dictate the level of hazard analysis and risk assessment needed. Other elements considered in design reviews include cost, legal, product liability, environmental, quality, compliance, and marketing to mention several. For instance, from a consumer products standpoint, product design reviews are used by manufacturers to identify and eliminate potential product defects or misuses in an effort to avoid product liability exposures.

Many design reviews include a compliance review aspect. If an office building is to be built, it must meet certain (state and city) building code requirements and standards for fire protection and life safety (SFPE, 2009). In addition, many organizations have internal requirements or insurance carrier requirements that go beyond compliance standards. A compliance review is used to verify that external and internal compliance specifications are incorporated into the design. Applicable standards promulgated by consensus groups such as ANSI, the American Society for Testing and Materials (ASTM), and the National Fire Protection Association (NFPA), nonconsensus bodies including Underwriters Laboratories (UL) and Factory Mutual (FM), and regulatory standards issued by federal and state OSHA should be included in a compliance review of a proposed design.

Compliance reviews are common, especially in the fire protection area. These are sometimes conducted by third party consultants and insurance brokers and carriers. However, designing to compliance does not ensure that all hazards are adequately controlled or that "error traps" are avoided.

In his book *Risk Assessment: Challenges and Opportunities*, Bruce Main suggests that formal safety analyses and risk assessments be performed separately from and prior to the engineering design review (Main, 2012). By doing so, specific safety guidance can be given to the design team for inclusion in to the design process. Obviously, any hazards recognized throughout the process should be addressed by the design safety and engineering teams. In many design reviews, a less formal identification of hazards is performed by the team during the design phase, which may be appropriate for relatively simple designs. However, as Main points out, a separate risk assessment should be made for more complex designs. Where potential risk is high, design safety reviews may be needed at each phase or stage (i.e., conceptual, preliminary, final design, and testing) as a sign-off or approval process from the SH&E, ergonomics, and compliance stakeholders.

10.6 HAZARDOUS ENERGY CONTROL

One of the prominent theories on hazardous energy control that has withstood the test of time was developed by Dr William Haddon, Jr., in the 1970s. Haddon's "Energy Release Theory" is based on an engineering approach that establishes a relationship between incident causation and risk control methods (Haddon, 1970). The use of Haddon's model should be considered in design safety reviews since it relates well to engineers and can be applied systematically. Haddon's "Energy Release Theory" includes 10 sequential control strategies similar to a hierarchy of control that should be considered in the design of new products and systems. The following are abbreviated descriptions of his 10 strategies:

1. *Prevent stored energy.* Prevent the marshaling of the form of energy in the first place, such as preventing the generation of thermal, kinetic, or electrical energy, or ionizing radiation that can be potentially released.
2. *Reduce stored energy.* Reduce the amount of energy marshaled by its amount and concentration, such as limiting the amount of chemicals stored, reducing the size of materials handled, or reducing the speed of vehicles.
3. *Prevent energy release.* Prevent the release of the energy by incorporating physical containment.
4. *Reduce rate of release.* Modify the rate or spatial distribution of release of energy from its source such as reducing compressed air pressure to 30 pounds per square inch (psi) or reducing the slope of warehouse ramps for forklifts.

ERGONOMIC REVIEW OF DESIGNS

5. *Separate energy release from humans and assets by space or time.* Separate, in space or time, the energy being released from that which is susceptible to harm or damage. This strategy eliminates the intersection (exposure) of energy and humans or assets. Examples include increasing the distance between the point of operation of a punch press and the operator or scheduling human interaction with machine when its functions are neutralized.
6. *Separate energy release from humans and assets by physical barriers.* Separate by interposition of a material "barrier" such as the use of insulation on electrical lines, machine guards, or welding curtains.
7. *Modify contact surfaces.* Modify appropriately the contact surface, subsurface, or basic structure, as in eliminating, rounding, and softening corners edges, and points with which people can come in contact.
8. *Strengthen susceptible structures.* Strengthen the structure, living or nonliving that might otherwise be damaged by the energy transfer such as the reinforcement of storage racks exposed to forklift damage.
9. *Increase detectability and prevention of harm.* Move rapidly in detection and evaluation of damage that has occurred or is occurring and counter its continuation or extension. Examples include fire alarms and sprinkler systems, proximity limit switches, or presence sensing devices.
10. *Prevent further damage.* After the emergency period following the damage energy exchange, stabilize the process. Examples include disaster recovery plans and emergency action and evacuation plans.

As indicated in Haddon's control strategies, the most effective control of potential energy release is accomplished when such measures are incorporated into the design. Special attention should be given to the potential for hidden energies in products and systems. Table 10.1 provides a simple list of energy types and hazards that should be investigated during a design review.

TABLE 10.1 Energy Types and Hazards

Energy	Hazards
Mechanical	Acute force to tissues, chronic stress to tissues
Physical	Noise, vibration, gravity, inertia, configuration
Chemical	Toxicity, caustic, acute, chronic
Electrical	Electrocution, explosion, fire, arc flash
Biological	Bacteria, blood-borne pathogens, microorganisms, animals, insects
Thermal	Extreme temperatures
Radiological	Ionizing, nonionizing

10.7 ERGONOMIC REVIEW OF DESIGNS

As discussed in *Chapter 16, Ergonomic Risk Assessment*, the concept of designing in ergonomic principles into the workplace is supported by recent standards such as *ANSI/ASSE Z590.3-2011, Prevention through Design: Guidelines for Addressing Occupational Hazards and Risks in Design and Redesign Processes* standard, and *ANSI/AIHA/ASSE Z10-2012, the Occupational Health and Safety Management Systems* standard. Certain industries such as automotive, aviation, and technology have incorporated ergonomic standards and guidelines within their products and system design specifications and procurement requirements to a certain degree.

However, there is a much room for improvement. According to Walt Rostykus, a noted ergonomist and researcher, only 5% of organizations consistently address ergonomics during design of new processes, equipment, and products (Rostykus, 2012). His research indicates that this small percentage of advanced organizations successfully incorporate ergonomics in the design phase by using three critical elements:

1. Use a formal ergonomic design review process that incorporates a phase-gate review and approval process during design.
2. Adherence to ergonomic design standards and guidelines (i.e., reach, force, work height, weight limits).
3. A system for holding engineers accountable for the risk level of their designs (level of musculoskeletal disorder risk factors).

Rostykus states that "the greatest value of good upstream design is the reduced cost of making changes. The cost of changing equipment and layout once it is in place is more than 1000 times the cost of making the change in the design phase."

Phase-gate or stage-gate reviews are an evaluation and approval process step between phases of a project. They require specific reviews from key departments or individuals to evaluate the proposed design to determine if it meets their approval for the design to proceed to the next phase. An example of the use of phase-gate type reviews is provided in Figure 10.2, Operational Risk Management within Life-Cycle Phases. SH&E professional as well as ergonomist should have an active role in "phase-gate reviews" for new designs.

Designing systems and products that are ergonomically conducive and free of error traps requires an effective ergonomics risk assessment of the proposed design. The ergonomic design review should include an assessment of risk factors such as forceful exertions, repetition and duration, static or awkward postures, excessive reach, contact stress, vibration, poor lighting, and cold temperatures. Specific tools such as the Ergonomic Risk Assessment Tool (ERAT) and other targeting tools discussed in Chapter 16 can be used to identify and eliminate ergonomic hazards during the preoperational stage, avoiding costly retrofitting later.

10.8 DESIGN REVIEW PROCESS

"Formal design safety reviews are effective processes for delivering inherently safer designs." That statement is taken from the opening sentence in ANSI Z590.3's *Addendum E – A Design Safety Review Guide (Informative)*. ANSI Z590.3 further defines that design safety reviews are an important management tool used to integrate safety into the design process for new facilities, expansions in existing buildings, new or modified processes and systems, equipment and machines, and products.

For a design safety review to be successful, potential hazards must be anticipated, identified, eliminated, and controlled to an acceptable risk level prior to being introduced to the workplace. This requires top management commitment and an established process for design safety reviews within the organization's operational risk management system. As outlined in ANSI Z590.3, key elements and steps for a formal design review process are summarized in Figure 10.3 and described in the following.

1. *Management policy* – A written management policy should be established and communicated providing direction on when, where, and how hazard analyses and risk assessments are performed, including the design phase, in the organization. The policy should outline roles and responsibilities, as well as accountabilities for design safety to include engineers, designers,

DESIGN REVIEW PROCESS 219

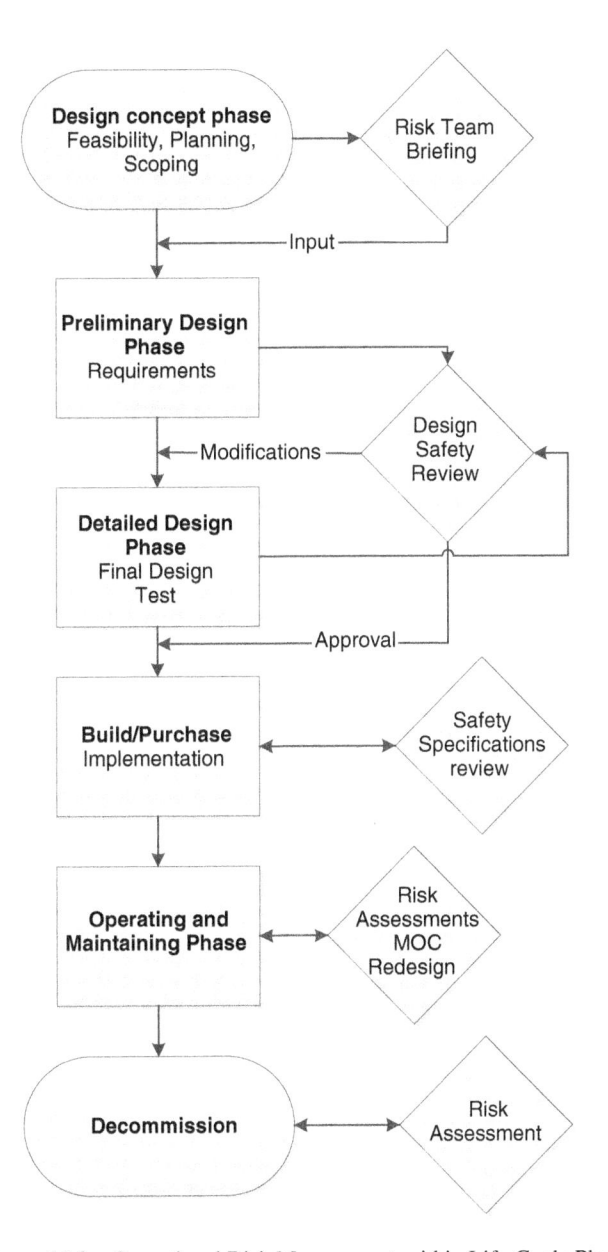

Figure 10.2 Operational Risk Management within Life-Cycle Phases

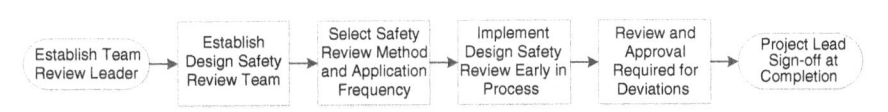

Figure 10.3 Design Safety Review Process Steps

production, maintenance, quality, legal, SH&E, human resources, procurement, and other involved parties.

2. *Establish a project design leader and team* – Management should designate a qualified leader and cross-functional team to perform the design safety review. Roles and responsibilities for all members should be defined and communicated for the project design review. The design safety review team should include members with expertise in applicable areas such as safety and health, ergonomics and human factors engineering, environmental safety, fire prevention and protection, and product liability prevention as appropriate for the project. In some cases, it may be necessary to include outside specialists to help in the design safety review. Effective communication between the safety review team and engineering/design team is a critical factor in the success of the design.

3. *Establish a design safety review process* – The organization should establish a documented process for conducting design safety reviews. The process should follow guidance from ANSI Z590.3, ANSI Z10, and other available standards and resources. For each project, an appropriate methodology and frequency should be determined based on the complexity and criticality of the design.

4. *Conduct hazard analysis and risk assessment early in the process* – The earlier a hazard analysis and risk assessment can be introduced into the design process the more effective it becomes. ANSI Z590.3 advises that safety reviews be performed early when design objectives are being discussed and defined. Depending upon the complexity and circumstances, the hazard analysis and risk assessment may be performed separately from the design review, with the findings and risk reduction recommendations incorporated into the design process. A analysis and assessment may include reviews of existing similar designs and literature; plan drawings and specifications; hazard checklists; applicable standards; discussions with manufacturers of components and materials; safety data sheets; loss experience related to similar designs; and existing controls and technology on similar designs. Not only should the design safety review address operational hazards and risk but should also consider hazards and risks resulting from nonroutine events such as emergency breakdowns, upsets and repairs, scheduled and nonscheduled service and maintenance, testing, adjusting, lubricating, and other maintenance-related activities.

5. *No deviation from standards without approval* – The design team should follow established safety standards and specification. Any requested deviation from the specifications should be reviewed for approval or disapproval by appropriate safety and management personnel.

6. *Design completion sign-off* – A written certification verifying that the design safety review has been completed should be signed by the project leader signifying a consensus among the safety team and engineering group. For this to occur, communication is needed between the design safety review team and engineering/design group throughout the process.

7. *Design safety review deliverables* – Depending on the scale and complexity of the design, specific deliverables resulting from the review may include modifications and markups of drawings and specifications; a risk-prioritized list of specific hazards and means for elimination or control; a list of design modifications necessary prior to approval; action item list with assigned responsibilities; and follow-up questions, concerns, or requests for additional information necessary to satisfy or complete the review and approve the design.

10.9 HAZARD ANALYSIS AND RISK ASSESSMENT IN DESIGN

When conducting a hazard analysis and risk assessment as part of a design review, it's vital that an agreed-upon process be followed. For those projects that are relatively simple, it may be appropriate

HAZARD ANALYSIS AND RISK ASSESSMENT IN DESIGN

for the safety professional to informally review a proposed design with the engineer to identify hazards and necessary controls. However, for more complex situations, management should establish a formal hazard analysis and risk assessment method that is incorporated into the design and redesign process. The following is an abbreviated outline adapted from ANSI Z590.3 for performing hazard analyses and risk assessments as part of a design safety review:

1. *Select risk assessment matrix* – The organization shall agree upon a selected risk assessment matrix and definitions for risk probability and severity appropriate for the hazards and risks involved. A clear understanding of the selected matrix and definitions is required by all personnel involved in the risk assessment process. As described in ANSI Z590.3, a matrix provides a method to categorize combinations of probability and severity of harm to establish a risk level for each hazard, acts as a communication tool for decisions makers in risk management efforts, and assists in comparing and prioritizing risks so that resources are effectively allocated.
2. *Select risk assessment technique* – Selection of a method appropriate for the design being reviewed will include consideration of the complexity and criticality of the design, scope and parameters of the assessment, and experience of the team. Team members should be adequately trained and familiar with the method used. In ANSI Z590.3, eight techniques are listed including Preliminary Hazard Analysis (PHA); Checklist Analysis; What-If/Checklist Analysis; Hazard and Operability Analysis (HAZOP); Failure Mode and Effects Analysis (FMEA); Fault Tree Analysis (FTA); and Management Oversight and Risk Tree (MORT). The standard makes note that PHA, FMEA, and What-If methods are sufficient to address most risk situations. For designs with potential human factors and ergonomics risk, a specific ergonomics risk assessment may be required.
3. *Establish parameters* – Determine the purpose and scope of the analysis and what is at risk in terms of people (employees, the public), processes, property, equipment, or the environment. The scope parameters and the design's boundaries are defined to include the design's applications, uses, potential misuses, operating phases, potential interfaces, and other potential boundaries.
4. *Identify potential hazards* – Those responsible for conducting design safety reviews should be adequately trained and experienced in hazard identification and control. Based on the design's complexity level, trained individuals or teams perform a systematic hazard analysis of the design to create a hazard inventory. The hazard identification process should involve brainstorming and include the following:
   (a) Reviews of available design documents and plans
   (b) Discussions with designers, engineers, operators, or users
   (c) Use of checklists for general and specific hazards concerning the design's technology, equipment, interactions with other systems, substances, and tasks
   (d) Review of potential hazardous energy release during its life span
   (e) Consideration of hidden hazards and potential "error-sensitive" situations that could lead to human error, omissions, and misuses
   (f) Investigation into similar designs, analyses, control measures, and lessons learned
   (g) Analysis of available historical loss for similar designs
   (h) Review of applicable industry standards, compliance standards, codes and regulatory requirements, and internal and customer requirements
   (i) Consideration of synergistic effects from combined hazards (e.g., noise and toluene, cold stress and hand vibration, heat stress, and lifting)
   (j) Analysis of hazardous atmospheres or environments created.

5. *Identify failure modes* – Following development of a hazard inventory, failure modes and foreseeable uses/misuses of the design for each hazard are identified. For each failure mode, the circumstances that could cause such "credible" scenarios to occur are determined.
6. *Evaluate existing controls* – For each hazard, an evaluation of the existing/planned controls is performed to determine whether they are adequate and effective. Control measures are evaluated for their maintainability and care required to maintain adequate control and the ability to defeat or bypass controls.
7. *Determine likelihood of occurrence* – For each hazard, the likelihood or probability of a hazardous event or exposure occurring is estimated. This includes an estimation of the frequency and duration of the exposure to harm or damage by the identified hazard. Likelihood estimation is generally related to an interval base such as a unit of time; activities, tasks, or events; units produced; or the life cycle of a facility, equipment, material, process, or product. The following factors are considered in the estimation:
    (a) Frequency of task or process performed
    (b) Duration of exposure and whether the exposure is continuous or intermittent
    (c) Number of people exposed
    (d) The occupational health and environmental exposures (which may require dose–response analyses and exposure assessments)
    (e) Potential production loss or business interruption.
8. *Assess severity of harm* – For each hazard, the worst "credible" consequences are estimated in determining severity of harm. This determination is made using historical and objective information such as
    (a) Likely number of injuries or illnesses and their severity level
    (b) Dollar loss value
    (c) Property or equipment damage value
    (d) Business interruption impact
    (e) Environmental impact
    (f) Loss of market share potential.
9. *Define initial risk level* – The selected risk assessment matrix is used to identify initial risk and help communicate risk levels to decision makers. To define the risk of a specific hazard, the estimates for its likelihood and severity are plotted in the matrix to arrive at a risk category.
10. *Select and implement risk reduction measures* – Using the hierarchy of controls, risk reduction measures are selected to avoid, eliminate, or reduce risks to an acceptable level. Based on risk prioritization, control measures and their action plan are developed and proposed. Alternate control strategies may be necessary in some cases. The proposed control measures should include estimated costs to assist decision makers. To ensure risk controls are properly implemented and effective, a documented method of tracking completion and testing effectiveness is needed.
11. *Assess residual risk* – Upon implementation of risk reduction measures, a documented assessment of remaining risk is performed to determine whether residual risk levels are acceptable. If they are not acceptable, further risk reduction measures are to be implemented where feasible. If it is not feasible to reduce risk to an acceptable level, the design shall require modification to allow acceptable risk levels or be discontinued. Completed risk assessments shall be shared and communicated with all stakeholders to aid in further risk assessment and control efforts.
12. *Documented risk acceptance and follow-through* – Management will sign off on completed design safety reviews and risk assessments indicating their acceptance. Follow-up activity is

HAZARD ANALYSIS AND RISK ASSESSMENT IN DESIGN 223

performed to ensure that all problems are resolved and that actions taken have not created new hazards. At this point, the design safety review is complete.

Figure 10.4 provides an example of a risk assessment process flowchart reprinted with permission from ANSI/ASSE Z590.3-2011.

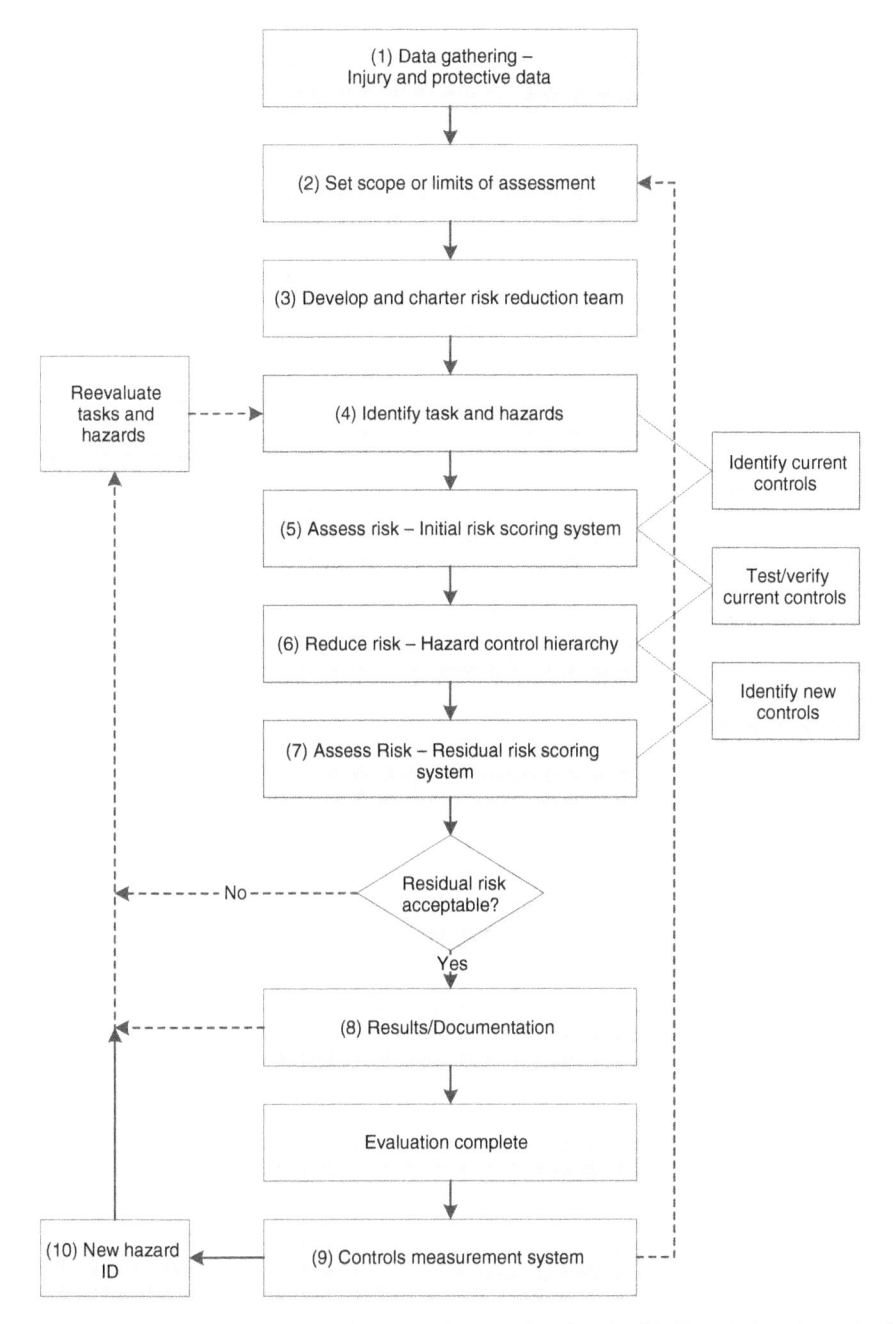

Figure 10.4 Example Risk Assessment Process *Source*: Reprinted with Permission from ANSI/ASSE Z590.3-2011 (Courtesy of the American Society of Safety Engineers)

The degree of involvement from SH&E in a design process will depend upon the criticality and complexity of the design. A single informal safety review may be adequate in simple designs. Other situations may require a series of hazard analyses and risk assessments at or between critical phases of a design.

As described by Thomas Hunter in his chapter *Integrating Concepts into the Design Process* in the book *Safety Through Design*, a series of design reviews may be necessary (Christensen & Manuele, 1999; Hunter, 1999). Hunter advises that a thorough study and identification of hazards and their control begins in the preliminary design phase and proceeds through final design and testing. At the preliminary stage, an initial design safety review of available data and literature, anticipated energy sources and controls, and integrity of materials is conducted. As a result of the first design review, questions and suggestions are generated for further study by the design team. Any identified hazards are eliminated or mitigated. As the project proceeds to the final design stage, a second design review is performed by SH&E stakeholders. Any new elements or changes incorporated into the design are reviewed to determine if new hazards have been created or introduced into the design. Once the design meets approval, it proceeds to the testing stage. A final design review is made to determine if all parameters have been met. After approval, the design moves to the implementation phase and on to the operational phase.

10.10 CONCLUSION

The greatest opportunity for reducing risk and improving safety within systems and products lies within the design phase. SH&E professionals must find ways to actively participate in the design and redesign process if progress is to be made. To summarize some of the salient points regarding design safety, the following are offered:

- A formal process for conducting design safety reviews should be established within an organization's operational risk management system.
- Design safety reviews, hazard analyses, and risk assessments should be incorporated early in the design process.
- Design safety reviews, hazard analyses, and risk assessments may need to be performed separately from the engineering review.
- Depending upon complexity and criticality of design, a series of hazard analyses and risk assessments may be necessary at critical phases of the design.
- Designs should be reviewed from an ergonomics and human factors standpoint to eliminate or reduce musculoskeletal disorder risk factors and error-provocative conditions.
- Design engineers should be held accountable for the risk level or safety quality of their designs.
- Organizations should ensure designs meet their acceptable risk levels.
- Safety and ergonomic specifications should be incorporated into the selection and procurement process for new designs and materials.
- The ANSI Z590.3 Prevention through Design standard should be used as a model by organizations developing design safety review protocols.

As with any process, follow-up and continual refinements of the design safety review process are required. Decision makers will require cost justification to accompany design safety review findings in many cases. Cost/benefit analysis and ROI calculations can be useful in showing the value of safety design recommendations as demonstrated in *Chapter 19, Business Aspects of Operational Risk Assessments*. The benefits derived by incorporating safety into design are many. These include reduced risk to people and the environment, fewer retrofits, reduced costs for ongoing risk controls, lower costs in energy use, reduced waste, improved operating efficiencies, lower project costs, reduced worker incident frequency and severity resulting in cost savings in risk financing, regulatory compliance, and reduce liability.

As decision makers become more aware of these benefits, the use of design safety reviews will increase. It's up to the SH&E professional to find ways to expand their influence and apply their skills at this most critical phase. With no doubt, there remains significant opportunity to fill the void that currently exists.

REVIEW QUESTIONS

1. Describe the concept of designing for safety (also known as prevention through design).
2. List some of the challenges and barriers to performing risk assessments at the design/redesign stage.
3. List several standards that include design safety in their requirements.
4. Explain the energy release theory and how it relates to the design safety process.
5. Explain how ergonomics impacts design safety.
6. Give four reasons why risk assessments and design safety reviews are valuable to an organization.
7. Describe the basic steps taken in a design safety review.

REFERENCES

Adams, P. S., 1999. Application in General Industry. Christensen, W.C., & Manuele, F.A. (Eds.), *Safety Through Design*. Itasca, IL: National Safety Council, NSC Press, pp. 155–169.

ANSI/AIHA/ASSE Z10-2012. *American National Standard – Occupational Health and Safety Management Systems*. Fairfax, VA: American Industrial Hygiene Association, 2012.

ANSI/ASSE Z590.3-2011. *Prevention Through Design: Guidelines for Addressing Occupational Hazards and Risks in Design and Redesign Processes*. Des Plaines, IL: American Society of Safety Engineers, 2011.

ANSI B11.0-2015. *Safety of Machinery – General Safety Requirements and Risk Assessments*. Houston, TX: B11 Standards, Inc., 2015.

ASSE. 2014. The American Society of Safety Engineers Job Board. Search performed November 26, 2014 at http://jobs.asse.org/jobseeker/search/results/ (accessed December 24, 2015)

Christensen, W. and F. Manuele, Co-editors. *Safety Through Design*. Itasca, IL: National Safety Council, 1999.

Haddon, W. J., Jr. May, 1970."On the escape of tigers: an ecological note." *American Journal of Public Health and the Nation's Health*, 60(12), 2229–2234. Available at: http://www.ncbi.nlm.nih.gov/pmc/articles/PMC1349282/pdf/amjphnation00041-0001c.pdf (accessed December 24, 2015)

Hunter, T. A., 1999. Integrating Concepts into the Design Process. Christensen, W.C., & Manuele, F.A. (Eds.), *Safety Through Design*. Itasca, IL: National Safety Council, NSC Press. pp. 73–87.

Main, B. W., *Risk Assessment: Challenges and Opportunities*. Ann Arbor, MI: Design Safety Engineering, 2012.

Manuele, F. A., "Prevention through design: addressing occupational risks in the design and redesign processes." *Professional Safety*, October 2008.

Manuele, F. A., *Advanced Safety Management: Focusing on Z10 and Serious Injury Prevention*, 2nd ed. Hoboken, NJ: Wiley, 2014.

Rostykus, W., "Five changing trends in managing workplace ergonomics." *Occupational Health & Safety*, 81(10), 76–80. October 2012. Available at http://ohsonline.com/articles/2012/10/01/five-changing-trends-in-managing-workplace-ergonomics.aspx (accessed December, 2015)

SFPE, *Guidelines for Peer Review in the Fire Protection Design Process*. Bethesda, MD: Society of Fire Protection Engineers, October 2009. Available at: http://www.sfpe.org/Portals/sfpepub/docs/pdfs/About-Us/peer_review_guidelines_2009.pdf (accessed December 24, 2015)

Walline, D. L., "Prevention through design: proven solutions from the field." *Professional Safety*, 59(11), 43–49. November 2014.

11

RISK ASSESSMENT AND THE PREVENTION THROUGH DESIGN (PtD) MODEL

GEORGI POPOV
School of Environmental, Physical & Applied Sciences, University of Central Missouri, Warrensburg, MO, USA

BRUCE K. LYON
Risk Control Services, Hays Companies, Kansas City, MO, USA

JOHN N. ZEY
School of Environmental, Physical & Applied Sciences, University of Central Missouri, Warrensburg, MO, USA

11.1 OBJECTIVES

- Introduce relationships between risk assessment methodologies and PtD model
- Review risk assessment techniques and demonstrate use in PtD model
- Demonstrate PtD principles/tools that can be used as a part of management practices and business process of the organization

11.2 INTRODUCTION

As discussed in Chapter 10, *Design Safety Reviews*, the use of risk assessment techniques in the design and redesign phase is one of the most effective ways to avoid and eliminate risk from being introduce into a system or product. While its use has increased during the past decade, design-phase risk assessments remain one of the most underutilized aspects of operational risk management. This represents an important opportunity for SH&E professionals that desire to enhance their organization's risk management process as well as their own value within the company.

There are in fact, a significant number of standards, guidelines, and initiatives for the practice of operational risk assessments available worldwide. And the number is growing. Since 2005, there have

---

Risk Assessment: A Practical Guide to Assessing Operational Risks, First Edition.
Edited by Georgi Popov, Bruce K. Lyon, and Bruce Hollcroft.
© 2016 John Wiley & Sons, Inc. Published 2016 by John Wiley & Sons, Inc.

been more than 35 such standards and initiatives that require or promote the use of risk assessments as outlined by Fred Manuele in Chapter 1, Addendum A. As noted in other chapters of this text, the requirement for employers to perform operational risk assessments has primarily occurred outside of the United States. However, in 2011, two important consensus standards were released by the American National Standards Institute (ANSI) providing requirements for the practice of risk management and risk assessment in the workplace.

In 2009, the International Standards Organization (ISO) published a series of standards for the principles, framework, and process for managing risk. This family of risk management standards referred to as ISO 31000 includes

- ISO Guide 73:2009, *Risk management – Vocabulary*
- ISO 31000:2009, *Risk management – Principles and guidelines*
- IEC/ISO 31010:2009, *Risk management – Risk Assessment Techniques.*

In 2011, the ANSI and American Society of Safety Engineers (ASSE) adopted the ISO 31000 standards in the ANSI Z690 series to include

- *ANSI/ASSE Z690.1-2011, Vocabulary for Risk Management (National Adoption of: ISO Guide 73:2009)*
- *ANSI/ASSE Z690.2-2011, Risk Management Principles and Guidelines (National Adoption of: ISO 31000:2009)*
- *ANSI/ASSE Z690.3-2011, Risk Assessment Techniques (National Adoption of IEC/ISO 31010:2009).*

The purpose of the ISO 31000/ANSI Z690 series is to establish principles and guidelines for the practice of risk management and risk assessment. These risk management standards are designed to

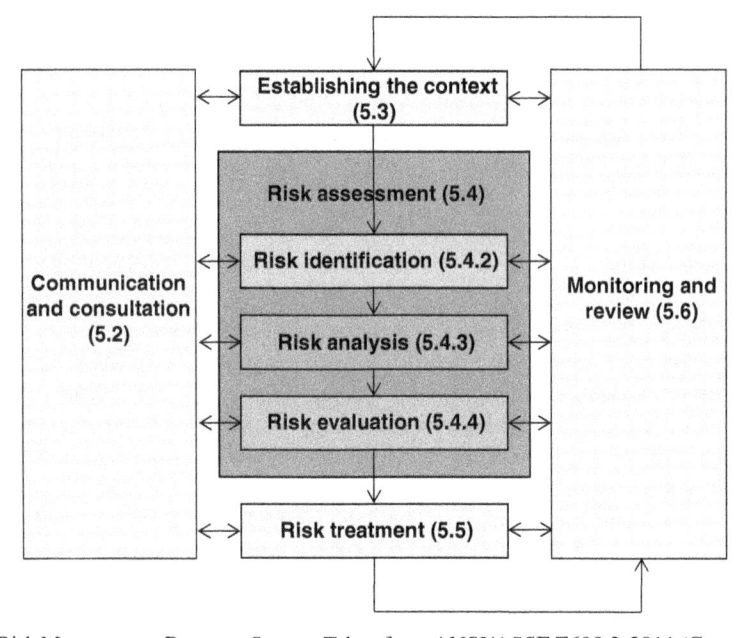

Figure 11.1 Risk Management Process. *Source*: Taken from ANSI/ASSE Z690.2-2011 (Courtesy of the American Society of Safety Engineers)

be applied to an entire organization, as well as specific processes, activities, or projects. The process of managing risk involves applying an internationally recognized process depicted in Figure 11.1 reprinted with permission from the ASSE. The steps in this process outlined in ISO 31000/ANSI Z690.2 include establishing communication (5.2) and context (5.3); risk assessment (5.4) that involves identification (5.4.2), analysis (5.4.3), and evaluation(5.4.4); treatment of risk (5.5); monitoring and reviewing risk (5.6); and recording and reporting the results appropriately.

11.3 THE CONCEPT OF PREVENTION THROUGH DESIGN (PtD)

As the highest rung on the hierarchy of controls ladder, avoidance of hazards and risk should be the absolute first choice. This is echoed in the Prevention through Design principles and initiatives discussed in this text. Fundamentally, and practically, it makes the most sense to avoid a problem rather than allow it to exist and try to manage it. This is the concept of PtD.

Although there were earlier efforts to establish prevention through design, in 1994 the ASSE released a position paper approved by the Board of Directors to promote gathering of knowledge and application of "Designing For Safety" concepts. This was followed by the National Safety Council in 1995 with the establishment of the *Institute for Safety through Design*. The Institute was formed to fulfill a need to integrate hazard analysis and risk assessment into the early stages of design so that hazards and risks could be avoided and minimized to an acceptable level. In 1999, the Institute published "*Safety through Design*," a significant work with 20 contributing authors edited by Fred Manuele and Wayne Christensen. The book provides a number of examples from various industries of safety being incorporated into the design process and the benefits derived. The research and work developed through the Institute and the National Safety Council was instrumental in the current Prevention through Design concepts.

In 2007, the National Institute for Occupational Safety and Health (NIOSH) launched the *Prevention through Design (PtD)* initiative. Like *Safety through Design*, the purpose of PtD is to "design out" or minimize hazards and risks. Experienced safety professionals are well aware that eliminating or avoiding hazards is the most effective way to reduce risk and control occupational injuries, illnesses, and fatalities.

In 2011, an important standard for the safety profession was released. The ANSI/ASSE Z590.3-2011 *Prevention through Design – Guidelines for Addressing Occupational Hazards and Risks in Design and Redesign Processes* standard is the first standard to address such needs in the design and redesign phase. Its purpose is to provide a framework for effectively preventing or minimizing work-related hazards and risks associated with the construction, manufacture, use, maintenance, and disposal of facilities, materials, equipment, and the service sector. One of the stated goals of the standard is to educate designers, engineers, machinery and equipment manufacturers, SH&E professionals, business leaders, and workers in the PtD principles so that they are applied to the design and redesign of new and existing facilities, processes, equipment, tools, and organization of work.

11.4 RISK ASSESSMENT PROCESS AND THE PTD MODEL

A review of the two standards and their structures reveals similarities and differences. The ISO 31000/ANSI/ASSE Z690.2-2011 *Risk Management Principles and Guidelines* standard includes three main sections:

- Risk Management Principles (Clause 3)
- Framework (Clause 4)
- Process (Clause 5).

Within the ANSI/ASSE Z590.3-2011 Prevention through Design standard, there are six primary sections:

- Roles and Responsibility (Section 4)
- Relationships with Suppliers (Section 5)
- Design Safety Reviews (Section 6)
- The Hazard Analysis and Risk Assessment Process (Section 7)
- Hazard Analysis and Risk Assessment Techniques (Section 8)
- Hierarchy of Controls (Section 9).

ISO 31000 (later adopted nationally as ANSI/ASSE Z690.2-2011) was not one of the standards referenced in the ANSI/ASSE Z590.3-2011 Prevention through Design standard; however, the principles align with the risk management process. PtD is a critical concept to the management of risk and should be integrated into an organization's risk management process. SH&E professionals should lead this effort and help to facilitate its integration.

Within the ISO 31000/ANSI Z690.2, Clause 5 defines the risk management process as shown in Figure 11.1. Central to this process is *Risk Assessment* (Section 5.4). According to ISO 31000/ANSI Z690.2, Risk Assessment is defined in the following.

- **Risk Assessment** – the overall process of risk *identification*, risk *analysis*, and risk *evaluation* (ISO 31000/ANSI Z690.2-2011).

Similarly, Section 7 of the ANSI/ASSE Z590.3 PtD standard defines process slightly different as the *Hazard Analysis and Risk Assessment Process*. The flowchart shown in Figure 11.2 is adapted from the standard and illustrates the process.

Obviously, the risk assessment process in the ANSI Z590.3 PtD standard includes a few additional and more detailed steps as compared to the ISO 31000/ANSI Z690.2 process, as does its definition for risk assessment indicated in the following.

- **Risk Assessment** – a process that commences with hazard identification and analysis, through which the probable severity of harm or damage is established, followed by an estimate of the probability of the incident or exposure occurring, and concluding with a statement of risk (ANSI Z590.3-2011, 3.10, p. 10).

Notice that the PtD standard refers to "hazard identification and analysis" where the ISO 31000/ANZI Z690.2 definition uses "risk identification" and "risk analysis." These differences are largely attributed to the context, scope, and purpose of each standard. ISO 31000/ANSI Z690.2 is much broader in its application addressing risk management concepts, while ANSI Z590.3 is more specifically focused on design level hazards and risks.

Although there are subtle differences, the authors of this text see a direct link between the ISO 31000/ANSI Z690 series and the ANSI Z590.3 PtD standard. Both standards provide sound guidance on risk assessment and fundamental techniques. ISO 31010/ANSI Z690.3-2011, *Risk Assessment Techniques*, covers more than 30 different risk assessment methods, while ANSI Z590.3 addresses eight common techniques.

The PtD model was developed based on selected risk assessment techniques discussed in both standards. The authors believe that the PtD standard concepts should be incorporated into Clause 5 of the ISO 31000/ANSI Z690.2 risk management process. Suggestions for ISO 31000/ANSI Z690.2 and ANSI/ASSE Z590.3-2011 integration are presented in Figure 11.3.

RISK ASSESSMENT PROCESS AND THE PTD MODEL

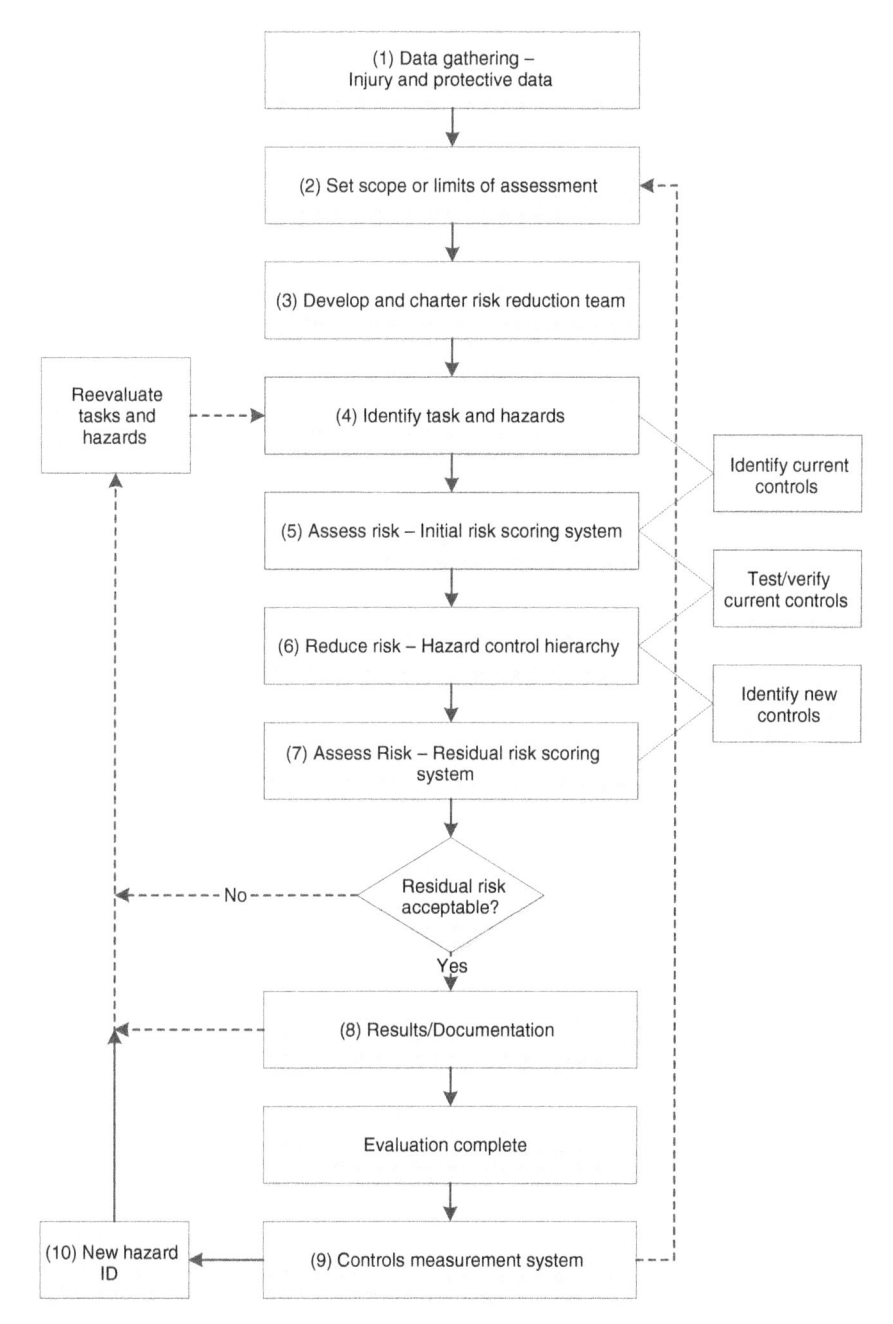

Figure 11.2 Hazard Analysis and Risk Assessment Process. *Source*: Taken from ANSI/ASSE Z590.3-2011 (Courtesy of the American Society of Safety Engineers)

As illustrated in Figures 11.3 and 11.4, the two standards are similar in the process steps with a few differences. In ANSI Z590.3 Section 7.8, the selection of a risk assessment matrix is listed as the second step in the process.

SH&E professionals are facing increased pressure to diversify and develop their skills in new risk assessment techniques. It is vital that professionals know how to develop tools and models to

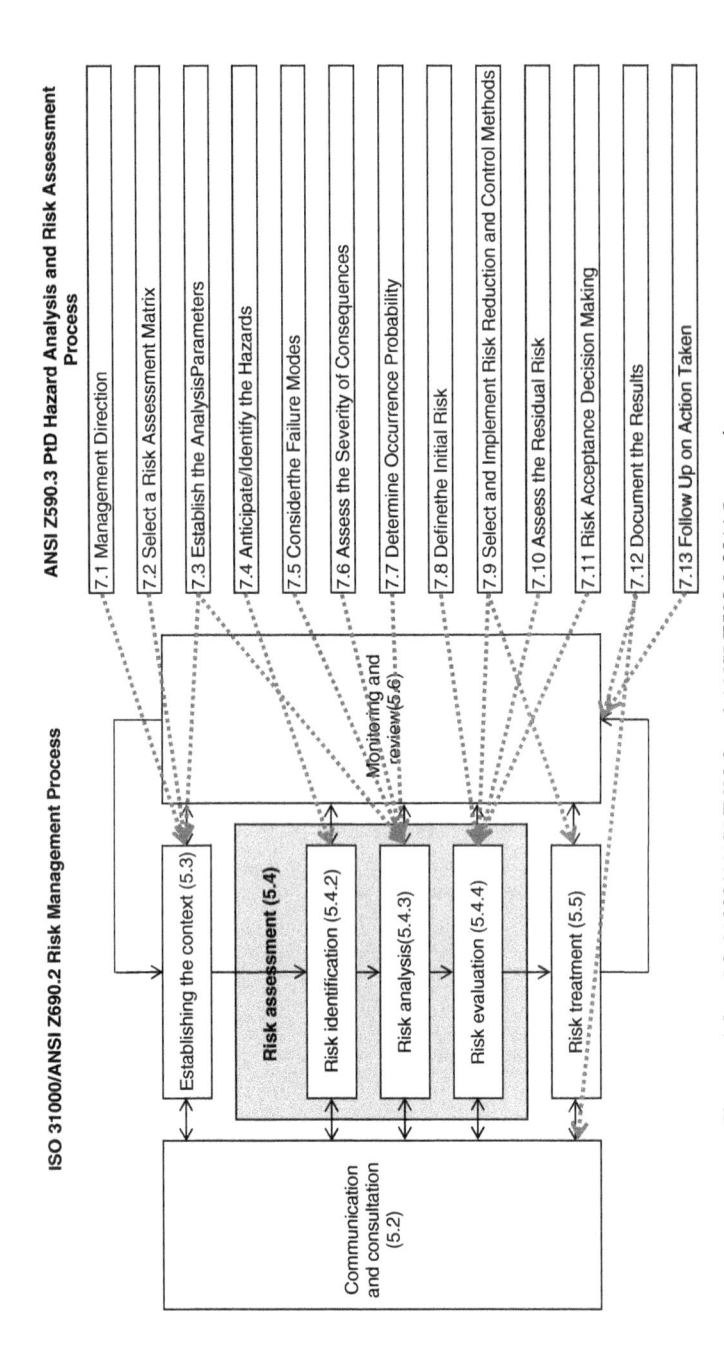

Figure 11.3 ISO 31000/ANSI Z690.2 and ANSI Z590.3-2011 Integration

RISK ASSESSMENT PROCESS AND THE PTD MODEL

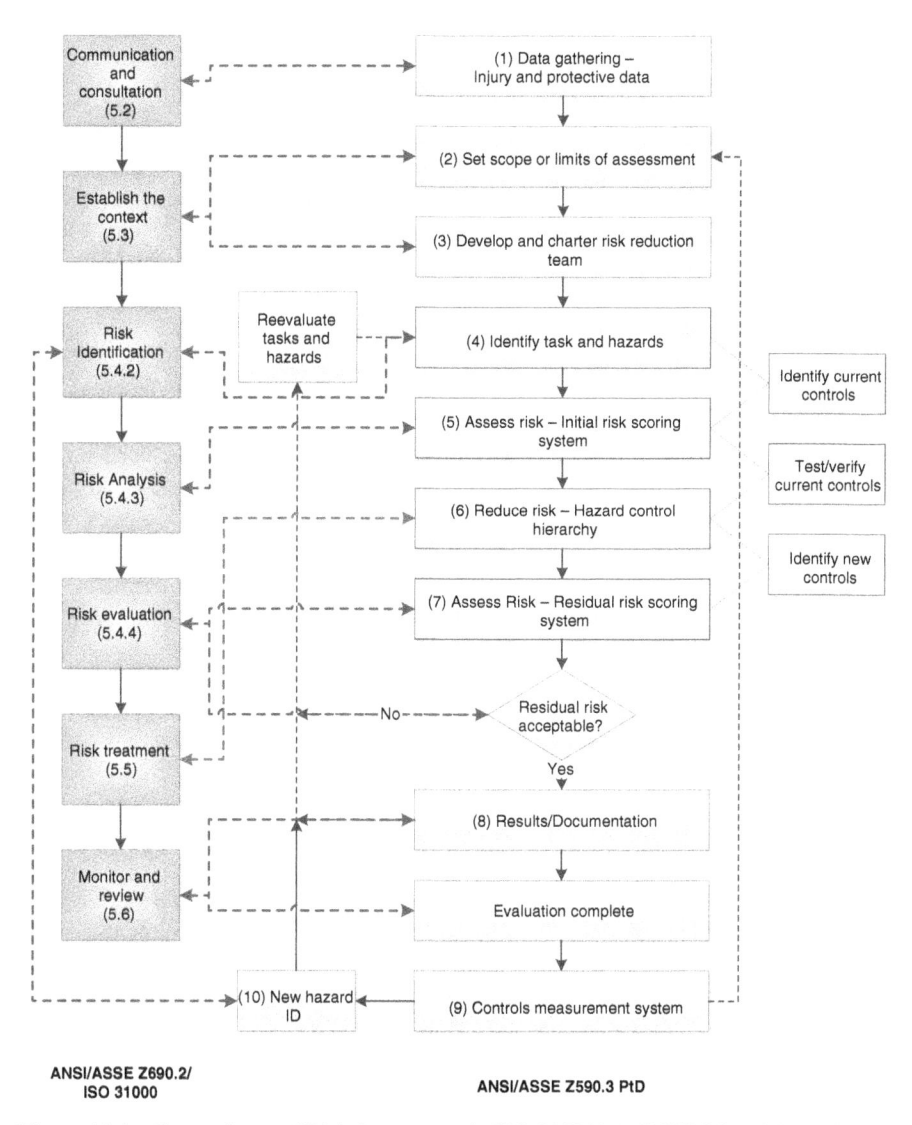

Figure 11.4 Comparisons of Risk Assessment in ISO 31000/ANSI Z690.2 and ANSI Z590.3

incorporate appropriate hazard identification and risk assessment techniques into the risk management process. Some of the newly developed tools are based on the recommended risk assessment techniques referenced in both standards. It is also noted that the PtD model follows the well-established Six Sigma practice of Define, Measure, Analyze, Improve, and Control (DMAIC) logic. Many risk assessment tools, including those described in the PtD standard, can be used to effective identify, assess, and manage risk of new products. The PtD model presents a logical step by step approach to conduct hazard analysis and risk assessment in all life cycles of a product or system. To demonstrate a practical application of the PtD model, the following case study is presented.

11.5 CASE STUDY

A small-size manufacturing company requested that a hazard analysis and risk assessment be performed on a new product. The company's products are intended for export to the European Union, which requires all products meet ISO standard requirements. However, the product is manufactured in the United States. For a number of beneficial reasons, senior management wanted to implement PtD principles. The author developed new tools and successfully implemented the new PtD model to evaluate the product.

The model allows different solutions to be evaluated and prioritized. The Excel-based tool (https://centralspace.ucmo.edu/handle/123456789/463) helped the SH&E professionals compare PtD-based design to an existing product that was not developed according to PtD principles. To satisfy the new product expectations and gain support for SH&E improvements, the team had to develop a new PtD-based risk assessment methodology.

Specifically, the purpose of this project was to determine the noise levels, hand and arm vibration risk, and potential particulate matter (PM) exposure from a normal production unit. The risk assessment evaluation included sound level meter, hand and arm vibration instruments, and PM measurement system. The evaluation was conducted during simulated work activities.

11.5.1 Methods

A new decision-making model was developed to evaluate a new product intended for export. This case study identified potential areas of SH&E professional involvement in the decision-making process. The authors developed a new PtD model that incorporates risk assessment, hierarchy of controls, and future state risk reduction. The model follows DMAIC logic. Separate tools were developed for each phase. For instance, brainstorming and preliminary hazard analysis (PHA) were used in the "Define" phase. A modified Bow-Tie diagram, risk assessment matrix, and failure mode effect analysis (FMEA) were used in the "Measure" phase.

Applicability of FMEA tools to prioritize the hazards and modify the procedures was utilized to demonstrate and quantify the risk reduction after the proposed SH&E improvements. Hand and arm vibration, noise levels, and air pollutants emissions were evaluated.

To demonstrate the applicability of the PtD model integration into ISO 31000/ANSI/Z690.2 risk management process, the authors evaluated two different products as discussed in the following case study.

11.5.2 Results

Utilization of hazard analysis and risk assessment tools described in both ISO 31010/ANSI Z690.3 and ANSI Z590.3 PtD including FMEA and a modified model were estimated to significantly reduce the risks of ergonomics injuries, noise levels, and air pollutants of the product evaluated in this case study. Similar benefits are possible for products manufactured in the United States and intended for the European Union market.

11.5.2.1 Hand and Arm Vibration Evaluation Three industrial cleaner vacuum units were evaluated utilizing the VibTrack/HAVSense system. HAVSense® is an autonomous vibration dosimeter that records the operator's exposure to hand and arm vibration. HAVSense provides monitoring that satisfies the requirements of the European Directive 2002/44/EC. (European Agency for Health and Safety at Work, 2002) The directive was issued in June 2002 and defines the minimum health and safety requirements regarding the exposure of workers to the risks arising from physical agents (vibration) and forms the 16th individual directive within the meaning of Article 16(1) of Directive 89/391/EEC. European directive 2002/44/EC lays down minimum requirements, in particular

Figure 11.5 HAVSense Sensor Inside Operator's Glove

the fixing of lower values for the daily exposure limit value (ELV) for vibrations. The European Directive acknowledges the possible damaging consequences of vibration for human health and lays down maximum levels of vibration exposure.

The HAVSense sits between the second and third fingers of either hand. The topside rests over the fingers. The underside rests under the fingers and is pressed by the fingers against the operating surface. The HAVSense was placed comfortably inside a protective glove for the data collection (see Figure 11.5).

Exposure data was downloaded directly to a computer via the docking station. To calculate the exposure, the team used the British Health and Safety Executive (HSE) hand–arm vibration exposure calculator. The assessment of the vibration exposure is calculated in relation to a standardized 8 h daily exposure value A(8). After establishing the A(8) value, this should be compared with the exposure action and limit values. Different units could be compared based on the daily exposure action value (EAV) and the daily ELV.

Exposure Action Value. Whenever an operator is subjected to vibration exposure A(8) exceeding the EAV at 2.5 m/s^2, the employer must carry out a risk assessment of the operation and introduce control measures. For more details, see Directive 2002/44/EC and Member State legislation.

Exposure Limit Value. In any event, workers shall not be exposed above the ELV (5.0 m/s^2).

Results revealed that the right hand of the operator is exposed slightly more than the left hand.

British HSE hand–arm vibration exposure calculations made by the survey team for left-hand exposure are shown in Table 11.1. (Health and Safety Executive A, n.d.; Health and Safety Executive B, n.d.).

HSE hand–arm vibration exposure calculations for right-hand exposure made by the survey team are shown in Table 11.2.

As indicated in the results, there was a difference in the vibration measurements for the right and left hands. The difference could be explained with the fact that the operator is right handed. In addition, the design of the units could be another contributing factor.

The operator should not operate/run the industrial vacuum cleaner more than 3.1 h based on right-hand exposure alone. Right-hand exposure is considered a worst-case scenario.

TABLE 11.1 Hand–Arm Vibration Sampling Data: Left Hand

	Vibration magnitude m/s² r.m.s	Exposure points per hour	Time to reach EAV 2.5 m/s² A(8) hours	minutes	Time to reach ELV 5 m/s² A(8) hours	minutes	Exposure duration hours	minutes	Partial exposure m/s² A(8)	Partial exposure points
Tool or process 1	5.3659	58	1	44	6	57	3	24	3.5	196
Tool or process 2	0						0			
Tool or process 3										
Tool or process 4										
Tool or process 5										
Tool or process 6										

Daily exposure m/s² A(8): 3.5 Total exposure points: 196

TABLE 11.2 Hand–Arm Vibration Sampling Data: Right Hand

	Vibration magnitude m/s² r.m.s	Exposure points per hour	Time to reach EAV 2.5 m/s² A(8) hours	minutes	Time to reach ELV 5 m/s² A(8) hours	minutes	Exposure duration hours	minutes	Partial exposure m/s² A(8)	Partial exposure points
Tool or process 1	7.9648	127	0	47	3	9	3	24	5.2	432
Tool or process 2	0						0			
Tool or process 3										
Tool or process 4										
Tool or process 5										
Tool or process 6										

Daily exposure m/s² A(8): 5.2 Total exposure points: 432

11.5.2.2 Noise Measurements The same units were evaluated for nose exposure. The sampling was conducted based on 2000/14/EC requirements.

OSHA sets legal limits on noise exposure in the workplace in the United States. These limits are based on a worker's time weighted average over an 8 h day. With noise, OSHA's permissible exposure limit (PEL) is 90 dBA for all workers for an 8 h day (Occupational Safety and Health Administration, n.d.).

The OSHA standard uses a 5 dBA exchange rate. This means that when the noise level is increased by 5 dBA, the amount of time a person can be exposed to the new noise level (now 95 dBA) to receive the same dose is cut in half (or 4 h).

The NIOSH has recommended that all worker exposures to noise should be controlled below a level equivalent to 85 dBA for 8 h to minimize occupational noise-induced hearing loss. (National Institute for Occupational Safety and Health, n.d.). NIOSH also recommends that the exchange rate be 3 dBA.

The British HSE Noise Regulations also require specific action at certain action values (Health and Safety Executive, 2005). These relate to the below:

Lower EAVs:

- Daily or weekly exposure of 80 dB
- Peak sound pressure of 135 dB.

Upper EAVs:

- Daily or weekly exposure of 85 dB
- Peak sound pressure of 137 dB.

There are also levels of noise exposure that must not be exceeded. These are called ELVs:

- Daily or weekly exposure of 87 dB
- Peak sound pressure of 140 dB.

None of the tested units exceeded the 85 dBA noise level.

11.5.2.3 PM Exposure Measurements Two units were tested for PM emissions using a DustTrak DRX PM Measurement system. The purpose of the test was to evaluate the PM levels approximately one (1) meter from the dust collection system.

11.5.3 Occupational Size-Selective Criteria and Particles Size Sampling

Occupational health and safety professionals have traditionally sampled for two particulate size fractions: total or respirable (Occupational Safety and Health Administration, n.d.):

Total particulate includes both respirable and nonrespirable particles, also known as Particulates not otherwise regulated (PNOR). OSHA PEL – **15 mg/m^3 (15,000 µg/m^3) TWA**.

Respirable particulate includes only the smaller particles than can penetrate to the alveolar or gas-exchange region of the lung. PNOR *respirable fraction* – OSHA PEL – **5 mg/m^3 (5000 µg/m^3) TWA**.

At this time, US OSHA and MSHA still use total and respirable particulate size fractions for regulatory standards and compliance monitoring.

The company's engineering unit designed a special filtering dust containment system to reduce the PM pollution and operators' exposure. The results are presented in Table 11.3 and Figure 11.6.

Figures 11.7 and 11.8 display the results. The used dust containment system is the most efficient. It provides reduced operator exposure, and it is more protective of the environment. The new unused dust containment system is also effective. However, it might be concluded that the collection efficiency increases with accumulation of the particles on the inner surfaces of the filtering bag.

TABLE 11.3 PM Exposure Measurements

	Respirable Particles 4 µg/m^3	Total PNOR µg/m^3
Used dust containment system	72.45	81.53
New dust containment system	91.8	107.12
No dust containment system	118.275	161.7
No bag	296.95	367.63

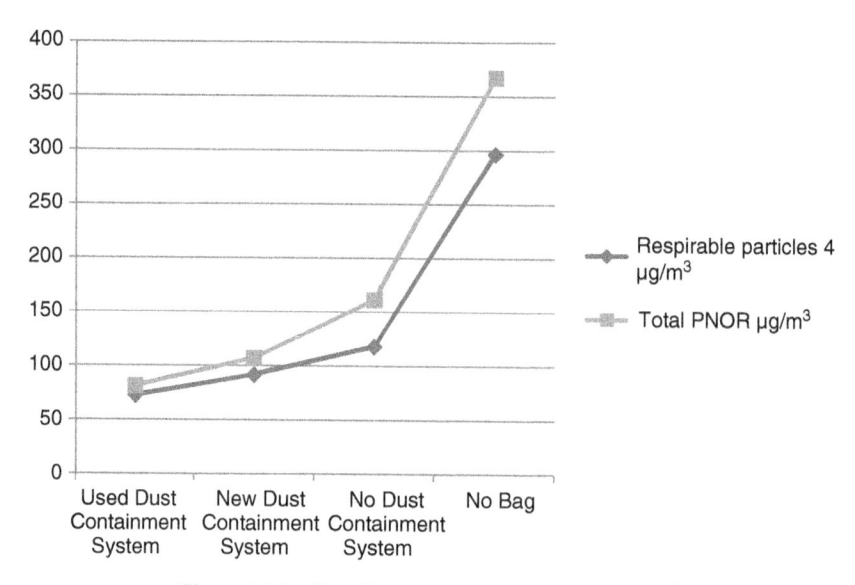

Figure 11.6 Dust Exposure Measurements in µg/m³

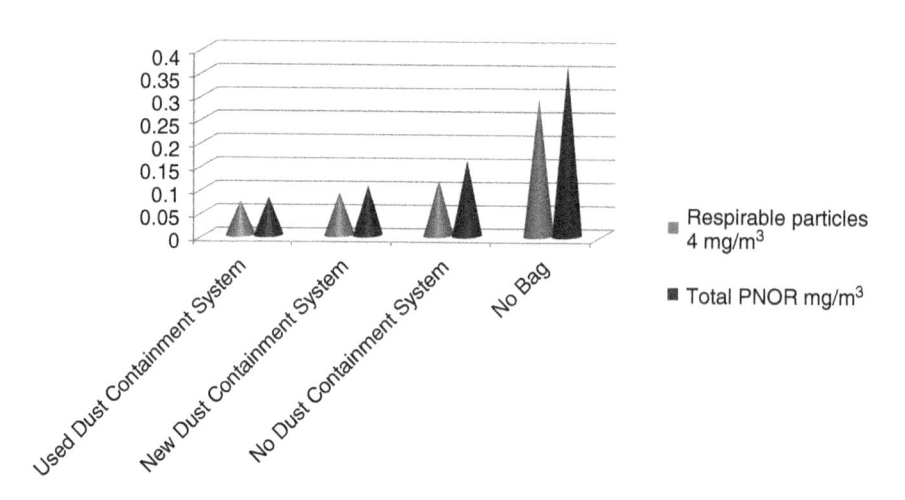

Figure 11.7 Dust Exposure Measurements in mg/m³

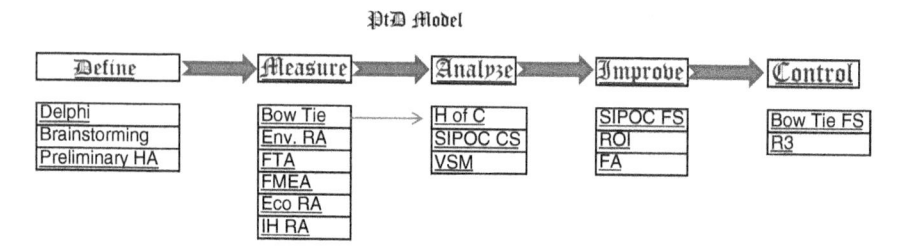

Figure 11.8 The PtD Model

CASE STUDY 239

PRELIMINARY HAZARD ANALISIS WITH TRACKING LOG

		Date:	07-08-2015
Project/Process			
Prepared by:	Dr. G. Popov		
Methods Used			
Hazardous Event	**System Effects**	**RF or RPN**	**Comments**
Wrist injury	Operator affected	27	
HAVS	Operator affected	24	
Noise	Operator and Community affected	9	

Figure 11.9 Current State PHA

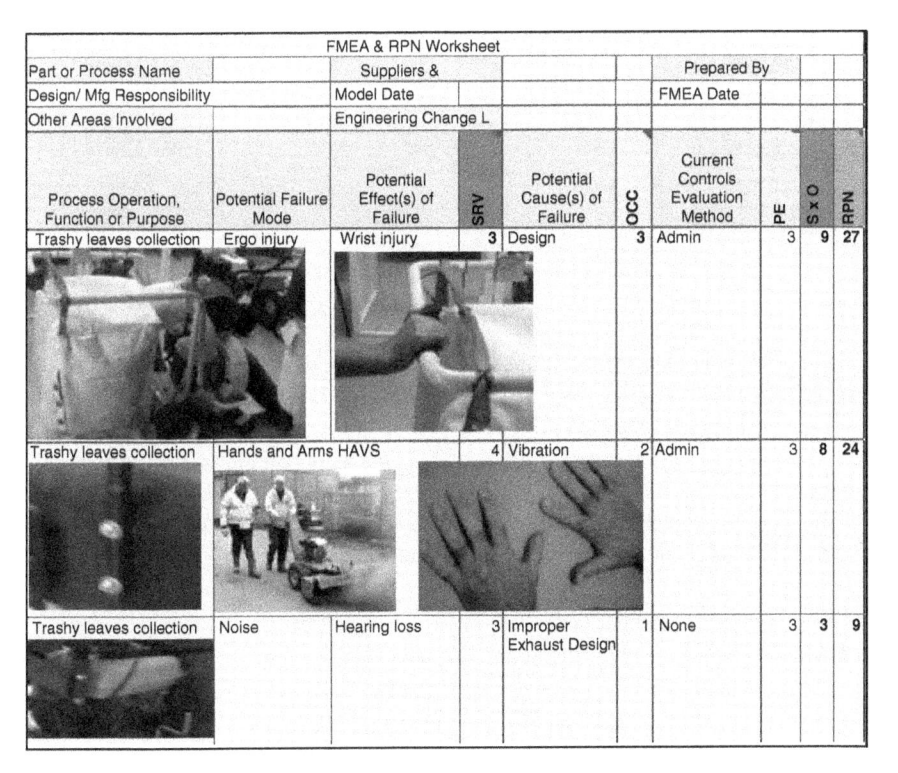

FMEA & RPN Worksheet

Part or Process Name		Suppliers &				Prepared By			
Design/ Mfg Responsibility		Model Date				FMEA Date			
Other Areas Involved		Engineering Change L							
Process Operation, Function or Purpose	Potential Failure Mode	Potential Effect(s) of Failure	SRV	Potential Cause(s) of Failure	OCC	Current Controls Evaluation Method	PE	S x O	RPN
Trashy leaves collection	Ergo injury	Wrist injury	3	Design	3	Admin	3	9	27
Trashy leaves collection	Hands and Arms HAVS		4	Vibration	2	Admin	3	8	24
Trashy leaves collection	Noise	Hearing loss	3	Improper Exhaust Design	1	None	3	3	9

Figure 11.10 FMEA and RPN Worksheet

The sampling results indicate potential operator exposure is well below occupational exposure limits.

After a careful evaluation of the results, the authors developed a new PtD model. The model follows DMAIC logic. (Popov & Zey, 2012). This PtD model is shown in Figure 11.8.

During the Define phase, PHA was performed for hand–arm vibration, noise, and PM exposures. The current state PHA example is presented in Figure 11.9.

Risk priority number (RPN) can be calculated using standard FMEA and RPN worksheet as shown in Figure 11.10, where

- SEV = Severity
- OCC = Occurrence/probability
- PE = Prevention effectiveness.

[See Chapter 8 FMEA for more detailed Severity, Probability, and Prevention Effectiveness rating scales.]

Based on the initial limited hazard analysis, a Bow-Tie analysis diagram was prepared. Figure 11.11 presents the current state risk assessment.

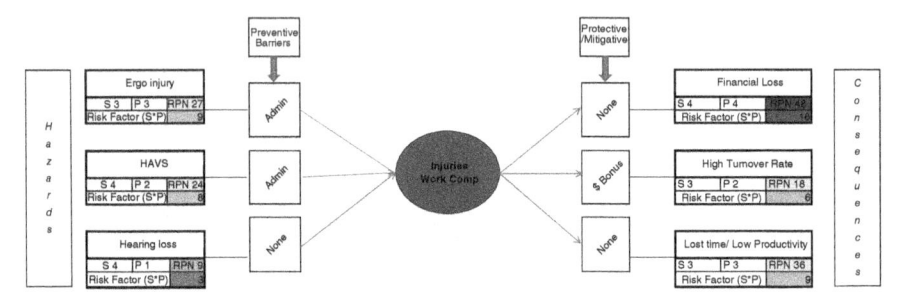

Figure 11.11 Current State Risk Assessment

The PtD hierarchy of controls (see Figure 11.12) was utilized to develop suggestions for engineering controls. A better handle design was suggested. Polyurethane dampers could reduce vibrations and a new muffler and lower RPMs could further reduce noise.

Based on PtD improvements, future state FMEA RPNs were calculated (see Figure 11.13).

Future state Bow-Tie risk analysis was prepared based on the future state RPNs (see Figure 11.14).

Additionally, residual risk and risk reduction were calculated. SH&E improvements resulted in a 55% (S × P) and 85% (RPN) risk reduction; see Figure 11.15, where

- Risk factor is severity times probability.
- RPN is severity times probability and prevention effectiveness.
- Simple mathematical equations can be applied to calculate risk reduction and residual risk. For instance, percent risk reduction is calculated based on the following formula:

$$\%\text{Risk Reduction} = \frac{\text{Risk Factor CS} - \text{Risk Factor FS}}{\text{Risk Factor CS}}$$

- Residual risk is simply 100%-% risk reduction.

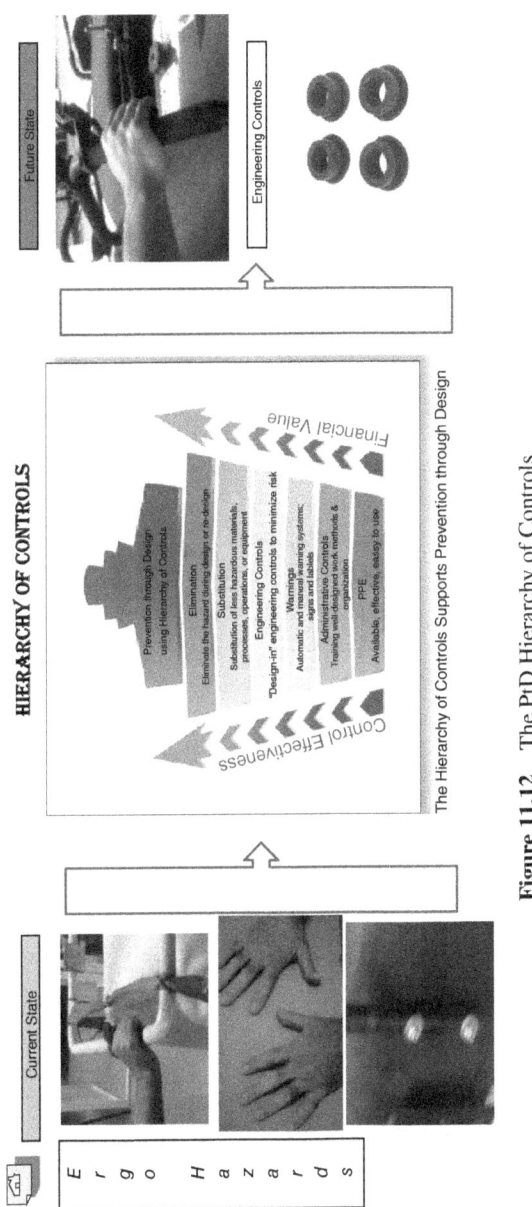

Figure 11.12 The PtD Hierarchy of Controls

FMEA AND RPN WORKSHEET

Part or Process Name		Suppliers and		Prepared By											
Design/ Mfg Responsibility		Model Date		FMEA Date											
Other Areas Involved		Engineering Change Level													
Process Operation, Function or Purpose	Potential Failure Mode	Potential Effect(s) of Failure	SEV	Potential Cause(s) of Failure	OCC	Current Controls Evaluation Method	PE	S×O	RPN	Recommended Action(s)	Area/Individual Responsible and Completion Date	Action Results Actions Taken	SEV 2	OCC 2	PE2	RPN 2
Trash/ leaves collection	Ergo injury	Wrist injury	3	Design	3	Admin	3	9	27	New Design	Eng. and Management	Improved wrist position	3	1	1	3
Trash/ leaves collection	Hands and arms	HAVS	4	Vibration	2	Admin	3	8	24	PU dampers	Eng. and management	Reduced exposure	3	1	1	3
Trash/ leaves collection	Noise	Hearing loss	3	Improper exhaust design	1	None	3	3	9	New muffler and low RPM	Eng. and management	Reduced noise levels	3	1	1	3

Figure 11.13 Future State FMEA

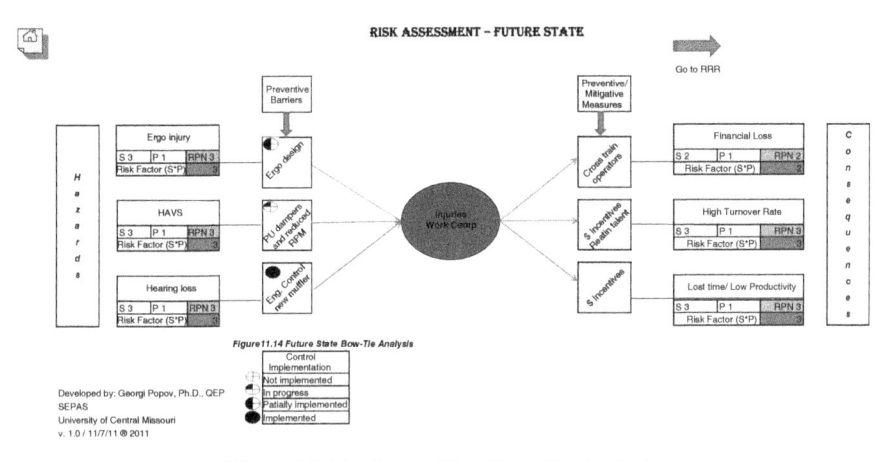

Figure 11.14 Future State Bow-Tie Analysis

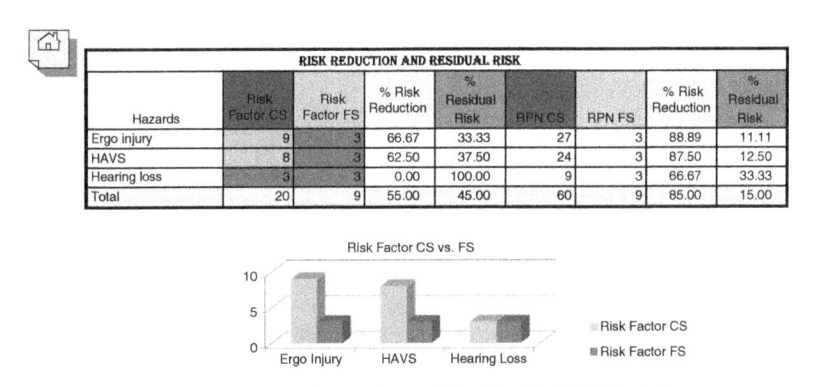

Hazards	Risk Factor CS	Risk Factor FS	% Risk Reduction	% Residual Risk	RPN CS	RPN FS	% Risk Reduction	% Residual Risk
Ergo injury	9	3	66.67	33.33	27	3	88.89	11.11
HAVS	8	3	62.50	37.50	24	3	87.50	12.50
Hearing loss	3	3	0.00	100.00	9	3	66.67	33.33
Total	20	9	55.00	45.00	60	9	85.00	15.00

Figure 11.15 Risk Reduction and Residual Risk Calculations

11.6 PTD AND THE BUSINESS PROCESS

As stated in the NIOSH PtD initiative "PtD's purpose is to promote this concept and highlight its importance in all business decisions." PtD integration into the business decision will be discussed in greater detail in Chapter 19, *Business Aspects of Operational Risk Assessment*. However, the authors strongly believe that SH&E professionals can play a significant role in new products development and business decision making through active participation in the process. PtD principles can be successfully integrated into the ISO 31000/ANSI Z690.2 risk management process. The case study described in this chapter illustrates how the process can lead to a decision by management to approve the new product design, which will result in reduced ergonomics injuries, reduced emissions, and improved operator productivity. These changes should also enhance the company's abilities to sell their products on the global market. In addition, the case study effectively demonstrates that risk reduction is a vital business benefit for the company.

It was concluded that PtD tools could be successfully incorporated in the risk management and business decision-making process. Such processes can be used effectively to develop and present business cases for environmental, health, and safety interventions.

11.7 SUMMARY

This chapter has provided an example of effectively integrating the Prevention through Design concept into the ISO 31000/ANSI Z690.2 risk management standard. There are many other specific risk assessment techniques and assessment methods used in numerous companies and industries, nonprofit organizations, and governmental agencies. However, PtD methodologies remain critical part of the risk management process. The case study and the PtD tool provided an example of hazard identification, initial risk assessment, hierarchy of controls, consequent risk analysis, and evaluation to manage risk to a tolerable level. The case study also demonstrated that the PtD tools could be successfully incorporated in the risk management and business decision-making process.

REVIEW QUESTIONS

1. Summarize the efforts to establish the Prevention through Design initiative.
2. Explain the concept of Prevention through Design
3. Describe differences and similarities between ISO 31000/ANSI/ASSE Z690.2-2011 *Risk Management Principles and Guidelines* standard and ANSI/ASSE Z590.3-2011 Prevention through Design standards.
4. Express your opinion about Risk Management Principles and PtD standard integration.
5. Identify a risk assessment method discussed in this chapter that you prefer and explain why.

REFERENCES

ANSI/ASSE Z590.3-2011. *Prevention Through Design: Guidelines for Addressing Occupational Hazards and Risks in Design and Redesign Processes*. Des Plaines, IL: American Society of Safety Engineers, 2011.

ANSI/ASSE Z690.1. *Vocabulary for Risk Management*. Des Plaines, IL: American Society of Safety Engineers, 2011.

ANSI/ASSE Z690.2. *Risk Management Principles and Guidelines*. Des Plaines, IL: American Society of Safety Engineers, 2011.

ANSI/ASSE Z690.3. *Risk Assessment Techniques*. Des Plaines, IL: American Society of Safety Engineers, 2011.

European Agency for Health and Safety at Work. 2002. Directive 2002/44/EC – Vibration. Information available at: https://osha.europa.eu/en/legislation/directives/exposure-to-physical-hazards/osh-directives/19 (accessed December 19, 2015) Information obtained February 21, 2013.

Health and Safety Executive. 2005. Noise at Work. Employers' Responsibilities – Legal Duties. Retrieved from http://www.hse.gov.uk/noise/employers.htm (accessed January 6, 2016).

Health and Safety Executive. n.d.-a. Hand–Arm Vibration Exposure Calculator. Information available at: http://www.hse.gov.uk/vibration/hav/readyreckoner.htm (accessed January 6, 2016).

Health and Safety Executive. n.d.-b. Exposure Points System and Ready-Reckoner. Information available at: http://www.hse.gov.uk/vibration/hav/readyreckoner.htm.

National Institute for Occupational Safety and Health (NIOSH). n.d. Noise and Hearing Loss Prevention. Workplace Health and Safety Topics. Retrieved from http://www.cdc.gov/niosh/topics/noise/ (accessed January 6, 2016).

NIOSH PtD Initiative. 2013. PtD Initiative (Ref: http://www.cdc.gov/niosh/topics/ptd/) (accessed January 6, 2016).

REFERENCES

Occupational Safety and Health Administration. n.d. Occupational Noise Exposure. OSHA Quick Takes. Information available at: http://www.osha.gov/SLTC/noisehearingconservation/ (accessed December 19, 2015).

Occupational Safety and Health Administration. n.d.-b Table Z-1. Limits for Air Contaminants. OSHA Quick Takes. Retrieved from: https://www.osha.gov/pls/oshaweb/owadisp.show_document?p_table=STANDARDS&p_id=9992.

Popov, G., Zey, J. N., 2012. Prevention Through Design (PtD): Combining Risk Assessment, Productivity and Sustainability. Presented at ASSE 2012, Session # 703. Proceedings Safety2012. ASSE.

12

INDUSTRIAL HYGIENE RISK ASSESSMENT

GEORGI POPOV AND STEVEN HICKS
School of Environmental, Physical & Applied Sciences, University of Central Missouri, Warrensburg, MO, USA

TSVETAN POPOV
Inspectorate Division, Organization for the Prohibition of Chemical Weapons (OPCW), The Hague, The Netherlands

12.1 OBJECTIVES

- Introduce industrial hygiene risk assessment
- Fundamental concepts
- Health risk assessments and prioritization
- Practical application
- Occupational health risk and PtD process alignment
- Summary
- Practice exercises/questions

12.2 INTRODUCTION

What is meant by the term "industrial hygiene risk assessment"? No universal methodology exists for assessing industrial hygiene (IH) risk that is applicable to all situations and all workplace health hazards. The conundrum of IH risk assessment reminds readers of a certain Donald Rumsfeld quote:

> Reports that say that something hasn't happened are always interesting to me, because as we know, there are known knowns; there are things we know we know. We also know there are known unknowns; that is to say we know there are some things we do not know. But there are also unknown unknowns – the ones we don't know we don't know. And if one looks throughout the history of our country and other free countries, it is the latter category that tend to be the difficult ones. (Rumsfeld, 2002)

Risk Assessment: A Practical Guide to Assessing Operational Risks, First Edition.
Edited by Georgi Popov, Bruce K. Lyon, and Bruce Hollcroft.
© 2016 John Wiley & Sons, Inc. Published 2016 by John Wiley & Sons, Inc.

This anecdote illustrates the depth and breadth of occupational risk and underscores that IH risk assessments not only serve to help evaluate known or anticipated exposures but also reveal health-related risks that were previously undiscovered. In other words, IH risk assessments address both "known unknowns" and "unknown unknowns."

IH, sometimes referred to as occupational health, is the science of anticipating, recognizing, evaluating, and controlling workplace conditions that may cause workers' injury or illness (OSHA, 1998). The American Industrial Hygiene Association (AIHA) describes the IH practice as the science and art devoted to the anticipation, recognition, evaluation, prevention, and control of those environmental factors or stresses, arising in or from the workplace, which may cause sickness, impaired health and well-being, or significant discomfort among workers or the citizens of the community (AIHA, 2015).

Being a very dynamic and complex discipline, it is difficult to describe where IH starts and ends in the occupational safety, health, and environmental (SH&E) field. With a great deal of overlap in the field, there are no clear professional boundaries for the IH practice. For instance, occupational ergonomics is considered within the IH domain, yet many safety and engineering professionals work in this area. Environmental concerns are often addressed by industrial hygienist. And some professionals are charged with coordinating all SH&E efforts.

In addition, there is considerable debate in the professional IH community regarding the practice of IH risk assessment. Health risk assessment, exposure assessment, exposure risk assessment and management (ERAM) are all terms used by IH professionals to refer to IH risk and its assessment. It should be noted that exposure assessment does not equal a true risk assessment. Exposure assessment can be defined as the process of estimating or measuring magnitude, frequency, and duration of exposure (the dose) to an agent considering the population exposed. It also should describe the sources, pathways, routes, and uncertainties in the exposure assessment. In fact, exposure assessment is aptly viewed as a part of the risk assessment process (Korchevskiy, Rasmuson, & Rasmuson, 2014).

New standards and a rapidly changing profession require proper alignment of traditional IH risk assessment models with the methodologies described in the standards. The risk assessment process is well defined in the American National Standard, *ANSI Z10-2012, Occupational Health and Safety Management Systems*, by the following statement taken from its *Appendix F, Risk Assessment*:

"Risk Assessment is the process to determine the level of risk based on the likelihood the hazard will cause an injury or disease and the severity of the injury or disease that may result. When a hazard is identified and discussion takes place concerning its potential for harm and the probability an injurious incident or exposure can occur, a risk assessment has been made." (ANSI, AIHA, & ASSE, 2012, page 47)

For the purposes of this chapter, the authors will primarily use definitions and risk assessment processes described in ANSI Z10, the International Standards Organization's ISO 31000/ANSI Z690 series and ANSI Z590.3-2011, Prevention through Design standards. Recent developments in standards emphasize the importance of risk assessment in operational risk management systems and in today's IH toolset. Unfortunately, there is no universal tool for IH risk assessment that can be applied indiscriminately to all situations (Jayjock, Lynch, & Nelson, 2000). In this chapter, some of the commonly used health risk assessment methods will be discussed. In addition, newer methodologies and modifications of the existing methods will be introduced.

12.3 FUNDAMENTAL CONCEPTS

Industrial hygienists and occupational health professionals conduct risk assessments on a regular basis using a variety of tools ranging from simple qualitative tools to sophisticated quantitative analysis. A variety of risk assessment methods are described in the latest editions of voluntary standards such as ISO 31010/ANSI Z690.3-2011, Risk Assessment Techniques and ANSI Z590.3.

An important development in consensus standards is the fact that hazard analyses and risk assessments are required. For instance, ANSI Z10-2012 includes a "shall" provision on risk assessment at paragraph 5.1.1. It says, "The organization shall establish and implement a risk assessment process(es) appropriate to the nature of hazards and level of risk." (ANSI, AIHA, & ASSE, 2012, para 5.1.1). The standard does not identify specific tools or methodologies for health risk assessments. However, Appendix F of the standard includes the following definition:

> "Risk assessment is the process to determine the level of risk based on the likelihood the hazard will cause an injury or disease, and the severity of the injury or disease that may result" (ANSI, AIHA, & ASSE, 2012, page 47).

This is a critical development, since the definition specifically includes the potential severity and likelihood factors of a disease. Some of work-related diseases or illnesses are easier to observe or diagnose. For instance, a 18.7 ml (milliliter) mean excess loss in forced vital capacity (FVC) in lung function, due to silica dust exposure, is an adverse health impact that may not be diagnosed as an illness, but it is an observable impairment of health and quality of life (Ehrlich et al., 2011). Decrease in lung function may lead to job transfer or a disability.

In 2002, AIHA published a white paper on risk assessment and risk management. The paper outlined AIHA's policy on the practice of human health risk assessment techniques in regulatory decision making, development of public health policy, and in the allocation of government and private sector resources to occupational and environmental issues (AIHA, 2002). In the same paper, AIHA "supports the use of an iterative approach to the assessment process in which relatively simple techniques are used initially to identify the potential magnitude of the health risk, followed by more sophisticated analyses as needed."

12.4 ANTICIPATING AND IDENTIFYING OCCUPATIONAL HEALTH RISKS

Within the risk assessment process, there are three consecutive components that are performed: (1) hazard/risk identification, (2) risk analysis, and (3) risk evaluation. Like other types of risk, an occupational health risk must be first identified or recognized so that proper analysis and evaluation can be performed.

Occupational health hazards are grouped into four category types: physical, chemical, biological, and ergonomic hazards (please refer to Chapter 5, Appendix B, Common Hazards and Descriptions). These categories are briefly described in the following:

1. **Physical health hazards** – include elements that present a health risk due to their "physical nature" such as noise, vibration, extreme temperature, ionizing and nonionizing radiation, dust, fibers, particulates, and nanomaterial.
2. **Chemical health hazards** – include substances (solid, liquid, gas, vapor, mists, fumes, or dusts) that have a toxic, flammable, corrosive, carcinogenic, or sensitizer nature that present an acute or chronic health risk.
3. **Biological health hazards** – include infectious diseases or pathogens (viruses, bacteria, fungi, and other living organisms) that can cause acute and chronic infections by entering the body either directly or through breaks in the skin. Exposures to biological health hazards can occur from plants and insects, animals, humans, blood and bodily fluids, stagnant water or wet conditions, and infectious wastes.
4. **Ergonomics health hazards** – include exposure to repetitive or cumulative trauma, biomechanical stress, excessive physical demands, and mental demands.

A fundamental approach to anticipating and identifying existing and potential health risks begins with a review of what is "known." Similar to other types of risk, occupational health hazards and their risks can be identified by reviewing available information of the operation or similar operations. Sources that are useful in identifying occupational health hazard/risk include the following:

- A review of the operations, processes, maintenance, or cleaning activities being performed
- Evaluation of materials, chemicals, and substances and their safety data sheets
- Consideration of hazards generated by interactions with other elements and by-products of production
- A review of health incident data history and previously identified health concerns
- The use of specific checklists and inventories of known hazards associated with the industry processes or job classifications
- Previous occupational health or IH studies.
- Capturing and interpreting worker complaints/concerns related to symptoms which may be associated with OSH risk

Upon completion, an inventory of recognized occupational health hazards is compiled and used in the risk analysis and evaluation.

12.5 DETERMINING OCCUPATIONAL HEALTH RISKS

To determine risk level, an evaluation of the potential health risk's severity and likelihood must be calculated. Several methods are available for making this determination and are briefly described in the following sections and examples.

12.5.1 Health Risk Rating Methodology

One technique used to identify the potential magnitude of the health risk is the "health risk rating" (HRR) methodology described in Chapter 6 of AIHA's Exposure Assessment Strategy (Ignacio & Bullock, 2006). Mulhausen, Damiano, and Pullen (2006) defined HRR as a function of the potential health effect caused by the agent and the potential exposure. HRR is calculated based on Equation 12.1 where Health Effect Rating is a semiquantitative numerical rating based on the scheme presented in Table 12.1:

$$\text{HRR} = \text{Health Effect Rating} \times \text{Exposure Rating} \tag{12.1}$$

TABLE 12.1 Health Effect Rating Scheme: AIHA Health Effects Rating

Category	Health Effect
4	Life-threatening or disabling injury or illness
3	Irreversible health effects of concern
2	Severe, reversible health effects of concern
1	Reversible health effects of concern
0	Reversible effects of little concern or no known or suspended adverse health effects

Source: Adapted from Ignacio and Bullock (2006). Copyright 2006 by the American Industrial Hygiene Association.

12.5.2 Exposure Rating Methodologies

Mulhausen, Damiano, & Pullen (2006b) and AIHA (2006) defined exposure rating as an estimate of the exposure level relative to the occupational exposure limit (OEL). Two different exposure rating categorization schemes may be suggested in this chapter. The first one is based on an estimate of the arithmetic mean of the exposure profile relative to the long-term average (LTA) OEL (illustrated in Table 12.2).

TABLE 12.2 Exposure Rating Categorization: Based on an Estimate of the Arithmetic Mean of the Exposure Profile

Exposure Rating Categorization Example	
Numerical Rating	LTA-OEL
4	>LTA-OEL
3	50–100% LTA-OEL
2	10–50% LTA-OEL
1	<10% LTA-OEL

Source: Adapted from Ignacio and Bullock (2006). Copyright 2006 by the American Industrial Hygiene Association.

The second exposure rating scheme is included in ANSI/ASSE Z590.3-2011 *"Prevention through Design: Guidelines for Addressing Occupational Hazards and Risks in Design and Redesign Processes"* standard (ANSI & ASSE, 2011). Table 12.3 depicts a more detailed exposure rating categorization scheme.

12.5.3 Health Effect and Exposure Methodology

HRR Matrix based on Equation 12.1 (page 250) is depicted in Figure 12.1. It is also included in the supplemental materials related to this chapter. The matrix is a 4 × 4 simple multiplication of Health Effect Rating × Exposure Rating. The matrix can be used for single agent's exposures.

12.5.4 COSHH Essentials Tool

One technique used to identify the potential magnitude of the health risk is the Control of Substances Hazardous to Health (COSHH) Regulations' COSHH Essentials online tool. COSHH Essentials was developed by the United Kingdom (UK) Health and Safety Executive (HSE), in collaboration with the Confederation of British Industry (CBI) and the Trades Union Congress (TUC), to help companies and small businesses comply with the COSHH Regulations (Health and Safety Executive, n.d.).

COSHH Essentials provides guidance on controlling the use of chemicals for a range of common tasks, for example, mixing or drying. COSHH Essentials online tool has two aspects:

1. Simple generic risk assessment producing advice on good control practice for common operations – "control guidance sheets"
2. Control guidance sheets for certain industry-specific tasks or processes.

The tool has the status of guidelines developed for use by small and medium-sized enterprises (SMEs). Occupational hygienists (OH) as they are called in the United Kingdom may not be hired as full time employees in such companies. Consequently, employers who can't afford full time OHs

TABLE 12.3 Exposure Rating Categorization Scheme

AIHA Exposure Categorization Scheme				
Exposure Category	Rule-of-Thumb Description	Qualitative Description	Recommended Statistical Interpretation	Notes
0	Exposures are trivial to nonexistent – employees have little to no exposure, with little to no inhalation contact	Exposures, if they occur, infrequently exceed 1% of the OEL	$X_{0.95} \leq 0.01 \times \text{OEL}$	1
1	Exposures are highly controlled – employees have minimal exposure, with little to no inhalation contact	Exposures infrequently exceed 10% of the OEL	$0.01 \times \text{OEL} < X_{0.95} \leq 0.1 \times \text{OEL}$	2
2	Exposures are well controlled – employees have frequent contact at low concentrations and rare contact at high concentrations	Exposures infrequently exceed 50% of the OEL and rarely exceed the OEL	$0.1 \times \text{OEL} < X_{0.95} \leq 0.5 \times \text{OEL}$	2–4
3	Exposures are controlled – employees have frequent contact at low concentrations and infrequent contact at high concentrations	Exposures infrequently exceed the OEL	$0.5 \times \text{OEL} < X_{0.95} \leq \text{OEL}$	2, 4
4	Exposures are poorly controlled – employees often have contact at high or very high concentrations	Exposures frequently exceed the OEL	$X_{0.95} > \text{OEL}$	4

Note: 1 – Category 0 was added to distinguish between highly controlled exposures and situations where exposures are either nonexistent or trivially low. It was included in the 1991 AIHA rating Scheme. 2 – "Infrequently" refers to an event that occurs no more than 5% of the time. 3 – "Rarely" refers to an event that occurs no more than 1% of the time. 4 – "High concentrations" are defined as concentrations that exceed the TWA OEL.

Source: Adapted from "ANSI/ASSE Z590.3-2011 Prevention through Design: Guidelines for Addressing Occupational Hazards and Risks in Design and Redesign Processes," 2011. Copyright 2011 by the American Society of Safety Engineers.

Health Risk Rating Matrix: Numerical Gradings

	Agent 1	Agent 2	Agent 3
Health Effect rating:	4	3	2
Exposure rating:	4	4	3
Total	16	12	6

RA Matrix	Exposure Rating			
	1	2	3	4
Health Effect Rating 4	4	8	12	16
3	3	6	9	12
2	2	4	6	8
1	1	2	3	4

Figure 12.1 *Health Risk Rating Matrix*

DETERMINING OCCUPATIONAL HEALTH RISKS

are unlikely to have skills in chemical risk assessment. COSHH essentials guidance is intended to inform SMEs management but not to limit occupational hygiene professionals (Health and Safety Executive, 2009).

The tool is heavily based on European Union (EU) Risk Phrases (R-Phrases) (Canadian Centre for Occupational Health and Safety, 2012). These phrases are used to describe the risks associated with chemical products. In the future, R-phrases will be discontinued and replaced by Hazard Statements (H Statements). According to EU-OSHA, many of the H Statements will be the same or similar to the R-Phrases (European Agency for Safety and Health at Work, n.d.). The COSHH Essentials online tool is based on the methodology presented in Figure 12.2.

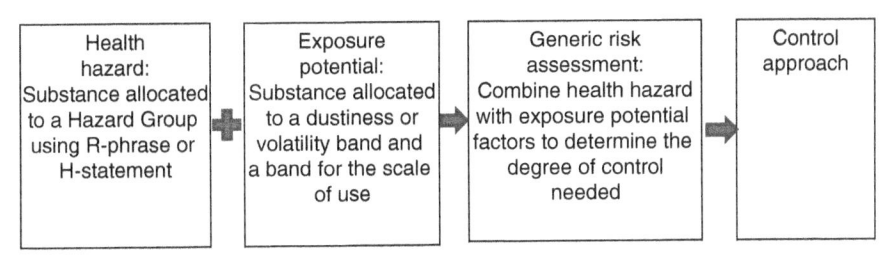

Figure 12.2 COSHH Essentials Methodology

For example, the COSHH Essentials tool will produce the following result for benzene (CAS Number 71-43-2), one of the substances used in the QAP case study illustrated in Figure 12.3.

According to the example, benzene belongs to the Hazard Group "E" (special cases) and, from a control standpoint, it is recommended to substitute the chemical with a less toxic material. However, one of the limitations of COSHH Essentials tool is that it does not sufficiently address the additive or synergetic effect of combined chemical exposures.

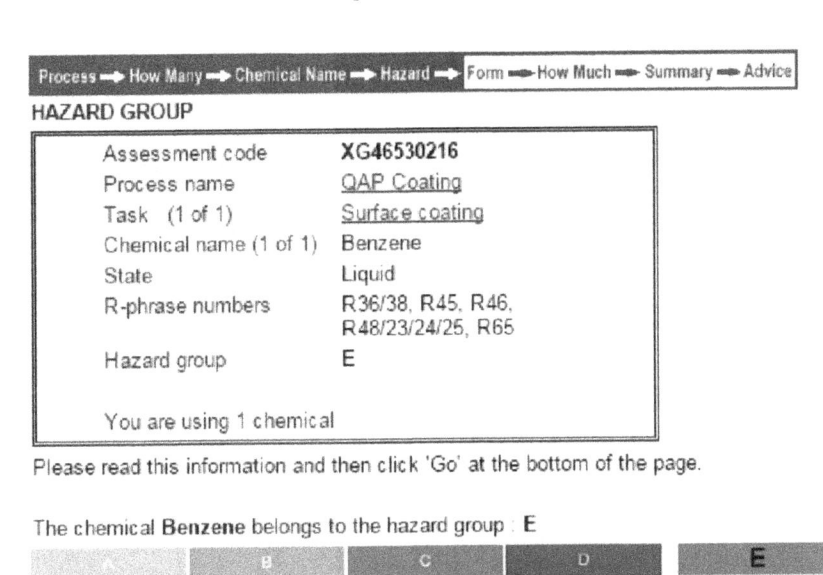

Figure 12.3 Benzene CAS Number 71-43-2, Hazard Group

12.5.5 OSHA's Calculation for Mixtures

The US Federal Occupational Safety and Health Administration (OSHA) suggests that in case of a mixture of air contaminants, which have additive toxicological effects on the same target organ(s), an employer shall compute the equivalent exposure as in Equation 12.2 (OSHA, n.d.a) where E_m is the equivalent exposure for the mixture, C is the concentration of a particular contaminant, L is the exposure limit for that substance specified in Subpart Z of 29 CFR Part 1910, and the value of E_m shall not exceed unity (1):

$$E_m = \left(\frac{C_1}{L_1} + \frac{C_2}{L_2} + \frac{C_3}{L_3}\right) + \cdots + \left(\frac{C_n}{L_n}\right) \qquad (12.2)$$

An example of benzene, toluene, ethyl benzene, and xylene (BTEX) mixture exposure is presented in Table 12.4.

TABLE 12.4 BTEX Exposure to the Mixture Example

	(PEL = 1)	(PEL = 100)	(PEL = 100)	(PEL = 100)	
Time (h)	Benzene	Toluene	Ethyl Benzene	Xylene	
2	0.5	75	75	75	
3	0.75	50	50	50	
3	0.75	77	25	55	
Sum	5.5	531	375	465	
TWA	0.6875	66.375	46.875	58.125	
PEL	1	100	100	100	
TWA/PEL	0.6875	0.66375	0.46875	0.58125	$\Sigma = 2.40125$

In the example provided previously, none of the individual substances exceeded the permissible exposure limit (PEL); however, E_m is greater than 1. Therefore, an overexposure to the mixture of chemicals appears to have occurred.

12.5.6 The ART Tool

"The Advanced REACH Tool (ART) version 1.5 incorporates a mechanistic model of inhalation exposure and a statistical facility to update the estimates with measurements selected from an in-built exposure database or the user's own data (Advanced Reach Tool, 2011). This combination of model estimates and data produces more refined estimates of exposure and reduced uncertainty." ART can be used to assess exposure to inhalable dust, vapors, and mists. Currently ART cannot be used for the assessment of fumes, fibers, gases, and dust resulting from emissions during hot metallurgical processes.

12.5.7 Stoffenmanager

Another widely accepted tool internationally is Stoffenmanager. The Stoffenmanager is a generic tool, initially developed for small- and medium-size business to support them in assessing, prioritizing, and controlling risks from chemicals at the workplace. It provides support for companies and EHS professionals in performing a risk assessment and controlling exposure by taking proper risk management measures. The Stoffenmanager is a free internet-based tool, and it is available on the website of the designer (https://stoffenmanager.nl/). The online tool provides different options for health risk assessments. The user can switch between "control banding" (CB) and "quantitative risk assessment" modules.

12.6 HEALTH RISK ASSESSMENTS AND PRIORITIZATION

The health risk assessment should focus on a probability of harm rather than the actual health outcomes, which may be diagnosed years after the exposure. Varieties of simple health risk assessment tools are available. For instance, qualitative risk assessment matrix, as described in ANSI Z10-2012 Appendix F, can be used for initial risk assessment. Numerical risk assessment method can also be utilized.

An example of a risk assessment matrix with numerical grading and scoring is presented in ANSI Z 590.3-2011. Combining the severity and occurrence probability values in the matrix yields a risk score. The matrix is not specifically designed for health risk assessments. However, it could be slightly modified and successfully used for chemical, physical, and biological exposures. An example of two-dimensional severity × probability/frequency risk assessment model is presented in Table 12.5.

TABLE 12.5 HRR Example

Process/ Operation	OSH Aspect/ Hazard Description	Potential Effects/ OSH Risk Description	Reference	Frequency	Severity	Risk Level (RL)
Spray paint mixing	Solvents exposure, skin exposed	Dermatitis	http://www.cdc.gov/niosh/topics/skin/	3	3	9
Spray paint application	Benzene: chemical exposure, VOCs generation	Leukemia, cancer, anemia	http://www.bt.cdc.gov/agent/benzene/basics/facts.asp	4	4	16
Clean spray paint booth	Disposal of contaminated containers and wastes. Solvents exposure	Leukemia, cancer, anemia. Dermatitis	http://www.epa.gov/dfe/pubs/auto/factsheet/spraybooth filters.pdf	3	4	12

The example is based on the QAP case study use of solvents containing 5% benzene. Examples of probability/frequency and severity rating scales are presented in Tables 12.6 and 12.7.

TABLE 12.6 Frequency Rating Scale Example

	Frequency (Routine Operations)
5	Frequent: exposure ≥ 8 or more hours per day
4	Probable/likely: exposure occurs on a regular basis (more than once per week)
3	Exposure likely to occur once per week
2	Exposure likely to occur < once per week
1	Exposure very unlikely to occur

Risk Level is simply the multiplication of the respective numerical values for severity and probability/frequency. Spray paint application is the highest risk in this example. Spray paint application is a continuous operation, and the operator is exposed to the benzene-containing mixture on a regular basis. Mixing of the solvents and cleaning of the spray booth are done once per week. Therefore, the frequency/probability of an exposure is rated at 3. As the matrix is completed, highest rated processes/operations or workstations can be easily identified.

TABLE 12.7 Severity Rating Scale

	Severity Rating
5	Death or multiple disabling illnesses
4	Permanent disability
3	DART illness
2	First aid
1	Fresh air/discomfort

The SH&E professional should identify appropriate risk reduction control measures based on ANSI/ASSE Z590.3-2011 "Prevention through Design: Guidelines for Addressing Occupational Hazards and Risks in Design and Redesign Processes" hierarchy of controls. Sometimes, multiple control measures are required to address the hazards and reduce the health risk to an acceptable level.

12.7 MODIFIED HRR/IH FMEA METHODOLOGY

The HRR methodology previously described does not include the AIHA's Exposure Categorization Scheme. Therefore, a modified HRR using a failure mode and effects analysis (FMEA) model that includes an *exposure* variable is presented in this example. The modified HRR/IH FMEA methodology assigns risk rating numbers to severity, probability, and exposure. An example case study follows, which illustrates how a SH&E professional used the worksheet within the risk assessment process and applied the HRR concept.

Case Study

A small company that produces foam mattresses imports a variety of chemicals used in the production process. Three chemicals were selected to demonstrate the applicability of the modified FMEA methodology. For instance, polymeric polyols are generally used to produce other polymers. They react with isocyanates to make polyurethanes used to make mattresses. Furthermore, boric acid is also added to the mix.

The process flow is presented in Figure 12.4.

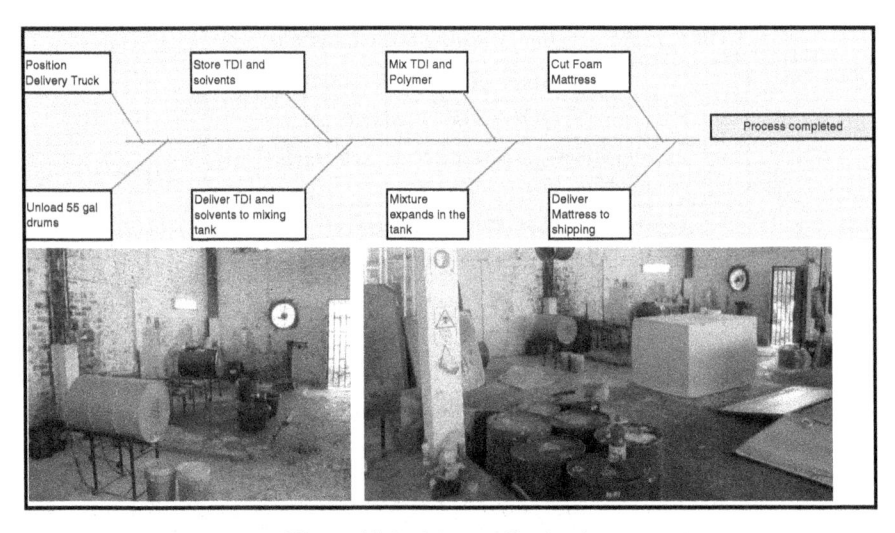

Figure 12.4 Mattress Production

RESULTS

- Processes: creation of the foam mattresses, cutting, shaping, wrapping in cloth
- Chemicals: TDI-80: toluene diisocyanate (TDI) (NIOSH SEQ 0621), polyol (polymer that reacts with TDI to create foam), and boric acid
- TDI-80: chemical formula: $CH_3C_6H_3(NCO)_2$
- OSHA PEL: C 0.02 ppm
- Target organs for all three chemicals: eyes, skin, respiratory system.

The production facility is located in South America. In the absence of an alternative to an AIHA accredited laboratories in the country, the potential worker health risk assessment had to be performed utilizing field detection methods. Direct-read wipes (DRW) for surface chemical detection were used for surface contamination evaluation. Experimental Microteq® TDI/MDI direct-read wipes were tested for the first time outside of the United States in 2008. Dräger TDI 0.02/a detector tubes were used for airborne TDI contamination measurements.

SAMPLING

Area samples were collected at key locations along work areas to evaluate the risk of exposure to employees who may be required to work in specific locations for long periods. In addition, area sampling was used to identify any potential "hot spots" and areas that may require special control measures. Nine (9) TDI colorimetric sampling tube samples were collected. Each detector tube sample was taken progressively farther from the TDI drums. In addition, 20 DRW surfaces were sampled for potential dermal exposure. DRW were taken at surfaces most likely to be contacted by employees. Colorimetric wipe samples may be a very useful screening tool for determining the probability of a dermal exposure.

RESULTS

Suspected areas of contamination were compared to a control area. The newly developed wipe samples measured concentrations at semiquantitative levels for surface detection and evaluation. Dräger TDI detector tubes showed presence of diisocyanates (Figure 12.5) in the same locations where surface contamination was determined.

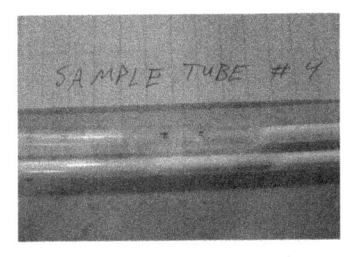

Figure 12.5 Confirmed Airborne TDI Concentrations of 0.1 ppm

Not surprisingly, TDI airborne concentration (Figure 12.6) was highest where high levels of surface contamination were detected.

Figure 12.6 Confirmed Presence of TDI in 50% of the Surfaces Sampled

A more detailed IH risk assessment tool was developed as a supplemental material. Some of the tools from Chapters 7 and 9 were used for this case study. Chapter 12 IH RA tool can be downloaded from https://centralspace.ucmo.edu/handle/123456789/464.

The following modified FMEA worksheet was developed to demonstrate the health risk assessment process. The model was used to calculate health risk priority number (HRPN) using Equation 12.3:

$$\text{HRPN} = \text{Severity} \times \text{Probability} \times \text{Exposure} \tag{12.3}$$

The following severity ratings could be suggested using the ratings presented in Table 12.8.

TABLE 12.8 Severity Ratings

	Severity Rating
5	Death or multiple disabling illnesses
4	Permanent illness or disability
3	Days away, restricted, and transfer (DART) illness
2	First aid
1	Fresh air/discomfort

With regard to probability or frequency ratings, two options might be suggested, as shown in Table 12.9.

TABLE 12.9 Probability or Frequency Ratings

	Probability of Developing an Occupational Disease Related to Chemical/Agent Exposure		Frequency (More Applicable to Routine Operations)
5	Extremely likely	5	Frequent: exposure ≥8 or more hours per day
4	Likely	4	Exposure occurs on a regular basis (more than once per week)
3	Somewhat likely	3	Exposure likely to occur once per week
2	Unlikely	2	Exposure likely to occur < once per week
1	Extremely unlikely	1	Exposure very unlikely to occur

AIHA's Exposure Categorization Scheme can be used to rank the exposure. It is well described in ANSI/ASSE Z590.3-2011 PtD standard. However, the authors would suggest adding a Category 5 for exposure to multiple chemicals. The AIHA scheme does not address multiple chemicals exposure and their additive effect. Therefore, a Category 5 was added in this case as presented in Table 12.10.

TABLE 12.10 Exposure Rating

	Numerical Rating	Exposure Rating	Recommended Follow Up/ Exposure Control
Acceptable Health Risk	0	<1% of OEL	No action necessary
	1	<10% of OEL	General hazard communication (aligned with GHS)
Uncertain Health Risk	2	10–50% of OEL	Chemical specific hazard communication (aligned with GHS)
	3	50–100% of OEL	Consider engineering controls, exposure surveillance, medical surveillance, work practices
Unacceptable Health Risk	4	>100% of OEL	Consider PtD hierarchy of controls. Start with avoid, eliminate, substitute. If not possible consider engineering controls, warnings, work practice controls, and as a last resort PPE
	5	Multiple chemical exposures, synergetic effect, Em >1	Immediate PtD controls or process shutdown, validate controls, and PPE selection. Use multiple controls if necessary

Note: Users are highly discouraged to use Category "0" for exposure as it will produce a "0" HRPN.

The three selected chemicals can be ranked using the modified FMEA and HRPN worksheet, as shown in Table 12.11. Severity, probability, and exposure 1–5 ratings were used in this example.

TDI severity level was rated 5 due to a possible death or multiple disabling illnesses as a result of overexposure. Probability of developing an occupational disease related to chemical/agent exposure was rated 4, since worker exposure was "likely." Finally, exposure was rated 5 due to exposures exceeding OEL and multiple chemicals exposure.

Poly(tetrahydrofuran) and *boric acid* severity levels were both rated 3 due to the lower toxicity of the chemicals. In addition, operators were using P 100 respirators, which offered some protection from boric acid particles. Boric acid is listed/regulated by OSHA, CAL OSHA, or ACGIH as "particulate not otherwise classified" or "nuisance dust" permissible exposure limit is 15 mg/m^3, total dust 5 mg/m^3, respirable dust. However, P 100 type respirators will not protect workers from TDI and poly(tetrahydrofuran). It could be argued that the exposure rating should be 5 for all three substances. Using such methodology allows SH&E professionals to quickly prioritize the hazards and urgently address the highest ranked chemical hazards. For instance, this practical example shows the need to apply risk reduction measures according to the PtD hierarchy of controls.

One possible option is to substitute TDI and poly(tetrahydrofuran) with a less toxic materials. To encourage transition to safer chemicals, OSHA has developed this step-by-step toolkit to provide employers and workers with information, methods, tools, and guidance on using informed substitution in the workplace (OSHA, n.d.b).

Process modification may be suggested to accommodate production of soy-based polyurethanes by nonisocyanate route (Javni, Hong, & Petrović, 2008). Other studies suggest using soy-based polyurethane (PU) foam product reinforced with carbon nanotubes (Liang & Shi, 2010).

Substitution of the identified extremely toxic chemicals, with less toxic materials, will lead to significant reduction of severity ratings. Probability of developing an occupational disease related to chemical exposure will also decrease significantly. However, the new chemical hazards should be evaluated carefully since they may present different hazards and/or affect target organs differently.

TABLE 12.11 Modified FMEA and HRPN Worksheet

FMEA and HRPN Worksheet

Part or process name	Foam mat.	Suppliers and Model date	Various	Prepared by FMEA date		GP
Design/Mfg responsibility			1/27/2015			
Other areas involved	Eng.	Engineering change				

Base case – current state

Process operation, function, or purpose	Potential exposure mode	Potential effect(s) of failure	SEV	Potential cause(s) of failure	PROB	Exposure control rating reasoning	EXP	S × P	HRPN
Foam mattresses production	TDI exposure	Eyes, skin, respiratory system affected	5	Operators unprotected or underprotected	4	Multiple chemicals Exp. – inadequate PPE	5	20	100
Foam mattresses production	Poly(tetrahydrofuran) exposure	Eyes, skin, respiratory system affected	3	Operators unprotected or underprotected	4	Multiple chemicals Exp. – inadequate PPE	5	12	60
Foam mattresses production	Boric acid exposure	Eyes, skin, respiratory system affected	3	Operators somewhat protected	3	Multiple chemicals Exp. – somewhat adequate PPE	5	9	45

12.8 CONTROL BANDING NANOTOOL

The addition of nanomaterials in the workplace presents new challenges and occupational health risks to consider. An excellent nanomaterials risk assessment tool was proposed by Paik, Zalk, and Swuste (2008) as a control banding (CB) Nanotool for risk prioritization and management. The following year, a slightly modified version was published by Zalk, Paik, and Swuste (2009). CB Nanotool 2.0 is available at http://www.controlbanding.net.

According to its authors, the CB Nanotool is a novel CB approach being used at the Lawrence Livermore National Laboratory (LLNL), by both experts and nonexperts, to assess risks associated with nanotechnology operations and prescribe appropriate engineering controls. Available at: http://controlbanding.net/Work.html. Other control banding approaches for nanomaterials are summarized by Brouwer (2012).

12.9 DERMAL RISK ASSESSMENT

Sahmel and Boeniger (2006) presented an interesting dermal exposure risk assessment methodology in Chapter 12 of AIHA's exposure assessment strategy. The authors use a rating scheme to rank the exposure profiles and priorities for further information gathering. Using the rating numbers, the dermal risk assessment methodology can be described as a semiquantitative method since it assigns approximate rating values rather than an exact measurement. The methodology is available in the supplemental materials of this chapter. The model is based on dermal exposure potential rating (DER) and dermal hazard rating (DHR) 1024 × 4 Dermal Risk Matrix.

DER is a multiplication of five rating factors that are multiplied together to estimate DER for similar exposure groups (SEGs) using Equation 12.4 where DCA = dermal contact area, DC = dermal concentration, DCF = dermal contact frequency, DRT = dermal retention time, and DPP = dermal penetration potential:

$$DER = DCA \times DC \times DCF \times DRT \times DPP \qquad (12.4)$$

All five factors are rated based on 1–4 scale. DHR is also 1–4 rating. The level of dermal risk is determined by finding the closest DHR and DER values on the vertical and horizontal axis. The convergence point of the two ratings determines the risk. An example of the matrix is depicted in Figure 12.7. The actual rating scales are included in the supplemental materials.

Dermal Risk Rating Matrix: Numerical Grading

	Agent 1	Agent 2	Agent 3
Dermal Hazard Rating (DHR): 4	1	3	2
Dermal Exposure Rating (DER): 1024	256	1024	256
Total	256	3072	512	Caution

RA Matrix — Dermal Exposure Rating

Dermal Hazard Rating	16	64	256	1024
4	64	256	1024	4096
3	48	192	768	3072
2	32	128	512	2048
1	16	64	256	1024

DER =	DCA	DC	DCF	DRT	DPP
Exp. PR =	4	2	4	4	4
DER =	512				

Dermal Hazard Rating
Note

Figure 12.7 Dermal Risk

12.10 OCCUPATIONAL HEALTH RISK AND PTD PROCESS ALIGNMENT

For many years, IH professionals have used the National Academy of Sciences (NAS) risk assessment paradigm (also adapted from EPA) as a framework for estimating risk from exposure to environmental chemicals (National Research Council, 1983). EPA's four-step risk assessment process is illustrated in Figure 12.8.

The 4 Step Risk Assessment Process

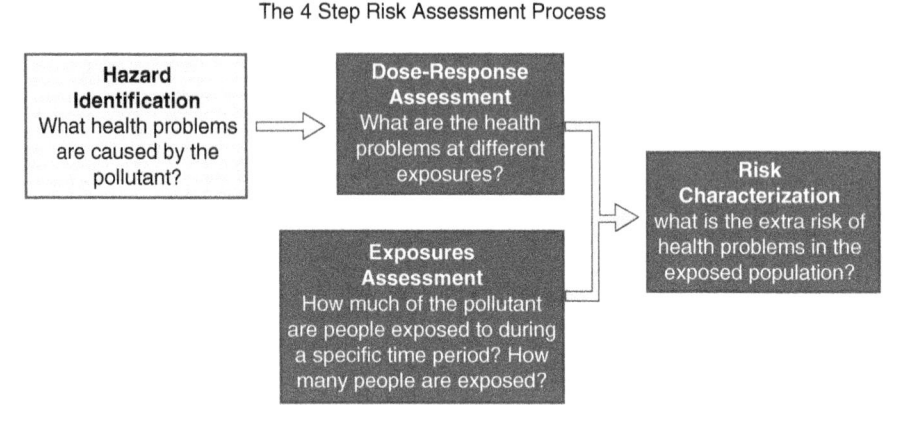

Hazard Identification — What health problems are caused by the pollutant?

Dose-Response Assessment — What are the health problems at different exposures?

Exposures Assessment — How much of the pollutant are people exposed to during a specific time period? How many people are exposed?

Risk Characterization — what is the extra risk of health problems in the exposed population?

Figure 12.8 EPAs Four-Step Risk Assessment Process. Available at: http://www.epa.gov/hsrb/new-orientation/019-risk.htm

Within this framework, quantitative health risk assessments serve as the cornerstone of health-based exposure limits. Exposure assessment is a key step of the risk assessment paradigm. AIHA's exposure assessment strategy committee developed excellent exposure assessment tools. The tools user manuals are available at https://www.aiha.org/get-involved/VolunteerGroups/Pages/Exposure-Assessment-Strategies-Committee.aspx

However, exposure assessment is just one of the steps of the health risk assessment process. For instance, Korchevskiy, A. Rasmuson, and Rasmuson (2014, February) state that evaluating health risks begins with exposure assessment. The authors of the article "New Information and New Models Are Transforming Asbestos Risk Assessment" also state that "sampling and measurement techniques involve various uncertainties and random characteristics, including the determination of SEGs, the extrapolation of daily measurements versus long-term exposures, the individual variability of results as they relate to dose–response relationships, and the high variability associated with analytical methodologies."

NAS risk assessment paradigm can be used to address occupational hazards, environmental hazards, natural hazards, and other stressors that may be present in the environment. As stated earlier NAS risk assessment paradigm was adopted by EPA, and it is widely accepted by IH professionals. The IH's anticipate, recognize, evaluate, and control hazards. Anticipation, recognition, and evaluation can be related to the EPA's risk assessment process, as shown in Table 12.12 (Jayjock, Lynch, & Nelson, 2000).

In a series of articles titled "Risk Assessment's New Era" published in the AIHA's *"The Synergist,"* the authors described the need to improve the risk assessment process used by IH professionals. (The Synergist, April, May, June/July, and August 2012). At the time NAS risk assessment paradigm was published, there were no voluntary Occupational Safety and Health (OSH) standards requiring risk assessment. Since then, significant number of standards, guidelines, and initiatives for the OSH practice requiring risk assessments were published worldwide. Since 2005, more than 35 such standards and initiatives that require or promote the use of risk assessments were promulgated. As noted

TABLE 12.12 Comparison of EPA's Risk Assessment Process and IH Principles

EPA's Risk Assessment Process	IH Principles
1. Hazard identification	Anticipation/recognition
2. Dose–response assessment 3. Exposure assessment 4. Risk characterization	Evaluation
Risk management	Control
Risk communication	Hazard communication

in other chapters of this text, the requirement for employers to perform risk assessments has primarily occurred outside of the United States. Since 2011, three important voluntary standards were published by the American National Standards Institute (ANSI) providing applicable requirements for the practice of health risk assessment and management. These standards are as follows:

- ANSI Z10-2012 – Occupational Health and Safety Management Systems
- ISO 31000/ANSI Z690-2011 Risk Management Standards Series
- ANSI Z590.3-2011 *Prevention through Design – Guidelines for Addressing Occupational Hazards and Risks in Design and Redesign Processes.*

The need to align NAS risk assessment paradigm with the new voluntary standards is apparent. Comparisons of Risk Assessment in ISO 31000/ANSI Z690 and ANSI Z590.3 was provided in Chapter 11 of this book. In Figure 12.9, the authors offer the following NAS risk assessment paradigm and ISO 31000/ANSI Z690 alignment.

Obviously, additional risk management or risk treatment steps are added to the NAS risk assessment paradigm. More detailed health risk assessment and management models can be developed and aligned with the PtD standard as illustrated in Figure 12.10.

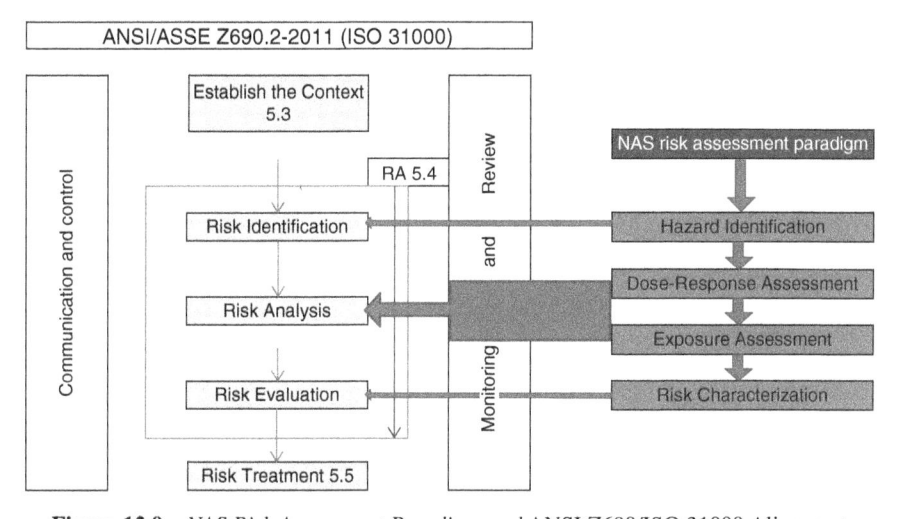

Figure 12.9 *NAS Risk Assessment Paradigm and* ANSI Z690/ISO 31000 Alignment

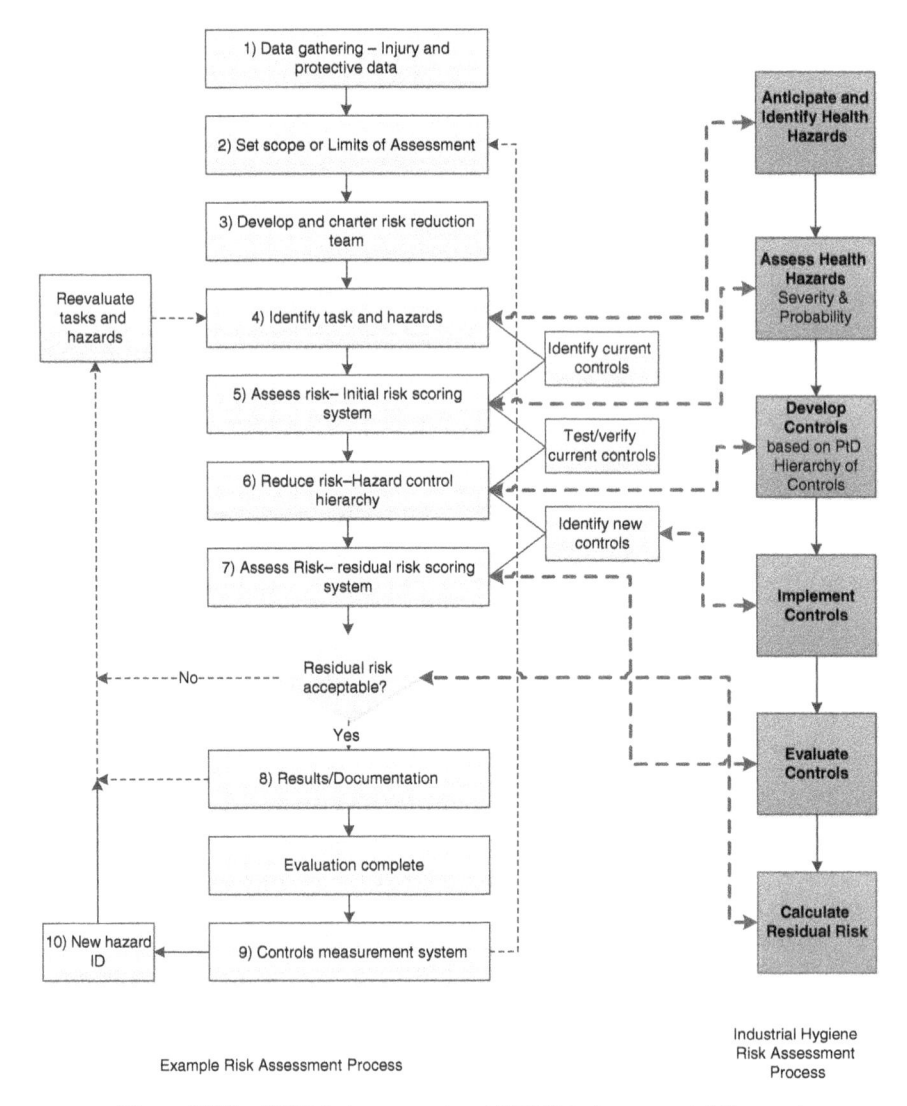

Figure 12.10 IH Risk Assessment and PtD Risk Assessment Alignment

12.11 SUMMARY

The tools and methods discussed in this chapter represent a summary of methods used in the assessment and management of IH and occupational health risks. By using these methodologies, occupational health risks can be systematical identified, analyzed, and evaluated so that risk can be rated and prioritized for risk reduction treatment.

The proposed use of FMEA for IH interventions relies on professional judgment and semiquantitative decision making in order to assess the significance of hazards and assign a HRPN to each task or process. Utilizing modified FMEA for IH risk assessments helps in prioritizing recommendations for reducing risks. Through these methods, the SH&E professional is better equipped to identify, measure, and manage "known" health risks, as well as previously "unknown" risks in the workplace.

REVIEW QUESTIONS

1. Compare some of the risk assessment terms used by industrial hygienists with those used by safety professionals.
2. State the primary purpose of an IH Risk Assessment.
3. Describe industrial hygiene's role in the risk assessment process.
4. Name three tools that have application in the IH risk assessment process.
5. Summarize some of the challenges associated with identifying and assessing industrial hygiene risks.
6. Explain the benefits of the IH risk assessment methods presented in this chapter.

REFERENCES

Advanced REACH Tool (ART). 2011. Retrieved from: https://www.advancedreachtool.com/ (accessed December 18, 2015).

AIHA. 2006. *A Strategy for Assessing and Managing Occupational Exposures*, 3rd ed. J. S. Ignacio, W. H. Bullockhttps://www.aiha.org/marketplace/Pages/Product-Detail.aspx?productid={96E7072A-4778-DE11-96B0-0050568361FD} (accessed December 18, 2015).

American Industrial Hygiene Association. 2002. AIHA White Paper on Risk Assessment and Risk Management. Retrieved from https://www.aiha.org/get-involved/VolunteerGroups/Documents/RISKVG-whitepaper02_Risk.pdf (accessed December 18, 2015).

American Industrial Hygiene Association. 2015. *Discover Industrial Hygiene*. Retrieved from https://www.aiha.org/about-ih/Pages/default.aspx (accessed December 18, 2015).

ANSI/AIHA Z10-2012. *Occupational Health and Safety Management Systems*. Fairfax, VA: The American Industrial Hygiene Association, 2012. ASSE is now the Secretariat. Available at: https://www.asse.org/cartpage.php?link=z10_2005 (accessed December 18, 2015).

ANSI/ASSE Z590.3-2011. *Prevention Through Design: Guidelines for Addressing Occupational Hazards and Risks in Design and Redesign Processes*. Des Plaines, IL: American Society of Safety Engineers, 2011.

Brouwer, D.H., 2012. "Control banding approaches for nanomaterials." *Annals of Occupational Hygiene*, 56(5), 506–514.

Canadian Centre for Occupational Health and Safety. 2012. *Risk Phrases Used in the Countries of European Union*. Retrieved from http://www.ccohs.ca/oshanswers/chemicals/risk_phrases.html (accessed December 18, 2015).

Ehrlich, R. I., Myers, J. E., te Water Naude, J. M., Thompson, M. L., & Churchyard, G. J., 2011. "Lung function loss in relation to silica dust exposure in South African gold miners." *Occupational and Environmental Medicine*, 68(2), 96–101. doi: 10.1136/oem.2009.048827.

European Agency for Safety and Health at Work. n.d. *What are the H Codes and How Can I Compare Them to the Familiar R-Phrases?* Retrieved from https://osha.europa.eu/en/faq/dangerous-substances-clp-reach/what-are-the-h-codes-and-how-can-i-compare-them-to-the-familiar-r-phrases (accessed December 18, 2015).

Health and Safety Executive. 2009. *The Technical Basis for COSHH Essentials: Easy Steps to Control Chemicals*. Retrieved from http://coshh-essentials.org.uk/assets/live/CETB.pdf (accessed December 18, 2015).

Health and Safety Executive. n.d. *COSHH Essentials*. Retrieved from http://www.hse.gov.uk/coshh/essentials/ (accessed December 18, 2015).

Ignacio, J. S., & Bullock, W. H. (Eds.). *A Strategy for Assessing and Managing Occupational Exposures*, 3rd ed. Fairfax, VA: AIHA Press, 2006.

Javni, I., Hong, D. P., & Petrović, Z. S. 2008. "Soy-based polyurethanes by nonisocyanate route." *Journal of Applied Polymer Science,* 108(6), 3867–3875.

Jayjock, M. A., Lynch, J. R., & Nelson, D. I. *Risk Assessment Principles for the Industrial Hygienist.* Fairfax, VA: AIHA Press, 2000.

Korchevskiy, A., Rasmuson, J., & Rasmuson, E. 2014. ""Miracle Mineral" risk assessment." *The Synergist,* 26–30.

Liang, K., & Shi, S. Q. 2010. "Soy-based polyurethane foam reinforced with carbon nanotubes." *Key Engineering Materials,* 419–420(2010), 477–480.

Mulhausen, J., Damiano, J., & Pullen, E. L. 2006. Defining and Judging Exposure Profiles. In Ignacio, J. S., & Bullock, W. H. (Eds.), *A Strategy for Assessing and Managing Occupational Exposures,* 3rd ed. Fairfax, VA: AIHA Press.

National Research Council. 1983. Risk Assessment in the Federal Government: Managing the Process. Retrieved from: http://www.nap.edu/read/366/chapter/1 (accessed December 18, 2015).

Occupational Safety and Health Administration. 1998. *Industrial Hygiene.* Retrieved from https://www.osha.gov/Publications/OSHA3143/OSHA3143.htm (accessed December 18, 2015).

Occupational Safety and Health Administration. n.d.a-a. *Occupational Safety and Health Standards: Toxic and Hazardous Substances.* Retrieved from https://www.osha.gov/pls/oshaweb/owadisp.show_document?p_table=Standards&p_id=9991 (accessed December 18, 2015).

Occupational Safety and Health Administration. n.d.b-b. *Transitioning to Safer Chemicals: A Toolkit for Employers and Workers.* Retrieved from https://www.osha.gov/dsg/safer_chemicals/index.html (accessed December 18, 2015).

Paik, S. Y., Zalk, D. M., & Swuste, P. (2008). "Application of a pilot control banding tool for risk level assessment and control of nanoparticle exposures." *Annals of Occupational Hygiene,* 52(6), 419–428.

Rumsfeld, D. H. 2002. *DoD News Briefing – Secretary Rumsfeld and Gen. Myers.* Retrieved from http://www.defense.gov/transcripts/transcript.aspx?transcriptid=2636 (accessed December 18, 2015).

Sahmel, J., & Boeniger, M. 2006. Dermal Exposure Assessments. In Ignacio, J. S., & Bullock, W. H. (Eds.), *A Strategy for Assessing and Managing Occupational Exposures,* 3rd ed. Fairfax, VA: AIHA Press.

Welcome to the Advanced Reach Tool 1.5. n.d. Retrieved from https://www.advancedreachtool.com (accessed December 18, 2015).

Zalk, D. M., Paik, S. Y., & Swuste, P. 2009. "Evaluating the control banding nanotool: a qualitative risk assessment method for controlling nanoparticle exposures." *Journal of Nanoparticle Research,* 11, 1685–1704.

13

MACHINE RISK ASSESSMENTS

BRUCE K. LYON
Risk Control Services, Hays Companies, Kansas City, MO, USA

13.1 OBJECTIVES

- Introduce machine hazards and safeguarding
- Review techniques used in machine risk assessments and risk reduction
- Examine the use of techniques and their strengths and limitations
- Provide guidance on the use of machine risk assessment techniques

13.2 INTRODUCTION

Machines perform many functions in the workplace that were, at one time, performed by people. The mechanical advantage provided by machines over humans has enabled industries to grow and expand and ultimately reduce the cost of production and services rendered. Designed to be powerful, quick, and tireless, machines are capable of performing repetitive tasks unachievable by people. Functions are performed daily by countless machines such as cutting, punching, bending, shaping, tempering, assembling, fusing, finishing, packaging, and handling of products and materials in almost all industries. These mechanical advancements have eliminated a number of hazards workers previously were exposed to and have saved many from occupational injury and death.

However, with all their benefits, machines present certain hazards and risks. Machine-related injuries make up the highest number of permanent, partial disabilities and rank third among all industrial accidents. Many of these incidents result from the lack of hazard recognition and control and the bypass of existing safeguards. On their website, the *Occupational Safety and Health Administration (OSHA) Machine Guarding eTool* says that "employee exposure to unguarded or inadequately guarded machines is prevalent in many workplaces. Consequently, workers who operate and maintain machinery suffer approximately 18,000 amputations, lacerations, crushing injuries, abrasions, and over 800 deaths/year. Amputation is one of the most severe and crippling types of injuries in

---

Risk Assessment: A Practical Guide to Assessing Operational Risks, First Edition.
Edited by Georgi Popov, Bruce K. Lyon, and Bruce Hollcroft.
© 2016 John Wiley & Sons, Inc. Published 2016 by John Wiley & Sons, Inc.

the occupational workplace, and often results in permanent disability" (OSHA, 2015). For these reasons, it has become increasingly important for SH&E professionals to be skilled in identifying and controlling machine-related hazards and risks.

13.3 MACHINE SAFETY STANDARDS

There are a number of machine-related safety compliance and consensus standards. In the United States, employers must comply with the OSHA standards as an absolute minimum. OSHA states that employers must provide a safe and healthy workplace for employees and provides specific standards for General Industry (29 CFR 1910) and Construction (29 CFR 1926). Specifically, machine safety standards fall into the General Industry standards in *1910 Subpart O, Machinery and Machine Guarding* and include

1910.211, Definitions

1910.212, General requirements for all machines

1910.213, Woodworking machinery requirements

1910.214, Cooperage machinery [Reserved]

1910.215, Abrasive wheel machinery

1910.216, Mills and calendars in the rubber and plastics industries

1910.217, Mechanical power presses. Includes general requirements in addition to specific requirements for construction, safeguarding, dies, inspection, maintenance, modification, operation, injury reporting, and presence sensing device initiation (PSDI)

- Appendix A, Mandatory requirements for certification/validation of safety systems for PSDI of mechanical power presses
- Appendix B, Nonmandatory guidelines for certification/validation of safety systems for PSDI of mechanical power presses
- Appendix C, Mandatory requirements for OSHA recognition of third-party validation organizations for the PSDI standard
- Appendix D, Nonmandatory supplementary information

1910.218, Forging machines

1910.219, Mechanical power-transmission apparatus.

In addition, *1910 Subpart R, Special Industries* addresses machine safety in:

1910.262, Textiles. Paragraph (c)(3) [reserved] contains a short statement on machine guarding requirements and a reference to 29 CFR 1910.219. [Related topic page]

1910.263, Bakery equipment. Paragraph (c) addresses general requirements for machine guarding.

1910.268, Telecommunications. Paragraph (b)(1)(v) addresses some general requirements for machine guarding.

The listed OSHA standards address machine safety during normal production operations. However, machines periodically require maintenance, service, and repair exposing maintenance workers and contractors to potential release of hazardous energy. For such machine maintenance and service activities, the *OSHA 1910.147, The Control of Hazardous Energy* (lockout/tagout) standard is in place to protect workers from the unexpected energization, start-up, or release of stored energy while performing these operations.

When it comes to consensus standards for machine safety, there are a number national and international standards and guidelines established. Although they are voluntary, consensus standards such

MACHINE SAFETY STANDARDS

as those published by the American National Standards Institute (ANSI) are used to support compliance standards and are often referenced by OSHA in their own standards and citations. Some of these machine safeguarding and risk assessment standards include the below:

- ANSI B11.0-2015. Safety of Machinery – General Requirements and Risk Assessment
- ANSI B11.TR3-2000. Risk Assessment and Risk Reduction – A Guide to Estimate, Evaluate and Reduce Risks Associated with Machine Tools
- ANSI B11.TR6-2010. Technical Report for Machines – Safety Control Systems for Machine Tools
- ANSI B11.19-2010. Performance Criteria for Safeguarding
- ANSI/RIA R15.06-2012. For Industrial Robots and Robot Systems – Safety Requirements
- NFPA 79-2012. Electrical Standard for Industrial Machinery
- ANSI/ASSE Z244.1-2003 (R2014). Control of Hazardous Energy – Lockout/Tagout and Alternative Methods
- ANSI/PMMI B155.1-2011. Standard for Packaging Machinery and Packaging-Related Converting Machinery – Safety Requirements
- SEMI S10-0307. Safety Guideline for Risk Assessment and Risk Evaluation Process
- MIL-STD-882E-2012. Standard Practice for System Safety
- CSA Z432-04. Safeguarding of Machinery – Occupational Health and Safety
- CSA Z434-03. Industrial Robots and Robot Systems – General Safety Requirements
- CSA Z460-05. Control of Hazardous Energy – Lockout and Other Methods
- ISO 12100:2010. Safety of machinery – General principles for design – Risk assessment and risk reduction
- EN 954-1:2000/ISO 13849-1:1999. Safety of machinery – Safety-related parts of control systems – Part 1: General principles of design
- ISO 13849-1:2006. Safety of machinery – Safety-related parts of control systems – Part 1: General principles of design
- 2006/42/EC. European Machinery Directive.

Of these standards, ANSI B11.0 is considered the centerpiece for machine safety and risk assessments. It is one of more than 30 standards and technical reports for metalworking machinery published by ANSI and B11 Standards, Inc., known as the B11 series. In the absence of machine-specific standards, ANSI B11.0, along with *ANSI B11.19 Performance Criteria for Safeguarding*, combine to form the foundation for the B11 series of machine-specific (Type-C) standards and for other industrial machinery lacking a machine-specific safety standard. The B11 standards and technical reports are organized with the ISO A-B-C level structure as described by ANSI B11.0-2015 as follows:

- ***Type-A standards*** (basis standards) give basic concepts, principles for design, and general aspects that can be applied to machinery.
- ***Type-B standards*** (generic safety standards) deal with one or more safety aspects or one or more types of safeguards that can be used across a wide range of machinery.
- ***Type-C standards*** (machinery safety standards) deal with detailed safety requirements for a particular machine or group of machines.

Since it applies to an array of machines and contains general requirements, ANSI B11.0 is considered a Type-A standard. Its scope states that it specifically applies to "new, existing, modified or rebuilt power driven machines, not portable by hand while working, that are used to process materials by cutting; forming; pressure; electrical, thermal or optical techniques; lamination; or a combination of these processes." In addition, the standard states that machinery *suppliers* and *users* are responsible for defining and achieving acceptable risk and that any risks associated with the operation, maintenance, dismantling, and disposal of machinery *shall* be reduced to an acceptable level. B11.0 includes a formal method to conduct and document the risk assessment process and also identifies some preparations that must be made before a risk assessment begins. The standard presents the basic risk assessment process in a step-by-step approach to assist in achieving this goal.

13.4 MACHINE HAZARDS

Machines require human control and interaction to perform their tasks. From a production standpoint, operators interact with their machines daily by performing such activities as setting up, adjusting, activating, loading and unloading, monitoring, unjamming or clearing, and shutting down machines. Periodically machines need ongoing maintenance, service, repair, or replacement. These nonproduction-related interactions pose certain hazards and risks to workers. Many of these hazards are derived from the energy sources that power machines and equipment. Hazardous energy sources associated with machines include electrical, gravity, extreme temperature, chemical, mechanical, pneumatic, hydraulic, and stored energy.

OSHA describes machine-related hazards and where they exist in the following excerpt taken from their website (OSHA):

Basics of Machine Safeguarding

Crushed hands and arms, severed fingers, blindness – the list of possible machinery-related injuries is as long as it is horrifying. There seem to be as many hazards created by moving machine parts as there are types of machines. Safeguards are essential for protecting workers from needless and preventable injuries.

A good rule to remember is: Any machine part, function, or process which many cause injury must be safeguarded. When the operation of a machine or accidental contact with it can injure the operator or others in the vicinity, the hazards must be either controlled or eliminated.

This manual describes the various hazards of mechanical motion and presents some techniques for protecting workers from these hazards. General information covered in this chapter includes – where mechanical hazards occur, the hazards created by different kinds of motions and the requirements for effective safeguards, as well as a brief discussion of non-mechanical hazards.

Where Mechanical Hazards Occur

Dangerous moving parts in three basic areas require safeguarding:

The point of operation: That point where work is performed on the material, such as cutting, shaping, boring, or forming of stock.

Power transmission apparatus: All components of the mechanical system which transmit energy to the part of the machine performing the work. These components include flywheels, pulleys, belts, connecting rods, couplings, cams, spindles, chains, cranks, and gears.

Other moving parts: All parts of the machine which move while the machine is working. These can include reciprocating, rotating, and transverse moving parts, as well as feed mechanisms and auxiliary parts of the machine.

In Annex B of the ANSI B11.0 standard, a list of machinery hazards can be found. Using the risk assessment process through all phases of a machine's life cycle, these potential hazard categories

MACHINE SAFEGUARDING 271

should be assessed and controlled to an acceptable level. Both the *supplier* (machine manufacturer) and the *user* have responsibilities in assessing and controlling machine risk. The following are hazard groups associated with machines adapted from the ANSI B11.0, Annex B:

- Chemical, material, or substance hazards (i.e., oil mist, metal dust, flammable liquids)
- Control systems failures or undesired machine behavior (i.e., unintended start-up, failure to stop)
- Electrical hazards
- Environmental hazards (i.e., emissions, releases)
- Ergonomics and human factor-related hazards (i.e., excessive reach, awkward postures, unclear controls or displays)
- Fire and explosion
- Fluids
- Extreme temperatures
- Radiation (i.e., UV light, lasers, infrared light, electromagnetic, ionizing radiation)
- Material handling hazards
- Mechanical hazards (i.e., moving or rotating parts, point of operation)
- Noise and vibration
- Slips and falls
- Ventilation and confined space.

13.5 MACHINE SAFEGUARDING

Safeguarding is essential for protecting employees from machine-related injuries. To use machines safely, appropriate safeguarding methods such as enclosures, fixed guards, barriers, and presence sensing devices are required to reduce machine-related risks to an acceptable level.

The term "safeguarding" is defined by ANSI B11.0-2015 as the "protection of personnel from hazards by the use of guards, safeguarding devices, awareness devices, and safeguarding measures." Further, the *ANSI B11.TR3 – 2000 Technical Report for Machine Tools* defines safeguarding to include "guards, safeguarding devices, awareness devices, safeguarding methods and safe work procedures."

As stated by OSHA "any machine part, function, or process that may cause injury must be safeguarded" (OSHA, 2007). The primary purpose of safeguarding is to detect or prevent inadvertent access to a hazard and reduce risk to an acceptable level. As in other areas of operational risk management, the hierarchy of control should be used in the selection of safeguarding methods for machines. The *ANSI B11.19-2003 Performance Criteria for Safeguarding* standard provides valuable guidance in the design, construction, installation, operation, and maintenance of the safeguarding used to protect employees from machine hazards.

Table 13.1 presents the hazard control hierarchy and types of safeguarding methods for machines reprinted with permission from *ANSI B11.0-2015*.

In order to effectively protect people from machine hazards, certain requirements must be met in the design and use of machine safeguards. OSHA defines the following minimum requirements for machine safeguards found on their website (OSHA).

1. *Prevent contact: The safeguard must prevent hands, arms, and any other part of a worker's body from making contact with dangerous moving parts. A good safeguarding system eliminates the possibility of the operator or another worker placing parts of their bodies near hazardous moving parts.*

TABLE 13.1 The Hazard Control Hierarchy

	Risk Reduction Measures	Examples	Influence on Risk Factors	Classification
Most preferred ↓	Elimination or substitution	• Eliminate pinch points (increase clearance) • Intrinsically safe (energy containment) • Automated material handling (robots, conveyors, etc.) • Redesign the process to eliminate or reduce human interaction • Reduced energy • Substitute less hazardous chemicals	• Impact on overall risk (elimination) by affecting severity and probability of harm • May affect severity of harm, frequency of exposure to the hazard under consideration, and/or the possibility of avoiding or limiting harm depending on which method of substitution is applied	Design out
	Guards, safeguarding devices, and complementary measures	• Barriers • Interlocks • Presence sensing devices (light curtains, safety mats, area scanners, etc.) • Two-hand control and two-hand trip devices	• Greatest impact on the probability of harm (occurrence of hazardous events under certain circumstances) • Minimal if any impact on severity of harm	Engineering controls
	Awareness devices	• Lights, beacons, and strobes • Computer warnings • Signs and labels • Beepers, horns, and sirens	• Potential impact on the probability of harm (avoidance) • No impact on severity of harm	Administrative controls
	Training and procedures	• Safe work procedures • Safety equipment inspections • Training • Lockout/tagout/verify	• Potential impact on the probability of harm (avoidance and/or exposure) • No impact on severity of harm	
Least preferred	Personal protective equipment (PPE)	• Safety glasses and face shields • Ear plugs • Gloves • Protective footwear • Respirators	• Potential impact on the probability of harm (avoidance) • No impact on severity of harm	

Source: Reprinted with permission from ANSI B11.0-2015 – Safety of Machinery.

2. *Secure:* Workers should not be able to easily remove or tamper with the safeguard, because a safeguard that can easily be made ineffective is no safeguard at all. Guards and safety devices should be made of durable material that will withstand the conditions of normal use. They must firmly be secured to the machine.
3. *Protect from falling objects:* The safeguard should ensure that no objects can fall into moving parts. A small tool which is dropped into a cycling machine could easily become a projectile that could strike and injure someone.
4. *Create no new hazards:* A safeguard defeats its own purpose if it creates a hazard of its own such as a shear point, a jagged edge, or an unfinished surface, which can cause a laceration. The edges of guards, for instance, should be rolled or bolted in such a way that they eliminate sharp edges.
5. *Create no interference:* Any safeguard that impedes a worker from performing the job quickly and comfortably might soon be overridden or disregarded. Proper safeguarding can actually enhance efficiency as it can relieve the worker's apprehensions about injury.
6. *Allow safe lubrication:* If possible, one should be able to lubricate the machine without removing the safeguards. Locating oil reservoirs outside the guard, with a line leading to the lubrication point, will reduce the need for the operator or maintenance worker to enter the hazardous area.

Risk reduction method concepts and characteristics are well defined in ANSI B11.0. The standard addresses the need to design machines for safe use and interface by operators and safe accessibility by maintenance and repair personnel. Safeguards and controls included in the standard cover a range of items that should be consulted by those involved in assessing machines, machine systems, and robotics.

13.5.1 Machine Safety Control Systems

As stated by ANSI B11.0, certain control systems on machines are designed to perform safety functions to reduce risk of harm. These control elements are referred to by the standard as the "safety-related parts of the control system" (SRP/CS). Machine safety control systems as defined by ANSI B11.0 include "sensors, manual input and mode selection elements, interlocking and decision-making circuitry and output elements to the machine actuators, operating devices and mechanisms." Such machine safety controls reduce certain risks related to the machine's function and are becoming increasingly complex.

For machines with safety control systems, it is important to conduct two separate risk assessments: (1) risks associated with the machine or process and (2) risks associated with the machine safety control systems. As indicated in an article written by Bruce Main and Fred Hayes published in ASSE Professional Safety, these two assessments should be performed separately. They point out that many machine hazards are beyond the reach of existing machine safety control systems and that both aspects must be assessed separately. Main and Hayes state the following:

> "ISO 13849-1 *[Safety-related parts of control systems standard]* only addresses the safety-related parts of the control system and only applies once the risk assessment for a machine has determined that a control system is needed as a risk reduction measure. Many hazards on a machine are unrelated to control systems (e.g., fall hazards from elevated work; slips and trips; ergonomic hazards from lifting/bending/twisting). PLs *[performance levels]* and categories have no meaning with these hazards, thus attempting to merge the machinery risk assessment and the control system specification is unwarranted." (Main & Hayes, 2014)

For those working with complex machinery and robotic systems, further study of the aforementioned article, other publications from Bruce Main, and the referenced standards is advised.

13.6 SELECTING MACHINES FOR ASSESSMENT

To properly select machines for risk assessment, an inventory of the various machines used by an operation is needed. Since most organizations have limited resources, it is prudent to follow some method of prioritization and selection of machines for risk assessment. Such selection methods should include factors such as exposure to the workforce (degree machines are used and population exposed), previous loss experience, maintenance demands, and known or perceived risk levels of machines.

Machines with higher frequency of use or exposure to the workforce and greater perceived risk should take priority. A mistake too often committed by organizations is waiting until an accident occurs before a risk assessment is conducted. A proactive plan should be developed to include all machines in a formal risk assessment process.

13.7 RISK ASSESSMENT OF MACHINES

A risk assessment seeks to prioritize identified hazards so that risks associated with each hazard can be eliminated or reduced. Where a hazard is identified, an assessment of risks associated with the hazard should be performed. Risk assessments should be performed for each machine initially, when changes or modifications occur, and when new hazards are recognized.

Machine manufacturers as well as the organizations that use them have responsibilities for assessing risk and safeguarding machines, as outlined in the ANSI B11.0 standard. This text will focus primarily on methods of risk assessment performed by the user.

There are a number of methods available for assessing machine-related risk. It should be noted that the real value of risk assessment is in the discipline and structure of performing the process. As stated in ANSI B11.0, it has been shown that various risk assessment methods lead to very similar results in both risk-level determination and control method strategies.

The risk assessment process is considered "scalable," meaning that it can be applied to a single hazard, multiple hazards, simple machines, or complex automated machines. This "scalable" process can be applied to new, existing, or modified machines; however, to be effective, it must incorporate the fundamental elements described in ANSI B11.0. The process steps are depicted in Figure 13.1 reprinted with permission from *ANSI B11.0-2015 – Safety of Machines*.

Using the ANSI B11.0 model, the following sequence of steps is briefly described for performing a machine risk assessment:

1. *Prepare for risk assessment*

    For any risk assessment undertaken, the first and foremost step is to clearly define the purpose, scope, and context. The "purpose" of the assessment will often dictate the needed scope and level of complexity (or simplicity). The scope should clearly define the limits of the assessment and include critical factors such as the assessment team (personnel with the technical competence and relevant skill), assessment method to be used, timeframe for completion, the machine to be assessed, phase of the machine's life cycle (i.e., design/redesign, construction, installation, operation, maintenance, decommission, dismantle), the component(s) to be assessed, operational state, acceptable risk levels, and other limits of the assessment.

    The machine's limits should be determined and considered such as
    - intended use
    - production rates
    - cycle times
    - forces exerted

RISK ASSESSMENT OF MACHINES

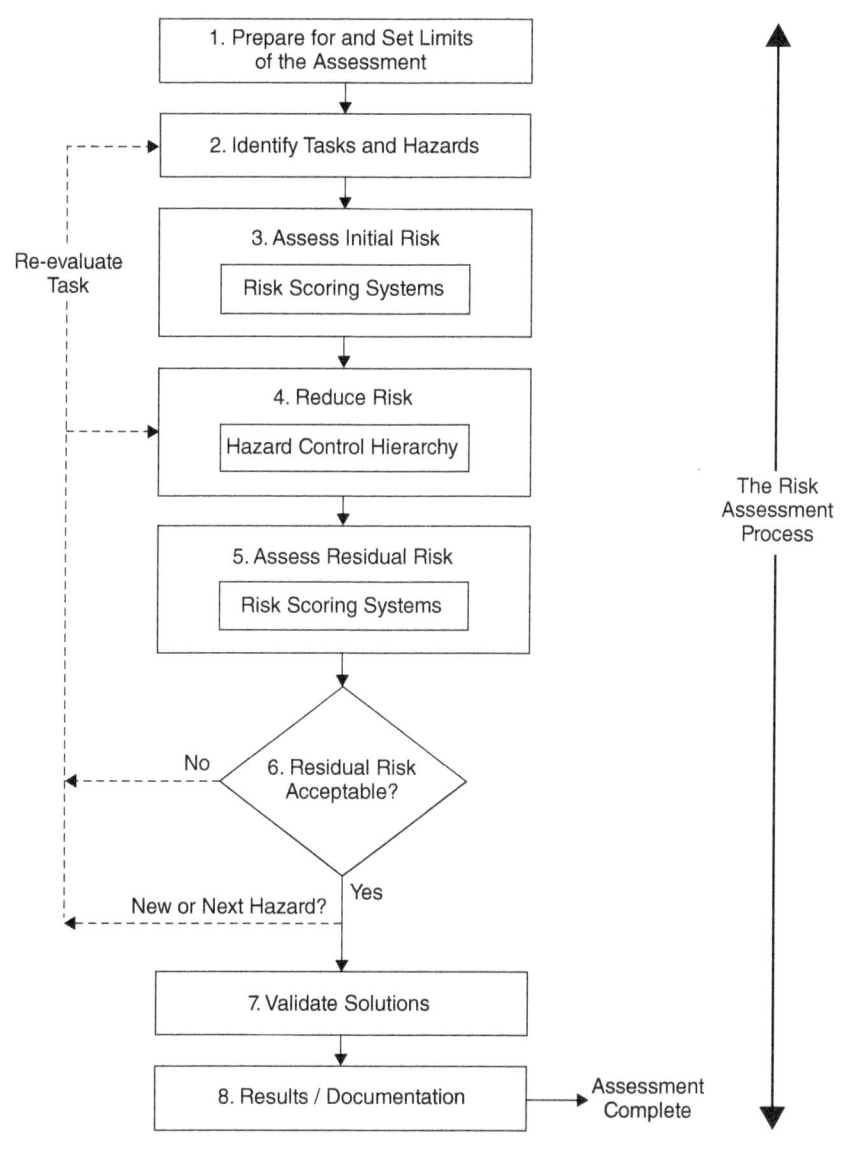

Figure 13.1 The Risk Assessment Process. *Source:* Reprinted with permission from ANSI B11.0-2015 – Safety of Machines

- materials used
- number of people involved in its use
- speed and range of movement
- operator–machine interfaces
- energy sources
- maintenance requirements
- environmental limits such as temperature, humidity, vibration, or other factors.

Once the purpose and scope are adequately defined, a plan of action to conduct the assessment is developed and communicated.

2. *Identify tasks and hazards*

In order to properly manage the risks, the anticipated tasks and associated hazards of the machine must be identified. This requires the assessment team to identify the following elements:

- Personnel exposed to the machine (i.e., operators, maintenance, engineering, contractors, installers, setup, material handlers)
- Tasks performed (i.e., operational modes, maintenance/service, troubleshooting, cleaning)
- Foreseeable hazards of tasks (i.e., mechanical, energy sources, unexpected starts, slips and falls, hot surfaces, combustible atmosphere, sharp edges, operational hazards).

To properly identify these elements, the assessment team needs to be technically qualified and knowledgeable of the machine's use and capabilities. A review of pertinent information such as machine specifications and operational instructions; accident history; applicable codes, regulations, and consensus standards; interviews with operators and maintenance; relevant checklists; and previous risk assessments should be performed by the team. Further identification and clarification of hazards and potential mishaps with the machine and surrounding operations can be accomplished through brainstorming.

3. *Assess initial risk*

For each hazard identified, risk is assessed. Risk related to the identified hazard is the result of both the "severity" of harm and the "probability" of the harm occurring. As in any risk assessment initiative, a risk scoring system or risk matrix must be selected and used. There are many qualitative and quantitative risk scoring systems are available. Which system is selected is largely based on personal, organization, or industry preference (Main, 2012). For example, the MIL-STD-882 Two-Factor Risk Scoring System (4 × 5) model shown in Table 13.2 is a qualitative system that can be used.

Using the selected scoring system, each risk is assessed before (initial risk level) and after any risk reduction (residual risk level) measures are implemented. To assess risk, severity is first estimated using the highest "credible" level of harm caused by the hazard. For instance, the severity level of a hazard capable of causing death or permanently disabling injuries would be considered "catastrophic" in Table 13.2 example scoring model. The probability of the occurrence is then assessed. Factors such as the frequency, duration, and extent of exposure; characteristics of the hazard; human error potential; and operator awareness and training are considered in determining probability. It should be noted that qualitative descriptions are most often used due to the difficulty in accurately predicting probability in quantitative terms. Based on the severity and probability estimations in the risk scoring system, a risk level is derived. This initial risk assessment level is then compared to the organization's established acceptable risk levels to determine if further risk reduction is required.

TABLE 13.2 MIL-STD-882 Two-Factor Risk Scoring System (4 × 5) Example

Probability	Severity			
	Catastrophic	Critical	Marginal	Negligible
Frequent	High	High	Serious	Medium
Probable	High	High	Serious	Medium
Occasional	High	Serious	Medium	Low
Remote	Serious	Medium	Medium	Low
Improbable	Medium	Medium	Medium	Low

4. *Reduce risk*

For those hazards with risk levels exceeding acceptable levels, additional risk reduction measures are needed. Using the hierarchy of control approach, feasible risk reduction measures are selected in their order of effectiveness beginning with "prevention through design" measures, engineering controls, administrative controls, and finally personal protective equipment (PPE) (refer to Table 13.1 Hierarchy of Machine Hazard Control from ANSI B11.0-2015). When selecting and implementing safeguards, it is critical to ensure that no new hazards are introduced. This is achieved by repeating the risk assessment process of the affected elements.

5. *Assess residual risk*

Following the selection of risk reduction measures, the team assesses the residual or remaining risk. Taking into account the effect of the selected risk reduction measures, the severity and probability risk factors are estimated to verify measures are effective and appropriate. (*Note: ANSI B11.0 emphasizes the importance of evaluating safeguarding measures for the potential to – intentionally or unintentionally – defeat or bypass selected safeguards or any incentives to bypass controls in order to meet production demands.*) Using the risk scoring system, the estimated severity and probability factors are combined to determine the residual risk level.

6. *Achieve acceptable risk*

Based on determined residual risk levels for the machine, a decision is made to accept or further reduce the risk for each hazard. It should be noted that acceptable risk is not a universal characteristic and that the same exposures may be judged differently by various organizations. Rather, acceptable risk is determined by each organization based on their culture and industry, among other factors.

7. *Validate solutions*

Once fully implemented and in place, the risk reduction measures are evaluated, tested, and verified as to their effectiveness. This validation process should include physical testing of the machine and safeguard devices and review of procedures. Such testing of machines must be performed safely, avoiding any exposure to harm.

8. *Document the process*

The final step in the process is documentation of the risk assessment. This includes documentation of the assessment steps, risk reduction measures, and results achieved, as well as the following:

- Machine information (identification, location, operation, limitations)
- Observations made during risk assessment (including photographs, digital video, dimensions, distances, stop times, and other relevant factors)
- Types of hazards (mechanical hazards such as moving parts or point of operation and exposure to hazardous energy – refer to ANSI B11.0-2015, Annex B – List of Machinery Hazards)
- Severity of harm from hazard
- Probability of harm from hazard
- Current safeguarding measures
- Initial risk level
- Additional safeguarding required (reference to ANSI B11.19 is recommended)
- Residual risk level.

An example of a risk assessment spreadsheet with columns for each item of data collected is provided in Figure 13.2.

Location:								
Date:	Assessor(s)							
Machine Identification:								
Machine Mode/Life Cycle:								
Observations:								
Exposure		Initial Assessment				Follow-up Assessment		
Activity	Hazard	Severity	Probability	Current Safeguards	Risk Level	Additional Safeguards	Severity	Probability	Risk Level

Figure 13.2 Sample Risk Assessment Spreadsheet for Documenting Data

13.8 ESTIMATING RISK

When determining risk levels for machine hazards, a standardized method should be applied to allow consistent comparison and prioritization of risks. The *ANSI B11.TR3-2000 Risk Assessment and Risk Reduction – A Guide to Estimate, Evaluate and Reduce Risk Associated with Machine Tools* technical report was specifically established for such machine safeguarding risk assessments. This qualitative method uses severity and probability of harm as the primary risk factors in estimating risk levels. As with all risk assessments, the ultimate purpose is to determine if the risk level is acceptable or if further reduction of risk is required. Following the TR3 technical report guidance, a brief description is provided of the risk estimation process.

1. *Estimate severity of harm* – First, the degree of injury or illness that could occur from the hazard is determined using the four categories as follows:

    A. Catastrophic – death or permanently disabling injury or illness
    B. Serious – severe debilitating injury or illness
    C. Moderate – significant injury or illness beyond first aid
    D. Minor – no injury or slight injury requiring first aid.

    Severity is estimated regardless of the safeguards in place. In making this determination, the worst-credible severity of harm that can occur is selected. For example, the severity level of a properly guarded table saw might be estimated as "catastrophic" due to the potential amputation of fingers or hands if contact is made. When the potential severity of a hazard falls between two risk ratings, the higher category should be selected.

2. *Estimate probability of harm* – The likelihood of the occurrence causing harm is then determined using the categories listed as follows:

    A. Very likely – near certain to occur
    B. Likely – may occur
    C. Unlikely – not likely to occur
    D. Remote – so unlikely as to be near zero.

TABLE 13.3 ANSI B11.TR3 Two-Factor Risk Model (4 × 4)

Probability of Occurrence of Harm	Severity of Harm			
	Catastrophic	Serious	Moderate	Minor
Very likely	High	High	High	Medium
Likely	High	High	Medium	Low
Unlikely	Medium	Medium	Low	Negligible
Remote	Low	Low	Negligible	Negligible

Source: Reprinted with permission from ANSI B11.0-2015 – Safety of Machinery.

The highest credible level of probability is estimated taking into account the frequency, duration, and extent of exposure, training, awareness, and presentation of the hazard. In estimating probability, ANSI B11.TR3 includes the following factors that are evaluated in the determination:

- Exposure to a hazard (frequency and duration, extent of exposure to the body, and number of people exposed)
- Personnel who perform tasks (their level of training, skill, and experience)
- Machine/task history (machine reliability, history of incidents and failures, perceived risk comparison to other machines)
- Workplace environment (housekeeping, layout/workflow, working surfaces, lighting, noise, vibration, ventilation, temperature, and humidity)
- Human factors (potential for errors and omissions, ergonomic risk factors, incentives to bypass safeguards, elements that reduce communication or clarity of tasks)
- Reliability of safety functions (control systems integral to the machine)
- Possibility to defeat or circumvent protective measures (degree of ability to bypass safeguards)
- Ability to maintain protective measures (sustainability and continued effectiveness, level of maintenance required).

Using the example of a properly guarded table saw, the probability of harm occurring might be rated "likely" considering the relative ease in removing the guard, accident history on table saws, possible incentive to remove guard for better visibility, operator training, and experience, among other factors.

The ANSI B11.TR3 Two-Factor Risk Model reprinted with permission from *ANSI B11.0-2015 – Safety of Machines* provided in Table 13.3 is used to estimate risk magnitude for each hazard. For risks estimated as "high," the greatest degree of risk reduction is required.

13.9 CASE STUDY

A metal fabrication company recognized the need to assess their risks associated with machines in their facility. The company defined the scope to include all fabrication machinery used in normal production operations. The purpose was to identify and prioritize machines that need additional risk reduction measures to achieve an acceptable risk level. A risk assessment team consisting of an experienced facilitator, production, maintenance, engineering, quality, and management was assembled. The team decided upon a risk assessment method that incorporated existing safeguarding protection and was trained in the method. The following data was collected and used in the risk assessment:

Machine risk assessment method – The organization had a new manufacturing facility that was experiencing a high incident rate related to machine operations. A risk assessment method was selected and a team trained to perform assessments of machines in the production process. The assessment tool used included three factors: severity level (S), likelihood level (L), and existing protection factor (P). An initial risk priority number (RPN) for each machine hazard was determined by multiplying severity level (S) with the product of likelihood (L) and protection factor (P) as shown in the formula $S \times (L \times P) = RPN$. The organization determined that a risk score of 15 or higher on the RPN scale shown in Table 13.7 was unacceptable and would require further risk reduction. Risks above 9 were considered a high priority for remedial action. For hazards with initial RPN score above 9, additional risk reduction measures were selected and calculated in a second risk assessment. The second RPN was calculated to indicate risk levels for the added controls. With this information, the company could prioritize efforts to reduce machine risks appropriately.

Machine – The machine assessed in this case study was a manually fed and operated nut welding machine. The machine, which used two spot welding guns and a fixture, was recently added to the operation. The machine assessed was in routine production mode.

Task – The tasks assessed were the loading and activating of the machine. For each part, two operators worked together placing ordinary steel parts into the fixture and two nuts in the location of the welding points of operation. Then the main operator would activate the welding machine using a two-hand control.

Affected personnel – Two employees were observed working together loading parts and activating the machine. (Other potentially affected personnel included welding technicians and maintenance personnel.)

Activity or task – The task involved placing metal pieces into the welding gun fixture along with two nuts to be welded; then the main operator activated the welding machine using two-hand controls; upon completion of the weld, the welded piece was removed from fixture and placed into bin.

Hazards – The identified hazards to operators included "caught-in" point of operation and puncture to hands, lacerations from handling parts, welding burns to hands, burns from welding sparks to body, flash burns to eyes, and struck by falling materials. (Further assessment of material handling and ergonomic risk factors were considered in a separate ergonomic risk assessment.)

Existing controls – A two-hand control system for activating the welding machine; partial barrier protection on sides and back of machine; use of cut/burn-resistant gloves and arm sleeves; ANSI Z87.1, Shade #2 protective eye wear and side guards; flame-resistant clothing; protective footwear; and a work procedure were in place.

Severity (S) level – For each hazard assessed, a risk rating was assigned that is most closely aligned with the most credible serious consequence. Table 13.4 was used.

Likelihood (L) level – For each hazard assessed, the likelihood of the occurrence was determined using Table 13.5.

Protection factor (P) level – For existing safeguards and control measures, Table 13.6 was used to determine protection level, which was factored into the RPN.

CASE STUDY

TABLE 13.4 Severity Levels

Score	Rating	Description
5	Catastrophic	• Fatality or permanent disability
4	Critical	• Lost time permanent impairment or multiple lost time
3	Substantial	• Lost time, full recovery
2	Marginal	• Medical treatment, OSHA recordable
1	Minimal	• First aid injury

TABLE 13.5 Exposure Levels

Score	Rating	Description
5	Very likely	• Hazard occurs continuously
4	Likely	• Hazard occurs daily
3	Possible	• Hazard occurs less than once a week
2	Unlikely	• Hazard occurs less than once a month
1	Very unlikely	• Hazard rarely occurs

TABLE 13.6 Protection Factor Level

Risk Formula	
Severity × (Likelihood × Protection Factor) = Risk	
Protection Factor (P)	Multiplier
Elimination	0.1
Substitution	0.4
Engineering – multiple	0.6
Engineering – single	0.7
Warning	0.8
Administrative	0.9
PPE	0.95
No controls	1

Risk calculation – For each hazard, the resulting scores for severity (S) multiplied with the product of likelihood (L) and protection factor (P) to calculate the RPN using the formula $S \times (L \times P) = RPN$.

Risk priority number – If the RPN was above 15, further risk reduction measures were required. For RPN scores between 9 and 14, further risk reduction was given high priority as described in Table 13.7.

Risk assessment worksheet – A portion of the resulting risk assessment is displayed in Figure 13.3.

Risk assessment matrix – The resulting RPNs are compared and prioritized according to the matrix in Figure 13.4.

TABLE 13.7 Risk Priority Number Scale with Action Requirements

Risk Level (RL)	Risk Scores	Actions
Very high	15–25	Operation not permissible; immediate action required
High	9–14	Remedial action to be given high priority
Moderate	5–8	Remedial action to be taken at appropriate time
Low	1–4	Remedial action discretionary

Machine Mode/Life Cycle: Nut Welding Machine – Operational Mode											
Observations: Two operators loading/activating machine; coordination and communication required. Sides and rear of machine were partially exposed											
Exposure				Initial Assessment				Follow-up Assessment			
Personnel	Activity	Hazard	Controls	S	L	P	RPN	Additional Controls	S	L	P	RPN
Operator	Place part and nuts in welding fixture and activate weld	Caught in weld point of operation	Two-hand control	5	4		14	Presence sensing device light curtain; pre start verification test; PPE	4	2	0.6	4.8
Assistant; other workers in area	Place part and nuts in welding fixture	Caught in weld point of operation; access from side and rear of machine	Partial barrier to side and rear; work procedure	5	4	0.7	14	Complete enclosure with interlocked barriers; presence sensing device light curtain; pre start verification test; PPE	4	2	0.6	4.8
Operator and assistant	Place part and nuts in welding fixture and activate weld	Burns from welding guns	PPE– burn resistant gloves and clothing	3	5	0.95	14.25	Presence sensing device light curtain; pre start verification test	3	2	0.6	3.6
Operator and assistant	Place part and nuts in welding fixture and activate weld	Welding flash and objects in eyes	PPE– ANSI Z87.1 eye protection – shade #2	4	4	0.95	15.2	Welding curtains	4	3	0.7	8.4

Figure 13.3 Nut Welder Machine Risk Assessment Example.

13.10 ASSESSMENT OF MACHINE MAINTENANCE AND SERVICE

Nonroutine work on machines presents significant risks to life and limb. Common to all operations is the need for periodic maintenance and service of facilities, systems, and equipment. As companies expand and develop, new operations and machines are often needed to replace older or inefficient equipment. In addition, unexpected situations such as equipment breakdowns, power failures, system

ASSESSMENT OF MACHINE MAINTENANCE AND SERVICE

			Severity (S)				
			Insignificant (1) Inconsequential with respect to injuries or illnesses, system loss or downtime, or environmental release	Negligible(2) First aid or minor medical treatment only, nonserious equipment or facility damage, chemical release requiring routine cleanup without reporting	Marginal(3) Medical treatment or restricted work, minor subsystem loss or damage, chemical release triggering external reporting requirements	Critical(4) Disabling injury or illness, major property damage and business downtime, chemical release with temporary environmental or public health impact	Catastrophic (5) One or more fatalities, total system loss, chemical release with lasting environmental or public health impact
			1	2	3	4	5
Likelihood (L)	Frequent (5) Likely to occur repeatedly	5	5	10	15	20	25
	Likely (4) Probably will occur several times	4	4	8	12	16	20
	Occasional (3) Could occur intermittently	3	3	6	9	12	15
	Seldom (2) Could occur, but hardly ever	2	2	4	6	8	10
	Unlikely (1) Improbable, may assume incident or exposure will not occur	1	1	2	3	4	5

Risk Matrix (modified from ANSI Z10)

Figure 13.4 Risk Assessment Matrix Example. Adapted from ANSI/ASSE Z10.2012

faults, and other emergencies that require immediate attention by internal or external resources can occur. These nonproduction and nonroutine situations present unusual and oftentimes significant risks to workers from the sudden release of hazardous energy. In such cases, the isolation and control of hazardous energy is required to perform these tasks safely.

Hazardous energy control in machine systems is most commonly referred to as lockout/tagout. The OSHA 1910.147, Control of Hazardous Energy standard covers the servicing and maintenance of machines and equipment in which the unexpected energization or start-up of the machines or equipment, or release of stored energy, could harm employees. It requires employers to establish an Energy Control Program with machine-specific lockout/tagout procedures for isolating specific energy sources in systems during maintenance and service activities.

However, in certain situations, there is a need for power to remain in the system to perform the required function. Activities such as die changing, jam clearing, lubrication, tool changes, minoring cleaning, adjustments, and setup, which are considered part of the production process, may inhibit the use of traditional lockout/tagout control measures. This presents a challenge in performing these tasks safely. Guidance for such situations is provided in the ANSI/ASSE Z244.1-2003 (R2014), *Control of Hazardous Energy Lockout/Tagout and Alternative Methods* standard, which was revised in 2014. The standard contains requirements for conducting risk assessments when the use of lockout/tagout is not practical, and an alternative method is needed. Selection of an appropriate alternative method is based on a risk assessment of the equipment or process and takes into consideration that existing safeguards may need to be removed or modified to perform the task. Figure 13.5 Decision Matrix for Safeguarding Hazardous Energy reprinted with permission from ANSI Z244.1 provides a decision process for determining the need for a risk assessment.

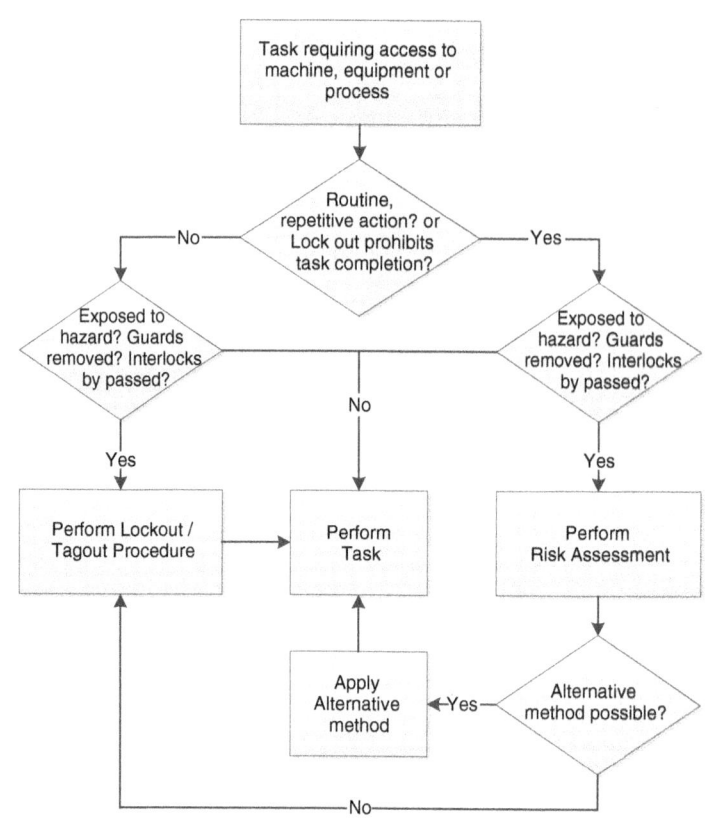

Figure 13.5 Decision Matrix for Safeguarding Hazardous Energy. *Source:* Taken from ANSI/ASSE Z244.1-2003 (R2014) (Courtesy of the American Society of Safety Engineers)

Although a specific methodology is not prescribed, ANSI Z244.1, Annex A provides guidance on the risk assessment and risk reduction processes. Those process steps for risk assessment and risk reduction of machine hazards and the use of alternative methods are outlined as follows.

13.10.1 Risk Assessment Process

The risk assessment process is an analytical process intended to identify, evaluate, and assess hazards and associated risks for the purpose of managing risks to acceptable level. The process involves the following:

1. *Identify all tasks* – All tasks and activities should be considered, including setup, installation, removal, maintenance, operating, adjusting, cleaning, troubleshooting, and programming.
2. *Identify hazards* – Hazardous energy releases such as electrical, gravity, mechanical, chemical, thermal, pneumatic, hydraulic, radiation, and other types of hazards such as human factors associated with each task should be considered.
3. *Assess potential consequences* – Assess the most severe injury that could occur with each task.
4. *Assess potential exposure to hazards* – Assess the potential exposure of all persons to the identified hazards. The number of persons exposed and the frequency, duration, and nature of exposure should be considered in the assessment.

5. *Assess probability of occurrence* – Estimate the probability of occurrence of the hazardous event by considering the following factors:
   - Existing safeguards, safety devices, and safety systems
   - Reliability, history, and failure mode
   - Operational/maintenance demands of task
   - Possibility of defeat or failure of safeguards
   - Accident history of task, activity, machine, equipment, or process
   - Competence of persons performing task
   - Working environment.
6. *Evaluate the risk* – Each identified hazard and task should be evaluated to determine the level of risk. This will determine whether the task is an acceptable risk.
7. *Achieve an acceptable level of risk* – If the level of risk is found to be acceptable, the process is complete. If the risk(s) is/are determined to be unacceptable, the risk reduction process should be implemented.

13.10.2 Risk Reduction Process

For those risks that are unacceptable, risk reduction measures are selected and implemented using the hierarchy of control. The process should be performed and documented by qualified and appropriate personnel. ANSI Z244.1, Annex A lists the following hierarchy of control choices in order of most effective to least effective as follows:

(a) Risk reduction by design
(b) Risk reduction by use of engineered safeguards
(c) Risk reduction by use of warning and alerting techniques
(d) Risk reduction by use of administrative controls
(e) Risk reduction by use of PPE.

Upon completion of the risk reduction measures, a second risk assessment should be performed. This assessment should take into consideration the risk control measures implemented and any new tasks or hazards generated. If the risks are judged acceptable, the process is complete. If residual risk remains unacceptable, further risk reduction followed by a risk assessment is applied until the risk is considered acceptable.

Further review of these processes should be conducted on a periodic basis or when changes or incidents occur to ensure control measures are effective and sustainable.

13.11 SUMMARY

Machine hazards are abundant in many industries and workplaces. It is foreseeable that the use of machines and automation will continue to increase, as will the number of regulatory and consensus standards addressing machine safety. The sheer number machines and magnitude of risk they can present signify a need for enhanced skills in machine risk assessment and safeguarding practices. It can be expected that organizations using machines will seek out those professionals that can help them better manage their machine-related risks.

REVIEW QUESTIONS

1. List five hazard types created by machines and safeguarding methods for their control.
2. What standard is considered the centerpiece for machine safety and risk assessment?
3. Regarding the hazard control hierarchy, explain how each of the following methods affects the severity of harm and the probability of occurrence:
   a. Elimination and substitution
   b. Guards and safeguarding
   c. Awareness devices
   d. Training and procedures
   e. Personal protective equipment.
4. List the six minimum requirements of the design and use of machine safeguards as defined by OSHA.
5. Explain why nonroutine work on machines presents significant risks to life and limb, and what methods are used to prevent exposure to these risks.
6. Risk assessments should be performed for each machine initially, when changes or modifications occur, and when new hazards are recognized. Describe the fundamental steps in performing a machine risk assessment.

REFERENCES

American National Standards Institute (ANSI): ANSI B11.0-2015. *Safety of Machinery – General Safety Requirements and Risk Assessments*. Houston, TX: B11 Standards, Inc., 2015.

ANSI B11.19-2010. *Performance Criteria for Safeguarding*. New York: ANSI, 2010.

ANSI B11.TR3-2000. *Risk Assessment and Risk Reduction – A Guide to Estimate, Evaluate and Reduce Risks Associated with Machine Tools*. New York: ANSI, 2000.

ANSI/ASSE Z244.1-2003 (R2014). *Control of Hazardous Energy – Lockout/Tagout and Alternative Methods*. Des Plaines, IL: American Society of Safety Engineers, 2014.

ANSI/ASSE/AIHA Z10-2012. *American National Standard – Occupational Health and Safety Management Systems*. Fairfax, VA: American Industrial Hygiene Association, 2012.

Main, B. W. *Risk Assessment: Challenges and Opportunities*. Ann Arbor, MI: Design Safety Engineering, Inc., 2012.

Main, B. W., Hayes C. F., July, 2014. "Control systems: insights on ISO 13849-1." *Professional Safety*, 59(7), 41–51.

MIL-STD-882E. *Standard Practice for System Safety*. Washington, DC: Department of Defense, 2012.

OSHA. Safeguarding Equipment and Protecting Employees from Amputations. Department of Labor, Occupational Safety and Health Administration. OSHA 3170-02R 2007. https://www.osha.gov/Publications/osha3170.pdf (accessed December 31, 2015).

OSHA. Machine Guarding eTool. Department of Labor, Occupational Safety and Health Administration, 2015. Available at https://www.osha.gov/SLTC/etools/machineguarding/ (accessed January 20, 2016).

OSHA. Basics of Machine Guarding. Department of Labor, Occupational Safety and Health Administration, n.d. https://www.osha.gov/Publications/Mach_SafeGuard/chapt1.html (accessed December 31, 2015).

OSHA. 29 CFR 1910.147. The Control of Hazardous Energy. Department of Labor, Occupational Safety and Health Administration. https://www.osha.gov/pls/oshaweb/owadisp.show_document?p_table=STANDARDS&p_id==9804 (accessed December 31, 2015).

APPENDIX 13.A
MACHINE SAFEGUARDS METHODS

Machine Guards			
Method	Function	Advantages	Limitations
Fixed guard – permanent part of the machine	Prevents contact with mechanical hazards	• Can be designed for specific applications • In-plant construction is often possible • Usually requires minimum maintenance	• May interfere with visibility • Suitable for high production/repetition operations • Can be limited to specific uses • Machine service and repair require guard removal, requiring other means of protection for personnel such as lockout/tagout
Interlocked guard – to disengage machine when guard is removed	Shuts off/disengages power, stops movement, and prevents start-up when the guard is open; should require the machine to be stopped before the worker can reach into the danger area	• Allows access to the machine for removing jams without time consuming removal of the fixed guards	• Requires careful adjustment and maintenance • May be easy to disengage or bypass
Adjustable guard – to accommodate various sizes of stock	Provides a barrier that may be adjusted to facilitate a variety of production operations	• Can be designed for specific applications • Can be adjusted to admit varying sizes of stock	• Protection may be incomplete at times • May require frequent service • Guard may be made ineffective by operator • May interfere with visibility
Self-adjusting guard – openings of these barriers are determined by the movement of the stock	Provides a barrier that moves according to the size of the stock entering the danger area	• Off-the-shelf guards are often commercially available	• Does not always provide maximum protection • May interfere with visibility • Requires maintenance and adjustment

Safeguarding Devices			
Photoelectric (optical) presence sensing device – uses a system of light sources and controls that can interrupt the machine's operating cycle	Machine will not start cycling when the light field is interrupted; when the light field is broken by any part of the operator's body during the cycling process, immediate machine braking is activated	• Can allow more freedom of movement for operator • Simplicity of use • Used by multiple operators • Provide passerby protection • No adjustment required	• Does not protect against mechanical failure • Limited to machines that can be stopped
Radiofrequency (capacitance) presence-sending device – uses a radio beam that is part of the machine control circuit	Machine cycling will not start when the capacitance field is interrupted; when the capacitance field is disturbed by any part of the operator's body during the cycling process, immediate machine breaking is activated	• Can allow more freedom of movement for operator	• Does not protect against mechanical failure • Antennae sensitivity must be properly adjusted and maintained • Limited to machines that can be stopped
Electromechanical sensing device – uses a probe or contact bar for a predetermined distance	Contact bar or probe travels a predetermined distance between the operator and the danger area; interruption of this movement prevents the starting of machine cycle	• Can allow access at the point of operation	• Contact bar or probe must be properly adjusted for each application and maintained
Types of Safeguarding Devices			
Pullback devices – uses a series of cables attached to the operator's hands, wrists, and/or arms	As the machine begins to cycle, the operator's hands are pulled out of the danger area	• Eliminates the need for auxiliary barriers or other interferences at the danger area	• Limits movement of operator • May create obstructions • Adjustments must be made for specific operations and each individual • Requires frequent inspections and maintenance • Requires close supervision of the operator's use of the equipment

Restraint (hold-back) device – uses cables or straps that are attached to the operator's hands and a fixed point	Prevents the operator from reaching into the danger area	• Little risk of mechanical failure	• Limits movement of operator • May obstruct work space • Adjustments must be made for specific operations and each individual • Requires close supervision of operator's use
Safety trip controls – uses pressure-sensitive body bar controls that deactivate the machine	Stops machine when tripped	• Simplicity of use	• Manually activated • May be difficult to activate controls because of their location • Only protects operator • May require special fixtures • May require a machine brake
Two-hand control – requires constant, concurrent pressure of both hand controls by the operator to activate the machine	Concurrent use of both hands is required, preventing the operator from entering the danger area	• Operator's hands are at a predetermined location • Operator's hands are free to pick up a new part after first half of the cycle is completed	• Requires a partial cycle machine with a brake • Some two-hand controls can be rendered unsafe by holding with arm or blocking, thereby permitting one-hand operation • Protects only the operator

Two-hand trip – requires concurrent application of both the operator's control buttons to activate the machine cycle, after which the hands are free	Concurrent use of two hands on separate controls prevents hands from being in the danger area when machine cycle starts	• Operator's hands are away from danger area • Can be adapted to multiple operations • No obstruction to hand feeding • Does not require adjustment for each operation	• Operator may try to reach into danger area after tripping machine • Some trips can be rendered unsafe by holding with arm or blocking, thereby permitting one-hand operation • Protects only the operator • May require special fixtures
Gate – a moveable barrier that protects the operator at the point of operation before the machine cycle can be started	Provides a barrier between danger area and operator or other personnel	• Can prevent reaching into or walking into the danger area	• May require frequent inspection and regular maintenance • May interfere with operator's ability to see the work

Source: Adapted from OSHA.

14

PROJECT-ORIENTED RISK ASSESSMENTS

BRUCE K. LYON

Risk Control Services, Hays Companies, Kansas City, MO, USA

14.1 OBJECTIVES

- Introduce project-oriented risk
- Review techniques used in construction projects and jobsites
- Review techniques used in maintenance and service
- Review techniques used for specific hazards
- Provide guidance on the use of project-oriented techniques

14.2 INTRODUCTION

With change there is risk. And one thing all projects tend to share throughout their life cycle is "change." Project-oriented tasks such as transportation, installation, construction, maintenance, service, repair, demolition, teardown, retrofitting, and other unique activities have a beginning and an end. And these tasks generally involve rapid and continuous change throughout the project creating certain challenges from an operational risk standpoint.

The Webster's Dictionary defines "project" as a plan, proposal, or assignment to conduct an important undertaking requiring concerted effort. Note the words "plan" and "concerted effort." The definition implies that there is a need to perform preplanning to adequately understand the objectives and scope, as well as the risks; communicate the plan to all affected parties; and coordinate the execution of the plan with stakeholders. Projects can be characterized as follows:

- Temporary or short-term activities with a specific beginning and end
- Not part of the normal production or routine work that takes place
- Dynamic in nature with the potential for the conditions and/or work demands to fluctuate
- Many times performed within other operations, activities, or harsh environments.

Risk Assessment: A Practical Guide to Assessing Operational Risks, First Edition.
Edited by Georgi Popov, Bruce K. Lyon, and Bruce Hollcroft.
© 2016 John Wiley & Sons, Inc. Published 2016 by John Wiley & Sons, Inc.

TABLE 14.1 Examples of Project-Oriented Tasks

Projects	Tasks
Setup	• Loading, transporting, and unloading (delivery) • Rigging, lifting, and erecting • Staging and preparation
Installation	• Assembly of systems, structures, and equipment • Utilities and power systems
Construction	• New facilities, expansions, and additions • Remodeling, modification, and repair of existing facilities or areas
Maintenance and service	• Shutdown • Preventive maintenance • Equipment service and adjustment • Disassembly and reassembly of equipment • Equipment breakdown repair • Return to service
Decommission	• Shutdown • Disassembly • Demolition and teardown • Removal, disposal, and cleanup

With project-oriented activities such as equipment maintenance, repair, installation, demolition, or construction, there is a greater risk of exposure to high energy sources. Energy sources include electrical, gravity, thermal, chemical, mechanical, pneumatic, hydraulic, and stored energy.

The Occupational Safety and Health Administration's (OSHA) standard 1910.147, *The Control of Hazardous Energy*, covers "the servicing and maintenance of machines and equipment in which the unexpected energization or startup of the machines or equipment, or release of stored energy, could harm employees" (OSHA, 1989). Employers are required to develop and implement an energy control procedure for safely performing maintenance and service on machines and processes using lockout and tagout measures. An example of a risk assessment approach for lockout/tagout is provided in Chapter 13, "Machine Risk Assessments."

Table 14.1 provides basic types of projects and examples of tasks that might be performed in projects.

For a project to be safely executed, a plan including a preproject hazard analysis and risk assessment appropriate for the scope and type of work is needed. Such a project plan requires effective communication among stakeholders regarding the project's scope, goals and objectives, timeline, personnel involved and their duties, the equipment and tools necessary, materials, and budget.

A number of hazard analysis and risk assessment tools can be used in project-related tasks. Some include checklists, job hazard analysis (JHA), preliminary hazard analysis (PHA), What-If hazard analysis, and failure mode and effects analysis (FMEA), which are covered in the chapters of this text. This chapter highlights several methods used for nonroutine activities including:

1. *Construction project risk assessment*
2. *Pretask hazard analysis*
3. *Maintenance task hazard analysis*
4. *Operating hazard analysis (OHA)*
5. *Specific hazard analysis examples:*

a. *Confined space pre-entry hazard analysis*
   b. *Fall hazard analysis.*

These represent a fraction of the available methods and variations that are used in the project-oriented tasks. No single tool can adequately address all types of tasks and their hazards. It is up to the user to determine which method(s) is most appropriate for the scope and complexity as well as expected risk levels of the project.

14.3 FATALITIES AND SERIOUS INCIDENTS

Project-oriented tasks have a higher risk of fatalities and serious incidents (FSI) according to studies cited by Fred Manuele (Manuele, 2014). His own study involved a review of more than 1800 incident investigation reports mostly for serious injuries and found that a large percentage of incidents involved:

- Unusual and nonroutine work
- Nonproduction tasks
- Facility modification or construction activities
- Shutdowns and startups for repair and maintenance tasks
- Exposure to high energy sources (e.g., electrical, steam, pneumatic, chemical)
- Upsets: situations going from normal to abnormal.

Taking these factors into consideration, it becomes obvious that such tasks present greater risk and have the potential for serious incidents and fatalities. To clarify, a serious incident can be defined as an unplanned event or series of events that result in death, permanent injuries or illnesses, substantial damage or loss of equipment or property, and harmful release to the environment. The following is an example of how one organization defines a serious accident:

- One or more fatalities
- Three or more personnel hospitalized, for other than observation, as a direct result of or in support of operations
- Property or equipment damage of $250,000 or more.

From a regulatory standpoint, OSHA has reporting requirements for *serious accidents* that involve a fatality or the hospitalization of three or more employees as described in OSHA, 29 CFR 1904.39 – *Reporting fatalities and multiple hospitalization incidents to OSHA* as follows (OSHA, 2014):

1904.39(a)

Basic requirement. Within eight (8) hours after the death of any employee from a work-related incident or the in-patient hospitalization of three or more employees as a result of a work-related incident, you must orally report the fatality/multiple hospitalization by telephone or in person to the Area Office of the Occupational Safety and Health Administration (OSHA), U.S. Department of Labor, that is nearest to the site of the incident. You may also use the OSHA toll-free central telephone number, 1-800-321-OSHA (1-800-321-6742).

Some state agencies have further definitions of a serious accident that also includes inpatient hospitalization for more than 24 hours for other than observation (regardless of number of employees), the loss of a body part, and serious disfigurement.

If the potential for serious incidents and fatalities is to be controlled, risk assessments must be recognized and established as the core of an operational risk management system (Manuele, 2012).

If the individuals performing the work have adequate knowledge and are able to correctly anticipate and recognize hazards and risks, their ability to perform work safely increases, and the risk of serious injuries and fatalities will decrease. As Manuele states, getting the required knowledge embedded into workers' minds requires a major, ongoing endeavor. Specifically directed communication and training must be crafted to achieve the awareness and knowledge required – and to achieve the necessary culture change. Empowering employees to competently assess risks and encouraging them to adopt a mind-set whereby identifying and analyzing hazards and their risks become integral to how they approach and think about work would be a major step forward in injury and fatality prevention. Having knowledge of hazard identification and analysis and risk assessments become rooted within an organization's culture is the type of innovative action needed to further reduce serious injury and fatality potential.

14.4 ERROR TRAPS IN NONROUTINE TASKS

Project-related activities such as equipment installation or removal, maintenance, service, and repair can be target-rich environments for "error traps." Error traps can be described as tasks or situations that have a high potential of producing human error. These error-provoking tasks are created by certain conditions and work demands that lead workers into the same type of error regardless of who is performing the work. The types of tasks that can create a higher likelihood of error are those that involve frequent removal and replacement of large numbers of varied components performed in tight, poorly lit spaces with less-than-adequate tools and usually under severe time pressure (Reason & Hobbs, 2003). These types of conditions and activities are oftentimes found in maintenance, repair, and other nonproduction-related activities.

It should be noted that human error is not a cause of incidents but rather the result of factors and conditions embedded in the work systems. Errors are shaped and provoked by upstream workplace and organizational factors (Reason). Identifying an error is just the beginning of the search for causes, not the end, and needs further explanation. Only by understanding the context that provoked the error can a recurrence be prevented. Further study of human error can be found in a number of resources including *The Field Guide to Understanding Human Error* (Dekker, 2006). The following is a short statement from Dekker that supports this notion:

> *Human error is not a cause of failure. Human error is the effect, or symptom, of deeper trouble. Human error is … systematically connected to features of people's tools, tasks, and operating systems. (p. 15)*

Error traps can be costly. They certainly impact quality and performance of operations; and, these error traps can also result in harm to those involved or affected by the task. By performing hazard analyses and risk assessments of tasks that have a history of human error or have a higher potential for error, error-provoking causes can be identified and eliminated or controlled.

14.5 MANAGEMENT OF CHANGE

When changes are made without proper assessment or management, unintended and often severe consequences can result. For this reason, standards and guidelines for operational risk management systems contain requirements for identifying and assessing risk and managing change. Several examples include

- ANSI/AIHA/ASSE Z10-2012, the Occupational Health and Safety Management System standard:
    - Risk Assessment – Section 5.1.1 and Appendix F

MANAGEMENT OF CHANGE 295

- Management of Change – Section 5.1.3 and Appendix H
- BS OHSAS 18001-2007, Occupational Health and Safety Management Systems:
  - Hazard Identification, Risk Assessment, and Determination of Controls – Section 4.3.1
- ILO OHSMS-2001 (ILO-OHS:2001):
  - Initial Review – Section 3.7
  - Continual Improvement – Section 3.16
  - Management of Change – Section 3.10.2
- OSHA VPP – 2008:
  - Hazard Analysis of Routine Jobs, Task, and Processes
  - Worksite Analysis
  - Hazard Analysis of Significant Changes
  - Pre-use Analysis.

Management of Change (MOC) is a process used to minimize the introduction of new hazards and risks into the workplace from changes made to the operation. These may include changes to structures, work flows, tools and equipment, technology, substances and materials, work practices and procedures, designs and specifications, organizations and staffing, and regulations and standards. The process is applied before and during the modification to ensure that

- hazards are identified, analyzed, and their risks assessed;
- proper decisions are made regarding avoidance, elimination, and control to achieve acceptable risk levels during the change process;
- new hazards are not knowingly created by the change;
- changes do not reintroduce previously controlled hazards;
- changes do not increase the potential or severity of an existing hazard.

The MOC process follows a continuous improvement cycle of planning, doing, checking, and acting (plan–do–check–act) used to understand and control the exposure to hazards and reduce risks from planned modifications. For demonstration purposes, the following example is provided:

A sheet metal fabrication operation is considering a new robotic welding machine to be installed in their fabrication and assembly process. As required by the MOC process, the production department initiates a formal "change request" using the established form that includes information on the specifications, potential hazards, and recommended controls. The form is submitted to the MOC committee for review. The MOC committee, which includes engineering, maintenance, quality, safety, and management, performs a review of the formal request according to the MOC criteria and procedures. The committee performs an analysis and risk assessment of the proposed changes. Based on the risk assessment and necessary control measures to achieve acceptable risk levels, a management decision is made to either "accept" the requested change, "accept with modifications," or "reject" the requested change.

Basic elements that should be considered in an MOC process include the following:

- A clearly defined purpose and scope
- Defined roles, responsibilities, and accountabilities
- Defined triggers for magnitude of change and degree of analysis required
- Established procedures for change requests
- Established change analysis and review procedures

- Criteria for acceptance, modification, and rejection
- Procedures for implementing, tracking, and verifying required actions
- Procedures for completion and closure of the project.

The process of identifying operational hazards and assessing their risks is fundamental to managing risk when change is introduced to the work environment. Managing change and assessing risk are necessary to properly inform decision makers in making decisions, especially before modifications are made to the workplace.

14.6 CONSTRUCTION PROJECT WORK

Anyone who has been involved with construction likely knows that construction-related activities, workplace environments, and conditions can create great risk. According to the Bureau of Labor Statistics' Census of Fatal Occupational Injuries (CFOI), more construction workers were killed on the job in 2012 than in any other industry (BLS, 2012). Of the 4175 worker fatalities that occurred in private industry in calendar year 2012, 806 or 19.3% were in construction. The leading causes of worker deaths on construction sites were falls, followed by struck by object, electrocution, and caught in/between (OSHA, 2014). As a result of this high hazard industry, OSHA has established standards in its 29 CFR 1926, Safety and Health Regulations for Construction. On its Construction Industry resource web page, OSHA describes construction work in the following statement:

> *Construction is a high hazard industry that comprises a wide range of activities involving construction, alteration, and/or repair. Examples include residential construction, bridge erection, roadway paving, excavations, demolitions, and large scale painting jobs. Construction workers engage in many activities that may expose them to serious hazards, such as falling from rooftops, unguarded machinery, being struck by heavy construction equipment, electrocutions, silica dust, and asbestos.*

To manage construction-related hazards and risks, it is necessary to conduct proper prejob planning, hazard analysis, and risk assessment. In the United Kingdom, the Health and Safety Executive (HSE) requires employers with five or more employees to conduct risk assessments. The HSE states that risk assessment in the construction industry is particularly important since construction activities tend to be "inherently dangerous and labor intensive" (HSE, 2014). For construction work, the HSE states on its website the following about legally required risk assessments in the United Kingdom:

> ***General assessment*** – *Employers are required to make an assessment of the health and safety risks to which employees and others are exposed on construction sites. The significant findings must be recorded where five or more people are employed.*
>
> ***Specific assessments*** – *Certain regulations require risk assessments for specific hazards and state in more detail what is required. These include work at height, hazardous substance (COSHH), manual handling, noise, vibration, and lead.*

There are many variations of hazard checklists, hazard analyses, and risk assessments used in construction-related activities. Two basic tools used in construction projects are provided in this section: (1) *construction project risk assessment* used at a project level and (2) *pretask hazard analysis* used for daily tasks.

14.7 CONSTRUCTION PROJECT RISK ASSESSMENT

Where construction work is to be performed, a risk assessment capable of adequately identifying and assessing the anticipated hazards and risks is necessary to properly manage operational risks. There are numerous available techniques for identifying, analyzing, and assessing construction-related hazards and risks, each of which may be useful in particular circumstances. Some of these techniques used in construction projects include checklists, brainstorming, What-If techniques, task analysis, fault and event tree analysis, hazard and operability studies (HAZOP), and FMEA. While there is no single method for any particular situation, stakeholders assessing risks should consider the nature and extent of construction activities and select the most appropriate combination of techniques to ensure that risks are adequately assessed.

The assessment should take into consideration the specific tasks and related risks of the planned construction activities in the project as well as the work environment in which it will take place. This is particularly relevant for projects where tasks and conditions will change during the course of the project. Such an assessment should be proportionate to the risks involved and consider the following factors:

- **Worksite environment and conditions** – site location, access and egress to/from site, weather and ground conditions of site, and risks related to other activities onsite or surrounding areas.
- **Tasks** – extent and complexity of task, duration, frequency, and potential for human error.
- **People** – number of workers involved, degree of exposure to risk, competence levels of the workers involved compared to competence levels required for tasks, and the levels of supervision required. Stakeholders should also consider risks to or created by those not directly involved in the work.
- **Equipment and structures** – suitability of existing structures for work at height (including the existence of unstable surfaces), the selection of equipment to be used, and any risks arising from pre- and postuse of the equipment (e.g., installing and dismantling scaffolding or using a mobile elevated platform or ladder on a busy road).

As part of the preproject planning process, a risk assessment covering the various expected project tasks should be performed by the stakeholders. Following the identification of construction activities, a risk assessment is performed. Steps for conducting a construction project risk assessment are described in the UK HSE booklet *Health and Safety in Construction* (2006) available for free download at http://www.hse.gov.uk/pubns/priced/hsg150.pdf. The fundamental steps listed and tables found in the booklet are reprinted later in accordance with the Crown's *Open Government License* found at www.nationalarchives.gov.uk/doc/open-government-licence/.

1) *Look for the hazards*
   a. *Consider the job, how it will be done, where will it be done and what equipment, materials and chemicals will be used.*
   b. *What are the hazards that could cause harm? Here are some examples that are regular causes of serious and fatal accidents or ill health:*
      i. *falling from an open edge or through a fragile surface;*
      ii. *being struck by site vehicles;*
      iii. *collapse of an excavation or part of a structure;*
      iv. *use of a vibrating hand tool;*
      v. *work with materials (e.g. lead, asbestos or solvents) that could be a health problem;*
      vi. *dust from cutting, grinding or drilling.*

2) *Decide who might be harmed and how*
   a. *Think about employees, the self-employed, employees of other companies working on the job, site visitors and members of the public who may be in the area or outside the site.*
   b. *Safe working often depends on co-operation between firms. Consider how they need to be taken into account in the assessment. Identify problems the work may cause for others at the site, or problems they may cause for those doing the work and agree necessary precautions. Tell the principal contractor or whoever is controlling the site what has been agreed.*
3) *Evaluate the risks and decide on action*
   a. *Where there is a risk that someone could be harmed consider:*
      i. *Can the hazard be removed completely? Could the job be done in another way or by using a different, less hazardous, material? If it can, change the job or process to eliminate the risk.*
      ii. *If the risk cannot be eliminated, can it be controlled? For example, while it may be necessary to apply a solvent-based material, the exposure of workers to hazardous vapors may be reduced by applying it by brush or roller rather than by spraying. If the precautions described in Section 3 have not been taken, is there an equivalent or better standard of protection? If not, more needs to be done.*
      iii. *Can measures be taken which will protect the whole workforce? For example, to prevent falls, guard rails at edges provide safety for everyone in the area.*
      iv. *Can the number of people at risk be reduced? For example, by reducing the size of the site workforce while cranes are in use for erecting structural frames etc., or by undertaking higher-risk tasks outside normal site working hours when only essential personnel will be present.*
4) *Record the findings*
   a. *Employers with five or more employees should record the significant findings of their assessment as an aid to controlling hazards and risks. No specific form is required providing that the information is recoverable.*
   b. *Employers should pass on information about significant risks and the steps they have taken to control the risks, even when they employ less than five people.*
5) *Review the findings*
   a. *Reviews are important. They take account of unusual conditions on some sites and changes in the way the job is done. Reviews allow lessons learned from experience to be taken into account. A new assessment is not always needed for every job, but if there are major changes, a new assessment will be needed. In other cases only the principal contractor will be in a position to do a full assessment. For example, it may be the potential interaction of two or more contractors that leads to increased risk; in such cases the principal contractor should take the lead.*

For evaluating and scoring construction project risk, a simple system can be used such as the HSE risk scoring system depicted in Table 14.2, HSE Severity Levels, and Table 14.3, HSE Likelihood Levels.

The HSE risk scores shown in Table 14.4, HSE Risk Assessment Matrix, provide a simple approach to prioritizing hazards/risks using the following risk levels and recommended actions:

TABLE 14.2 HSE Severity Levels

Severity Level	Description
High	Fatality; major injuries or illness causing long-term disability
Medium	Injury or illness causing short-term disability
Low	Other injuries or illnesses

SAFE WORK METHODS

TABLE 14.3 HSE Likelihood Levels

Likelihood Level	Description
High	Certain or near certain to occur
Medium	Reasonably likely to occur
Low	Very seldom or never occurs

TABLE 14.4 HSE Risk Assessment Matrix

Risk Assessment Matrix		Likelihood		
		High	Medium	Low
Severity	High	3	3	2
	Medium	3	2	1
	Low	2	1	1

Construction Project Risk Assessment				
Date:		Assessors:		
Company:				
Project Description:				
#	Hazard	Existing Controls	Risk Level	Additional Controls Needed

Figure 14.1 Construction Project Risk Assessment Example

1. Low risk requires no action.
2. Medium risk requires action to reduce risk level.
3. High risk requires immediate action.

Two example forms used in assessing construction project risk are provided in Figures 14.1 and 14.2. The two examples show how variations in format and risk categories can be applied to fit the needs of the stakeholders and specific project.

14.8 SAFE WORK METHODS

Following the risk assessment, it may be necessary for the stakeholders to develop safe work methods for certain hazards as part of an overall risk reduction strategy. Such method statements include

Project Risk Assessment

Operation/Location: Street Light Work – Carondelet St between Canal St. and Common St

Assessor(s): Bruce Lyon, CSP, PE Hays Companies

Date: November 12, 2013

Likelihood	Severity			
	Low	Moderate	High	Very High
Very Likely	4	5	6	6
Likely	3	4	5	6
Unlikely	2	3	4	5
Very Unlikely	1	2	3	4

#	Hazard	Likelihood	Severity	Risk Rating	Existing Controls	Additional Controls Required
1	**Electrical Hazard** Operating JLG boom lift near overhead power lines and street car cables	Very Likely without controls	**Death** if contact with power lines	6	Unit placed at curb side	Trained Spotter to ensure safe distance maintained from electrical lines while moving / elevating boom. (10 to in. for 50V; 15 to in. for >50V <200kV... ANSI/SIA A92.5)
2	**Tip over** Moving JLG lift with boom/operator elevated	Likely without controls	Disability	5	Move slow; Fall protection system	Ensure operator is trained on JLG lift; keep boom lowered when moving
3	**Struck by** live traffic - crew crossing street with live traffic from cars, street car	Likely without controls	Disability	5		High visibility vests on crew working near street; traffic control – Local Police
4	**Struck by/ caught in** scissor lift positioned in front of the doorway with pedestrian traffic	Likely without controls	Medical	4		Spotter during scissor lift movement (raising and lowering) to ensure safe distance maintained from scissors and potential falling objects
5	**Falls** climbing over scissor lift rails to enter lift	Likely without controls	Medical	4		Ensure operators are trained on scissor lift; following instructions for mounting lift (thru open gate)
6	**Manual material handling** – setup and breakdown, loading	Likely without controls	Disability	4	Carts, dollies, straps, and other handling aids	Ensure all crew members are properly trained in safe lifting, team lifting, and use of material handling aids. (Observed crew carrying too many boxes blocking view – tripping hazard)
7	**Falls** – stepping off/on elevated slanted tail gate lift on trailers	Likely without controls	Medical	4		Stair platforms with rails to ascend and descend from trailers

Figure 14.2 Project Risk Assessment Example

safe work instructions, procedures, permit systems, and other written prepared steps to conduct work safely. The HSE says this about safe work methods (HSE, 2006):

> *A safety method statement describes in a logical sequence exactly how a job is to be carried out in a safe manner to avoid risks. It includes all the risks identified in the risk assessment and the measures needed to control those risks. This allows the job to be properly planned and resourced.*
>
> *The arrangements for carrying out demolition or dismantling must be recorded in writing before the work begins. This is usually achieved by means of a method statement that can be generated from a risk assessment. Such statements are prepared for many higher risk construction activities e.g. roof work.*

Safe work method statements are most often found in the construction industry and are particularly helpful for higher-risk, complex, or unusual work such as steel and formwork erection, demolition, or the use of hazardous substances. These methods provide documented guidance to employees about how the work should be performed and the precautions to be taken and should be incorporated into the construction project safety and health plan. As in any risk reduction effort, the hierarchy of controls should be used in selecting the most effective control measures to manage risks. Safe work methods are considered an administrative level control and should be used in conjunction with appropriate design and engineering-type controls.

14.9 PRETASK HAZARD ANALYSIS

Certain construction tasks require more specific attention during the life of the project to properly manage their hazards and risks. One simple technique used by construction workers is the task hazard analysis or pretask hazard analysis. (Pretask hazard analysis is also covered in Chapter 5, Fundamental Techniques.)

The task-based approach begins with a job, breaks it down into specific steps, identifies hazards associated with each step, and determines necessary control measures to eliminate or mitigate the hazard. In this chapter, the tool is applied to construction projects and used as part of the preplanning, analyzing, and briefing process.

The *ANSI/ASSE A10.1-2011, Pre-Project and Pre-Task Safety and Health Planning*, standard provides guidance on performing hazard analyses of tasks within construction and demolition operations. The stated purpose of ANSI A10.1 is to assist construction owners, project constructors, and contractors in making preproject and pretask safety and health planning a standard part of their planning processes. Incorporated into the standard are requirements for performing task hazard analyses. Section 3.21 of ANSI A10.1 defines task hazard analysis as:

> *The process of analyzing work tasks to identify potential hazards and determine how to address them so that the task can be completed safely. The process includes the evaluation of the task, affected workers, tools, materials, equipment and identification of appropriate hazard controls. The process is also commonly referred to as an Activity Hazard Analysis or Job Hazard Analysis.*

For any sizable construction project, there should be a preproject safety and health plan that includes policies and procedures for managing the operational risks, including hazard analysis, risk assessment, and control. Hazard analysis and risk assessment should, at a minimum, be employed prior to complex tasks that require specific expertise or potentially hazardous activities such as

- elevated work (tasks where workers could fall six feet or more to lower levels);
- exposure to high energy sources such as electrical, mechanical, stored energy, etc., such as equipment maintenance or utility work;
- critical lifts with cranes or helicopters;
- tunneling, excavation, and trenching;
- confined space entry;
- work that involves hazardous substances;
- routine work that becomes nonroutine as a result of breakdowns, weather changes, and other emergencies.

Where project work is of a general nature, it may be reasonable to have a more generic hazard analysis and risk assessment that can be reviewed on a periodic basis. However, there should be provisions for adding site-specific information should changes in work environment, tasks, or hazards present themselves.

A pretask hazard analysis (or task hazard analysis) should be conducted whenever work is to be carried out for a task that is new, revised, or complex and for tasks without a written job safety procedure to ensure all hazards are correctly identified and controlled. As part of managing project safety, ANSI A10.1 requires task hazard analyses as stated:

Section 8.1.1. As part of the planning process, each contractor shall perform a task hazard analysis before beginning work on any complex task or potentially hazardous task. All task hazard analyses shall occur as close to the time the task will begin as is feasible.

The purpose of a pretask hazard analysis is to analyze work tasks to identify potential hazards and determine proper control measures to allow safe completion of work. The process involves identification and analysis of anticipated tasks and related hazards, affected workers, tools, equipment, materials, workplace conditions, and environment.

The process begins with a review of available project documents, specifications, schedules, site assessments, drawings, plans, and other related documents to identify complex or hazardous tasks anticipated on the project. Tasks that have a potential for greater risk include working from heights; exposure to high energy sources; working in excavations, trenches, or confined spaces; working under suspended objects or around moving equipment; and working with or around hazardous substances. In addition, the potential for unexpected occurrences or interruptions such as equipment breakdowns, weather conditions, or other emergencies can complicate tasks creating new risks or increase existing ones.

A pretask hazard analysis is similar to a JHA, with its three-column format listing sequential task steps, associated hazards, and control measures. However, it differs in that it is performed by the work crew immediately before the task; often incorporates targeted checklists to aid in the identification of hazards; and includes a pretask briefing with the crew. To ensure completeness, as well as ownership by stakeholders, it is vital that all affected workers be involved in the analysis and development of the safe work procedures. The following are basic steps in performing a pretask hazard analysis:

1. **Task steps**: Using a three-column-type analysis worksheet as depicted in Figure 14.3 *Pretask Hazard Analysis Example*, the stakeholders (work crew to perform the task) discuss and record the basic steps of the task in the sequence they will occur. Each step is briefly and clearly described in the first column as an "observable action" such as "enter confined space" or "climb scaffold." The number of steps should be kept manageable and accurately identify the potential hazards of the task. Any anticipated deviations from standard procedures should be included in the analysis as these tend to create new hazards.
2. **Hazards**: For each task, the stakeholders "brainstorm" and identify the potential hazards that are recorded in the second column. A hazard checklist is used to assist in identifying hazard categories such as falls from heights, release of energy sources, struck by, or caught in exposures.
3. **Control measures**: For each hazard, a control measure or safe work procedure is determined using the risk control hierarchy to eliminate or mitigate the risks and recorded in the third column. Stakeholders should first consider the possibility of elimination of hazards, substitution of less hazardous elements, or engineering methods before settling on administrative procedures and protective equipment.
4. **Pretask briefing**: Upon completion of the analysis, a documented pretask briefing with crew members should be held to review the task steps, hazards, and safe work procedures. The briefing should verify that all necessary safety systems such as safe work permits, protective equipment, and emergency resources are in place. Any specific training required to perform the work safely should be provided to affected workers prior to the task.

THE USE OF CHECKLISTS

Pre task Hazard Analysis		
Project:		
Task:		
Performed by:		
Date/Time Start:	Date/Time Finish:	
#	Step	Hazards	Control Measures

Hazards

__ Fall from Heights	__ Cranes and Rigging	__ Explosives
__ Confined Spaces	__ Moving Machinery	__ Power Tools
__ Excavations and Trenches	__ Hazardous Materials	__ Uneven Terrain
__ High Energy Sources	__ Toxic Gas	__ Slips, Trips, and Falls
__ Heavy/Moving Equipment	__ Dust/Fumes	__ Congested Work
__ Fire Hazards	__ Heat Stress	__ Working Alone
__ Overhead Utilities	__ Cold Weather	__ Poor Lighting
__ Underground Utilities	__ Material Handling	__ Suspended Loads
__ Vehicle Traffic	__ Noise/Vibration	__ Other _____

Controls

Permits	Inspections	Systems	PPE
__ Confined Space	__ Tools/Equipment	__ Training	__ Fall Protection System
__ Scaffolds	__ Cranes/Rigging	__ Procedures	__ Hard Hat __ Face shield
__ Excavation	__ Electrical	__ Meetings	__ Gloves __ Arm __ Body
__ Lockout/Tagout	__ Scaffolds/ladders	__ Observations	__ Eye __ Hearing
__ Hot Work	__ Housekeeping	__ Work method	__ Foot __ Respiratory
__ Other	__ Other	__ Other	__ Other

Pre task Safety Briefing

Task assignments and discussion:

Hazard review:

Control measures required:

Safe work methods:

Emergency procedures:

Work Crew Signatures and Date:

Figure 14.3 Pretask Hazard Analysis Example

There are many examples of pretask/task hazard analysis used in various industries. A quick search on the internet can produce a number of examples that follow the basic steps and concept presented here.

14.10 THE USE OF CHECKLISTS

Likely the most common and simplistic tool used in hazard identification and analysis is the *checklist*. As discussed in Chapter 6, checklists typically consist of specific hazards, safety elements, and "yes/no" questions derived from published standards, codes, and industry practices for a specific application. They are relatively quick and easy to use but rely upon the quality and completeness of the checklist. As Main (2004) points out, checklists only guide the user to consider the items on the list, making it a somewhat "static" assessment tool. It does not prompt the user to identify new or unlisted hazards. If pertinent items or questions are omitted from the checklist, the analysis may miss existing hazards or hazard scenarios. It's up to the user's skill and experience to identify existing or anticipated hazards in the task or workplace.

Checklists are used by individuals or teams to identify deviations and resulting hazards associated with a task, process, or system. For a checklist to be effective, it must target specific concerns, standards, or practices of process or system being analyzed. Checklists should be selected and/or developed by someone experienced in the operation, potential deviations, types of hazards, and controls. When conducting preproject planning, specific checklists addressing the anticipated tasks and activities should be developed and used. There are many resources that have established checklists available including governmental and regulatory agencies, industry practices and associations, the insurance industry, safety and health professional associations, universities, and consulting firms to mention a few.

14.11 MAINTENANCE AND SERVICE WORK

Common to all operations is the need for periodic maintenance and service of facilities, systems, and equipment. And as companies grow and develop, new structures and additions as well as equipment are installed, while obsolete assets are decommissioned and dismantled or demolished. In addition, unexpected and abrupt situations occasionally occur such as equipment breakdowns, power failures, system faults, natural and man-made emergencies, and other unwanted events, which require immediate attention by internal or external resources. Such activities are considered nonproduction and nonroutine and oftentimes present significant hazards and risks to an organization. Unfortunately, these activities are frequently overlooked from a formal hazard analysis and risk assessment standpoint.

Several examples of hazard analysis and risk assessment of maintenance-related activities are available on the HSE website. The following HSE example illustrates how maintenance-related activity can be analyzed and assessed prior to the job (HSE, 2013). It is reprinted in accordance with the Crown's Open Government License at www.nationalarchives.gov.uk/doc/open-government-licence/.

ABC Engineering manufactures parts for the motor industry. The company employs 40 people on a site built in the 1970s. The managing director reviewed the company's health and safety arrangements and found that although risk assessments for the production, storage and distribution of products were done and the necessary risk control measures had been put in place, no assessment of the risks had been done and recorded for maintenance work in the factory. The MD told the maintenance manager (the 'fitter') to do this risk assessment and to put its findings into practice. Where possible, maintenance work at the factory is done in-house by the fitter. His main job is to support production by, for example, maintaining plant, machinery and tools and undertaking minor jobs on the building fabric. The company also uses outside contractors, for example for most building repairs, detailed repairs to machinery, most electrical work and work on the LEV system. The fitter's job includes the selection of contractors and, with the works manager, the oversight of their work. The fitter works out of a small workshop, which has some basic engineering machinery, a welding kit and secure storage for solvents and flammables. His work, however, takes him to all parts of the factory.

Following the guidelines for conducting risk assessment, the fitter identified the hazards by:

- *walking around all the areas where he and contractors may go and noting those things that might pose a risk*

Maintenance Task Hazard Analysis

Company/Location:						
Date:		Personnel Involved:		Approved by:		
Maintence Task Description:						
Hazard	Personnel Exposed	Current controls	Additional controls needed	Action By	Target Date	Completion Date

Figure 14.4 Maintenance Task Hazard Analysis

- talking with the safety representative, supervisors and other members of staff to learn from their knowledge and experience of particular jobs and to listen to their concerns and opinions on health and safety issues
- looking at the accident book to learn what had previously resulted in accidents or near misses.

As he identified the hazards he also thought about who could be harmed by them and how accidents might happen. He noted what was already being done to control the risks. Where he considered existing controls not to be good enough he noted what further actions were required. Putting the risk assessment into practice he set out what actions needed to be taken, who would do them and by when. The fitter discussed the findings with the safety representative, supervisors and managing director and placed a copy of the risk assessment on the notice board. The fitter decided to review the risk assessment whenever there were any significant changes to the workplace or one-off jobs needed to be done.

For those interested in further explanation, the HSE website provides a detailed scenario in its *example risk assessment for maintenance work in a factory* available at http://www.hse.gov.uk/risk/casestudies/pdf/factory.pdf.

Although the HSE refers to this example as a *risk assessment*, the authors consider the term *hazard analysis* more accurate since risk estimation is not conducted. Figure 14.4 shows an example of a maintenance task hazard analysis form that can be used.

14.12 OPERATING HAZARD ANALYSIS

As described in the US Department of Defense, Standard Practice for System Safety, MIL-STD-882E-2012, Operating and Support Hazard Analysis (O&SHA), also known as OHA, is a method performed to identify hazards and assess risks introduced by operational and support activities such as installation, maintenance, repair, service, and other support tasks. Even though it is referred to as an *analysis*, the technique contains the elements of a *risk assessment*.

The OHA is used to evaluate the adequacy of operational and support procedures, facilities, processes, and equipment used to mitigate risks associated with identified hazards. According to the MIL-STD-882E standard, an OHA is used to identify measures needed to eliminate or control risks and should consider the following:

- Planned system configurations
- Facility/installation interfaces to the system
- Planned operation and support environments
- Supporting tools or other equipment
- Operating and support procedures

- Task sequence, concurrent task effects, and limitations
- Human factors, regulatory, or contractually specified personnel requirements
- Potential for unplanned events, including hazards introduced by human errors
- Past evaluations of related legacy systems and their support operations.

An example of the OHA is found in the Federal Transit Administration (FTA) Office of Safety and Security 2001 publication entitled "Hazard Analysis Guidelines for Transit Projects." The FTA guidelines indicate that the OHA is used to identify and analyze hazards associated with personnel and procedures during production, installation, testing, training, operations, maintenance, and emergencies. The procedure is described as follows:

The OHA will be conducted on all tasks and human actions, including acts of omission and commission, by persons interacting with the system, subsystems and assemblies, at any level. When the OHA indicates a potential safety hazard, it will be made known to the responsible engineer to initiate a design review or a system safety working group action item. The OHA will be reviewed on a continuous basis to provide for design modifications, procedures, testing, etc. that do not create hazardous conditions.

Figure 14.5 demonstrates an example of the OHA form adapted from the FTA guidelines. The severity categories (Table 14.5), probability levels (Table 14.6), and risk assessment matrix (Table 14.7) used by the OHA are taken from the MIL-STD-882E standard. The following steps for filling in the columns are described in the following:

1. **Task description** – Briefly describe task being performed.
2. **Hazard description** – Describe human action or condition that could lead to an incident.
3. **Probability of occurrence** – Enter the level of probability of such an occurrence using the probability levels in Table 14.6.
4. **Potential cause** – Enter the most likely primary and secondary causes, including those induced by hardware, software, procedures, and the environment.
5. **Effect on personnel or system** – Describe the effect that the action or condition may have on people, equipment, facilities, and entire system in terms of system safety and operational impact (i.e., delay, injury, damage, fatality, etc.).

System: Preventive Maintenance Procedure Subsystem: Pump OHA No.: 1			**OPERATING HAZARD ANALYSIS (OHA)**				Prepared by: Date: Reviewed by: Date: Approved by: Date:	
General Description			Hazard Cause/Effect				Corrective Action	
Task Description	Hazard Description	Probability of Occurrence	Potential Cause	Effect on Personnel or System	Hazard Risk Index	Possible Controlling Measures	Resolution
Preventive maintenance on pumping system that serves tunnel	Failure to perform power plug reversal; failure to test pump after reversal	Occasional	(1) Inattention or distraction; (2) unaware of potential hazard	Pump failure; tunnel flooding from pump failure	1C	(1) <u>Design change</u>: Install automatic transfer switch set to monthly or specified number of hours <u>Procedure</u>: Add steps to test pump after plugs have been reversed	1D Transfer Switch installed (10/6/13) PM testing procedure revised, approved with training (9/5/13)

Figure 14.5 Operating Hazard Analysis. *Source:* Adapted from the Federal Transit Administration (2000)

OPERATING HAZARD ANALYSIS

TABLE 14.5 Severity Categories from MIL-STD-882E

\multicolumn{3}{c	}{Severity Categories}
Description	Severity Category	Mishap Result Criteria
Catastrophic	1	Could result in one or more of the following: death, permanent total disability, irreversible significant environmental impact, or monetary loss equal to or exceeding $10 M
Critical	2	Could result in one or more of the following: permanent partial disability, injuries or occupational illness that may result in hospitalization of at least three personnel, reversible significant environmental impact, or monetary loss equal to or exceeding $1 M but less than $10 M
Marginal	3	Could result in one or more of the following: injury or occupational illness resulting in one or more lost work day(s), reversible moderate environmental impact, or monetary loss equal to or exceeding $100K but less than $1 M
Negligible	4	Could result in one or more of the following: injury or occupational illness not resulting in a lost work day, minimal environmental impact, or monetary loss less than $100K

Source: MIL-STD-882E (2012).

TABLE 14.6 Probability Levels from MIL-STD-882E

\multicolumn{4}{c	}{Probability Levels}
Description	Level	Mishap Result Criteria	Fleet or Inventory
Frequent	A	Likely to occur often in the life of an item	Continuously experienced
Probable	B	Will occur several times in the life of an item	Will occur frequently
Occasional	C	Likely to occur sometime in the life of an item	Will occur several times
Remote	D	Unlikely but possible to occur in the life of an item	Unlikely but can reasonably be expected to occur
Improbable	E	So unlikely it can be assumed that occurrence may not be experienced in the life of an item	Unlikely to occur but possible
Eliminated	F	Incapable of occurrence. This level is used when potential hazards are identified and later eliminated	Incapable of occurrence. This level is used when potential hazards are identified and later eliminated

Source: MIL-STD-882E (2012).

6. **Hazard Risk Index** – Using the risk assessment matrix in Table 14.7, determine the Hazard Risk Index rating based on the hazard's severity category from Table 14.5 and probability level from Table 14.6.
7. **Possible control measures** – Describe actions that can be taken or procedural changes to prevent the anticipated hazard from occurring.
8. **Resolution** – Describe changes made or steps taken to eliminate or control the hazard.

TABLE 14.7 Risk Assessment Matrix from MIL-STD-882E

SEVERITY / PROBABILITY	Catastrophic 1	Critical 2	Marginal 3	Negligible 4
Frequent A	High	High	Serious	Medium
Probable B	High	High	Serious	Medium
Occasional C	High	Serious	Medium	Low
Remote D	Serious	Medium	Medium	Low
Improbable E	Medium	Medium	Medium	Low
Eliminated F	Eliminated			

Source: MIL-STD-882E (2012).

14.13 ANALYZING SPECIFIC HAZARDS

Project-related work can involve specific types of hazards that require specific analysis and assessment. Some of these project-related hazards may include critical lifts by cranes, release of hazardous energy, falls from elevations, excavation and trench collapse, exposure to hazardous atmospheres or substances, contact with electrical overhead power lines, arc flash, and heavy equipment use. For illustration purposes, two examples of specific hazards are presented in this section including (1) confined space pre-entry hazard analysis and (2) fall hazard assessment.

14.14 PRE-ENTRY HAZARD ANALYSIS

Maintenance and repair work is sometimes required within confined spaces such as tanks, vessels, silos, storage bins, hoppers, vaults, pits, manholes, tunnels, equipment housings, ductwork, and pipelines. Confined spaces present unique and severe risks that lead to a number of FSI each year. Such incidents often involve more than one fatality resulting from ill-prepared attendants or rescuers attempting to rescue entrants that are incapacitated or trapped in a confined space.

In OSHA 1910.146 (OSHA, 1993), Permit-required Confined Space standard, OSHA defines confined space as a space that:

1. is large enough and so configured that an employee can bodily enter and perform assigned work; and
2. has limited or restricted means for entry or exit (for example, tanks, vessels, silos, storage bins, hoppers, vaults, and pits are spaces that may have limited means of entry); and assigned work; and
3. is not designed for continuous employee occupancy.

OSHA's definition of a "permit-required confined space (permit space)" is a confined space that has one or more of the following characteristics (risks):

1. Contains or has a potential to contain a hazardous atmosphere

PRE-ENTRY HAZARD ANALYSIS

Confined Space	Building/Location	Type of Space	Potential Hazards	Type of Entry
Electrical vault	Mechanical building #1 – Northwest corner	Partially enclosed vault with single access	Potential electrical energy	Permit reclassified to non permit with LOTO and air monitoring clearance
Nitrogen storage tank	Main production building – South end.	Enclosed 1000 gallon tank with one manhole at bottom and vent	Atmospheric hazards – oxygen deficient	Permit – alternative entry with air monitoring and continuous forced ventilation
Crawl space	Under main production building	Enclosed tunnel with two entries	Configuration	Permit
Air handling plenum	Above main production building	Enclosed tunnel with three entries	Falls; mechanical energy	Permit
Powder coat oven	Main production building – finishing department	Partially enclosed natural gas fired oven with steps to walk in entry (2)	Atmospheric hazards – natural gas; high temperature	Permit reclassified to non permit with isolation of gas line and LOTO, temperature cool down and air monitoring clearance

Figure 14.6 Confined Spaces and Hazard Inventory Example

2. Contains a material that has the potential for engulfing an entrant
3. Has an internal configuration such that an entrant could be trapped or asphyxiated by inwardly converging walls or by a floor which slopes downward and tapers to a smaller cross-section or
4. Contains any other recognized serious safety or health hazard.

The OSHA confined space entry standard requires an initial evaluation of the workplace to identify confined spaces and determine if they are "permit-required" spaces. Figure 14.6 illustrates an example of a *confined spaces and hazard inventory* resulting from an initial evaluation.

If entry into a "permit-required" confined space is necessary, a pre-entry hazard analysis is required to identify and control hazards as part of the entry permit system. Specifically, OSHA 1910.146 says the employer must:

(d)(2) Identify and evaluate the hazards of permit spaces before employees enter them.

(d)(3) Develop and implement the means, procedures, and practices necessary for safe permit space entry operations ...

Permit spaces that have as their only hazard an actual or potential hazardous atmosphere may use "alternate entry procedures" according to OSHA 1910.146(c). Alternate entry procedures do not require the use of an entry permit; however, in order to use alternate entry procedures, the employer must demonstrate that

- the only hazard posed by the permit space is an actual or potential hazardous atmosphere;
- that continuous forced air ventilation alone is sufficient to maintain that permit space safe for entry;
- all other requirements in 1910.146(c) have been met.

TABLE 14.8 Confined Space Hazards

Atmospheric Hazards: • Oxygen below 19.5% • Oxygen above 23.5% • Toxic • Flammable or explosive
Physical Hazards: • Mechanical • Electrical • Thermal (heat stress; extreme cold) • Engulfment or entrapment • Struck by or against • Falls • Biological (animals, insects, etc.) • Noise or vibration • Chemical residues • Poor visibility • Congested configuration
Hazards Created by Work or Conditions: • Fumes from hot work • Exposure to chemicals or substances used • Substances entering/leaking into space from pipes or openings • Manual handling or cumulative trauma • Noise or vibration • Welding flash • Forced air ventilation failure or power failure

Confined Space Identification	Hazard(s)	Hazard Elimination Method	Potential Hazardous Atmosphere	Ventilation Equipment Required
CS #1 Elec. vault	Electrical energy release (440 V)	Elec. vault LOTO procedure	Potential oxygen deficient or CO_2 buildup from work performed	Forced air ventilation at entrance
CS #4 Powder coat oven	Mechanical overhead conveyor; natural gas; temperature	Powder coat LOTO procedure; double block and bleed gas line; cool down to 85 °F	Welding fumes from welding work	Welding fume extractors; forced air ventilation at entrance
CS #8 Mechanical space sub-basement	Potential atmospheric hazard	Forced air ventilation	Oxygen deficiency or CO_2 buildup	Forced air ventilation at entrance

Figure 14.7 Example of a Hazard Analysis for Alternative Entry Procedure Form

A list of confined space hazards is provided in Table 14.8, while Figure 14.7 provides a sample alternative entry hazard analysis form.

Further review of the 1910.146 standard is advised for development of a confined space entry program and compliance with the Permit-required Confined Space standard.

14.15 FALL HAZARD ASSESSMENT

Falls from elevations are a leading cause of workplace fatalities. Construction and maintenance projects can involve work that presents such exposures. A fall hazard analysis or assessment for elevated work should be a standard practice.

The following instructions for conducting an assessment of fall hazards are adapted from US Department of the Navy guidance documents, *Department of the Navy Fall Protection Guide for Ashore Facilities*, August 2013, and the *OPNAV Instruction 3500.39C*, *Operational Risk Management (ORM) Fundamentals*, July 2010 (Naval Safety Center, 2013).

1. **Conduct Hazard Survey.**
   (a) Assessor shall determine whether the walking or working surfaces on which employees are to work have the strength and structural integrity to support the workers safely. Employees shall not be permitted to work on those surfaces until it has been determined that the surfaces have the requisite strength and structural integrity to support the workers and equipment related to their task(s).
   (b) Once it has been determined that the surface is safe for employees to work on, then it should be determined if a Fall-Hazard exists at the work location. A Fall-Hazard Survey will help identify potential Fall-Hazards at the workplace. The gathered information will provide documentation to assist in the development of viable solutions to protect personnel exposed to Fall-Hazards. Understanding work procedures and how a person conducts the required task is very important in the selection and development of the most appropriate Fall Protection method. A Fall-hazard survey will help to identify options for fall hazard elimination and/or selection of other control measures.
   (c) The survey information, required for identification of Fall-Hazards at existing buildings or facilities should include the following:
   - Interview of workers and their supervisors
   - Work-paths and movement of the workers
   - Range of mobility in each fall-hazard zone
   - Location and distances to obstructions
   - Potential anchorage location, if a Fall-Hazard cannot be eliminated or prevented
   - Available clearance and total fall distance
   - Number of personnel exposed to Fall-Hazards
   - Frequency and duration of exposure
   - Lock-Out/Tag-Out hazards
   - Potential severity of the fall
   - Access or egress to fall-hazard area
   - Condition of floors and other surfaces
   - Review of any fall mishap reports at the facility

- Identify the presence of any:
  - Hot objects, sparks, flames, and heat-producing objects
  - Electrical and chemical hazards
  - Sharp objects
  - Abrasive surfaces
  - Moving equipment and materials
  - Impact of weather factors
  - Any other maintenance or work environment issues or conditions.

2. **Assess Risk.**
   (a) For each hazard identified, determine the associated degree of risk in terms of probability and severity. The result of the risk assessment is a prioritized list of hazards, which ensures that controls are first identified for the most serious threat to mission or task accomplishment. The hazard list is intended for use as a guide to the relative priority of risks involved and not as an absolute order to follow.
   - **Severity.** This is an assessment of the potential consequence that can occur as a result of a hazard and is defined by the degree of injury, illness, property damage, loss of assets (time, money, personnel), or effect on the mission or task. Consideration must be given to exposure potential. For example, the more resources exposed to a hazard, the greater the potential severity. Severity categories are assigned Roman numerals according to the following criteria in Table 14.9.
   - **Probability.** This is an assessment of the likelihood that a potential consequence may occur as a result of a hazard and is defined by assessment of such factors as location, exposure (cycles or hours of operation), affected populations, experience or previously established statistical information. Probability categories are assigned a letter according to the following criteria in Table 14.10.

3. **Complete Risk Assessment.**
   (a) Combine the severity with the probability to determine the risk assessment code (RAC) or level of risk for each hazard, expressed as a single number. Although not required, the use of a matrix, as illustrated in Figure 14.8 is helpful in identifying the RAC. In some cases, the worst credible consequence of a hazard may not correspond to the highest RAC for that hazard. For example, one hazard may have two potential consequences. The severity of the worst consequence (I) may be unlikely (D), resulting in a RAC of 3. The severity of the lesser consequence (II) may be probable (B), resulting in a RAC of 2. Therefore, it is important to consider less severe consequences of a hazard if they are more likely than the worst credible consequence, since this combination may actually present a greater overall risk.

4. **Select Risk Control Measures.**
   (a) The preferred order of control measures for Fall-Hazards is:
      (1) Elimination – Removal of the hazard from a workplace. This is the most effective control measure (e.g., lowering various devices or instruments installed at high locations, such as meters or valves, to the height level of the individual; instead of servicing such devices or instruments at heights).
      (2) Prevention – (traditional) – The isolation or separation of the hazards from the general work areas (e.g., same level barriers such as guardrails, walls, covers or parapets.)

FALL HAZARD ASSESSMENT

(3) Engineering Controls – Where the hazard cannot be eliminated, isolated, or separated, engineering control is the next-preferred measure to control the risk (e.g., design change or use of various equipment or techniques, such as aerial lift equipment or movable and stationary work platforms).

(4) Administrative Controls – This includes introducing new work practices that reduce the risk of a person's falling (e.g., erecting warning lines or designated areas, restricting access to certain areas, or posting of warning signs).

(5) Personal Protective Systems and Equipment – These shall be used after other control measures (such as eliminating or isolating Fall-Hazards) are determined not to be practical, or when secondary systems are needed (e.g., when it is necessary to increase protection by employing a backup system).

Control measures are not mutually exclusive. There may be situations where more than one control measure should be used to reduce the risk of a fall.

The Site-Specific Fall Survey Report Checklist in Figure 14.9 taken from the Department of the Navy document, and the Fall Hazard Analysis form example in Figures 14.10 and 14.11, can be used in performing fall hazard assessments.

TABLE 14.9 Severity Categories from OPNAV Instruction 3500.39C

Category	Description
I	Loss of the ability to accomplish the mission. Death or permanent total disability. Loss of a mission-critical system or equipment. Major facility damage. Severe environmental damage. Mission-critical security failure. Unacceptable collateral damage
II	Significantly degraded mission capability or unit readiness. Permanent partial disability or severe injury or illness. Extensive damage to equipment or systems. Significant damage to property or the environment. Security failure. Significant collateral damage
III	Degraded mission capability or unit readiness. Minor damage to equipment, systems, property, or the environment. Minor injury or illness
IV	Little or no adverse impact on mission capability or unit readiness. Minimal threat to personnel, safety, or health. Slight equipment or system damage but fully functional and serviceable. Little or no property or environmental damage

Source: OPNAV INSTRUCTION 3500.39C (2010).

TABLE 14.10 Probability Categories from OPNAV Instruction 3500.39C

Category	Description
A	Likely to occur, immediately or within a short period of time. Expected to occur frequently to an individual item or person or continuously over a service life for an inventory of items or group
B	Probably will occur in time. Expected to occur several times to an individual item or person or frequently over a service life for an inventory of items or group
C	May occur in time. Can reasonably be expected to occur sometime to an individual item or person or several times over a service life for an inventory of items or group
D	Unlikely to occur but not impossible

Source: OPNAV INSTRUCTION 3500.39C (2010).

Risk Assessment Matrix			Probability			
			Frequency of occurrence over time			
			A Likely	B Probable	C May	D Unlikely
Severity Effect of hazard	I	Loss of mission capability, unit readiness, or asset; death	1	1	2	3
	II	Significantly degraded mission capability or unit readiness; severe injury or damage	1	2	3	4
	III	Degraded mission capability or unit readiness; minor injury or damage	2	3	4	5
	IV	Little or no impact to mission capability or unit readiness; minimal injury or damage	3	4	5	5
Risk Assessment Codes (RAC)						
1 – Critical 2 – Serious 3 – Moderate 4 – Minor 5 – Negligible						

Figure 14.8 Risk Assessment Matrix. *Source:* OPNAV INSTRUCTION 3500.39C (2010)

Site-specific Fall Survey Report Checklist	
General Information	
Activity	Date
Location	Area
Conducted by	Approved By
Fall Hazard #	
Survey Information	
Major Fall Hazard Zone or Type:	Work Location:
Personnel Interviewed:	Guiding Documents:
	Work Type:
Distance of Personnel from Fall Hazard in Feet:	
Distance to Ground Below in Feet:	
Number of Personnel Exposed to Fall Hazard:	
Location or Distance to Obstructions in Feet:	
Suggested Anchorage:	
Frequency/Duration of Fall Exposure: ___/___	Exposure Risk:
Potential Severity of Fall:	Obstructions in Fall Path:
Access to Fall Hazard Area:	Condition of Work Surface:
Previous Fall Incidents?	Lockout Tagout Hazards?
Other Hazards	**Suggested Fall Protection Solutions**
Hotwork	Guardrails:
Chemical	Horizontal Life-line
Electrical	Portable System
Sharp Objects	Overhead Beam Strap
Abrasive Surfaces	Self-retracting Lanyard
Weather	Energy Absorbing Lanyard
Other Hazards	Maintenance Stand or workplatform
	Restraint System
Anchorage Locations	Positioning System
	Aerial Lift/Work Platform
Can Rescue Be Performed if Required?	Horizontal or Single Anchor Vertical Lifeline System
Is there a Rescue Plan prepared?	Type of Rescue Self Assisted
Are end-users trained in fall arrest systems?	Other:
Do Swing Fall Hazards Exist?	Additonal Information:

Figure 14.9 Site-Specific Fall Survey Report Checklist

FALL HAZARD ASSESSMENT 315

Fall Hazard Analysis 1st page			
Company:	Location:		
Date Assessed:	Assessors:		
Procedures Reviewed:	Site Marked / Entry Control:		
Fall Hazard Checklist			
1. Area restricted to prevent unauthorized entry?		__Yes	__No
2. Are fall prevention systems in place?		__Yes	__No
3. All slip and trip hazards removed or controlled?		__Yes	__No
4. Visual warnings of fall hazards installed?		__Yes	__No
5. Can the distance a worker could fall be reduced?		__Yes	__No
6. All floor opening covered or protected?		__Yes	__No
7. Does the site contain any other hazards?		__Yes	__No
8. Is the space a designated Permit-required confined space?		__Yes	__No
9. Have anchor points been designated and load tested?		__Yes	__No
Analysis Information			
Initials	Hazard Factors	Recommendations	
	Total potential fall distance:		
	Number of workers involved:		
	Frequency of task:		
	Obtainable anchor point strength:		
	Required anchor point strength:		
Additional Requirements			
Initials	Environmental Conditions	Recommendations	
Initials	Structural Alternations Required	Remarks	
Initials	Task Modifications Required	Remarks	

Figure 14.10   Fall Hazard Analysis Form Example – Page 1

Fall Hazard Analysis
2nd page

Work Task Sketch (with breakdown of vertical and horizontal movement)

Training Requirements

Initials	Training Element	Remarks

Personal Protective Equipment (PPE) Requirements

Initials	PPE	Remarks

Authorization

I certify that I have conducted a Fall Hazard Analysis of the above designated location and have detailed the findings of the analysis on this form.

Name:	Signature:
Title:	Date:

Analysis Form Retention		Attachments	
Permanent retention:	Location:	__Yes	__No
Date Filed:	Filed by:		

Figure 14.11 Fall Hazard Analysis Form Example – Page 2

14.16 SUMMARY

This chapter has provided a primer in hazard identification, analysis, and risk assessment for project-related tasks. There are many other specific hazard analysis and assessment methods used in various industries, military branches, and governmental agencies. However, the concepts remain true to the risk management process: identify hazards, analyze and evaluate their risk, and reduce risk to an acceptable level. To reduce serious injuries and fatalities in the workplace, it is vital that nonroutine tasks be included in the assessment and control of operational risks.

REVIEW QUESTIONS

1. Describe the four primary characteristics that all "projects" share.
2. List five different project categories that occur in the workplace. Provide example tasks within each of these types of projects.
3. Describe the concept of Management of Change (MOC), and list four standards and/or guidelines that include MOC requirements in their guidance.
4. Explain the purpose of a pretask hazard analysis and/or risk assessment.
5. For construction projects, list seven high-risk activities where pretask hazard analyses and risk assessment should be performed.
6. Explain how maintenance and service work activities can present risk to workers.
7. List several hazard analysis and risk assessment methods mentioned in this chapter that are used for project-related tasks.

REFERENCES

American National Standards Institute (ANSI): ANSI/ASSE A10.1-2011. *Construction and Demolition Operations, Pre-Project & Pre-Task Safety and Health Planning.* Des Plaines, IL: American Society of Safety Engineers, 2011.

ANSI/ASSE/AIHA Z10-2012. *American National Standard – Occupational Health and Safety Management Systems.* Fairfax, VA: American Industrial Hygiene Association, 2012.

BLS. *Census of Fatal Occupational Injuries Summary.* Washington, DC: Bureau of Labor Statistics, US Department of Labor, 2012. USDL-13-1699: released August 22, 2013. Also at http://www.bls.gov/news.release/cfoi.nr0.htm (accessed December 21, 2015).

BS OHSAS 18001:2007. *Occupational Health and Safety Management Systems – Requirements.* London: British Standards Institution, 2007.

Dekker, S., *The Field Guide to Understanding Human Error.* Burlington, VT: Ashgate, 2006.

Federal Transit Administration. *Hazard Analysis Guidelines for Transit Projects.* Cambridge, MA: Department of Transportation, 2000. Also available at http://www.fta.dot.gov/documents/HAGuidelines.pdf (accessed December 21, 2015).

HSE. *Health and Safety in Construction.* The Health and Safety Executive, 2006. http://www.hse.gov.uk/pubns/priced/hsg150.pdf (accessed December 21, 2015).

HSE. *Example Risk Assessment for Maintenance Work in a Factory.* The Health and Safety Executive, 2013. http://www.hse.gov.uk/risk/casestudies/factory.htm (accessed December 21, 2015).

HSE. *Risk Assessments.* The Health and Safety Executive, 2014. http://www.hse.gov.uk/construction/safetytopics/admin.htm (accessed December 21, 2015).

ILO-OHS:2001. *Guidelines on Occupational Health and Safety Management Systems*. Geneva, Switzerland: International Labour Organization, 2001. http://www.ilo.org/wcmsp5/groups/public/@ed_protect/@protrav/@safework/documents/normativeinstrument/wcms_107727.pdf (accessed December 21, 2015).

Main, B. W., *Risk Assessment: Basics and Benchmarks*. Ann Arbor, MI: Design Safety Engineering, Inc., 2004.

Manuele, F. A., *On the Practice of Safety, 4th ed.* Hoboken, NJ: Wiley, 2012.

Manuele, F. A.,. *Advanced Safety Management: Focusing on Z10 and Serious Injury Prevention, 2nd ed.* Hoboken, NJ: Wiley, 2014.

MIL-STD-882E. *Standard Practice for System Safety*. Washington, DC: Department of Defense, 2012.

Naval Safety Center. *Department of the Navy Fall Protection Guide For Ashore Facilities*. Norfolk, VA: Department of the Navy, 2013. Available at: http://www.public.navy.mil/comnavsafecen/Documents/OSH/FP/Ashore_Fall_Protection_Guide_Aug_13.pdf (accessed December 21, 2015).

OPNAV INSTRUCTION 3500.39C. *Operational Risk Management Fundamentals*. Washington, DC: Department of Defense, 2010. http://doni.daps.dla.mil/Directives/03000%20Naval%20Operations%20and%20Readiness/03-500%20Training%20and%20Readiness%20Services/3500.39C.pdf (accessed December 21, 2015).

OSHA. *The Control of Hazardous Energy, 29 CFR 1910.147*. Washington, DC: Department of Labor, Occupational Safety and Health Administration, 1989. https://www.osha.gov/pls/oshaweb/owadisp.show_document?p_table=STANDARDS&p_id=9804 (accessed December 21, 2015).

OSHA. *Permit-Required Confined Space Entry, 29 CFR 1910.146*. Washington, DC: Department of Labor, Occupational Safety and Health Administration, 1993.

OSHA. *Commonly Used Statistics*. Department of Labor, Occupational Safety and Health Administration, 2014. Available at: https://www.osha.gov/oshstats/commonstats.html (accessed December 21, 2015).

OSHA. *Reporting Fatalities and Multiple Hospitalization Incidents to OSHA, 29 CFR 1904.39*. Washington, DC: Department of Labor, Occupational Safety and Health Administration, https://www.osha.gov/pls/oshaweb/owadisp.show_document?p_table=standards&p_id=12783 (accessed December 21, 2015).

Reason, J. and A. Hobbs. *Managing Maintenance Error: A Practical Guide*. Burlington, VT: Ashgate, 2003.

VPP - OSHA's Voluntary Protection Program, Policies and Procedures, Department of Labor, Occupational Safety and Health Administration, 2008, https://www.osha.gov/OshDoc/Directive_pdf/CSP_03-01-003.pdf (accessed December 21, 2015).

15

FOOD PROCESSING RISK ASSESSMENTS

GEORGI POPOV
School of Environmental, Physical & Applied Sciences, University of Central Missouri, Warrensburg, MO, USA

BRUCE K. LYON
Risk Control Services, Hays Companies, Kansas City, MO, USA

YING ZHEN
EHS Manager, Belzona Inc., Miami, FL, USA

15.1 OBJECTIVES

- Introduce the potential risk in food processing
- Review risk assessment techniques and tools used in the food industry
- ISO 31000/ANSI Z690.2 and Food Safety Management Systems Integration

15.2 OVERVIEW

The food industry is unique from other industries, in that it must consider risk to both employee health and safety and public safety and health. The high degree of exposure, the potential severity level of food-borne illnesses, and the legal ramification that can require an organization to properly assess hazards and manage food safety. Therefore, a clear understanding of the food processing steps, and characteristics of products being handled, is needed to efficiently prevent potential hazards in the food chain. This chapter will explain how fundamental risk assessment principles such as the hazard analysis and critical control points (HACCP) are applied in food hazard identification and how to successfully implement Prevention through Design (PtD) principles in the food industry. In addition, it will summarize opportunities for successful integration of food safety assessments and the American National Standard, ANSI/ASSE Z690.2-2011, Risk Management standard (nationally adopted from ISO 31000).

Risk Assessment: A Practical Guide to Assessing Operational Risks, First Edition.
Edited by Georgi Popov, Bruce K. Lyon, and Bruce Hollcroft.
© 2016 John Wiley & Sons, Inc. Published 2016 by John Wiley & Sons, Inc.

15.3 INTRODUCTION TO FOOD RISK

Thousands of years ago, ancient people learned that spoiled or contaminated food made people sick. Throughout history, various methods were implemented to preserve food and to reduce the threat of food-borne illnesses caused from biological, chemical, and physical contamination. The use of refrigeration, filtration, and pasteurization technologies has led to the development of food preservation practices used in today's modern food industry that begins with farming and livestock, raw material production, procurement and handling, manufacturing and processing, distribution, and finally the consumption of food products.

Almost every person relies on the national and international food supply system (Roberts, 2001). However, these developments have increased the risk of food-borne illnesses. A simple mistake or cross-contamination in one step of food manufacturing may cause a large number of people to get sick in different locations at the same time, which makes food safety one of the most important risk management topics in the 21st century. Despite of the scientific advances and control methods used, food-related diseases affect tens of millions of people and kill thousands each year. World Health Organization (WHO) claims "food safety is an increasing important public health issue." Food-borne diseases are widespread, which not only threatens public health but also significantly reduces the economic productivity.

According to WHO's estimation, food-borne and waterborne diarrheal diseases kill approximately 2.2 million people annually (World Health Organization, 2011). About 13 million children under the age of five die each year from infections and malnutrition, most often attributed to contaminated food (World Health Organization, 2007). According to CDC's research and analysis based on the information from multiple surveillance systems and other sources, food-borne diseases cause approximately 76 million illnesses, 325,000 hospitalizations, and 5000 deaths in the United States every year (Mead, 2000). In addition, CDC estimates that one out of six Americans (or 48 million people) gets sick from food poisoning each year; 128,000 of them are hospitalized, and 3000 die from food-borne diseases (US CDC, 2011).

Due to the specialty and severity of food risk, food processing risk assessment is a necessary and fundamental component to food safety and risk management.

15.4 RISK ASSESSMENT TECHNIQUES IN THE FOOD INDUSTRY

A significant number of national and international standards for food safety are available. Such standards mandate risk assessment in order to control certain parameters and help guarantee the safety of the food supply. With the numerous methods available for assessing risk, it is important to have some understanding of the different methodologies, their applications, advantages, and disadvantages to make the best selection. This chapter offers a simplified risk assessment method that could be applied in the food industry.

Hazard Analysis Critical Control Point (HACCP) is a management system extensively used by advanced food companies to identify, analyze, and control biological, chemical, and physical hazards through the whole food production process to achieve food safety (US FDA, 2011). HACCP provides a structure for hazard identification and placement of controls at critical points in a process to effectively prevent hazards from occurring.

The HACCP concept originally started from production process monitoring used during World War II because traditional "end of the pipe" testing on artillery shell's firing mechanisms could not be performed, and a large percentage of the artillery shells made at the time were either duds or misfiring (MIL-STD-105). The HACCP approach was further developed by Pillsbury Corporation with the US National Aeronautics and Space Administration (NASA) in the 1960s to ensure the safety of food and drinks for space flights. During the 1970s, HACCP was widely applied in the industry. In 1994, the organization of the international HACCP Alliance was established for the US meat and poultry industries. The National Advisory Committee on Microbiological Criteria for Foods (NACMCF) provided information for international standards on the development and implementation of HACCP principles. The General Accounting Office (GAO) endorsed HACCP as a scientific, risk-based system to protect public health. Then in 1996, the Food Safety and Inspection Service (FSIS) of the United States published a final rule of HACCP. In 2005, ISO issued ISO 22000, "Food Safety Management System – Requirements for Organizations in Food Chain," which is a complete food safety and quality management system. The standard included all HACCP principles and incorporated the prerequisite programs, such as good manufacturing practice (GMP) and Sanitation Standard Operation Procedures (SSOP) (Standard Kalite, n.d.).

The HACCP system is a science-based methodology that identifies specific hazards and measures the control effectiveness to ensure food safety. HACCP, similarly to a safety risk assessment, is a tool to assess food safety hazards and identify control measures that will focus on *prevention* rather than end-product testing. HACCP principles rely on prevention to achieve the main food safety objective – the minimum frequency or concentration of a hazard in food at the time of consumption that provides or contributes to the "*appropriate level of protection*" (ALOP).

Other critical risk assessment tools include failure mode and effects analysis (FMEA) and fault tree analysis (FTA). FMEA, discussed in Chapter 8, is a methodology to analyze the potential failure modes that may occur in a product or process; assess the severity, likelihood, and detectability level of each failure mode; prioritize all failure modes based on risk level; and provide control measures to eliminate or reduce the harm (Crow, 2002). FTA is a systematic methodology that is widely used in system reliability, maintainability, and safety analysis. It is used to identify potential causes of system failures by top-down deductive analysis (Pilot, 2002).

This chapter demonstrates a theoretic application of a simplified risk assessment methodology, including controls and preventive measures applicable in the food industry. The utilization of FMEA and FTA methodologies and the implementation of HACCP principles may reduce the probability of physical, chemical, and biological hazard contamination. Other identified data demonstrates that risk assessment tools will contribute to the implementation of the food safety system in food industry to minimize risk, improve productivity and quality of products, and reduce unnecessary waste.

15.5 FOOD SAFETY-RELATED HAZARDS

The food industry is different from other industries in that it requires an excellent understanding of the characteristics of products being handled as well as the process itself to efficiently prevent the development of potential hazards and to control the ones that exist. Three categories of hazards are related to food safety: biological hazards, chemical hazards, and physical hazards.

15.5.1 Biological Food Hazards

Biological food hazards include bacterial pathogens, viruses, and parasites. CDC estimated top five pathogens contributing to domestically acquired food-borne illnesses are presented in Table 15.1.

CDC also estimates the top five pathogens contributing to domestically acquired food-borne illnesses resulting in death presented in Table 15.2.

More details about biological microorganisms that frequently cause food-borne illnesses and deaths are discussed in the following:

- *Listeria monocytogenes* is one of the most virulent food-borne pathogens. It causes the highest fatality rate among food-borne bacterial pathogens. *L. monocytogenes* frequently result in septicemia, meningitis, encephalitis, and many other illnesses. The main sources of *L. monocytogenes* are raw milk, ice cream, raw meat, and seafood. It can survive at temperatures as low as 0 °C.

TABLE 15.1 Top Five Pathogens Contributing to Domestically Acquired Food-Borne Illnesses and Estimated Number of Illnesses

Pathogen	Estimated Number of Illnesses
Norovirus	5,461,731
Salmonella, nontyphoidal	1,027,561
Clostridium perfringens	965,958
Campylobacter spp.	845,024
Staphylococcus aureus	241,148

Source: US CDC (2011).

TABLE 15.2 Top Five Pathogens Contributing to Domestically Acquired Food-Borne Illnesses Resulting in Death and Estimated Number of Deaths

Pathogen	Estimated Number of Deaths
Salmonella, nontyphoidal	378
Toxoplasma gondii	327
Listeria monocytogenes	255
Norovirus	149
Campylobacter spp.	76

Source: US CDC (2011).

- The infection of *Escherichia coli O157:H7* leads to hemorrhagic diarrhea, abdominal cramps, and even kidney failure, especially in young children and the elderly. It is transmitted via the fecal–oral route, and the main source of infection is undercooked food, such as ground beef, unpasteurized milk, vegetables, and water.
- *Salmonella typhi* normally causes diarrhea, and the infection can be very serious to small children and the elderly. The main food sources are meats, poultry, eggs, and milk. Hepatitis A and Norwalk viruses are representations of enteric viruses associated with food. Undercooked meat or contaminated ready-to-eat foods may be infected by parasites (Roberts, 2001).

15.5.2 Chemical Food Hazards

Chemical food hazards are the chemical substances or compounds that exist in food, which will cause health problems to a sensitive population or even the general public by consumption. Different from biological hazards that always have a quick response, chemical hazards can either cause acute food-borne illnesses or result in chronic illness. Hazardous chemicals in food may be the product's ingredients, intentionally added or unintentionally added to food. Naturally occurring chemicals include shellfish toxins, mycotoxins, scombrotoxin (histamine), ciguatoxin, toxic mushroom species, and many other chemical substances or mycotoxins. Paralytic shellfish poison (PSP) and diarrhetic shellfish poison (DSP) are two types of shellfish toxins. PSP toxins are neurotoxic alkaloids that can block the entrance of sodium ions into nerve cells, and people may die because the muscles of respiration lose control. DSP causes slight sickness in the gastrointestinal (GI) tract system. Mycotoxin is the metabolic product of fungus. The common kinds of mycotoxins include aflatoxins, trichothecene mycotoxin, ochratoxins, saxitoxins, and grayanotoxins. They exist in different kinds of food and have various symptoms, but they all threaten human and animals' safety and health. Further, there are added chemicals in the process. Agricultural chemicals such as pesticides, fungicides, fertilizers, and growth hormones are added to facilitate the growth of raw material. Food additives are added as preservatives, flavor enhancers, nutritional additives, or color enhancements. The application of additives should follow the allowable limits under GMPs. Chemicals added into food products unintentionally are also a threat to customers, such as cleaners and sanitizers (US FDA, 1999, p. 366).

15.5.3 Physical Food Hazards

Physical food hazards are objects (foreign or naturally occurring) in food that may cause injuries or illnesses. The main materials considered as physical hazards include glass fragments, wood, stone, metal fragments, insulation, bone, plastic, jewelry, and many others. Physical food hazards are usually not as harmful as others. However, they can lead to life-threatening events for young children and the elderly.

15.6 TECHNIQUES FOR ASSESSING FOOD RISK

As previously described in this text, the ISO 31000 series (ISO Guide 73:2009, ISO 31000:2009, and IEC/ISO 31010:2009) is a family of standards relating to risk management codified by the International Organization for Standardization. The main goal of this series is to provide principles and generic guidelines on risk management. The ISO standards were nationally adopted by the American National Standards Institute in 2011. ISO 31000/ANSI/ASSE Z690.2-2011, Risk Management Principles and Guidelines, standard provides a universally recognized paradigm for practitioners and

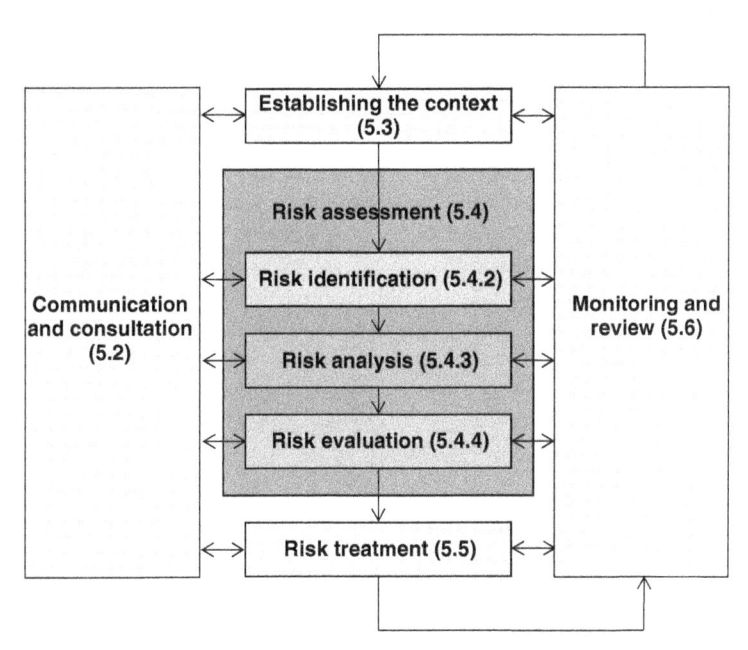

Figure 15.1 Risk Management Process. *Source:* Reprinted with Permission from ISO 31000/ANSI/ASSE Z690.2-2011 (Courtesy of the American Society of Safety Engineers)

companies employing risk management processes. Its purpose is to be applicable and adaptable for "any public, private or community enterprise, association, group or individual" (ISO, 2013). The following process for managing risk (Clause 5) illustrated in Figure 15.1 was selected for detailed evaluation of the applicable risk assessment tools.

In 2011, ANSI released its PtD standard, ANSI/ASSE Z590.3-2011. The authors believe that multiple risk assessment techniques listed in both ISO 31010/ANSI Z690.3, *Risk Assessment Techniques*, and ANSI Z590.3 are applicable to the food industry. Table A.1 in ISO 31010/ANSI Z690.3-2011 lists 31 tools used to assess risk and includes HACCP. According to Table A.1, HACCP is "strongly applicable" for *risk identification*, *consequence*, and *risk evaluation*; however, HACCP is "not applicable" for determining *probability* and *level of risk* (ISO 31010/ANSI Z690.3-2011).

15.7 HAZARD ANALYSIS AND CRITICAL CONTROL POINTS

For some time, the application of HACCP in food processing has been recommended by the US Food and Drug Administration (FDA). The reason for its use in food safety is that the HACCP methodology enables organizations to implement an effective management system used to identify, prevent, and control food hazards within a process. HACCP provides a scientific safety assurance theory that *prevents hazards from entering the process* rather than inspecting and evaluating the products by end testing (US FDA, 1997, p. 363):

1. *Preliminary tasks* – Similarly to other risk assessment methods, the first steps in developing an HACCP plan are to establish the scope and context of the assessment and assemble an HACCP team. Both the establishment of context and risk assessment team are discussed in other chapters throughout this text.
2. *Context* – The first and foremost important step is the establishment of the agreed upon purpose and the scope of the assessment effort. This is generally developed with management to ensure it aligns with the organization's business objectives and to assure management support. Key elements of the context include the purpose statement, limits of the assessment, targeted process, individuals involved, resources available, and risk criteria to be used.
3. *Team* – The team generally consists of knowledgeable and experienced personnel and employees who are involved in the operation. Team members should have specific knowledge and expertise appropriate to the product and process as well as the HACCP methodology. An experienced facilitator or leader is necessary, and it is considered a very good management practice to form a multidisciplinary team and include individuals from areas such as finance, risk management, engineering, production, sanitation, quality assurance, and food microbiology. As stated earlier, the team should include local personnel involved in the operation familiar with the variability and limitations of the operation. Furthermore, this practice promotes a sense of ownership among those who must implement the plan. According to the FDA, the HACCP team may need assistance from outside experts who are knowledgeable in the potential biological, chemical, and/or physical hazards associated with the product and the process. However, a plan that is developed totally by outside sources may be erroneous, incomplete, and lacking in support at the local level (US FDA, 1999).
4. *Targeted process* – The targeted process is then defined and described using various tools and methods such as existing process flowcharts and diagrams, descriptions of food processing activities and tasks, operating instructions of processing equipment, and other available information related to the targeted process. Several key actions are performed in this step including the following:
   - *Describe the food and its distribution.*
   - *Describe the intended use and consumers of the food.*

INTEGRATION OF RISK ASSESSMENT METHODS 325

- *Develop a flow diagram that describes the process.*
- *Verify the flow diagram.*

    To visually demonstrate the process, a flow diagram should be developed and used. A flow (or process) diagram is a schematic and systematic presentation of the sequence and interactions of steps. For effective use in risk assessment, flow diagrams must be clear, accurate, and sufficiently detailed (ISO, 2005).

5. *HACCP principles* – Following the completion of the proceeding tasks, the seven principles of HACCP are applied in the following order (US FDA, 1997):
    1. Conduct a hazard analysis.
    2. Identify critical control points (CCPs).
    3. Establish critical limits for each CCP.
    4. Establish CCP monitoring requirements.
    5. Establish corrective actions.
    6. Establish record-keeping procedures.
    7. Establish procedures for ensuring the HACCP system is working as intended.

15.8 INTEGRATION OF RISK ASSESSMENT METHODS

The following scenario of a frozen salmon processing operation without sufficient control measures is presented to demonstrate integration of various risk assessment methods including HACCP, ANSI Z590.3 PtD, and ISO 31000/ANSI Z690.2 principles:

Fish bone diagram – To aid in the understanding of the process, a simple process map using a fish bone diagram of the salmon processing is pictured in Figure 15.2.

The use of a process map or diagram helps the communication of the process steps, potential hazards, and necessary control points: in essence a "road map" for the process being assessed. In addition to the process map, a sequence of risk assessment techniques is presented as shown in Figure 15.3.

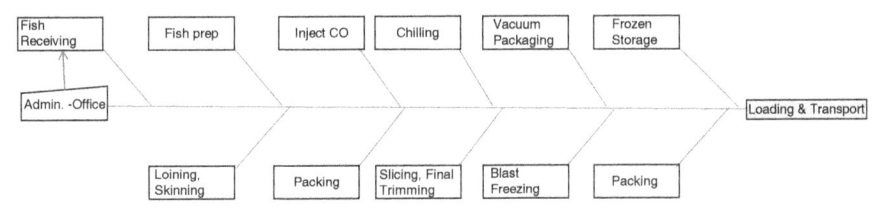

Figure 15.2 Salmon Processing Simplified Process Map

Detailed descriptions of the tools are presented in the following:

(A) *Hazard/risk identification* – ISO 31000/ANSI/ASSE Z690.2-2011 states that the goal of risk identification is to generate a comprehensive list of risks. There are a variety of risk assessment tools and techniques that can be used at this stage:
    (1) *Preliminary hazard analysis* (PHA) – After completing the preliminary tasks discussed earlier, the HACCP team begins with identifying hazards/risks and performs a PHA. Risk identification involves the identification of hazards/risk sources and can be successfully aligned with hazard analysis step (Step 1) of the HACCP principles. FDA states that the purpose of the food safety hazard analysis process is to develop a list of hazards, which

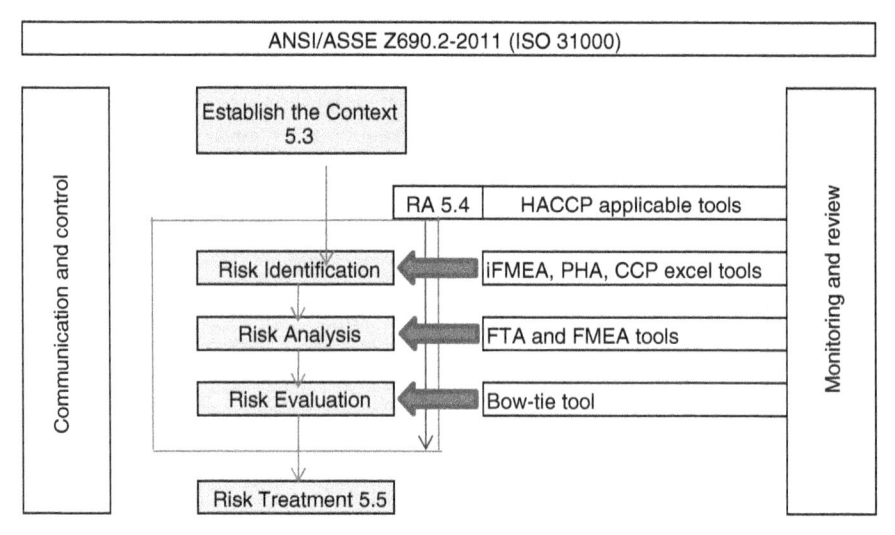

Figure 15.3 ANSI Z690.2, Risk Management Process and Assessment Tools Alignment

are of such significance that they are reasonably likely to cause injury or illness if not effectively controlled (US FDA, 1997). Furthermore, hazards that are not likely to occur would not require further consideration within an HACCP plan.

(2) *Simple hazard assessment* – The FDA suggests a simple hazard assessment form presented in Table 15.3. The simple form allows each step to be briefly described, potential hazards associated with each step, justification for including the hazard exposure, and a determination of whether the hazard will be included in the HACCP plan.

TABLE 15.3 Sample Hazard Analysis Form

Step	Potential Hazard(s)	Justification	Hazard to be Addressed in Plan? Y/N	Control Measure(s)
Fish receiving	Enteric pathogens: for example, *Salmonella*, verotoxigenic *Escherichia coli*	Enteric pathogens have been associated with outbreaks of food-borne illness from improperly processed fish	Y	TBD

(3) *Initial failure mode and effects analysis (iFMEA)* – To assist in the identification of hazards and potential failures, an iFMEA can be used. Using an iFMEA has some advantages such as assigning/calculating probability and level of risk during the risk analysis step, which the HACCP is not able to provide. The iFMEA form, presented in Table 15.4, can be described as a preliminary step of an FMEA.

(B) *Risk analysis* – The risk analysis step involves "developing an understanding" of the risks. As the second step with *risk assessment*, risk analysis provides the necessary input to the risk evaluation step and assists in making proper decisions regarding risk treatment. CCP, FTA, and FMEA, among others, can be used during the risk analysis step.

INTEGRATION OF RISK ASSESSMENT METHODS 327

TABLE 15.4 Sample iFMEA Form

iFMEA Worksheet		
Part or Process Name	Frozen Salmon Process	Suppliers & Plants Affected
Design/Mfg Responsibility		Model Date
Other Areas Involved		Engineering Change Level
Process Operation, Function or Purpose	Potential Failure Mode	Potential Effect(s) of Failure
Salmon receiving	Chemical, Biological cont.	Contaminated product
Loining, skinning and trimming	Physical contamination	Physical Hazards to customers
Inject CO	Chemical contamination	Contaminated product
Chilling	Biological contamination	Contaminated product
Metal Detection 1	Physical contamination	Contaminated product
Metal Detection 2	Physical contamination	Contaminated product

(1) *CCP* – CCP identification is the foundation of the HACCP process and is used as a decision-making tool to determine the critical steps or points within a process. A CCP is defined as a point in the food processing line where potential hazards present a potential for occurrence and a severity level that requires control measures to eliminate or reduce their risk to operate at an acceptable level. CCP can be applied to prevent, eliminate, or reduce the occurrence or the severity of food hazards. The identification process is based on the knowledge of the production process, characteristics of the food products, and the potential food safety hazards.

(2) *Critical control point decision tree (CCP DT)* – To help identify CCP within a process, the authors developed a CCP DT as an Excel form with step directions. A flowchart describing the steps is presented in Figure 15.4.

For each procedure that may cause or have potential hazards, the first consideration is (Q1) whether any controls exist for the identified hazard. If any controls exist for the hazard, an evaluation of the control effectiveness is performed (Q2). This involves determination whether the control(s) will eliminate the occurrence or reduce the risk of occurrence to an acceptable level. If the risk is reduced to an acceptable level, the step is considered a CCP and included in the HACCP plan. If the control does not reduce risk of occurrence sufficiently, then the severity of the hazards is evaluated. If no health threat exists from the identified hazard, it is not a CCP and the process stops. However, if the contamination presents a risk to consumer health, then the subsequent step CCP is considered. If no efficient subsequent step exists, the evaluated step is considered a CCP and included in the HACCP.

For this process, if (Q1) preventive measures do not exist for an identified hazard, the team determines whether control measures are needed. If control is not needed, the step is not a CCP. If control is necessary, the step needs to be modified with a preventive measure and placed back into the evaluation cycle. The Excel-based worksheet is presented in Figure 15.5.

(3) *FMEA and risk priority number (RPN)* – After the determination of the most hazardous processes and possible consequences, a detailed analysis of the process or operation can be performed. The risk analysis step will involve analyzing severity of consequences and probability of occurrence. As stated in ISO 31010/ANSI/ASSE Z690.3-2011 HACCP (CCP) is "not applicable" for probability assessments. Therefore, other tools of a more

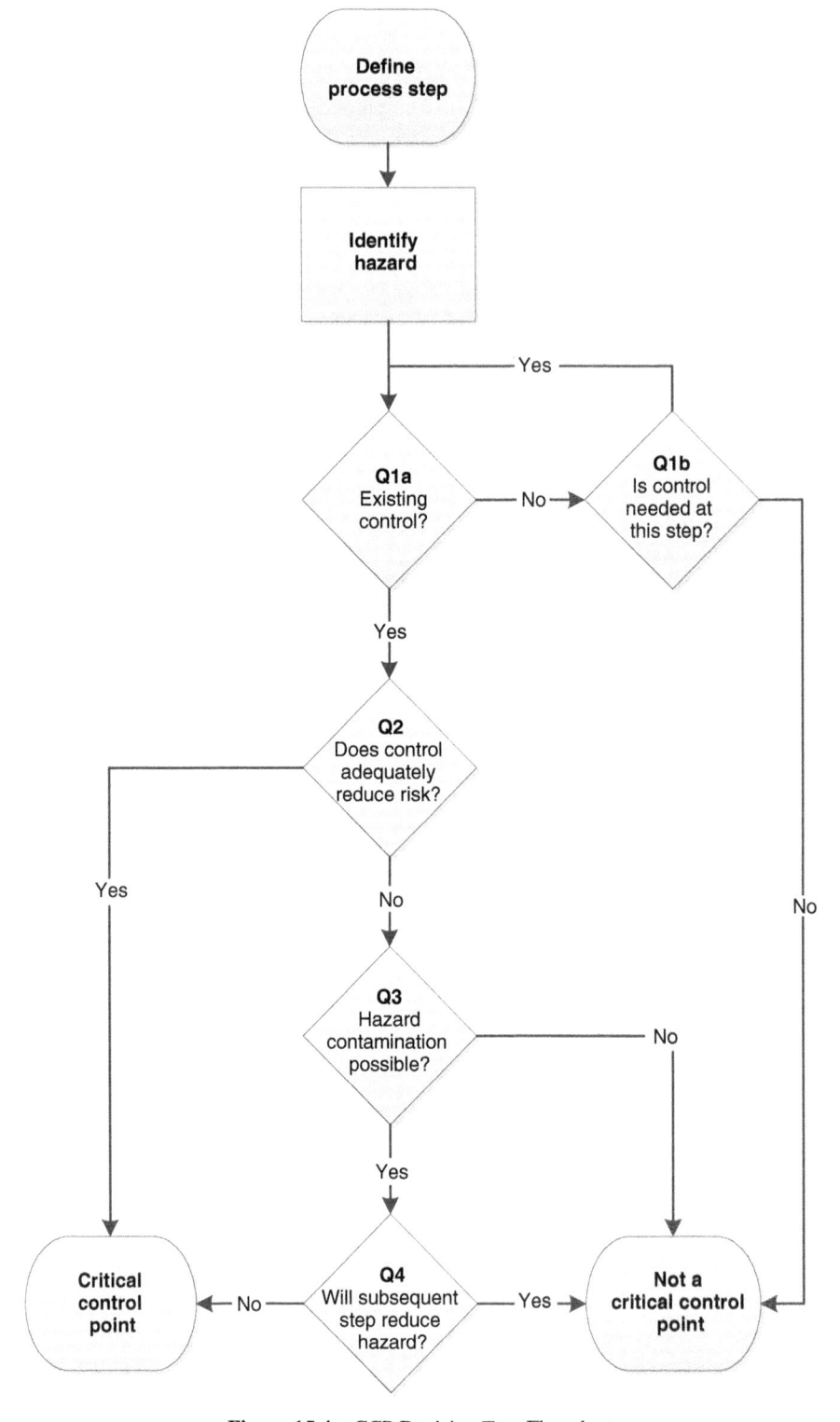

Figure 15.4 CCP Decision Tree Flowchart

CCP Decision Tree Form							
Product or Process Name:							
Developed by:							
Date:							
		Q1a	Q1b	Q2	Q3	Q4	CCP #
Process/Step	Hazard	Q1a. Do preventive measures exist for the identified hazard(s)?	Q1b. Is control at this step necessary for safety?	Q2. Does this step eliminate or reduce the likely occurrence of a hazard(s) to an acceptable level?	Q3. Could contamination with identified hazard(s) occur in excess of acceptable levels or could these increase to unacceptable levels?	Q4. Will a subsequent step eliminate hazard(s) or reduce the likely occurrence to an acceptable level?	
		If No – go to Q1b	If No – not a CCP	If No – go to Q3	If No – not a CCP	If No–CCP	
		If Yes – go to Q2	If Yes – modify step, process or product and return to Q1a	If Yes – CCP	If Yes – go to Q4	If Yes–not a CCP	

UCM CCP Form

Approved: ________ Date: ________

Developed by: Dr. Georgi Popov, Ph.D., QEP

Figure 15.5 CCP Decision Tree Identification Tool

specialized nature may have application in areas of food safety management. For instance, severity/probability/likelihood of food contamination can be assessed using an FMEA or FTA methodology.

FMEA is often applied to assist hazard analysis in food processing. Its application in food safety may provide detailed analysis of each process/operation based on the occurrence, the severity, and the detection of hazards. Based on the experience of the food safety risk assessment team members, an FMEA methodology can be used as a continuation of the CCP hazard analysis to facilitate the risk assessment and prioritization of the CCPs. The evaluation system leads to an RPN, which can be used to rank the hazards and quickly select the highest ranked hazards. The analysis of the recommended actions includes an evaluation of the effectiveness of controls related to the hazards. This can also be used as a recommendation for critical control actions. The HACCP team can compare the RPNs of different operations and prioritize the hazards. The priority of operations that requires critical controls will be listed from the highest RPN to the lowest. The organization can allocate their finances wisely with the RPNs. According to the information FMEA can bring to the hazard analysis, it is an effective tool to be used in the food processing. An example of FMEA and RPN worksheet is presented in Figure 15.6.

FMEA AND RPN WORKSHEET									
Part or Process Name		Suppliers & Departments				Prepared By:	Dr. G. Popov		
Design Responsibility		Model Date				FMEA Date			
Other Areas Involved		Engineering Change Level							
Assess Risk - Initial Scoring System: Current State									
Process Operation, Function or Purpose	Potential Exposure Mode	Potential Effect(s) of Exposure	SEV	Potential Cause(s) of Exposure	OCC	Current Controls Evaluation Method	PE	S x O	RPN

Figure 15.6 Example of FMEA and RPN Worksheet

(4) *FTA* – As described by ANSI Z590.3, FTA is a top-down, deductive logic model that traces the failure pathways for a predetermined, undesirable condition or event, called the TOP event. FTA generates a fault tree (a symbolic logic model) entering failure

probabilities for the combinations of equipment failures and human errors that can result in the incident or exposure. Each immediate causal factor is examined to determine its subordinate causal factors until the root causal factors are identified (ANSI Z590.3-2011). An example of FTA is presented as follows:

FTA example: The bacterial species *Vibrio parahaemolyticus*-related illnesses have been associated with consumption of raw, undercooked fish or oysters. What is the likelihood of *Vibrio* infection if an employee did not wash hands properly (probability of contamination, 25%), used contaminated cutting board (probability of contamination, 12%), and used contaminated knife (probability of contamination, 17%)? These events are mutually exclusive. Therefore, "OR" gate FTA is applicable for the *V. parahaemolyticus* contamination example in Figure 15.7. This example is for educational purposes and does not represent actual conditions. In fact, CDC Foodborne Diseases Active Surveillance Network (US CDC, 2014a) reports suggest that the probability of *V. parahaemolyticus*-related illnesses is much lower than represented. [There were only 242 confirmed *V. parahaemolyticus* infections in the United States during 2013. Data for 2013 are preliminary (US CDC, 2014b).]

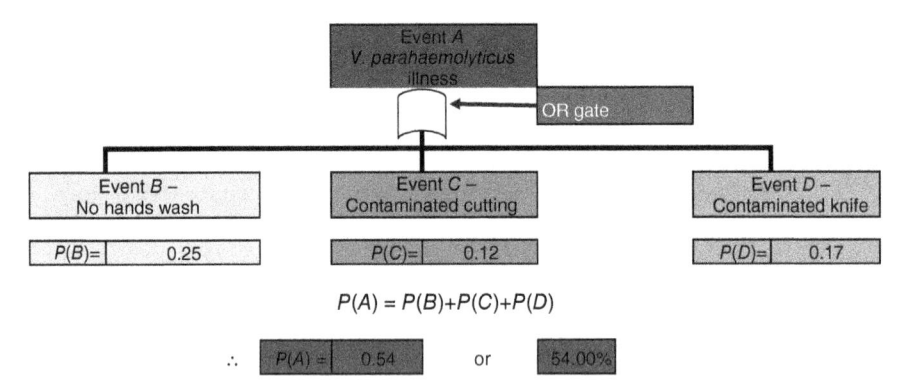

Figure 15.7 FTA "OR" Gate Probability Example

(C) Risk evaluation – The purpose of risk evaluation is to support the decision-making process. It is generally based on the outcomes of risk analysis and the priority for environmental, health, and safety interventions. Risk evaluation involves determining the level of risk found during the risk assessment process with risk criteria established when the CCPs were considered. Decisions should be made in accordance with legal, regulatory, and other requirements. In some cases, the risk evaluation can lead to further analysis. For instance, an FMEA performed for a process may require a more detailed FMEA for each step of the process. Performing detailed FMEA for specific steps of the process may be time consuming; however, such decisions will be influenced by the organization's risk acceptance and the risk criteria or "as low as reasonably practicable" ALARP that have been established.

(1) *Bow-Tie analysis* – The conventional Bow-Tie methodology (covered in Chapter 9) can be used for risk evaluation. ISO 31010/ANSI Z690.3 defines Bow-Tie analysis as a simple diagrammatic way of describing and analyzing the pathways of a risk from causes to consequences. The focus of the Bow Tie is on (1) the prevention barriers between the causes and the risk and (2) mitigative barriers between the risk and consequences. However, the "conventional" Bow-Tie analysis is not a quantitative tool. A conventional Bow-Tie analysis is presented in Figure 15.8.

INTEGRATION OF RISK ASSESSMENT METHODS 331

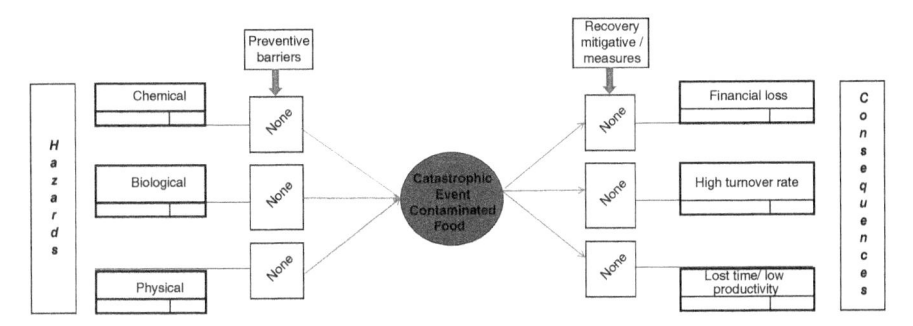

Figure 15.8 Example of "Conventional" Bow-Tie Analysis

(2) *Modified Bow-Tie analysis* – Dr. Popov developed a modified Bow-Tie analysis where the hazards and consequences are semi-quantitatively defined based on the severity and probabilities of occurrence. A conventional risk assessment matrix from the ANSI Z590.3 PtD standard can be used to produce a risk factor (which is Severity × Occurrence) or an RPN from the FMEA. A combination of both is presented in Figure 15.9.

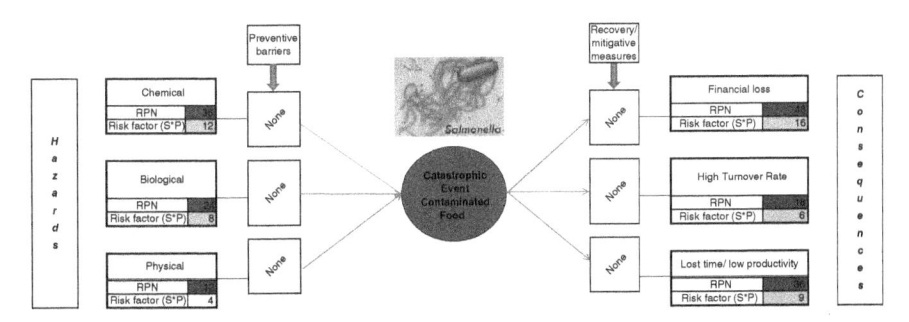

Figure 15.9 Example of Modified Bow-Tie Analysis

(3) *Process map* – As mentioned previously, a scenario of salmon processing without sufficient control measures is used to demonstrate a food risk assessment process. The process incorporates a combination of methods to minimize and control food-related hazards. The simplified process map of the salmon processing presented in Figure 15.10 displays the key steps.

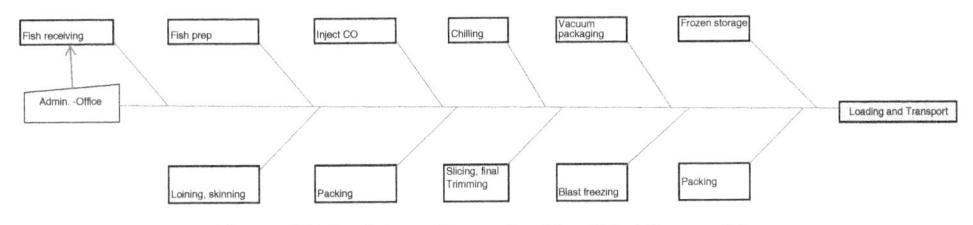

Figure 15.10 Salmon Processing Simplified Process Map

(4) *CCP DT* – The frozen salmon processing is evaluated according to the HACCP management system resulting in six CCPs identified. These identified CCPs represent the key points that may cause defects in product quality or food safety. An example of partially completed CCP DT form is presented in Figure 15.11.

CCP DECISION TREE FORM

Product or Process Name: Frozen Salmon Processing
Developed by:
Date:

FMEA

Process/Step	Hazard	Q1a. Do preventive measures exist for the identified hazard(s)?	Q1b. Is control at this step necessary for safety?	Q2. Does this step eliminate or reduce the likely occurrence of a hazard(s) to an acceptable level?	Q3. Could contamination with identified hazard(s) occur in excess of acceptable levels or could these increase to	Q4. Will a subsequent step eliminate the hazard(s) or reduce the likely occurrence to an acceptable level?	CCP #
		If No—go to Q1b	If No—not a CCP	If No—go to Q3	If No—not a CCP	If No—CCP	
		If Yes—go to Q2	If Yes—modify step, process or product and return to Q1a	If Yes—CCP	If Yes—go to Q4	If Yes—not a CCP	
	chemical and biological contamination						
Salmon receiving		Yes, BBT control, potable water		Yes			CCP1
Loining, skinning and trimming	physical hazards	Yes, X-ray detection		Yes			CCP2
Inject CO	Chemical contamination	Yes, needle inspection and sanitation		Yes			CCP3
Chilling	Biological contamination	Yes, temp≤3.3C, curing time 24hrs		Yes			CCP4
Metal Detection on individual package	Physical contamination	Yes, metal detection on individual package		Yes			CCP5
Metal Detection on individual box	Physical contamination	Yes, metal detection on individual box		Yes			CCP6

UCM CCP Form

Approved: Ying Zhen **Date:** Feb. 13, 2011

Developed by: Dr. Georgi Popov, Ph.D., QEP

Figure 15.11 CCP Decision Tree Form

INTEGRATION OF RISK ASSESSMENT METHODS 333

(5) *Detailed FMEA and RPN* – Hazards identified in the CCP DT are transferred to FMEA tool for further risk analysis and evaluation. The FMEA form provides a detailed analysis of each operation based on the occurrence, the severity, and the detection of hazards, which is a similar theory to the hazard analysis of CCP. Evaluation of the CCP and resulting risks, which includes severity and probability ratings and detectability rating, leads to RPNs to prioritize hazards. For the example presented, severity, probability/occurrence, and detection are evaluated on a 1–5 scale as follows:

- Severity (SEV):
  1. Insignificant
  2. Negligible
  3. Marginal
  4. Critical
  5. Catastrophic.
- Probability/occurrence (OCC):
  1. Unlikely
  2. Seldom
  3. Occasional
  4. Likely
  5. Frequent.
- Detection (DET):
  1. Detectable
  2. Likely to be detected
  3. Occasional detection
  4. Rarely
  5. Unlikely.

$$RPN = SEV \times OCC \times DET$$

The higher the RPN, the greater the risk. Therefore, if it is unlikely to detect a chemical, physical, and biological contamination, the "detection" level is estimated to be unlikely and scored a 5. Conversely, if contamination is detectable, the "detection" level is scored a 1 as presented in the FMEA example in Figure 15.12.

Figure 15.12 FMEA Detailed Analysis Based on Identified CCPs

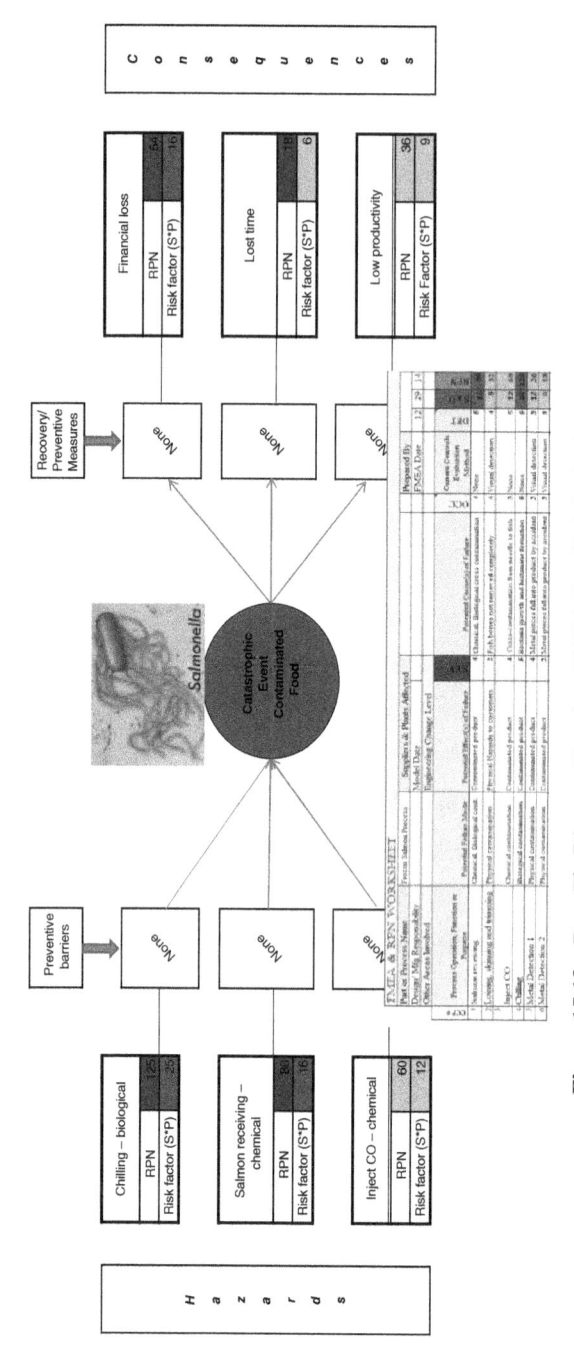

Figure 15.13 Bow-Tie Diagram Based on FMEA Detailed Analysis

Safety → EFFICIENCY → Revenue

FMEA & RPN WORKSHEET

Part or Process Name	Frozen Salmon Process	Suppliers & Plants Affected		Prepared By				
Design/Responsibility		Model Date		FMEA Date	12	29	14
Other Areas Involved		Engineering Change Level						

CCP #	Process Operation, Function or Purpose	Potential Failure Mode	Potential Effect(s) of Failure	SEV	Potential Cause(s) of Failure	OCC	Current Controls Evaluation Method	DET	S x O	RPN	Recommended Action(s)	Area/Individual Responsible & Completion Date	Action Results Actions Taken	SEV 2	OCC 2	DET 2	S x O	RPN 2
1	Salmon receiving	Chemical, Biological cont.	Contaminated product	4	Chemical, Biological cross contamination	4	None	5	16	80	Fish sampling	Plant Manager		4	2	1	8	8
2	Loining, skinning and trimming	Physical contamination	Physical Hazards to customers	2	Fish bones not removed completely	4	Visual detection	4	8	32	X-ray bone detection	Plant Manager		3	2	2	6	12
3	Inject CO	Chemical contamination	Contaminated product	4	Cross-contamination from needle to fish	3	None	5	12	60	Needle inspection and sanitation	Plant Manager		4	2	1	8	8
4	Chilling	Biological contamination	Contaminated product	5	Bacteria growth and histamine formation	5	None	5	25	125	Temp. Control	Plant Manager	temp≤3.3C	5	1	5	5	25
5	Metal detection 1	Physical contamination	Contaminated product	4	Metal pieces fall into product by accident	3	Visual detection	3	12	36	Metal detection	Plant Manager		4	1	1	4	4
6	Metal detection 2	Physical contamination	Contaminated product	2	Metal pieces fall into product by accident	3	Visual detection	3	6	18	Metal detection	Plant Manager		2	1	1	2	2

RPN also defined in: The Six Sigma Handbook, Revised and Expanded By Thomas Pyzdek

Developed by: Georgi Popov, Ph.D., QEP
v. 1.0 / 10/7/09 ® 2009

Figure 15.14 FMEA Assessments of CCPs

The analysis includes an evaluation of the effectiveness of each control and provides guidance for critical control actions. Comparison of RPNs allows an organization to prioritize and allocate resources to operations with the greatest needs. Considering the additional information and visualization FMEA can bring to the hazard analysis, it can be an effective tool in food processing safety. Following the analysis, the top three ranked hazards are transferred to the Bow-Tie analysis form as shown in the current state (before appropriate controls) Bow-Tie analysis in Figure 15.13.

(6) *Current state Bow Tie/FMEA* – Preventive controls and protective measures are then identified for each CCP. Using the FMEA, the resulting estimated RPNs for the identified process operations were improved significantly after implementing the recommended CCP controls (see recommended action column in Figure 15.14).

Although the *severity* of the potential failure did not decline, the identified control reduced the *probability* of occurrence and improved detectability of the failure, which ultimately reduced the RPN or priority score. The FMEA tool not only showed the necessity of controls but also prioritized the control actions. According to FMEA RPNs, bacterial growth during chilling was the most urgent possible failure to be addressed. Maintaining temperature below 3.3 °C is considered a critical control measure as shown in Figure 15.14.

Proper temperature control can reduce the probability of bacterial growth from 5 to 1. The chemical contamination of the fish was the second highest concern of the food poisoning. Fish sampling cannot reduce the severity and the occurrence, but the ability of detection was obviously improved, which made the threat drop from 5 to 1.

After the controls were identified, HACCP team can add the control measures to the process map. A simple process map with the suggested improvements is presented in Figure 15.15.

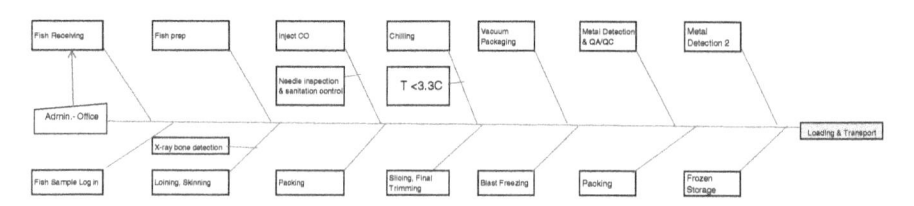

Figure 15.15 Process Map with Added Control Measures

According to the analysis, all the CCPs should be supervised strictly by the measures of the control limits if available. The same top three ranked hazards and recommended actions are transferred to the future state (after controls were implemented) Bow-Tie analysis form shown in Figure 15.16.

(7) *Future state Bow Tie/FMEA* – The risks associated with the hazards have been reduced due to implementation of proper controls. To maintain the benefits of the safety and the efficiency improvements, engineering controls and administration controls should be applied at the same time. The individuals responsible for implementation of controls should make sure all the CCP controls are verified as effective and are consistently applied and enforced (i.e., temperature control for specific operations and necessary equipment retrofits or updates for accurate results). Stakeholders should work with plant management to make sure the operation remains effective and include proper documentation, training, and supervision.

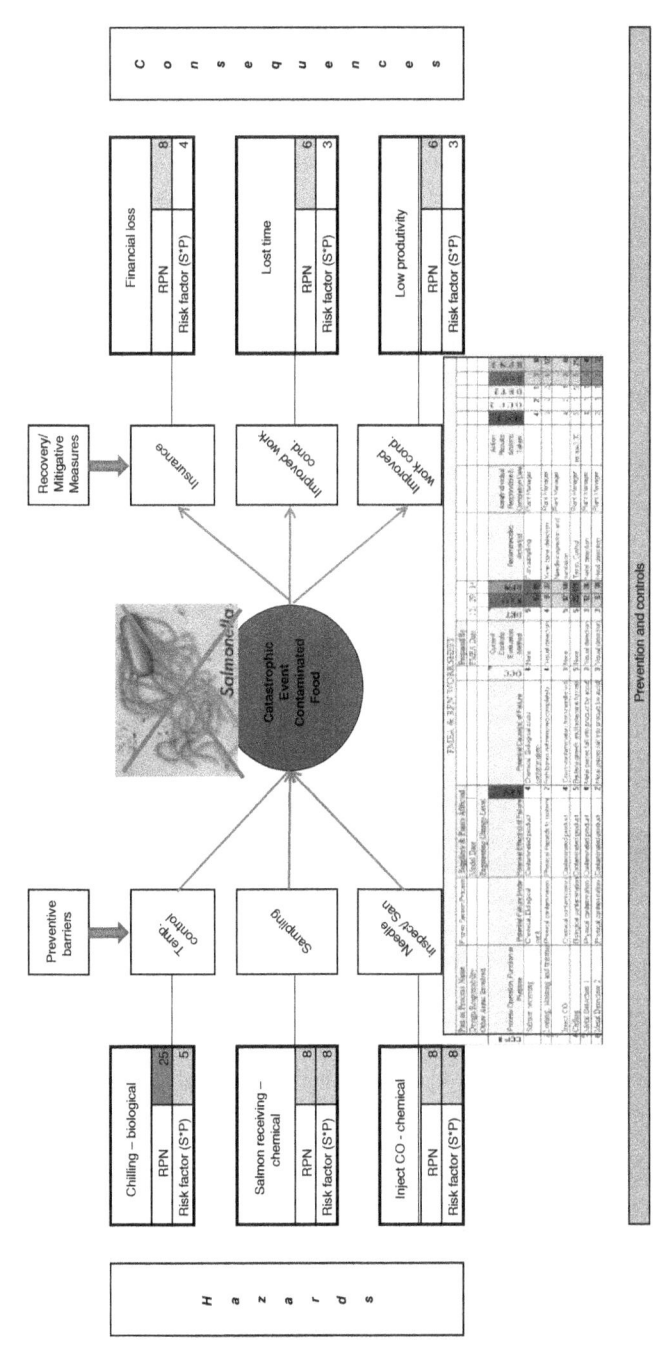

Figure 15.16 Bow-Tie Diagram Based on FMEA Recommended Actions

15.9 PTD AND HACCP INTEGRATION

Worker safety and consumer safety should both be considered during the analysis. It is essential to include safety, health, and environmental professionals in a HACCP as well as the development of workplace design solutions that will benefit both workers and the public. Integrating HACCP and PtD principles will help to modify the production process to eliminate or control biological, chemical, or physical hazards in food products, as well as protect workers from injuries and illnesses. The integration of HACCP and PtD principles helps ensure end products are safe for consumption and that risks to workers remain acceptable as presented in Figure 15.17.

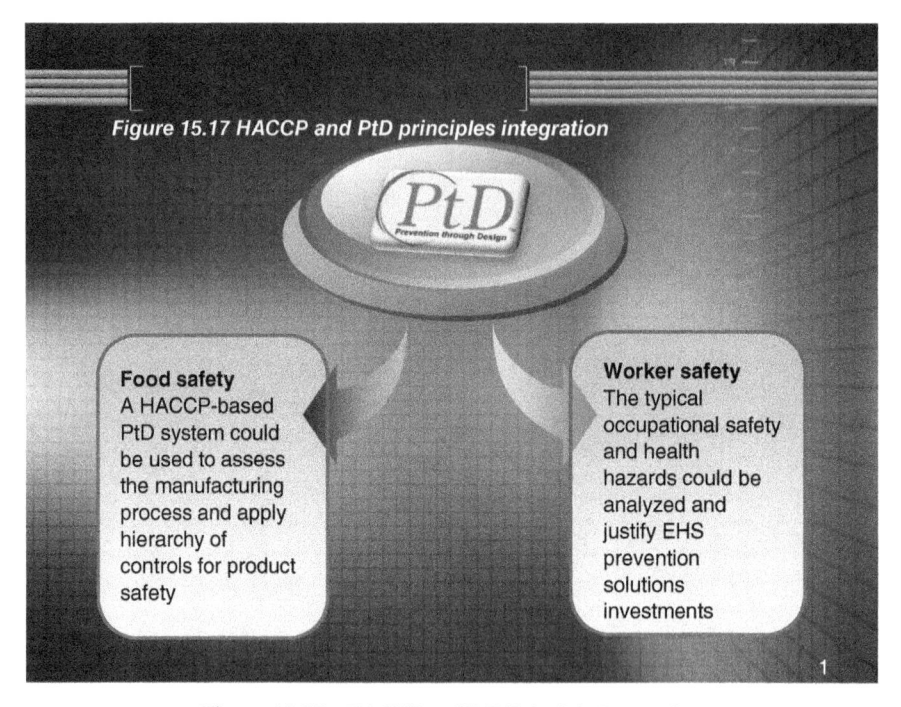

Figure 15.17 HACCP and PtD Principle Integration

PtD principles should be used in the food industry to anticipate and identify potential food safety hazards and eliminate or reduce the risks to an acceptable level. An HACCP and PtD integration approach is presented in Figure 15.18.

For instance, in the example presented in this chapter, some of the identified health and safety hazards for *workers* processing salmon fish included the following:

- Cuts, lacerations, or pinprick
- Extremely cold temperatures
- Ergonomic injuries due to long-time standing, awkward postures, repetitive motions, and lifting.

From a consumer food safety standpoint, product temperatures were identified as a CCP. For instance, a critical control measure identified in the salmon process included maintaining temperatures below 3.3 °C. It should be recognized that some control measure for food safety may increase risk to worker safety and health (i.e., lower temperatures to reduce food spoilage increases occupational health risk to workers). Other examples include cutting hazards and ergonomic injuries due to

CONCLUSIONS 339

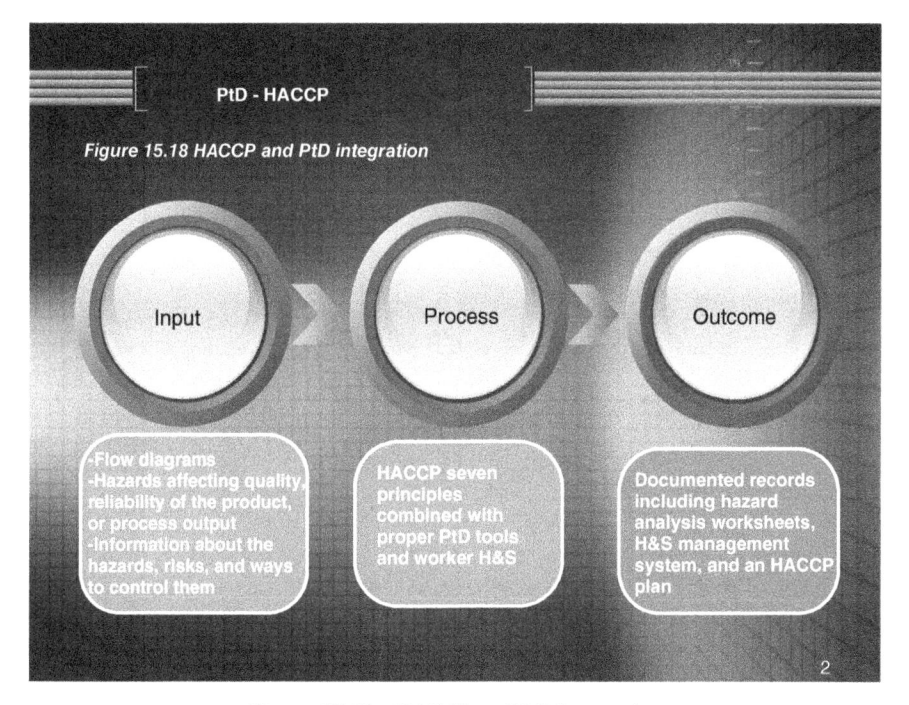

Figure 15.18 HACCP and PtD Integration

long-time standing, awkward postures, and lifting. For ergonomic-related risks, one or more of the ergonomic risk assessment tools described in Chapter 16 can be used.

Experienced and knowledgeable SH&E professionals and HACCP teams should address both food safety and worker safety hazards during the design phase. For instance, an automated salmon skinning machine may be a good application of PtD hierarchy of controls. It will nearly eliminate cutting hazards and significantly reduce ergonomic injuries. Different risk assessment and business tools can be used to evaluate the value of such control measures. For instance, select ergonomic risk assessment tools could be used to evaluate the process from an operator prospective, and business tools should be used to evaluate financial and nonfinancial benefits. During the design and redesign process, significant consideration should be given to protecting workers while addressing food safety concerns using HACCP and PtD concepts.

15.10 CONCLUSIONS

In conclusion, the combination of risk assessment methodologies suggested in ANSI Z590.3 PtD and ISO 31010/ANSI/ASSE Z690.3 and implementation of CCP controls, designed to minimize food-related illnesses, will improve food quality and worker safety and reduce financial losses. Risk and its corresponding risk factor scores or RPNs can be significantly reduced by implementing appropriate controls, as demonstrated in the examples presented.

The risk assessment process should be used for continuous improvement within an operational risk management system. Such a process should include an appropriate combination of methodologies and an effective HACCP plan. Such integration can lead to greater food safety, worker safety, and health, as well as financial and nonfinancial benefits to an organization.

REVIEW QUESTIONS

1. Describe the uniqueness of food industry.
2. Name three standards or guidelines related to food safety.
3. State the primary purpose of HACCP plans.
4. Briefly describe some of the risk assessment techniques used in the food industry.
5. Name three unique food safety hazards.
6. Describe the applicability of PHA, FMEA, and Bow-Tie analysis to food safety risk assessment.
7. Explain PtD and HACCP integration process.

REFERENCES

ANSI/ASSE Z590.3-2011. *Prevention Through Design: Guidelines for Addressing Occupational Hazards and Risks in Design and Redesign Processes*. Des Plaines, IL: American Society of Safety Engineers, 2011.

ANSI/ASSE Z690.2. *Risk Management Principles and Guidelines*. Des Plaines, IL: American Society of Safety Engineers, 2011.

ANSI/ASSE Z690.3. *Risk Assessment Techniques*. Des Plaines, IL: American Society of Safety Engineers, 2011.

Crow, K. Failure Modes and Effects Analysis (FMEA), 2002. Retrieved March 20, 2011 from http://www.npd-solutions.com/fmea.html (accessed December 31, 2015).

ISO 22000 Food Safety Management Systems. Requirements for any organization in the food chain, 2005, September 1. Retrieved January 4, 2015 from http://www.iso.org/iso/home/store/catalogue_tc/catalogue_detail.htm?csnumber=35466 (accessed December 31, 2015).

Mead, P. V., 2000 "Food-related illness and death in the United States." *Journal of Environmental Health*, 62(7), 9; Retrieved from EBSCOhost.

Pilot, S. What Is a Fault Tree Analysis? – Use a General Conclusion to Determine Specific Causes of a System Failure, 2002, March. Retrieved from http://asq.org/quality-progress/2002/03/problem-solving/what-is-a-fault-tree-analysis.html (accessed December 31, 2015).

Roberts, A. C. *The Food Safety Information Handbook*, 1st ed. Westport State: An Imprint of Greenwood Publishing Group Inc., 2001.

Standard Kalite. HACCP history [Web log message], n.d. Retrieved March 15, 2015 from http://www.standardkalite.com/haccp_iso22000_history.htm (accessed December 31, 2015).

United States. Centers for Disease Control and Prevention. CDC 2011: Estimates: Findings, 2011. Retrieved March 15, 2015 from http://www.cdc.gov/foodborneburden/2011-foodborne-estimates.html (accessed December 31, 2015).

United States. Centers for Disease Control and Prevention. 2014a. Foodborne Diseases Active Surveillance Network (FoodNet). Retrieved March 15, 2015 from http://www.cdc.gov/foodnet/index.html (accessed December 31, 2015).

United States. Centers for Disease Control and Prevention, 2014b. Incidence and trends of infection with pathogens transmitted commonly through food – foodborne diseases active surveillance network, 10 U.S. sites, 2006–2013. *Morbidity and Mortality Weekly Report (MMWR)*, 63(15), 328–332. Retrieved from http://www.cdc.gov/mmwr/preview/mmwrhtml/mm6315a3.htm?s_cid=mm6315a3_w (accessed December 31, 2015).

United States. Food and Drug Administration. 1997. Hazard Analysis and Critical Control Point Principles and Application Guidelines. Retrieved on December 28, 2014 from http://www.fda.gov/Food/GuidanceRegulation/HACCP/ucm2006801.htm (accessed December 31, 2015).

REFERENCES

United States. Food and Drug Administration. Hazard Analysis & Critical Control Points (HACCP), April 27, 2011. Retrieved from http://www.fda.gov/food/foodsafety/hazardanalysiscriticalcontrolpointshaccp/default.htm (accessed December 31, 2015).

United States. Food and Drug Administration. Annex 5: HACCP Guidelines, 1999. Retrieved March 15, 2015 from http://www.hospitalityguild.com/downloads/HACCP_Guidelines_FDA.pdf (accessed December 31, 2015).

World Health Organization. Food safety and Foodborne diseases, 2007. Retrieved from http://www.who.int/foodsafety/areas_work/foodborne-diseases/en/.

World Health Organization. Food Safety, 2011. Retrieved from http://www.who.int/foodsafety/en/ (accessed December 31, 2015).

16

ERGONOMIC RISK ASSESSMENT

BRUCE K. LYON
Risk Control Services, Hays Companies, Kansas City, MO, USA

GEORGI POPOV
School of Environmental, Physical & Applied Sciences, University of Central Missouri, Warrensburg, MO, USA

16.1 OBJECTIVES

- Introduce practical ergonomic risk assessment methods
- Review fundamental ergonomic risk assessment techniques
- Present a model for practical ergonomic risk assessment
- Examine the use of techniques, their strengths, and limitations

16.2 INTRODUCTION

Ergonomic risk factors exist in almost all industrial and office work settings. Where there are manual tasks, ergonomic risks can generally be found; however, in many cases they are not recognized or assessed. Such risks can negatively impact employee safety and health, quality of products and services, production efficiency, employee morale, and overall safety culture.

The cost of ergonomic risks can be significant. Work-related musculoskeletal disorders (WMSDs) or soft tissue injuries such as damage to nerves, tendons, muscles, and supporting structures are a common health concern in the workplace. These disorders can cause fatigue, discomfort, pain, swelling, numbness or tingling, and permanent tissue damage. According to the Bureau of Labor Statistics, WMSDs represent a third of all workplace disabling incidents in the United States. In addition, WMSDs make up over 40% of all workers compensation costs (Liberty Mutual Research Institute for Safety, 2013) and costs US businesses $20 billion a year (U S Department of Labor, BLS, 2012). These numbers indicate a need for better application of ergonomic principles in workplace and task design. Safety professionals skilled in practical ergonomic risk assessment and application

---

Risk Assessment: A Practical Guide to Assessing Operational Risks, First Edition.
Edited by Georgi Popov, Bruce K. Lyon, and Bruce Hollcroft.
© 2016 John Wiley & Sons, Inc. Published 2016 by John Wiley & Sons, Inc.

of ergonomic principles are needed in the workplace and will have an advantage when it comes to opportunities in the profession.

This chapter presents a practical risk assessment model and simplified ergonomic risk assessment tool (ERAT) that can be used to help manage ergonomic risks and reduce WMSDs.

16.3 ERGONOMICS AND DESIGN

According to the Human Factors and Ergonomics Society, and International Ergonomics Association, "ergonomics (or human factors) is the scientific discipline concerned with the understanding of interactions among humans and other elements of a system, and the profession that applies theory, principles, data, and other methods to design in order to optimize human well-being and overall system performance" (HFES, 2014). Notice the use of the word "design" in the definition. From an occupational standpoint, the following definition is offered:

> *Ergonomics* – The applied science of designing workplace demands and environment to accommodate human capabilities and limitation for well-being and optimum performance.

Some consider "human factors" to be more focused on the cognitive aspects of user interface, such as design of displays and controls and the usability of products and systems. A simple definition of applied human factors engineering is "the designing of systems with the user in mind." The Human Factors and Ergonomics Society has defined it as follows:

> Human Factors is concerned with the application of what we know about people, their abilities, characteristics, and limitations to the design of equipment they use, environments in which they function, and jobs they perform.

The concept of designing in safety and ergonomic principles into the workplace continues to be emphasized and reinforced by recent standards. The *ANSI/ASSE Z590.3-2011, Prevention through Design Guidelines for Addressing Occupational Hazards and Risks in Design and Redesign Processes* standard is a blueprint for safety through design. In ANSI Z590.3's scope, it states "This standard provides guidance for a life-cycle assessment and design model that balances environmental and occupational safety and health goals over the life span of a facility, process, or product." In Z590.3's foreword, the National Institute for Occupational Safety and Health (NIOSH), instrumental in the development of the standard, states that "one of the best ways to prevent and control occupational injuries, illnesses and fatalities is to design out and minimize hazards and risks early in the design process." This concept is emphasized throughout the Prevention through Design standard. Its stated goals include:

- Achieve acceptable risk levels
- Prevent or reduce occupationally related injuries, illnesses, and fatalities
- Reduce the cost of retrofitting necessary to mitigate hazards and risks that were not sufficiently addressed in the design or redesign processes.

The *ANSI/AIHA/ASSE Z10-2012, the Occupational Health and Safety Management Systems* standard establishes requirements for identifying, preventing, and controlling hazards and risks "associated with new processes or operations at the design stage." To prevent or limit the introduction of new hazards and risks into the work environment, the ANSI Z10 standard provides provisions for (1) risk assessment, (2) design review, (3) management of change, and (4) procurement. Regarding design review, the standard states that "The design review should consider all aspects including design, construction, operation, maintenance, and decommissioning."

Unfortunately, in many organizations, ergonomic principles are not considered or incorporated into the design or redesign of their work systems. As a result, inherent hazards become deeply embedded in their operations, workstations, and methods that are difficult and expensive to eliminate. In the Occupational Safety and Health Administration publication, *Ergonomics: The Study of Work* (OSHA publication 3125, revised in 2000), it states the following regarding the need for ergonomic principles and risk assessment in the workplace:

> If work tasks and equipment do not include ergonomic principles in their design, workers may have exposure to undue physical stress, strain, and overexertion, including vibration, awkward postures, forceful exertions, repetitive motion, and heavy lifting. Recognizing ergonomic risk factors in the workplace is an essential first step in correcting hazards and improving worker protection.

16.4 ERGONOMIC HAZARDS

Workplace ergonomic hazards can contribute to a number of negative effects. The most common and recognized impact is biomechanical and musculoskeletal stress to the worker. Workers exposed to ergonomic hazards such as prolonged repetitive motion, excessive force, awkward or static postures, compression, vibration, and extreme temperature are at risk of developing WMSDs. The accumulation of stress (also known as cumulative trauma) to the soft tissues such as muscles, nerves, and blood vessels over time can cause serious and even permanent damage to the individual. The overall costs of WMSDs are quite significant as mentioned earlier in this chapter.

Ergonomic hazards also contribute to "single-event" incidents such as soft tissue strains, slips and falls, and "struck against" objects. These hidden hazards can go unrecognized within the design of work areas, equipment, tools, or methods eventually causing workers to be injured from the exposure. For illustration purposes, consider the following examples:

Example #1

A fast-food chain selects a low-cost floor tile for their kitchen environments in spite of the fact that the tile has a relatively low coefficient of friction and is not recommended for environments with grease or water exposure. In addition, the company does not provide or enforce the use of nonslip sole footwear and as a result, many employees wear shoes that are not compatible with the environment. The kitchen designs are somewhat tight and congested with workers frequently moving back and forth to perform their duties. These physical conditions along with worker activities create ergonomic hazards that present a high risk of slip and fall incidents.

Example #2

An automotive repair shop has a number of repair tasks that are performed daily on various makes and models of automobiles. Due to costs, the company has a limited number of hand tools and sizes available. The wide variety of vehicle models serviced makes it difficult for workers to find the proper type and size of tool for every job. As a result, ergonomic mismatches can occur, such as ill-fitting tools that slip causing "struck against object" or muscle strain incidents.

Ergonomic hazards also include psychosocial stress or job stress. Psychosocial stress refers to the interaction of the worker and their work conditions. The term "work organization" is often associated with this exposure. In the NIOSH publication, *Stress ... At Work*, job stress is defined as harmful physical and emotional responses that occur when requirements of the job do not match the capabilities, resources, or needs of the worker. According to the NIOSH publication, work conditions that can contribute to job stress are listed as follows:

- Design of tasks – Heavy workload, infrequent rest breaks, long work hours, and shift work; hectic and routine tasks that have little inherent meaning, do not utilize workers' skills, and provide little sense of control
- Management style – Lack of participation by workers in decision making, poor communication in the organization, lack of family-friendly policies
- Interpersonal relationships – Poor social environment and lack of support or help from coworkers and supervisors
- Work roles – Conflicting or uncertain job expectations, too much responsibility, too many "hats to wear"
- Career concerns – Job insecurity and lack of opportunity for growth, advancement, or promotion; rapid changes for which workers are unprepared
- Environmental conditions – Unpleasant or dangerous physical conditions such as crowding, noise, air pollution, or ergonomic problems.

Ergonomic hazards related to system or product usability (poor "human factors engineering") can contribute to human error. Human error or omission can be initiated when workers interface with work systems that contain ergonomic hazards or latent conditions in work stations, controls and displays, equipment, or methods. Unfortunately, these latent conditions lay hiding in work systems contributing to worker mistakes, omissions, shortcuts, and workarounds until an injury or damage occurs.

16.5 ERGONOMIC RISK FACTORS

Indicators of ergonomic hazards are referred to as ergonomic risk factors or stressors. These risk factors are conditions of a job process, workstation, or work method that contribute to the risk of developing WMSDs or other ergonomic-related injuries. These risk factors include forceful exertions, high repetition and duration, static or awkward postures, contact stress, vibration, poor lighting, and cold temperatures. Table 16.1 provides a review of workplace-related ergonomic risk factors.

Ergonomic risk factors tend to be synergistic in their effect. Individually, an ergonomic risk factor may not contribute a notable risk; however, when two or more risk factors are combined, the risk of biomechanical stress leading to a WMSD can be significantly increased. For example, tasks with repetitive motion such as keying on a computer are common in office work environment. Repetitive keying by itself does not generally create a high risk for biomechanical stress. However, when a second risk factor is introduced such as wrist deviation or cold temperature, the combined effect can restrict blood flow and cause internal contact stress on soft tissues elevating the risk of WMSD. This is sometimes referred to as "multiple causation," the combined effect of several risk factors in the development of WMSDs.

16.6 ESTABLISHING AN ERGONOMICS ASSESSMENT PROCESS

A standardized method of assessing ergonomic risks should be established, including appropriate ERATs, user training. The purpose of this standardized method is to provide a defined and systematic process for stakeholders to:

- Identify existing jobs/tasks with ergonomic risks
- Assess, prioritize, and track ergonomic risk factors
- Select and implement effective control measures
- Prevent new ergonomic risks from being introduced into the workplace
- Support continuous improvement with team-based problem solving.

ESTABLISHING AN ERGONOMICS ASSESSMENT PROCESS

TABLE 16.1 Work-Related Ergonomic Risk Factors

Risk Factor	Description
Forceful exertions	Exerting force to complete a motion, which requires an application of considerable contraction forces by the muscles, causing rapid fatigue. Motions include lifting, lowering, pushing, pulling, carrying, gripping, pinching, and tossing
Repetitive motions	Performing repeated identical motions or multiple tasks where the motions of each task are very similar and involve the same muscles groups and tissues. Repetitive motions include assembly-type tasks, keying, sorting, cutting, and finishing work such as sanding and buffing and other similar activities
Awkward postures	Positions of the body (limbs, joints, back) that deviate significantly from the neutral position while job tasks are being performed. Muscles do not work as efficiently in awkward postures and must exert more physical effort to accomplish the task. Awkward postures include elevated reach, shoulder abduction, head tilt, deviated wrist, and bent waist
Static postures	Physical exertion held in the same posture or position throughout the exertion causing muscle fatigue. Also referred to as static muscle loading; examples include gripping tools, holding arms out or up to perform tasks, or standing in one place for prolonged periods
Contact stress	Internal and external contact stress occurs from compression of soft tissues and blood vessels causing blood flow restriction and damage. Contact stress can occur from leaning on table edges, gripping handles, contact with hard surfaces, sitting in unpadded chairs, and internal rubbing of soft tissues against tendons and bones. Can also result when using a part of the body as a hammer or striking instrument
Vibration	Vibration inhibits the blood supply to the affected body part and can lead to numbness, tingling, and damage of soft tissues. Localized vibration occurs when a specific body comes in contact with a vibrating object such as powered hand tools or equipment. Whole-body vibration occurs when standing or sitting in a vibrating environment (truck or heavy equipment operator) and when using heavy vibrating equipment such as a jackhammer
Cold temperature	Exposure to excessive cold while performing work tasks causing restricted blood flow and loss of dexterity and sensitivity. The risk of soft tissue damage is increased when exposure to cold is combined with other risk factors, such as vibration or contact stress
Duration and pace	Extended length of exposure time or elevated pace/cycle times without adequate recovery time between tasks leading to fatigue and stress
Inadequate lighting	Too little or too much lighting, indirect or direct glare causing eye strain, or awkward postures. Can also lead to human error-related incidents

Like other occupational safety and health initiatives, an ergonomic risk assessment process should be applied as a collaborative effort with full management support and active employee involvement to be effective and successful.

Assessment of ergonomic risk is best accomplished within an organization when there is a well-defined plan for consistent application. A written ergonomics process or management system outlining an organization's ergonomic risk assessment process should be developed and communicated to stakeholders to provide clear guidance. Management systems standards such as the *ANSI/AIHA/ASSE Z10-2012 Occupational Health and Safety Management Systems* provide a framework to consider. Ergonomics management systems should include the following considerations:

16.6.1 Scope and Context

First, a clear and concise scope should be determined and established. Based on need, the organization should consider the context of the process in terms of who will be involved and affected, what will be included, the time period and time requirements necessary, and how the process will be applied. The scope should encompass design, redesign, and management change efforts as well as risk assessments of new equipment, workstations, tools, and work methods. ANSI/ASSE Z590.3 Prevention through Design offers such guidance and is strongly recommended.

16.6.2 Goals and Objectives

Goal setting determines the "target(s)" the organization is moving toward and needs to have specific, measurable elements. For example, an organization may want to set specific goals for each operation to develop and train a cross-functional ergonomics task force to assess and improve three (3) high-risk jobs per quarter. Such a goal is specific, measurable, actionable, realistic, and time oriented (SMART).

Goals and objectives must be realistic and achievable by the responsible parties. Some goals may seem achievable at first; however, as the process unfolds, obstacles may arise. Organizations should revisit goals and objectives and make refinements as the process moves forward. As with any initiative, the organization must effectively communicate with stakeholders throughout the process.

16.6.3 Responsibilities

After the goals are set, the organization must clearly define and communicate roles, responsibilities, and accountabilities for all stakeholders. These responsibilities should be specific and measureable to allow for accountability and aimed at achieving the stated goals.

16.6.4 Training

Effective ergonomics training for stakeholders is essential. It should outline and describe the overall process, sequence of steps, specific ergonomic risk factors, hierarchy of controls, and problem-solving methods. Ultimately, the training is to enable stakeholders in performing their assigned tasks effectively by providing the necessary knowledge, skills, and tools.

Training content must be customized and appropriate to each group and include stakeholder participation and interaction. It has been the authors' experience that in general, scientific and medical terms do not find favor with those not familiar with ergonomics. Keeping terms simple, clear, and understandable helps improve comprehension and participation. The *ANSI/ASSE Z490.1-2001, American National Standard Criteria for Accepted Practices in Safety, Health, and Environmental Training* standard can be consulted for more guidance in this subject.

Organizations should determine the types of ergonomics training required for each level, who will participate, specific learning objectives, the timeframe, and frequency of training.

16.6.5 Ergonomics Team

Establishing a well-trained ergonomics team is vital to the success of the ergonomic risk assessment process. The ergonomics team along with the coordinator or leader performs essential functions to drive the improvement process. To increase participation and ownership, as well as leverage necessary skills, a cross-functional ergonomics team should be considered. Members of the ergonomics team should including operators (users of the system being assessed), maintenance, engineering, quality, SH&E, department managers, human resources, and plant management should be considered.

The ergonomics coordinator's role is to facilitate the process, lead, and direct the team in their efforts. This position requires an ability to collaborate with others, mentor team members, and provide technical resources as needed.

16.7 ASSESSING ERGONOMIC RISK

Standard risk assessment techniques are generally not designed to detect ergonomic risk factors to the level necessary. Tools discussed in this text such as job hazard analysis (JHA), preliminary hazard analysis (PHA), failure mode and effects analysis (FMEA) are used to identify and evaluate multiple types of hazards and risks on a broader scale. Ergonomic hazards are sometimes recognized in JHAs, PHAs, and FMEAs; however, these hazard analysis and risk assessment methods are unable to pinpoint, analyze, and assess detailed ergonomic risk factors. For this, a more targeted risk assessment tool is needed.

The American Industrial Hygiene Association (AIHA) Ergonomics Committee in their Ergonomics Reference Document says the following about ergonomic assessment:

> *Some ergonomics risks (such as force, frequency, and posture) are very easy to spot and fix, while some may require more detailed review by trained eyes. There are a number of good resources available from OSHA or state consultation programs, insurance companies, trade organizations and other local companies who may have similar issues. In addition, participatory ergonomics programs, where employees can get involved in identifying hazards and recommending solutions, have proven very successful. There are free resources on the OSHA website, including ergonomics eTools, which can be helpful* (AIHA, 2014). For more information, visit https://www.osha.gov/SLTC/ergonomics/index.html (OSHA, 2013).

TABLE 16.2 Partial List of Ergonomic Risk Assessment Tools

Assessment Tool	Application	Body Segment	Type	Complexity
OSHA Screening Tool	General industry repetitive tasks	Whole body	Qualitative checklist	Low
WISHA Checklist	General industry repetitive tasks	Whole body	Qualitative checklist	Low
ACGIH TLV for Lifting	General industry lifting tasks	Trunk	Qualitative table	Moderate
ERAT	General industry repetitive tasks	Upper extremities	Qualitative spreadsheet	Moderate
RULA	General industry repetitive tasks	Upper extremities Trunk	Semiquantitative worksheet	Moderate
REBA	General industry repetitive asks	Whole body	Semiquantitative worksheet	Moderate
Snook Tables	General industry manual handling	Trunk	Semiquantitative table	Moderate
ACGIH TLV for Hand Activity	Office settings hand work	Upper extremities	Semiquantitative worksheet	Moderate
NIOSH Revised Lifting Equation	General industry manual handling	Trunk	Quantitative formula	Moderate
Strain Index	General industry hand work	Upper extremities	Quantitative worksheet	Moderate
Energy Expenditure Prediction Program University of MI	General industry manual handling	Upper extremities trunk	Quantitative software	High
3D Static Strength Prediction Model University of MI	General industry manual handling	Trunk lower extremities	Quantitative software	High

Upon recognizing the potential for an ergonomic hazard, a tool designed to specifically detect and assess ergonomic risk factors is required. ERATs, of which there are many, are used to detect and assess existing risk factors on a microlevel. The systematic sequence of applying a standard risk assessment followed by an ergonomic assessment is useful in identifying, assessing, and controlling ergonomic risks to the degree necessary.

There are many ERATs to choose from as shown in Table 16.2. They range in application, body segment, ease of use, skill level, time requirements, software or special equipment, as well as other variables.

Summaries of ergonomic assessment tools are available from many sources. Some of these include the following:

- AIHA Ergonomics Committee Ergonomic Assessment Toolkit. The toolkit is available at https://www.aiha.org/get-involved/VolunteerGroups/Pages/Ergonomics-Committee.aspx
- Cornell University Ergonomics Web "Workplace Ergonomics Tools." Available at http://ergo.human.cornell.edu/cutools.html
- Ministry of Labour, Ontario, Canada, "Part 3C: MSD Prevention Toolbox; More on In-depth Risk Assessment Methods." Available at https://www.labour.gov.on.ca/english/hs/topics/pains.php
- University of South Florida, College of Public Health, Thomas E. Bernard, PhD, Ergonomics. Available at http://personal.health.usf.edu/tbernard/ergotools/.

Organizations that have recognized a need and have sufficient internal resources are more likely to have defined and in some cases mature processes, with specific tools and support for assessing ergonomic risks in their workplace. However, for those that lack the experience, knowledge, and skill of ergonomic assessment methods, it can be overwhelming. The challenge for such organizations is to select a tool that can be used by stakeholders with limited ergonomics training and experience that provides reliable and useful results in a timely manner.

16.8 ERGONOMICS IMPROVEMENT PROCESS

Similar to initiatives involving quality, *Lean*, and operational risk management systems, the ergonomics process should be managed as an ongoing, integrated, and sustainable process of continuous improvement (Rostykus, 2005). Once the infrastructure is in place, an operation can initiate the ergonomics improvement process and begin ergonomics risk assessments.

The process steps for conducting continuous improvement in work systems using ergonomic risk assessments generally follow the described ergonomics improvement process model in Figure 16.1. Each step is briefly described as follows:

16.8.1 Identify Jobs

To begin, the most obvious place to start is with loss history. Jobs or tasks that have a history of musculoskeletal disorder (MSD) incidents in both frequency and severity should be made a priority. This information can be derived from workers' compensation insurance claims data, incident investigation reports, incident analysis data, first aid logs, and OSHA incident data. Even though the OSHA 300 log no longer contains a column for MSDs, it is important that SH&E professionals track and measure ergonomics-related incidents.

Often times, information from employees can be most helpful in identifying high-risk jobs. Employee feedback, reported concerns, and discomfort surveys should be used in selecting problem areas. When asked, it is likely that employees will report on a large number of situations that they

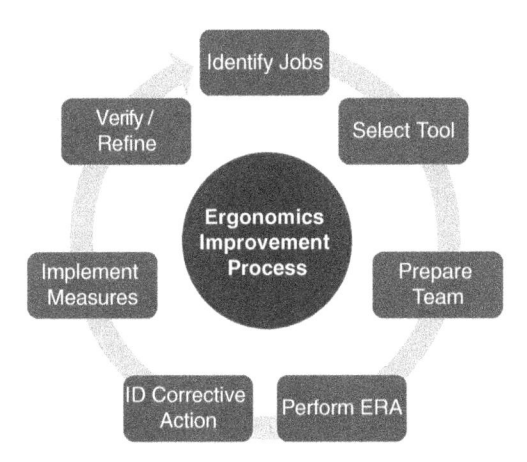

Figure 16.1 Ergonomics Improvement Process Model

perceive as stressful. Communication of some sort should result rather quickly on each report or employees might perceive that this effort is nothing more than another flavor of the month.

Workplace observations by the ergonomics team or other trained personnel are a good source of information. Modifications made to chairs, tools, and other workplace items are a "red flag" indicating a need of further investigation. One particular method used by the authors involves a sequential series of techniques to identify and prioritizing jobs by perceived risk:

(1) First, the team uses "brainstorming" to collect a population of jobs and tasks with perceived ergonomic risk. The brainstorming technique is a common tool used for quickly gathering ideas and data. It is important to set the ground rules with the team for the brainstorming session. Brainstorming can be conducted in structured or nonstructured fashion. A structured brainstorming session requires every person to offer an idea as their turn arises. An "unstructured" brainstorming approach relies on the group to provide ideas as they come to mind, which may create a more relaxed atmosphere, but risks domination by more vocal members of the group. Either way, brainstorming should promote the following:
   - An environment free of judgment of ideas/items – no criticism or favoritism
   - Creative thinking
   - Quantity over quality (at this stage).
(2) Once the list is populated, the team members individually select their top three to five jobs and write them down on index cards. The cards are collected and the numbers tallied to identify the team ranking of jobs. A white board, flip chart, or projected Excel spreadsheet is used to list the jobs by priority ranking. This technique is sometimes referred to as "The Priority Path" (GOAL/QPC, 1995).
(3) From there, the jobs at the top of the list (5–15 jobs) are selected for placement on the risk assessment matrix. It is critical that the team clearly understands the severity and likelihood categories used in the risk assessment matrix. Each job is evaluated by the team to determine a consensus on its severity level and likelihood level and plotted on the matrix accordingly. This provides an ergonomic risk ranking of jobs that the ergonomics team can begin addressing.

16.8.2 Assessment Tools

Many ERATs are available (see a partial list in Table 16.2) that have general or specific applications. Selecting the appropriate tool to fit the scope and purpose of the assessment as well as the application and target body segment should be taken into consideration. In addition, the complexity or "degree of difficulty" in using the selected tool and the skill level of personnel using the tool should be considered. All ERATs require some level of training and experience to be used properly and effectively. The end users' skill level should be considered in the tool selection.

There are three types of ERATs: qualitative, semiquantitative, and quantitative. Each are briefly described in the following:

1. *Qualitative tools* – Qualitative tools are most commonly used and are good for screening jobs. Most are manual or Excel based. Some of these include the OSHA Basic Screening Tool, NIOSH Checklists, Washington Industrial Safety and Health Act (WISHA) Checklists, and American Conference of Governmental Industrial Hygienists (ACGIH) Threshold Limit Value (TLV) for Lifting. There are also many tools developed by insurance carriers and consulting firms that can be used for basic screening and postural assessment. The ERAT discussed in this chapter is one example.

2. *Semiquantitative tools* – Semiquantitative tools generally require more expertise in their application. Typically, semiquantitative tools are targeted for specific applications and risk factors such as lifting, hand work, or body region. These types of tools include Rapid Upper Limb Assessment (RULA), Rapid Entire Body Assessment (REBA), Snook Tables, ACGIH TLV for Hand Activity, and Utah Back Compression among others.

3. *Quantitative tools* – A third group of tools are called quantitative assessment tools, which are typically used to perform more in-depth analysis. As a result, quantitative tools require a higher degree of user training, knowledge, and skill. Quantitative tools include NIOSH Revised Lifting Equation, University of Michigan's 3D Static Strength Prediction Model, University of Michigan's Energy Expenditure Prediction Program (EEPP), Strain Index (SI), Dynamic Work Analysis, and Static Work Analysis to mention a few.

16.8.3 Assessment Team

Ergonomic risk assessments are best performed by a cross-functional team of trained and qualified members. Such teams should include users/operators, production, quality, skilled trades, maintenance, and SH&E professionals. Safety committees that have a diverse and experienced group can be used in this process. A qualified ergonomics coordinator that is proficient in workplace ergonomics and risk assessment to guide the team is essential.

The assessment team should be properly trained and knowledgeable of the jobs and tasks being performed as well as the ability to identify ergonomic hazards and risk factors. In addition, team members should have some training and experience in problem-solving skills, ergonomics principles, and the hierarchy of controls.

16.8.4 Performing the Assessments

Ensure there are adequate resources and time allotted to perform assessments. Equipment such as digital cameras and measuring tapes can be useful in gathering and documenting data for analysis. Other instruments such as light meters, sound level meters, infrared thermometers, force gauges, and goniometers may be needed. During the assessment, it is advised that the assessor verify that the selected job is being performed at its typical capacity or rate and representative of its normal operation.

Certain workplaces such as manufacturing have work involving varied or long-cycle tasks presenting a unique challenge in observing and analyzing the complete job. In such cases, an observation plan with work flow diagram should be reviewed prior to the assessment. The work can then be split into segments or tasks much like a JHA with each task assessed individually.

The ergonomic risk assessment process involves three basic steps:

(1) Identify and collect data:
   a. Obtain operator input and involvement.
   b. Observe task cycles and identify task details.
   c. Collect digital video and photos of tasks, operator interactions for further analysis.
   d. Collect data regarding work area, materials, tools, and environment.
(2) Analyze data:
   a. Analyze collected data to understand the nature and types of ergonomic risk.
   b. Use selected ERAT inputs to determine level of risks.
(3) Evaluate:
   a. Evaluate severity, likelihood, and exposure/duration of each risk.
   b. Input resulting risk factor in the assessment tool.

Upon completion of the assessment, risk factors are entered into the ERAT according to its defined risk criteria. The resulting risk priority number (RPN) or action level value is used to determine whether the assessed job's ergonomic risk level is at an acceptable level or if it requires additional controls. For example, the ERAT tool featured in Figure 16.3 specifies three different levels (*Action Level 3* – high risk, *Action Level 2* – moderate risk, and *Action Level 1* – low risk) and indicates whether additional action is required or desired.

16.8.5 Identifying Corrective Measures

The ultimate goal of ergonomic risk assessment is to help organizations understand ergonomic risk levels and minimize risks to a tolerable level. Jobs with high-risk scores require immediate corrective measures to reduce the identified risk factors. The ergonomics team should be involved in identifying and developing effective ergonomics solutions. Again, brainstorming can be used to identify potential improvements and controls that will reduce ergonomic risk. The team must understand and use the *hierarchy of controls* concept to select the most effective controls measures. This often requires some use of cost/benefit analysis and return of investment calculations to select feasible control measures and help persuade decision makers.

16.8.6 Implementing Measures

Before any changes are initiated, affected workers should be made aware of what will take place, why it is necessary, and when it will take place. Any special training needed by workers to use the controls effectively should be provided prior to the implemented changes. If ergonomic changes are made without the user's input or knowledge, the likelihood of the change to not be accepted is increased dramatically. Workers should be included in the process of selecting ergonomic interventions to create buy-in and ownership.

Many ergonomic improvements will require engineering, maintenance, and production level staff to complete. Ensuring that task assignments, target dates, needed resources, and other items are communicated and tracked is important.

16.8.7 Verify and Refine

Soon after implementation, the ergonomics team and engineering personnel should meet with affected operators to verify the measures are working properly. There may be need for fine-tuning certain applications or complete overhaul in rare cases. Operator feedback or concerns about ease of operation, comfort, speed, etc. should be collected to identify any needed adjustments or corrections.

After a sufficient break-in period, a more in-depth review of the new control measures and their effectiveness should be made. This step will involve ergonomics team members and engineering using the selected ERAT to measure risk levels following the successful implementation of control measures.

16.8.8 Communicate Results

Results from the assessment must be tracked, measured, and communicated for the process to be effective (Lyon, 1997). The organization's ergonomics team should establish short-term and long-term metrics that demonstrate the effectiveness of the ergonomic risk assessment process. Many of these metrics are already measured and tracked by organizations in their various departments such as production, quality, or human resources. Some of the most common measures include result-based metrics that provide a long-term measure of performance:

- Number of ergonomic-related incidents
- Number of ergonomic-related lost time incidents
- Number of ergonomic-related lost days, restricted days, or transferred days
- Reduction of piece rework, rejection rate, and other quality-related measures
- Reduction in nonvalue added tasks
- Reduction in waste and scrap
- Increase in production rate values.

In addition, it is important that action or activity-based metrics be included in the process. Action-based metrics are typically "short-term" measures that can help communicate the impact made in workplace improvements and risk reductions. Examples of action-based metrics may include the following:

- Percent of jobs/tasks assessed for ergonomic risk
- Percent of jobs/tasks with reduced risk
- Percent of ergonomic risk reduction or reduction in risk levels (RPN)
- Number of employees trained
- Number of employee-generated improvements.

As with any successful risk assessment model, the ergonomic risk assessment process should involve stakeholders from beginning to end. As goals are achieved, the organization should celebrate, recognize, and reinforce involved stakeholders.

16.9 ERAT: A PRACTICAL ASSESSMENT TOOL

One of many quick screening and assessment tools available is the ERAT. ERAT was developed by the authors (Lyon et al., 2013) and is based on the draft ANSI Z365 Ergonomics Checklist (ANSI/ASC Z365-1998) that was part of the *"Management of Work-Related Musculoskeletal Disorders Accredited Standards Committee Z365 Working Draft"* withdrawn in 2003 (National Safety Council, 2002).

The relatively simple tool provides a standardized method of quickly identifying, assessing, and scoring ergonomic risks to upper extremities in most work environments. It is Excel based and has an *Initial Assessment (IA) worksheet*, a *Post-Controls Assessment (PCA) worksheet* (used after the

Initial Assessment and implementation of control measures to establish a current risk factor score), and a *Hierarchy of Ergonomic Risk Controls*. One of the advantages of simplified tools such as ERAT is the relatively small amount of training time required to learn how to use the tool. The authors have trained supervisors, lead persons, and operators in the ERAT in less than 4 h with users demonstrating adequate levels of competence in performing ergonomic risk assessments. A review of the checklist, explanation of risk factors with examples of how they are scored, live demonstrations of the tools use, followed by hands-on application is highly recommended.

ERAT is limited to the assessment of ergonomic risk factors impacting the upper extremities and is appropriate for office work and assembly-type work environments. The Initial Assessment (IA worksheet) is used to develop a baseline risk score and action level with possible control measures to reduce risk factor scores. The PCA worksheet is used after the Initial Assessment and implementation of control measures to establish a current risk factor score.

The assessment worksheets include six categories of risk factors:

(1) Repetition (R)
   a. Every few minutes
   b. Every few seconds
(2) Lifting (L)
   a. 5–15 lb
   b. 15–30 lb
   c. 30–50 lb
   d. Over 50 lb
(3) Push/pull force (F)
   a. Easy
   b. Moderate
   c. Heavy
(4) Carrying loads more than 10 ft (C)
   a. 5–15 lb
   b. 15–30 lb
   c. Over 30 lb
(5) Postures (P)
   a. Head tilt (forward, backward, or to the side)
   b. Shoulder reaching (arm extension)
   c. Flying elbow (shoulder abduction; raised elbow)
   d. Bent wrist (wrist deviation including ulnar, radial, flexion, or extension) or pinch grip
   e. Bend or twist waist (back flexion or extension, bending to the side, twisting at waist)
(6) Environmental factors (E)
   a. Noise (noticeably interfering with communication or comfort)
   b. Lighting/glare (inadequate or excessive light for task, direct or indirect glare interfering with task, noticeable eye strain caused by lighting)
   c. Impact/compression (impact to soft tissues from tools, handles, parts; using hand as a hammer; hard edges causing compression to soft tissues on arms, legs, or other body parts)
   d. Vibration/power tools (whole body vibration from equipment, hand–arm vibration from power tools)

e. Excessive pace (noticeable difficulty in maintaining pace; lack of control over pace; lack of recovery time)

f. Extreme temperature (cold temperature affecting hands, feet, or other body parts; heat stress).

For each category, the duration of the task is entered in the worksheet, which presents an individual risk factor score. Individual scores are added in the spreadsheet for a total risk score as represented by the formula in the following:

$$R + L + F + C + P + E = \text{Total risk score}$$

Instructions for using the tool follow:

For each row that applies in the initial assessment, the assessor scores the task based on the duration and observed risk factors. For individual scores values of 2 or 3, control measures may be needed and suggested in the Comments section on the worksheet. After scoring all six risk factor categories, add the individual scores in the total column. If the total sum of ergonomic risk factors is equal to or less than 10 (Action Level 1), the risk level is consider minimal and the need for further evaluation may be required; for scores between 11 and 22 (Action Level 2), the risk is considered moderate to high requiring intervention in the near future; and for scores exceeding 22 (Action Level 3), the risk is considered extremely high with immediate intervention required.

As with many assessment tools, it is suggested that video analysis be used to document and study task repetition, postures, lifts, pulls, pushes, carries, and other factors covered by the assessment tool.

To provide an illustration, an assessment of a pork processing task using the ERAT is described in the following:

16.9.1 ERAT Example: Pork Processing Belly Grader

16.9.1.1 Task Description Two operators standing at the end of an incoming conveyor approximately waist high, manually grab each pork belly with one hand and place on an adjacent weighing scale, visually grade the belly, and then toss it (with one arm) into one of several bins approximately 6–8 ft away. The work pace for each worker is approximately 10 bellies a minute equating to approximately 60 per hour or 480 bellies per 8 h shift. Pork bellies weigh between 11 and 14 lb each, with average around 13 lb.

16.9.1.2 ERAT Initial Assessment Scores

- *Repetition score* = 3 (cycle every few minutes with duration more than 4 h)
- *Lift score* = 1 (average weight is 13 lb each, with duration more than 4 h)
- *Push/pull score* = 3 (repetitive "one arm" throwing motion is consider physically demanding, with duration more than 4 h)
- Postures
  - *Head tilt score* = 1 (periodic head tilt forward approximately 45°)
  - *Shoulder reaching score* = 2 (repetitive shoulder extension postures performed more than 4 hs)
  - *Flying elbow score* = 2 (repetitive elevated elbow/shoulder abduction postures performed more than 4 h)
  - *Bent wrist/pinch grip score* = 2 (repetitive bent wrist and pinch grip during grasping of bellies performed more than 4 h)

- ○ *Bending/twisting score* = 1 (periodic twisting at waist while tossing bellies to bins)
- Environment
  - ○ *Noise score* = 2 (Hearing Conservation area with protection used)
  - ○ *Lighting score* = 2 (light levels and quality inadequate for inspection tasks)
  - ○ *Excessive pace score* = 2 (physical effort required at a 10 per minute pace performed for more than 4 h)
  - ○ *Extreme temperature score* = 2 (cold ambient temperatures require use of protective gloves and clothing).

The Initial Assessment of this task has a score of 23, which is categorized as *Action Level* 3 or extremely high and requires immediate intervention as illustrated in Figure 16.2.

Job Task: Belly Grader	Evaluator: Bruce Lyon; Georgi Popov			Dept: Pork Processing Line	Date: 1/9/2012	
Risk factor	Duration of task			Score	Comments	
	<1 h	1–4 h	>4 h / N/A			
Repetition	Enter only one score					
Every few minutes	0	0	1		0	
Every few seconds	0	1	3		3	10 bellies/min
Lifting	Enter only one score					
5–15 lb	0	0	1		1	8 h shift; 12–14 lb each
15–30 lb	1	1	2	N/A		
30–50 lb	2	2	3	N/A		
Over 50 lb	3	3	3	N/A		
Push/pull force	Enter only one score					
Easy	0	0	1	N/A		
Moderate	0	1	2	N/A		
Heavy	1	2	3		3	Throwing bellies ~8 ft. to bin
Carry > 10 ft	Enter only one score					
5–15 lb	0	0	1	N/A		
15–30 lb	0	1	2	N/A		
Over 30 lb	1	2	3	N/A		
Postures	Score each					
Head tilt	0	1	2		1	
Shoulder reaching	0	1	2		2	Extreme reach
Flying elbow	0	1	2		2	Extreme elbow elevation
Bent wrist/pinch grip	0	1	2		2	Pinch grip
Bend or twist waist	0	1	2		1	
Environmental	Score each					
Noise	0	1	2		2	> 85 dBA
Lighting/glare	0	1	2		2	Shadows and glare
Impact/compression	0	1	2	N/A		
Vibration/power tool	0	1	2	N/A		
Excessive pace	0	1	2		2	Physically demanding
Extreme temperature	0	1	2		2	40°F
			Action level total		23	
Action level 1	Total score of 10 or less: may require further analysis					
Action level 2	Total score of 11–22: requires intervention in near future					
Action level 3	Total score of 23–36: requires immediate intervention					

Figure 16.2 Ergonomics Risk Assessment Tool: Initial Evaluation

Method	Phase/Application	Control Examples	Effectiveness
Avoidance	Conceptual Stage Design/Redesign	Prevent entry of hazard into workplace by design through selection of technology and work methods	High
Elimination	Existing Processes Redesign	Eliminate hazard by changes in design, equipment and methods	High
Substitution	Existing Processes	Substitute materials, sizes, weights and other aspects to a lower hazard severity or likelihood	Moderately High
Engineering Controls	Existing Workstations Redesign	Reduce hazard by changes to workplace, tools, equipment, fixtures, adjustability, layout, lighting, work environment	Moderate
Administrative Controls	Practices and Procedures	Reduce exposure to hazard by changes in work practices, training, job enlargement, job rotation, rest breaks, work pace	Moderately Low
Personal Protective Equipment	Workers	Reduce impact of hazard to employee by use of protective equipment and materials such as vibration attenuation gloves	Low

Figure 16.3 Hierarchy of Ergonomic Risk Controls

16.9.1.3 Ergonomics Risk Controls Selection and Implementation Following the Initial Assessment, the ergonomics team at the pork processing facility worked together to form appropriate risk controls using the Hierarchy of Ergonomic Risk Controls principles (Figure 16.3).

The resulting interventions involved engineering controls including in-line weighing scales placed in a new conveyor system, placement of chutes and bins beneath the conveyor system eliminating the need for the excessive handling and throwing tasks. The engineering control measures also helped to eliminate or reduce a number of extreme, repetitive postures.

16.9.1.4 Post-Control Assessment Following the implementation of controls, a second assessment using the PCA worksheet was performed with the resulting risk score shown in Figure 16.4. The PCA score was reduced from 23 (Action Level 3 – high risk) to 9 (Action Level 1 – low risk).

CONCLUSION

Job Task: Belly Grader	Evaluator: Bruce Lyon; Georgi Popov	Dept: Pork Processing Line	Date: 3/5/2012		
Risk factor	**Duration of task**			**Score**	**Comments**
	<1 h	1–4 h	>4 h / N/A		
Repetition				Enter only one score	
Every few minutes	0	0	1	N/A	
Every few seconds	0	1	3	3	No change made
Lifting				Enter only one score	
5–15 lb	0	0	1	0	Engineering controls reduce handling by ~50%
15–30 lb	1	1	2	N/A	
30–50 lb	2	2	3	N/A	
Over 50 lb	3	3	3	N/A	
Push/pull force				Enter only one score	
Easy	0	0	1	1	Engineering controls reduce handling by ~50%
Moderate	0	1	2	N/A	
Heavy	1	2	3	N/A	
Carry > 10 ft				Enter only one score	
5–15 lb	0	0	1	N/A	
15–30 lb	0	1	2	N/A	
Over 30 lb	1	2	3	N/A	
Postures				Score each	
Head tilt	0	1	2	0	Engineering controls reduce extreme postures significantly
Shoulder reaching	0	1	2	0	
Flying elbow	0	1	2	0	
Bent wrist/pinch grip	0	1	2	1	
Bend or twist waist	0	1	2	0	
Environmental				Score each	
Noise	0	1	2	2	No change made
Lighting/glare	0	1	2	0	Improved lighting quality
Impact/compression	0	1	2	N/A	
Vibration/power tool	0	1	2	N/A	
Excessive pace	0	1	2	0	Reduced handling
Extreme temperature	0	1	2	2	No change made
			Action level total	**9**	
Action level 1	Total score of 10 or less: may require further analysis				
Action level 2	Total score of 11–22: requires intervention in near future				
Action level 3	Total score of 23–36 requires immediate intervention				

Figure 16.4 Ergonomic Risk Assessment Tool: Post-Controls Evaluation

16.10 CONCLUSION

Ergonomic hazards and latent conditions to some degree exist in almost all work systems and can be missed in standard risk assessment efforts. Many organizations uncertain of the ergonomic risk levels in their operation or the impact it has on the business have a significant opportunity for improvement. For an organization to improve its ability to manage ergonomic hazards, an effective ergonomic risk assessment process is needed to reduce risks to an acceptable level.

For organizations that have yet to implement an ergonomics process, it is an opportunity for SH&E professionals to help the business decision makers understand and recognize the value of reducing ergonomic risk and guide them in establishing a process. Keeping it simple, yet effective is a key to success.

REVIEW QUESTIONS

1. Provide a summary of the types of hazards created by systems and products that are not ergonomically designed.
2. Briefly describe the ergonomic risk factors commonly included in ergonomic risk assessments.
3. Describe the ergonomic improvement process and how it fits within an operational risk management system.
4. Explain the limitations that general hazard analysis and risk assessment tools such as failure mode and effects analysis, preliminary hazard analysis, and fundamental risk assessments have regarding ergonomic risks.
5. List six ergonomic risk assessment tools commonly used.
6. Compile a list of departments/positions that should be included in an ergonomic risk assessment.

REFERENCES

AIHA. AIHA Ergonomics Committee Ergonomics Reference Document. Falls Church, VA: American Industrial Hygiene Association, 2014. Also at: https://www.aiha.org/get-involved/VolunteerGroups/Pages/Ergonomics-Committee.aspx (accessed December 31, 2015).

ANSI/AIHA/ASSE Z10-2012. Occupational Health and Safety Management Systems Standard. Fairfax, VA: The American Industrial Hygiene Association, 2012.

ANSI/ASC Z365-1998. Control of Cumulative Trauma Disorders. Itasca, IL: National Safety Council, 1998.

ANSI/ASSE Z490.1-2001. Criteria for Accepted Practices in Safety, Health, and Environmental Training Standard. Des Plaines, IL: American Society of Safety Engineers, 2001.

ANSI/ASSE Z590.3-2011. Prevention Through Design: Guidelines for Addressing Occupational Hazards and Risks in Design and Redesign Processes Standard. Des Plaines, IL: American Society of Safety Engineers, 2011.

BLS. Nonfatal Occupational Injuries and Illnesses Requiring Days Away from Work. Washington, DC: Bureau of Labor Statistics, US Department of Labor, USDL 12-2204: released November 8, 2012. Available at http://www.bls.gov/news.release/archives/osh2_11082012.pdf (accessed December 31, 2015).

GOAL/QPC. Coach's Guide to the Memory Jogger. Lawrence, MA, 1995.

HFES. Definitions of Human Factors and Ergonomics. Santa Monica, CA: The Human Factors and Ergonomics Society, 2014. Available at: http://www.hfes.org/web/educationalresources/hfedefinitionsmain.html (accessed December 31, 2015).

Liberty Mutual Research Institute for Safety. 2013 Liberty Mutual Workplace Safety Index. Hopkinton, MA: Liberty Mutual Research Institute for Safety, 2013. Available at https://www.google.com/#q=2013+liberty+mutual+workplace+safety+index (accessed December 31, 2015).

Lyon, B. K., 1997. "Ergonomic benefit/cost analysis: communicating the value of enhancements." *Professional Safety*, 42(3), 33–36.

Lyon, B. K., Popov, G. and Hanes, K., 2013. "Improving ergo IQ: a practical risk assessment model." *Professional Safety*, 58(12), 26–34.

National Safety Council. Management of Work-Related Musculoskeletal Disorders. (ANSI Accredited Standards Committee Z365 Working Draft). Itasca, IL: National Safety Council, August, 2002. Retrieved from http://www.humanics-es.com/z365_finldrft.pdf (accessed December 31, 2015).

NIOSH. Stress ... At Work. DHHS (NIOSH) Publication No. 99-101: released in 1999. Cincinnati, OH: National Institute for Occupational Safety and Health. Available at http://www.cdc.gov/niosh/docs/99-101/ (accessed December 31, 2015).

Occupational Safety and Health Administration. Analysis Tools. Washington, DC: U.S. Department of Labor, Occupational Safety and Health Administration, 2013. Available from https://www.osha.gov/dts/osta/oshasoft/index.html (accessed January 21, 2016).

OSHA. Ergonomics Program Management Guidelines For Meatpacking Plants, OSHA 3 123. Washington D.C.: Occupational Safety and Health Administration, 1993. Available at: https://www.osha.gov/Publications/OSHA3123/3123.html (accessed December 31, 2015).

OSHA. Ergonomics: The Study of Work, OSHA 3125. Washington, DC: Department of Labor, OSHA, 2000. Available at https://www.osha.gov/Publications/osha3125.pdf (accessed December 31, 2015).

Rostykus, W. G. Managing Ergonomics as a Process. ASSE Professional Development Conference Proceedings. New Orleans, LA, June 2005.

APPENDIX 16.A

SAMPLE ERGONOMIC RESPONSIBILITIES FOR INVOLVED STAKEHOLDERS

Role	Responsibilities
Senior management	• Establish goals, objectives, and scope of ERA process • Communicate with stakeholders monthly • Enable operations to accomplish objectives • Participate in assigned ergonomics training • Review and track progress • Provide visible support and reinforcement of process
Operations manager	• Participate in assigned ergonomics training • Enable and support stakeholders • Hold stakeholders accountable and ensure objectives are met • Identify ergonomics coordinator • Review and track progress • Provide visible support and reinforcement of process
Ergonomics coordinator	• Participate in assigned ergonomics training • Develop and coordinate plan • Verify elements of process are implemented and maintained • Ensure 3 assessments of priority jobs are successfully completed quarterly • Review and track progress • Provide leadership and direction to ergonomics team • Report progress monthly to operations management • Participate in monthly ergonomics team meetings

Role	Responsibilities
Ergonomics team	• Participate in assigned ergonomics training • Assess 3 priority jobs each quarter • Identify risk reduction measures and use cost/benefit analysis • Develop selected measures and assign implement • Confirm implement and follow-up • Review and track progress • Assist in incident analysis and corrective actions
Engineers	• Participate in assigned ergonomics training • Participate in all ergonomics team meetings • Work with team to identify engineering solutions • Work with team to develop cost/benefit analysis • Apply ergonomics guidelines in new designs, equipment, and workplace changes
Maintenance	• Participate in assigned ergonomics training • Participate in all ergonomics team meetings • Participate in solution development and cost/benefit analysis • Provide insight in maintenance and service needs • Help implement selected solutions
Supervisors	• Participate in assigned ergonomics training • Enable and ensure employees have assigned training • Assist ergonomics team in identifying priority jobs • Assist in defining task requirements in priority jobs • Participate in incident analysis • Reinforce and recognize safe work practices • Mentor and coach employees • Communicate with employees
Employees	• Participate in assigned ergonomics training • Help identify ergonomic risk factors using the defined assessment tool • Follow ergonomics guidelines and safe work instructions • Report discomfort, symptoms immediately • Participate in incident analysis • Provide input and feedback on control measures
Medical care and workers comp	• Participate in assigned ergonomics training • Assess and treat employees with reported symptoms • Inform management/ergonomics coordinator of symptoms, concerns, disorders • Participate in incident analysis • Manage and track incidents • Report trends to management and ergonomics coordinator • Help identify proper modified duties for employees with restrictions

APPENDIX 16.B
SAMPLE ERGONOMICS TRAINING FOR INVOLVED STAKEHOLDERS

Role	Training	Objectives	Frequency
Senior management	*Ergonomics management system* *2 h*	• Understand the purpose, scope, and goals of process • Overview of ergonomics process • Responsibilities and accountabilities for senior management • Resources necessary and available • Time and budget requirements • Managing and reinforcing the process steps • Process metrics and methods tied to goals	• Initially • 2 years • Changes
Operations management	*Plant ergonomics process* *3 h*	• Understand the purpose, scope, and goals of process • Understand basic ergonomics principles and risk factors • Understanding of plant ergonomics process steps • Responsibilities and accountabilities for operations management • Resources necessary and available • Time and budget requirements • Managing and reinforcing the process steps • Process metrics and methods tied to goals	• Initially • 2 years • Changes
Ergonomics coordinator Ergonomics team members	*Ergonomics team and risk assessment* *16 h initial* *4 h refresher*	• Understand the purpose, scope, and goals of process • Responsibilities and accountabilities for team • Understand ergonomics principles, risk factors, and control measures • Understand ergonomics process steps • Understand how to use ergonomics risk assessment tool • Ergonomic incident analysis and root cause • Problem solving and solution building • Resources necessary and available • Time and budget requirements • Process metrics and methods tied to goals	• Initially • 2 years • Changes

Role	Training	Objectives	Frequency
Engineers Maintenance	*Design guidelines for ergonomics* *16 h initial* *4 h refresher*	• Understand the purpose, scope, and goals of process • Responsibilities and accountabilities for designers and engineers • Understand engineering ergonomics principles, risk factors, and control measures • Understand ergonomics process steps • New workstation design guidelines • Problem solving and solution building • Resources necessary and available • Time and budget requirements • Process metrics and methods tied to goals	• Initially • 2 years • Changes
Supervisors	*Ergonomics process for supervisors* *2 h*	• Understand purpose, scope, and goals of process • Understand basic ergonomics principles and risk factors • Understanding of ergonomics process steps • Responsibilities and accountabilities for supervisors • Reinforcing the process steps • Process metrics and methods tied to goals	• Initially • 2 years • Changes
Employees	*Ergonomics process for employees* *2 h*	• Understand the purpose, scope, and goals of process • Understand basic ergonomics principles and risk factors • Responsibilities and accountabilities for employees	• Initially • 2 years • Changes

APPENDIX 16.C

ERGONOMIC RISK ASSESSMENT TOOL (ERAT) – INITIAL ASSESSMENT

Job Task:	Evaluator:				Dept:	Date:
Risk Factor	Duration of Task				Score	Comments
	<1 h	1–4 h	>4 h	N/A		
Repetition	Enter only one score					
Every few minutes	0	0	1			
Every few seconds	0	1	3			
Lifting	Enter only one score					
5–15 lb	0	0	1			
15–30 lb	1	1	2			
30–50 lb	2	2	3			
Over 50 lb	3	3	3			
Push/pull force	Enter only one score					
Easy	0	0	1			
Moderate	0	1	2			
Heavy	1	2	3			
Carry > 10 ft	Enter only one score					
5–15 lb	0	0	1			
15–30 lb	0	1	2			
Over 30 lb	1	2	3			
Postures	Score each					
Head tilt	0	1	2			
Shoulder reaching	0	1	2			
Flying elbow	0	1	2			
Bent wrist/pinch grip	0	1	2			
Bend or twist waist	0	1	2			
Environmental	Score each					
Noise	0	1	2			
Lighting/glare	0	1	2			
Impact/compression	0	1	2			
Vibration/power tool	0	1	2			
Excessive pace	0	1	2			
Extreme temperature	0	1	2			
	Action level total				0	
Action Level 1	Total score of 10 or less: may require further analysis					
Action Level 2	Total score of 11–22: requires intervention in near future					
Action Level 3	Total score of 23–36: requires immediate intervention					

APPENDIX 16.D

ERGONOMIC RISK ASSESSMENT TOOL (ERAT) – POST-CONTROL ASSESSMENT

Job Task:	Evaluator:				Dept:	Date:
Risk Factor	Duration of Task				Score	Comments
	<1 h	1–4 h	>4 h	N/A		
Repetition	Enter only one score					
Every few minutes	0	0	1			
Every few seconds	0	1	3			
Lifting	Enter only one score					
5–15 lb	0	0	1			
15–30 lb	1	1	2			
30–50 lb	2	2	3			
Over 50 lb	3	3	3			
Push/pull force	Enter only one score					
Easy	0	0	1			
Moderate	0	1	2			
Heavy	1	2	3			
Carry > 10 ft	Enter only one score					
5–15 lb	0	0	1			
15–30 lb	0	1	2			
Over 30 lb	1	2	3			
Postures	Score each					
Head tilt	0	1	2			
Shoulder reaching	0	1	2			
Flying elbow	0	1	2			
Bent wrist/pinch grip	0	1	2			
Bend or twist waist	0	1	2			
Environmental	Score each					
Noise	0	1	2			
Lighting/glare	0	1	2			
Impact/compression	0	1	2			
Vibration/power tool	0	1	2			
Excessive pace	0	1	2			
Extreme temperature	0	1	2			
	Action level total				0	
Action Level 1	Total score of 10 or less: may require further analysis					
Action Level 2	Total score of 11–22: requires intervention in near future					
Action Level 3	Total score of 23–36: requires immediate intervention					

APPENDIX 16.E

HIERARCHY OF ERGONOMIC RISK CONTROLS

Control Method	Phase/Application	Control Examples	Effectiveness
Avoidance	Conceptual stage design/redesign	Prevent entry of hazard into workplace by design through selection of technology and work methods	*High*
Elimination	Existing processes redesign	Eliminate hazard by changes in design, equipment, and methods	*High*
Substitution	Existing processes	Substitute materials, sizes, weights, and other aspects to a lower hazard severity or likelihood	*Moderately high*
Engineering controls	Existing workstations redesign	Reduce hazard by changes to workplace, tools, equipment, fixtures, adjustability, layout, lighting, work environment	*Moderate*
Administrative controls	Practices and procedures	Reduce exposure to hazard by changes in work practices, training, job enlargement, job rotation, rest breaks, work pace	*Moderately low*
Personal protective equipment	Workers	Reduce impact of hazard to employee by use of protective equipment and materials such as vibration attenuation gloves	*Low*

17

ASSESSING OPERATIONAL RISKS AT AN ORGANIZATIONAL LEVEL

BRUCE K. LYON
Risk Control Services, Hays Companies, Kansas City, MO, USA

17.1 OBJECTIVES

- Introduce organizational risks
- Examine elements within organizational risks
- Provide guidance on conducting organizational risk assessments

17.2 INTRODUCTION

The world is ever changing, presenting new and often times complex risks to businesses every day. In the globally connected and interdependent market place, risks are no longer isolated to a single location or entity. Risks that threaten the market, supply or distribution chains for an organization, affect all parties. Some risks are known and accounted for in an organization's risk management plan. But many risks remain hidden and unquantified, creating uncertainty and a potential for disruption and loss to an organization. Some of these risks if left untreated can destroy key assets of an organization, damage its reputation, or prevent the successful achievement of business goals and objectives.

The previous chapters in this text have addressed risk assessment methods for functions within an organization such as its processes, systems, machines, workstations, and tasks. This chapter takes a broader view of risks at a more strategic level, specifically risks that are capable of impacting an organization's ability to stay in business. If done well, an organizational risk assessment (ORA) gives management a clear picture of internal and external exposures to the organizations and enables decision makers in making better decisions.

An ORA is a broader, more holistic approach. Further analysis and assessment are often required using other targeted methods covered in this text such as Bow-Tie Analysis to help illustrate and communicate risks to management (Chapter 9). The bottom line is management needs adequate information to make the best decisions concerning risks and may require more detail for certain aspects of the organization.

Risk Assessment: A Practical Guide to Assessing Operational Risks, First Edition.
Edited by Georgi Popov, Bruce K. Lyon, and Bruce Hollcroft.
© 2016 John Wiley & Sons, Inc. Published 2016 by John Wiley & Sons, Inc.

17.3 RISKS TO AN ORGANIZATION

The purpose of an ORA is to identify plausible risks scenarios capable of business interruption, damage to an organization's reputation, or other catastrophic level consequences. In essence, it is the management of risks to the organization itself. This is sometimes referred to as total risk or enterprise risk management (ERM). The concept of ERM was developed in the 1990s and generally refers to the management of all risks, both negative and positive from internal and external sources. From their website, *RIMS, the Risk Management Society*™ defines ERM in the following statement (RIMS, The Risk Management Society, 2015):

> *Enterprise Risk Management ("ERM") is a strategic business discipline that supports the achievement of an organization's objectives by addressing the full spectrum of its risks and managing the combined impact of those risks as an interrelated risk portfolio. ERM represents a significant evolution beyond previous approaches to risk management in that it:*
>
> 1) *Encompasses all areas of organizational exposure to risk (financial, operational, reporting, compliance, governance, strategic, reputational, etc.);*
> 2) *Prioritizes and manages those exposures as an interrelated risk portfolio rather than as individual "silos";*
> 3) *Evaluates the risk portfolio in the context of all significant internal and external environments, systems, circumstances, and stakeholders;*
> 4) *Recognizes that individual risks across the organization are interrelated and can create a combined exposure that differs from the sum of the individual risks;*
> 5) *Provides a structured process for the management of all risks, whether those risks are primarily quantitative or qualitative in nature;*
> 6) *Views the effective management of risk as a competitive advantage; and*
> 7) *Seeks to embed risk management as a component in all critical decisions throughout the organization.*

Risks come from internal sources as well as those outside the organization. For example, the US Department of Homeland Security document entitled *Risk Management Fundamentals* published in April 2011, describes the sources of risk in the statements in the following:

Internal Sources of Risk

Risks impacting organizational effectiveness arise from both internal and external sources. Examples of internal sources are issues such as financial stewardship, personnel reliability, and systems reliability. Organizations across government and the private sector are all subject to these types of internal risks. These internal risks have the potential to derail effective operations and adversely affect mission accomplishment. A comprehensive approach to risk management serves to identify weaknesses and assists in creating internal systems and processes that minimize the potential for mission failure.

External Sources of Risk

Many organizations have additional risks to manage that are caused by external factors. Examples include global, political, and societal trends, as well as hazards from natural disasters, terrorism, malicious activity in cyberspace, pandemics, transnational crime, and manmade accidents. It is these hazards and threats that caused the Nation to make a significant commitment in homeland security, and it is important that the risks from external threats remain at the forefront of consideration for homeland security organizations.

In this text, the term "operational risk" is used to describe undesired risks resulting from hazards in the workplace. However, organizations face risks that are derived from sources other than those presented by workplace hazards. The Institutes, an organization providing professional education for the risk management and property–casualty insurance industry, categorizes risk into four groups or "quadrants of risk." These quadrants of risk are described in the following.

- *Hazard risks* – arise from property, liability, or personnel loss exposures and are generally the subject of insurance
- *Operational risks* – arise from people or failure in processes, systems or controls, and information technology, which generally fall outside of hazard risks
- *Financial risks* – arise from the effect of market forces on financial assets or liabilities and include market risk, credit risk, liquidity risk, and price risk
- *Strategic risks* – arise from trends in the economy and society, including changes in the political, economic, and competitive environments, as well as demographic shifts.

The Institute states that these risk classifications are general and not intended to cover every risk faced by an organization. Many risks overlap, and it is advisable that an organization develop its own risk classifications to suit its specific needs or situation.

To be clear, not all risks are negative. Some are desirable and necessary for the success of an organization. Investments, opportunities for growth through acquisitions and mergers, new product lines and services, expansion into global markets, and development of technology all present risks that have a potential "upside" as well as downside. All things involve risk. The decision to take a specific risk is based on an organization's desire to achieve an objective, the perceived likelihood and consequence of the risk, and management's risk tolerance level.

17.4 ORGANIZATIONAL RISK MANAGEMENT

Risks come in all sizes ranging from those that have the potential to impact an entire organization, individual divisions, facilities, systems or processes, operations and projects, down to individual workers. An operational risk management system (ORMS) should encompass all levels, especially at the organizational level, and requires senior management leadership, commitment, and involvement in the process. The American National Standard, ANSI/ASSE Z690.2-2011, Risk Management Principles and Guidelines (National Adoption of ISO 31000:2009) in its introduction summarizes the importance of managing risk within an organization in the following:

Organizations of all types and sizes face internal and external factors and influences that make it uncertain whether and when they will achieve their objectives. The effect this uncertainty has on an organization's objectives is "risk."

All activities of an organization involve risk. Organizations manage risk by identifying it, analyzing it and then evaluating whether the risk should be modified by risk treatment in order to satisfy their risk criteria. Throughout this process, they communicate and consult with stakeholders and monitor and review the risk and the controls that are modifying the risk in order to ensure that no further risk treatment is required. This standard describes this systematic and logical process in detail.

Risk assessment is at the heart of the risk management process. This is illustrated in Figure 17.1 reprinted with permission from the ISO 31000/ANSI/ASSE Z690.2-2011 standard. When considering the various process elements of risk management, it becomes apparent how significant the "risk assessment" component is to the process.

Most organizations purchase insurance or self-insure to cover their property (buildings, equipment, vehicles, aircraft, materials, product, and other assets) and casualty (worker compensation for employee injuries and illnesses, general liability, products liability, employer practices liability, directors and officers, etc.) losses. Many organizations consider the purchasing of insurance as their "risk

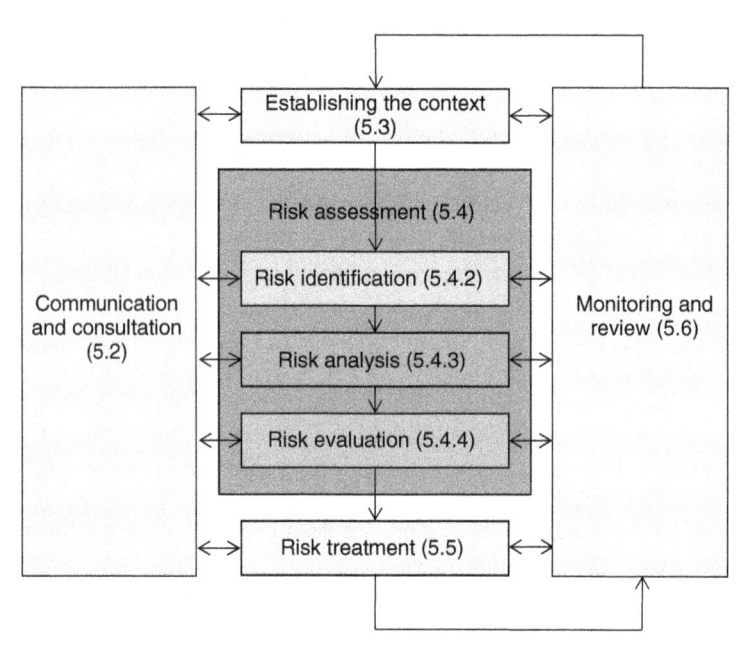

Figure 17.1 Risk Management Process. *Source:* Reprinted with permission from ISO 31000/ANSI/ASSE Z690.2-2011 (Courtesy of the American Society of Safety Engineers)

management plan." Insurance is only *one* option in the risk treatment (5.5) phase of risk management and does not treat all risks.

To truly manage risk, risk must be assessed so that it can be treated and managed to an acceptable level. As illustrated in Figure 17.1, a major component of the process involves risk identification, analysis, and evaluation, otherwise known as risk assessment. An ORA follows this process to manage risks that are capable of significant impacting the viability of an organization. The ISO 31000/ANSI Z690 Risk Management Standard series is the "blue print" for managing risk and is highly recommended by the authors to risk managers, safety professionals, insurance professionals, and other stakeholders.

17.5 KEY DEFINITIONS IN ORGANIZATIONAL RISK

Several definitions are important to review in the context of this chapter. The American National Standard, *ANSI/AIHA/ASSE Z10-2012, Occupational Health and Safety Management Systems*, defines an "organization" as follows (ANSI/AIHA/ASSE Z10-2012):

> **Organization** - *A public or private company, corporation, firm, enterprise, authority, or institution, or part or combination thereof, whether incorporated or not, that has its own management functions. This can consist of one or many sites or facilities.*

Organizations perform functions that produce goods or services to gain profit and market share (for-profit organization) or to benefit society in a tangible way (nonprofit organization). If an organization is unable to achieve its objectives, it is not sustainable. Operational risks are one several types of risk that organizations face. It is defined in the following:

> **Operational risks** *are defined as the risk of loss resulting from inadequate or failed processes, people and systems or from external events. This definition includes legal risk, but excludes strategic and reputational risk (Global Association of Risk Professionals).*

There are four primary causes of operational risk that can occur when there are operational failures due to the following: (1) people/human factors, (2) processes, (3) systems, or (4) external events. Generally, these represent macrolevel losses that have a greater potential for impacting the organization's ability to achieve its business goals. Examples may include business interruption, natural disasters, security breaches, environmental releases, fires and explosions, product recalls and liability, and loss of reputation to mention a few.

Terms that are used with organizational risk include assets, vulnerabilities, threats, exposures to hazards, and business continuity. "Business continuity" is defined as "the strategic and tactical capability of the organization to plan for and respond to incidents and business disruptions in order to continue business operations at an acceptable predefined level." *(ISO 22301:2012)*. Several other key terms are defined in the following to provide meaning for their use in the context of organizational risk:

Asset - *Something valuable that an entity owns, benefits from, or has use of, in generating income or to provide a service to society. Examples include employees and management, customers and vendors, property and buildings, liability, income, technology and information, and reputation. (Adapted from Businessdictionary.com)*

Vulnerability - *Degree to which an asset is susceptible to harm, degradation, or destruction by being exposed to a hazard (adapted from Businessdictionary.com). A weakness of an asset that can be exploited by one or more threat agents. Vulnerability refers to the security flaws in a system that allow an attack to be successful (Rausand, 2011).*

Threat - *Often used in security related concerns, a threat is an indication of an approaching or imminent menace. A threat is a negative event that can cause a risk to become a loss, expressed as an aggregate of risk, consequences of risk, and the likelihood of the occurrence of the event. A threat may be a natural phenomenon such as an earthquake, flood, storm, or a man-made incident such as fire, power failure, sabotage, etc. (Adapted from Businessdictionary.com.) Anything that might exploit a vulnerability of an asset. Examples include arson, theft, cyber-attacks, sabotage and terrorism (Rausand, 2011).*

Exposure - *State or condition of being unprotected and open to damage, danger, risk of suffering a loss in a transaction, or uncertainty. Examples of exposure to hazards include natural hazards, fire and explosion, spills or releases, process breakdowns, utility failures, transportation or distribution disruption, human error, intentional acts, regulatory and liability. (Adapted from Businessdictionary.com)*

Hazard - *Insurance context: Condition or situation that creates or increases chance of loss in an insured risk, separated into two kinds (1) Physical hazard: physical environment which could increase or decrease the probability or severity of a loss. It can be managed through risk-improvement, insurance policy terms, and premium rates. (2) Moral hazard: attitude and ethical conduct of the insured. It cannot be managed but can be avoided by declining to insure the risk. Workplace safety context: Dangerous event or situation that may lead to an emergency or disaster. It could also be a biological, chemical, or physical agent in (or a property of) an environment that may have an adverse health effect, or may cause injury or loss. (Adapted from Businessdictionary.com)*

As previously defined, risk assessment at the organizational level is a systematic process for identifying, analyzing, and evaluating risks or events that could affect the achievement of objectives. Assets or "things of value" within an organization can be affected when risk events intersect and disrupt the organization's ability to function normally.

17.6 ASSESSING ORGANIZATIONAL RISK

When setting out to define an assessment process for an organization, a firm understanding of relevant business goals and objectives is necessary. The assessment scope should reflect these objectives to ensure the resulting risk assessment plan is relevant to the organization's needs. Business objectives are defined at various levels of an organization such as by division, location, operation, or other segment. An analysis of an organization's strengths, weaknesses, opportunities, and threats otherwise known as "SWOT" is often times conducted to identify critical aspects of risk from both internal and external sources. Objectives are commonly laid out in an organization's annual reports, business unit strategic plans, presentations to analysts, functional unit charters, project/investment plans, and management documents (PricewaterhouseCoopers, 2008).

In the "Horizon Scan 2015 Survey Report" (BCI, 2015), a survey of business continuity professional worldwide was conducted by the *Business Continuity Institute* (BCI) and the *British Standards Institute* (BSI). According to the pole, the top threats (risks) facing organizations are (1) cyberattack, (2) unplanned IT and telecommunication outages, (3) data breach, (4) interruption of utilities, (5) supply chain disruption, (6) security incidents, (7) adverse weather, (8) human illness, (9) acts of terrorism, and (10) fire. Other risk trends and concerns highlighted in the report include the loss of key employees, new regulations and increased regulatory scrutiny, prevalence of Internet-dependent services, increasing supply chain complexity, political change, changing consumer attitudes and behavior, and social unrest. Notice that many of these risks or threats are from external sources.

TABLE 17.1 Business Assets and Risks Example

Assets at Risk	Risks to Assets

• Reputation and good will
• Market share
• Quality of products and services
• Opportunity for growth
• Customers
• Financial stewardship and credit
• Operational stability
• Physical assets
• People
• Key employees
• Ethics and business practices
• Legal and regulatory compliance
• Supply chain
• Distribution
• Environment
• Security
• Information
• Technology
	• Internal
	  ○ Fraud and unethical practices
	  ○ Fiduciary risk
	  ○ Legal liability
	  ○ Employer practices
	  ○ Product failures/liability
	  ○ Business interruption
	  ○ Loss of key employees
	  ○ Catastrophic or high-profile incidents
	  ○ Workplace violence and security
	  ○ Fire and explosion
	  ○ Sabotage and vandalism
	  ○ Noncompliance
	• External
	  ○ Interruption to utility supply
	  ○ IT/telecommunication outage
	  ○ Transport network disruption
	  ○ Natural disasters
	  ○ Terrorism
	  ○ Kidnap and ransom
	  ○ Cyber attack
	  ○ Data breaches
	  ○ Supply chain disruption
	  ○ Contractors and vendors
	  ○ Market conditions
	  ○ Loss of large customer
	  ○ Pandemic
	  ○ Social/civil unrest
	  ○ War/conflict

An example list of "assets at risk" and the "risks to assets" is presented in Table 17.1. This list is not considered complete; however, it provides a starting point. Risks can also be further defined and broken into categories such as accidental, intentional or deliberate, natural, man-made, or other desired groupings.

Several recent events illustrate the scale and impact these risk can have on an organization. For instance, in November of 2014, Sony, a global electronics and entertainment corporation, experienced a significant cyberattack (Richwine, 2014). The attack resulted in the release of sensitive and confidential data, scripts from upcoming movie projects, and films that were then downloaded illegally from the Internet. The attack caused an IT outage forcing employees to work without computers for a period of time. The estimated cost to Sony was around $100 million. A second example, the Ebola breakout, began in 2014 and continued into 2015. According to the US Centers for Disease Control and Prevention (2015), the Ebola epidemic was the largest to date, causing more than 8000 deaths and 20,000 infections over a 10-month period. A recent World Bank (United Nations, 2014) report estimates that the economic cost of the outbreak may exceed $500 billion. It is also highly likely to depress economic growth in the worst-affected countries (McKenna, 2015).

Risks such as these can threaten an organization's existence. However, organizational risks can be avoided and managed if properly assessed. For example, creating an effective business continuity and disaster recovery plan is a proactive risk mitigation measure that an organization can take to minimize disruption and damage in such events. To determine what risks or threats face an organization and the measures needed, an ORA process is needed to map out the top risks.

As in more focused, localized, or specific tasks and operations, a means of assessing and managing broader organizational risks is necessary. An organizational-level risk assessment is a coordinated effort of identifying critical process and assets in an organization, potential exposures and consequences, and needed controls.

There are a number of methods used to analyze risks at an organizational level, most of which are based on scenario analysis. One of those methods, listed in *ISO 31010/ANSI Z690.3-2011* (ANSI/ASSE Z690.3-2011), *Risk Assessment Technique Annex B*, is Business Impact Analysis or Assessment (BIA). As described by the standard, the BIA method "analyzes how key disruption risks could affect an organization's operations and identifies and quantifies the capabilities that would be needed to manage it." The process steps for BIA are similar to other ORA methods and include the following:

1. Determine key processes and outputs of the organization to determine their criticality.
2. Determine consequences of disruption on identified critical processes in financial and/or operational terms for a defined unit of time.
3. Identify interdependencies with key internal and external stakeholders within the supply chain, production, and distribution.
4. Determine current state of controls and resources and additional controls and resources necessary to manage risk to an acceptable level.
5. Identify existing and/or needed alternatives and backup resources for disruptions of critical processes/assets.
6. Determine maximum acceptable outage (MAO) time for each process based on the identified consequences and critical functions necessary. The MAO represents the maximum period of time the organization can tolerate the loss of capability.
7. Determine recovery time objectives (RTO) for any specialized equipment or technology. The RTO represents the time period the organization expects to recover essential specialized equipment or technology.
8. Verify preparedness level of critical processes to manage a disruption.

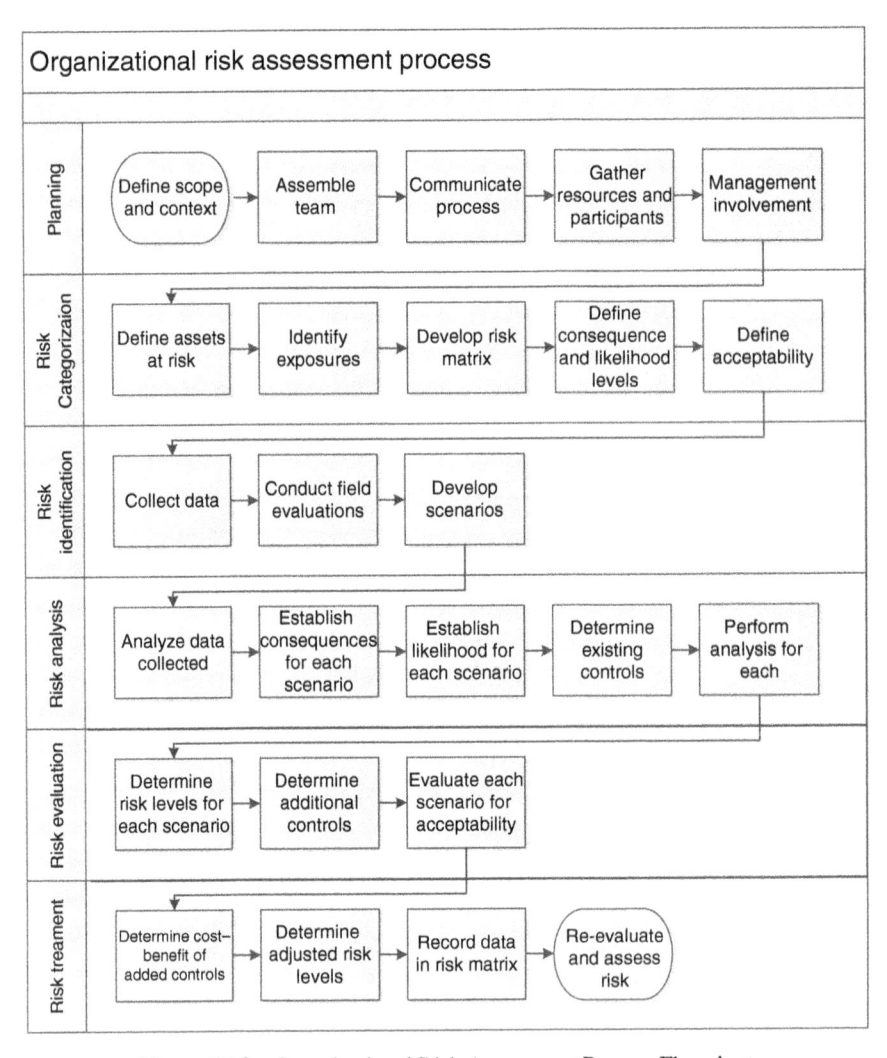

Figure 17.2 Organizational Risk Assessment Process Flowchart

The BIA steps are expanded into the following ORA method described in the following section. Figure 17.2 illustrates the sequence of steps used in the ORA method described in this chapter.

(1) *Plan ORA* – The process begins with establishing the scope and context, the risk assessment team, and the organization's key participants.
   a. *Assemble team* – A qualified facilitator and team are selected to conduct the ORA. The team should be knowledgeable in the risk management process, risk assessment techniques, gathering data, interviewing personnel, and conducting physical surveys. Experienced consultants, insurance professionals, and other outside team members may facilitate or be involved in the process depending upon the complexity and context of the ORA. Team members will need access to the appropriate individuals in the organization. This will likely include those with knowledge of the operations, financials, supplier relationships, customer demands, insurance program, legal status, human resources and employment practices, quality assurance programs and products liability, environmental regulations, facilities maintenance and fire protection, occupational safety and health, security, IT technology and cybersecurity,

ASSESSING ORGANIZATIONAL RISK

transportation and distribution, and other pertinent areas. The organization's risk manager and safety, health, and environmental manager are key components in an ORA.

b. *Establish ORA perimeters* – The team sets perimeters for the ORA including the purpose and scope, responsibilities for team members, and general approach to the ORA. These perimeters are set in conjunction with management and the ORA team and should include an outline of the organization's assets to be analyzed. Supporting documents and data such as annual reports, lists of major facilities, loss prevention reports, loss experience, etc. should also be included. Coordination is required to standardize data gathering, report writing, scenario development, and other aspects of the risk assessment. The ORA worksheets to collect data from divisions, facilities, or operations are developed. Possible questions to consider in determining the ORA perimeters are presented in Table 17.2.

c. *Communicate process* – The ORA team, risk manager, and senior management meet to review the proposed plan and reach agreement on the approach. Senior management representatives may include the COO, CFO, corporate legal, administration, human resources, and other management representatives.

d. *Gather participants and resources* – Key participants in the ORA will generally include the risk management department and a primary contact for each affected division or major asset category to provide necessary information and resources for the assessment. An orientation meeting with the key participants is conducted to review the goals of the ORA, methodology to be used, resource requirements, and time frames and commitments. Agreement on the scope, data needed, and format of the ORA is made. The ORA worksheets to collect data from divisions, facilities, or operations are shared with the group.

e. *Obtain management agreement* – An ORA is a big undertaking and requires complete support by senior management. It is important to know what management expects from the ORA. If expectations do not match the established purpose and scope, then a mutual revision of the process is needed. For instance, if management expects the ORA to provide information to assist in determining insurance deductible levels, and the scope of the ORA only seeks to identify and assess catastrophic events, it may be necessary to revise the scope of the project or adjust management expectations. Either way, it is important to reach a clear agreement with management on the ORA.

TABLE 17.2 Questions to be Asked of Management

- How does Management view the risk management function?
- What risk potentials or concerns does Management worry about most?
- What is the CEO's, CFO's, or organization's tolerance for risk?
- What is a major or significant event, and how is it measured?
- What are the critical processes, functions, or assets of the organization?
- What business dynamics are occurring that will change the organization's risk profile?
- What critical risk management business decisions does Management anticipate in the next two years?
- What information does Management need to make those decisions?
- How variable has the organization's risk been historically?
- Who in senior management is going to champion the risk assessment process and initiate the communication to the operating groups?
- What are the best sources of information for risk identification and analysis within the company?
- How does Management measures risk (change in earns per share, debt/equity ratio increase, falling short of analysts' earnings, etc.)?

(2) *Risk categorization* – Following the planning stage, the ORA team defines the specific "exposures" which pose risk to company assets and the potential severity of consequences.

  a. *Define "assets at risk"* – Specific assets to be included in the ORA previously discussed in the planning phase are formalized. The broad categories of assets at risk are defined and may include categories such as employees, customers/public, contractors, buildings, computers, telecommunications, transportation equipment, environment, and financial assets.

  b. *Define hazard exposures* – Similar to defining the assets at risk, define the specific hazard exposures to be evaluated. They should be defined within broad categories such as natural hazards, fires and explosions, machinery breakdown, releases and spills, transportation accidents, human error, deliberate acts, and errors and omissions. Within each of these broad categories, specific exposures should also be identified. For example, within natural exposures category, the more specific exposures of hurricane, tornado, rainstorm, flood, earthquake, subsidence/landslide, lightning, freezing, and ice would be included.

  c. *Define consequences* – Considering the assets at risk, the team defines the types of consequences that can occur as a result of the exposures. Both insurable and noninsurable consequences are included. Codes are assigned to each consequence type as illustrated in Table 17.3.

  d. *Prepare risk profile matrix* – To help communicate risks to stakeholders and decision makers, a risk profile matrix listing the identified assets and their exposures is prepared as illustrated in Figure 17.3.

  e. *Define risk levels* – The team defines severity of consequence levels and likelihood or probability levels to be measured for each of the hazard exposures. The matrix can be expanded into submatrices based on company structure or the type of consequence to be measured if desired. A simple set of risk level definitions is provided in Tables 17.4 and 17.5.

TABLE 17.3 Consequence Code Table Example

Code	Consequence
AL	Aircraft liability
BI	Business interruption
BM	Boiler and machinery breakdown
CP	Civil and criminal penalties
DO	Director and officer liability
EE	Extra expense
EN	Environmental liability (off-site)
FL	Fiduciary liability
OC	On-site contamination
PD	Property damage
PI	Damage to public image/goodwill
PL	Product liability
PR	Political risk
RC	Regulatory consequence
RD	Loss of research data
TP	Third party liability
WC	Fatalities and serious incidents
$	Financial impact

ASSESSING ORGANIZATIONAL RISK

Exposures	Hazard Exposure	Reputation/goodwill	Mgmt and employees	Customers	Suppliers	Distributors	Inventory	Property	Key equipment	IT and communication	Technology, R & D	Cash and finacials	Community
Natural	Hurricane												
	Tornado												
	Earthquake												
	Extreme weather												
Release	Chemical spill												
	Atmospheric release												
	Radioactive release												
Fire	Fire sprinker damage												
	Explosion/implosion												
Operations	Human error												
	Machinery breakdown												
	Process breakdown												
	Structural failure												
Management	Legal/liability												
	Non-compliance												
	Employment practices												
	Errors and omissions												
Supplier	Key supply failure												
	Supply quality failure												
	Delivery delay												
Distribution	Internal interruption												
	External interruption												
	Loss of distributor												
	Distribution delay												
Product	Design flaw												
	Misuse												
	Failure to perform												
Deliberate	Cyber attack												
	Sabotage/vandalism												
	Terrorism/threat												
	Theft/fraud												
	Kidnap/ransom												
	Strike/civil unrest												
Change	Regulatory												
	Societal												
	Technology												

Assets at Risk

Figure 17.3 Risk Profile Matrix

 f. *Prepare organizational risk matrix* – Using the defined risk levels, an organizational risk matrix is developed as shown in Figure 17.4.

 g. *Define acceptable risk level* – The organization's acceptable risk level is defined by management and included into the risk matrix. For example, management may decide that property damage losses less than $1 million, business interruption losses of less than 1 week are acceptable levels of risk. Acceptable risk levels for each consequence type should be established. The number of scenarios and the assessment effort expands significantly as the consequence cutoff levels are lowered. MAO time for critical functions and RTO for needed specialized equipment and technology is defined by the team with the help of management and other stakeholders.

(3) *Risk identification and data collection* – The purpose is to identify scenarios and collect data, which will help quantify probability and severity.

TABLE 17.4 Severity of Consequence Levels Example

Severity of Consequence Level	
Catastrophic	Total shutdown of business, communication, or operations; loss of a facility or building; loss of most or all key employees; or other loss that results in more than $25 million
Critical	Partial shutdown of business, communication, or operations; interruption of business for more than 1 week; damage to the organization's reputation; loss of vital or sensitive information; or other loss that results in more than $10 million
Marginal	Minor shutdown of a facility, operation, distribution, or service; partial interruption of business for 1 day; damage to the organization's reputation; limited number of assets damaged; or other loss that results in more than $1 million
Minor	Little or no impact on operations, distribution, or service; minimal damage to the organization's reputation; or loss less than $1 million
Negligible	No impact on operations, distribution, or service; no damage to organization's reputation; limited loss less than $100,000

TABLE 17.5 Likelihood of Occurrence Levels Example

Likelihood of Occurrence Level	
Very Likely	Expected to occur within the time frame contemplated by the objective or likely to occur once in less than 1 year
Likely	Likely to occur within the time frame contemplated by the objective or likely to occur once in 1–9 years
Moderate	Moderate likelihood that the threat will occur or likely to occur once in 10 years or more
Unlikely	Unlikely to occur within the time frame contemplated by the objective; likely to occur once in 100 years or more
Remote	Remotely possible for event to occur; not likely to occur more than once in 1000 years or more

   a. *Collect data* – The ORA team collects data according to the defined methodology gathering available data by document reviews, physical observations, and interviews with key personnel. The data is gathered and documented on the collection forms.
   b. *Scenario development* – Using the data collected, the team brainstorms to develop plausible worst-case scenarios. For each scenario, a description of the hazard exposures, the assets affected, and resulting consequences are developed and entered into the ORA worksheet. The likelihood of occurrence and the severity of the consequences are quantified as much as possible. Each scenario will include existing control strategies and a list of proposed preventive and mitigating alternatives. An estimate of the cost for needed preventive and mitigating alternatives is desirable to allow for cost–benefit calculations. There are numerous computer models and computerized databases that can be used in certain scenarios such as the natural hazards information system, or CAMEO/ALOHA gas dispersion models, vapor cloud explosion models, natural hazards models, and others.
(4) *Risk analysis* – The team analyzes the hazards to determine the risks for each scenario using the information gathered.
   a. *Analyze data* – An ORA risk assessment worksheet (such as those presented in Figures 17.5, 17.6, and 17.8) is used to document and analyze the data. The analyzed data is also logged

ASSESSING ORGANIZATIONAL RISK 381

		Severity				
		Negligent 1	Minor 2	Marginal 3	Critical 4	Catastrophic 5
Likelihood	Very likely 5	6	7	8	9	10
	Likely 4	5	6	7	8	9
	Moderate 3	4	5	6	7	8
	Unlikely 2	3	4	5	6	7
	Remote 1	2	3	4	5	6

Figure 17.4 Organizational Risk Matrix Example

into an organizational risk register or profile. For example, the asset category "computers and telecommunications" might be broken up into subcategories. Separate matrices for property damage, injuries and fatalities, and business interruption, as well as automated techniques such as computer and database programs can be applied. The decision will depend on the amount of data to be analyzed and the organization's needs.

b. *Establish consequence measures* – The team defines the organizations severity of consequence levels based on risk tolerance. Table 17.4 provides an example of a severity of consequence levels table. Consequences for each scenario defined earlier are translated into estimated dollar amounts, which can be aggregated and compared. For example, 1 week of downtime may be converted to $5 million and subsequently added to a $2 million property damage loss and $1 million extra expense loss to give a total loss of $8 million.

c. *Establish likelihood measures* – The team determines likelihood of occurrence based on historical data, frequency of exposure, duration of exposure, and other available data. Table 17.5 provides an example of a likelihood of occurrence table. For each scenario, a likelihood of occurrence measure is established using the defined likelihood levels.

d. *Determine existing controls* – Existing control strategies for the identified hazard exposures are entered into the ORA risk assessment worksheet. Risk control and risk financing strategies designed to prevent or mitigate the hazard exposures in the scenario are considered in determining the "current state" severity and likelihood risk levels.

e. *Analyze scenarios* – For each scenario, the team performs analyses using the selected ORA risk assessment worksheet. The analyses include the following:

- The cumulative losses for each "asset at risk"
- Scenarios grouped by the type of hazard, which initiated loss
- A summary of loss scenarios that results in a particular type of consequence, such as business interruption
- Individual scenarios ranked from highest to lowest depending on their severity
- Individual scenarios ranked based on their likelihood of occurrence

Organizational Risk Assessment Worksheet

Company: ABC Products	**Scenario Name:**	**Major Hurricane**
Division: Small Motor Division	Major class storm enters harbor, causing major flooding (loss of containment), loss of power, and possibly damage to processes for approximately 1 month. The Boston area averages a tropical storm event approximately every 1.75 years. According to NOAA, the Commonwealth has a 6 – 30% chance of a tropical storm or hurricane affecting the area each year	
Location: Boston, MA – Plant #2		
Hazard: Hurricane		
Category: Natural disaster		
Scenario #: 2		
Date: August 14		
By: J.Smith; K. Boyd; G. Murry		

Asset at Risk	Consequence	Estimated $ Loss
Owned property	Property damage	$22,000,000
	Material and finished product loss	$4,000,000
Production and distribution	Business interruption	$5,000,000
	Lost sales	$1,000,000
Environment	Environmental liability	$1,500,000
	Extra expense	$1,000,000
	Regulatory non-compliance	$500,000
	Total estimated loss	**$35,000,000**

Existing Control Strategies	Needed Control Strategies	Further Control Strategies
Process designed for high winds; emergency response plan; $2M BI and $10M property insurance limits	Increased insurance limits; include backup production arrangements in business continuity plan and disaster recovery plan; environmental management plan	Relocate critical facilities inland away from flood zones. Further assess natural disaster risks, business continuity and disaster recovery plans at all facilities

With Existing Controls		With Needed Controls		Further Control Strategies	
Total loss	$35M	Estimated total loss	$10M	Estimated total loss	
Severity	Very High	Severity	Moderate	Severity	
Likelihood	High	Likelihood	High	Likelihood	
Risk level CS	9	Risk level FS	7	Risk level FS2	
Control costs	$500K	Control costs	$1.5M	Control costs	
		Benefit–/cost	25	Benefit–/cost	

Note: Selected controls add cost of $1M; however expected loss is reduced by $25M producing a benefit–/cost ratio of 25. It was not considered feasible to relocate the plant at this time; however a formal proposal to further reduce risk is required within 6 months

Figure 17.5 Hurricane Scenario

- Combination of the probability of occurrence and the severity of the loss using a coding system and a weighing of the probability, severity, and cost of preventive measures. Multiplying probability by severity produces annualized risk. Loss should be segregated first using a risk decision matrix and then employ annualized risk concepts. Without this step, too much effort will be placed on high frequency, low severity losses. Risk should be evaluated at the upper end of the risk decision matrix first.

(5) *Risk evaluation* – The team evaluates risk levels for each scenario to develop a strategy for eliminating, reducing, or transferring unacceptable risk.

    a. *Evaluate risk levels* – For each scenario, the team evaluates risk levels for each hazard exposure to determine the needed for additional control strategies.

    b. *Evaluate additional controls* – For each scenario, the team evaluates the current control measures to determine if additional measures are required. Control strategies such as avoidance, elimination, substitution, mitigation, risk transfer, and risk financing options

ASSESSING ORGANIZATIONAL RISK

Organizational Risk Assessment Worksheet

Company: ABC Products	Scenario Name:	Cyberattack
Division: Corporate	\multicolumn{2}{l|}{Major cyber attack causing sensitive data breach and collapse of all internal and external communication systems via the intranet and website. Confidence in the organization's security is damaged with clients and prospects as well as general public due to the breach of personal data. Several lawsuits result.}
Location: Corporate Office – Network		
Hazard: Cyber Attack		
Category: Deliberate Acts		
Scenario #: 5		
Date: August 14		
By: J. Smith; K. Boyd; G. Murry		

Asset at Risk	Consequence	Estimated $ Loss
IT and Communication	Business interruption	$25,000,000
	Loss of sensitive data	$5,000,000
	Liability	$4,000,000
	Total estimated loss	**$34,000,000**

Existing Control Strategies	Needed Control Strategies	Further Control Strategies
Nominal IT security system and staff. Insurance coverage limits for minimal amounts. Controls have not kept up with the increase in reliance on IT communication	Implement robust cyber security system and additional staff based on vulnerability assessment. Increase Insurance coverage limits	Continue to assess and update IT security and cyber attack vulnerability

With Existing Controls		With Needed Controls		Further Control Strategies	
Total loss	$34M	Estimated total loss	$5M	Estimated total loss	
Severity	Very high	Severity	Moderate	Severity	
Likelihood	High	Likelihood	Moderate	Likelihood	
Risk level CS	9	Risk level FS	6	Risk level FS2	
Control costs	$1M	Control costs	$3M	Control costs	
		Benefit-/cost	14.5	Benefit-/cost	

Notes: Install robust IT security system with a cost of $2M reducing likelihood and severity of a cyber attack by $29M

Figure 17.6 Cyber Attack Scenario

(i.e., commercial insurance, deductible level, self-insurance and retention, etc.) and their feasibility are considered along with estimated costs, effectiveness, and reliability. Recommended control strategies to eliminate or reduce risk to an acceptable level are made and entered into the worksheet. Cost/benefit analyses are performed to help select the appropriate controls. For each risk category a priority list of recommendations is established. Alternative strategies for each type of risk are also identified. For example, it may be possible to significantly reduce the overall fire risk by completing a small number of major recommendations, which will qualify the organization for highly protected risk insurance coverage at a lower rate.

c. *Evaluate acceptable risk* – The team evaluates scenarios and their current risk control strategies and suggested additional strategies to determine acceptable risk levels. Risks that remain unacceptable will be further assessed and reviewed by management to determine actions needed.

Organizational Risk Assessment Worksheet

Company: ABC Products	**Scenario Name:**	**Hazardous Waste Release**
Division: Chemical Division	Chemical waste sent to hazardous waste disposal site where it is improperly handled and stored. A significant release into soil and city water supply occurs. Treatment Storage Disposal Facility (TSDF) is out of compliance.	
Location: Modesto, CA – Process Plant		
Hazard: Hazardous Waste		
Category: Environmental release		
Scenario #: 15		
Date: August 14		
By: J. Smith; K. Boyd; G. Murry		

Asset at Risk	Consequence	Estimated $ Loss
Community environment	Environmental liability off-site	$15,000,000
	Reputation and goodwill	$5,000,000
Liability/compliance	Environmental remediation (third party)	$5,000,000
	Total estimated loss	**$25,000,000**

Existing Control Strategies	Needed Control Strategies	Further Control Strategies
Insurance for third party liability from $25,000,000 to $75,000,000 does not provide on-site/off-site cleanup and business interruption	Reduce hazardous waste generated. Due diligence on waste handlers and TSDFs to ensure properly licensed, insured, and in compliance	Move towards elimination of all hazardous waste by substituting non-hazardous chemical use

With Existing Controls		With Needed Controls		Further Control Strategies	
Total loss	$25M	Estimated total loss	$8M	Estimated total loss	
Severity	Very high	Severity	Moderate	Severity	
Likelihood	Likely	Likelihood	Moderate	Likelihood	
Risk level CS	9	Risk level FS	6	Risk level FS2	
Control costs	$500K	Control costs	$1M	Control costs	
		Benefit–/cost	15	Benefit–/cost	

Notes: Reduced hazardous chemical use, due diligence of TSDF/waste handling companies, and increased environmental liability coverage reduce risk significantly. Benefit ($17M in savings) divided by added cost ($1M in due diligence) equals a return of $17 for each $1 spent

Figure 17.7 Hazardous Waste Release Scenario

(6) *Risk treatment* – The team develops action plans for the implementation of additional control strategies.

  a. *Determine cost–benefit analysis* – Depending on the complexity of the risk and proposed control strategies, a cost–benefit analysis is used in the final selection of strategies. The estimated total dollar cost of each risk control strategy (installation, hardware, software, administrative, maintenance, or insurance), and its estimated benefits in dollars savings are calculated by a simple "benefit divided by cost" equation. Strategies are then ranked according to their cost–benefit analysis for implementation.

  b. *Determine future state risk levels* – Use the selected control strategies, each hazard exposure's risk level is adjusted to reflect the estimated "future state" and entered into the worksheet.

  c. *Develop risk register* – A risk register or profile is created to list assets at risk, hazard exposures, consequences, and risk levels.

d. *Complete risk matrix* – The team enters the risk levels in the matrix using color-coding to highlight different risk levels.
   e. *Present findings to management* – A concise executive summary followed by a more detailed management-oriented summary of findings, conclusions, and recommendations is prepared and presented to management. Supportive data may be necessary for more complex risks. The goal of the presentation is to inform management in making the best decisions regarding risk management of the organization.
(7) *Reassess Risks* – The ORA team continues to reassess risks to as conditions change, new risks introduced, and control measures change to ensure risks remain acceptable to the organization.

Case Study

To demonstrate how an ORA process may be employed, the following "condensed" case study is presented. The previously illustrated risk matrix (Figure 17.4), risk level definitions (Tables 17.4 and 17.5), and hazard/asset matrix (Figure 17.3) are used in this example.

An astute risk manager of a large international manufacturing company raised concerns about potentially significant operational risks that the organization faced. The company's SH&E leader recommended that a broad, company-wide risk assessment be conducted to identify, assess, and prioritize risks needing further control in an effort to properly manage their risks.

To identify and understand its operational risks, the company formed a team of experienced and knowledgeable stakeholders to conduct an ORA to identify potential major risks and needed control strategies. The team consisted of the risk manager, SH&E leader, corporate legal, operations director, corporate physical and IT security, human resources director, key insurance representatives, and an experience risk management consultant.

The organization defined a *major event* as one that could result in a $10 million uninsured loss, interruption to business of the corporation or major supplier for more than 1 day, or significantly damages the image and reputation of the company. The risk matrix in Figure 17.4 was used. The organization determined that risks in the "very dark gray" (8 through 10) were unacceptable and required immediate action. Risks in the "dark gray" (7) were considered high priority for treatment within 6 months, and risks in the "medium gray" (6) required a formalized action plan proposal to management within 6 months.

The team performed a review of the organization's critical "assets at risk" and the risk assessment method to be used. A structured brainstorming process was used to identify potential hazard exposures or threats. An assets and hazard exposures matrix (shown in Figure 17.3) was developed to help determine major event scenarios.

The list of potential threats was limited to plausible worst-case scenarios that could result in a *major event*. Fifteen major event scenarios were selected. Data collection worksheets were used by the team to identify critical assets at risk, resulting consequences and their potential magnitude (dollar loss estimate), existing control, needed controls, and possible future controls strategies for each scenario. A final list of major exposure events were identified and assessed. Each major event scenario was documented in an ORA worksheet as shown in Figures 17.5–17.7.

The team completed each of the major event scenario risk assessments and compiled the results in a risk register (partial register is presented in Figure 17.8).

Scenarios with a current risk level (Risk Level CS) of 8 or higher are considered in the unacceptable red zone and required to have risk control and financing strategies to reduce the future state risk level (Risk Level FS).

An executive summary and management report was developed and presented to senior management following the completion of the ORA. Action plans for immediate, short-term, and long-term risk management strategies were developed as a result of the presentation. The organization was better informed of real risks to the company allowing it to make business decisions with confidence.

Organizational Risk Register

Major Event #	Division or Location	Scenario Name	Conseq. Codes	Risk Level CS	Existing Controls	Needed Controls	Risk Level FS
1	Customer Products	Fire	BI, PD, RC, PI, SFI, EL, OC	6	Fire Suppression Systems; Emergency Response Plan team; Process controls to prevent	Secondary controls in some places; Segregated diking for several parts of plant	4
2	Small Motor Divison Boston Plant #2	Major Hurricane	BI, RC, OC, EL		Designed for high winds; Emergency Response; $2M BI and $10M Property insurance limits	Increased insurance limits; Backup Production Arrangements in BCP and DRP; Environmental Management Plan	7
3	Products Group Accounts Receivable	Tornado	AR	6	Evaluate distributors' financial ability to pay should they not receive revenue from current crop	Insurance of A/R for catastrophic loss. Awareness of sales revenue / exposure by agric. region.	5
4	Corporate	Employment Practices	$, PI, DO	6	Seek diverse slate of internal, external candidates. Ensure equal pay for equal work. Create an environment that promotes a diversed workplace.	EPL coverage; compensation to reward hiring and retention of employees.	3
5	Company-wide	Cyber Attack	BI, PI, L		Nominal IT security system and staff. Insurance coverage limits for minimal amounts. Controls have not kept up with the increase in reliance on IT communication.	Implement robust cyber security system and additional staff based on vulnerability assessment. Increase Insurance coverage limits.	6
6	Overseas Group	E & O Currency Control	AR, $	6	None	Political risk insurance. Conduct risk analysis review on budgeted exposure Group-wide by country to preestablish limits of risk acceptable in each country.	4
7	Public Affairs	EPA Non-compliance	RC, PI	7	Continue dialogue with EPA on risk mitigation strategies	Increase political pressure to ensure favorable outcome; bring in specific commodity groups to bring pressure to bear on Agency	5
8	Agricultural Products	Overspray	TP, EE, EN, PL	4	Product Stewardship Programs	Commodity Group Partnering	3
9	Products Group	Fleet Accident	TP, PI, FSI	5	Strict policy regarding abuse of alcohol and company vehicles. MVR checks for all company vehicle drivers.	Incorporate ANSI Z15 Fleet Safety Management System and Interactive Driver Training	3
10	Agricultural Products	EPA Non-compliance	EL, $	6	Follow legislation as it develops, try to plan products around regulatory requirements.	Add additional screening parameters to look at molecule properties similar to those required by regulators.	5
11	Distribution Services	Fire	PD, BI, $, EE, PI	7	Fire suppression system throughout, Emergency response plan and team, Plant safety and fire prevention prog., Building and operation segregation, Process controls, Water treatment plant	Implement modifications identified from process safety management review.	6
12	Customer Products	Failure to Perform	TP, CP, PI, RC	7	Quality systems- ISO at 3 plants, Quality control lab at all plants, Cross contamination policy and procedure at all plants. Separation of processes.	ISO system at all plants, Cross contaminatioon testing on all batches before being shipped. setting acceptable limits of cross contamination.	6
13	Agricultural Products	Process Explosion	FSI, PI, PD, BI, RC	7	Employee training. Experienced employees in test pit. Documented test firing procedures.	Remote test firing facility; explosion containment. Use field service technicians to test boilers.	5
14	Products Group	Oil Release	EL, OC, RC, PI, BI	5	Oil tanks are above ground on concrete pads and include secondary containment. SPCC plans in place at all facilities.	Provide alarm system (monitored) on secondary containment	4
15	Chemical Division Modesto, CA	Hazardous Waste Release	EL, TP, OC, RC		Insurance for third party liability $25,000,000 to $75,000,000, does not provide onsite/offsite cleanup and business interruption.	Reduce hazardous waste generated. Due diligence on waste handlers and TSDFs to ensure properly licensed, insured, and in compliance.	6

Figure 17.8 Organizational Risk Register

17.7 SUMMARY

The need for organizations to understand the broader range of risks that threaten their existence is real. This requires a coordinated effort among key stakeholders to identify their critical operations and assets and the types of potential risks that they face. The ORA should be a beginning step to more in-depth and detailed analyses and assessments of these critical functions.

Organizations that successfully identify, assess, and manage plausible risks that are capable of major disruption or severe damage will succeed. Safety and risk professionals able to guide their organizations in such efforts will undoubtedly increase their own value.

REVIEW QUESTIONS

1. Describe the differences between organizational risk assessments and operational risk assessments.
2. List internal sources of risk to an organization.
3. List external sources of risk to an organization.
4. Identify the four quadrants of risk and list three types of risk within each quadrant.
5. Provide 10 examples of common assets at risk within an organization.
6. List 10 common risks to assets at an organizational level.
7. Briefly describe the process steps within an organizational risk assessment.
8. Provide a list of suggested team members by position and title that would be valuable to an organization risk assessment.

REFERENCES

ANSI/ASSE/AIHA Z10-2012. *American National Standard – Occupational Health and Safety Management Systems.* Fairfax, VA: American Industrial Hygiene Association, 2012.

ANSI/ASSE Z690.2-2011. *American National Standard – Risk Management Principles and Guidelines.* Des Plaines, IL: American Society of Safety Engineers, 2011.

ANSI/ASSE Z690.3-2011. *American National Standard – Risk Assessment Techniques.* Des Plaines, IL: The American Society of Safety Engineers, 2011.

Business Continuity Institute. Horizon Scan 2015 Survey Report. Business Continuity Institute in conjunction with the British Standards Institution, 2015. Available at http://www.bcifiles.com/BCIHorizonScan2015.pdf (accessed December 31, 2015).

BusinessDictionary.com, 2015. Available at: http://www.businessdictionary.com/ (accessed December 31, 2015).

Centers for Disease Control and Prevention. 2014 Ebola Outbreak in West Africa. CDC (online), 2015. Available from http://www.cdc.gov/vhf/ebola/outbreaks/2014-west-africa/ (accessed December 31, 2015).

Global Association of Risk Professionals. Operational Risk Management. No date provided. Available at: http://www.garp.org/media/665968/icbrr-operational0711preview.pdf (accessed December 31, 2015).

ISO 22301:2012. *Societal security – Business Continuity Management Systems – Requirements.* International Organization for Standardization, 2012.

McKenna, M. The Long Tail of Ebola: Depressing African Economic Progress. Wired (online), 2015. Available from: http://www.wired.com/2015/01/long-tail-ebola-depressing-african-economic-progress/.

PricewaterhouseCoopers. A Practical Guide to Risk Assessment: How Principles-Based Risk Assessment Enables Organizations to Take the Right Risks. 2008. Available at: https://web.actuaries.ie/sites/default/files/erm-resources/risk_assessment_guide_pwc_Dec08.pdf.

Rausand, Marvin. *Risk Assessment: Theory, Methods, and Applications*. Hoboken, NJ: John Wiley & Sons, Inc., 2011.

Richwine, L. Cyber Attack Could Cost Sony Studio as Much as $100 million. Reuters (online), 2014. Available from: http://www.reuters.com/article/2014/12/09/us-sony-cybersecurity-costs-idUSKBN0JN2L020141209 (accessed December 31, 2015).

RIMS. The Risk Management Society, 2015. https://www.rims.org/ERM/Pages/WhatisERM.aspx (accessed December 31, 2015).

The Institutes. Risk Management Principles and Practices, Quadrants of Risk: Hazard, Operational, Financial, and Strategic. Retrieved March 18, 2015 from: http://www.theinstitutes.org/comet/programs/arm/assets/arm54-chapter.pdf (accessed December 31, 2015).

United Nations. Ebola: World Bank Reports Economic Impact in Worst-Hit Countries to Exceed $500 million in 2014, 2014. UN News Service (online). Available from: http://www.un.org/apps/news/story.asp?NewsID=49490# (accessed December 31, 2015).

US Department of Homeland Security. Risk Management Fundamentals, Homeland Security Risk Management Doctrine, April 2011. Available at https://www.dhs.gov/xlibrary/assets/rma-risk-management-fundamentals.pdf (accessed December 31, 2015).

18

RISK ASSESSMENT APPLICATIONS IN LEAN SIX SIGMA AND ENVIRONMENTAL MANAGEMENT SYSTEMS

GEORGI POPOV

School of Environmental, Physical & Applied Sciences, University of Central Missouri, Warrensburg, MO, USA

18.1 OBJECTIVES

- Introduction
- Overview and Background
- ISO 14001 Implementation Procedures and Practical Application
- EMS and Implementation of Lean Six Sigma practices
- Practice Exercises/Questions

18.2 INTRODUCTION

Safety, health, and environmental (SH&E) professionals can play a significant role in the development of new environmental management system (EMS) models and implementation of Lean Six Sigma (LSS) practices designed to minimize variation and improve productivity and sustainability efforts of an organization. In this chapter, readers will learn how LSS principles are combined with risk assessment tools, variation reduction, and sustainability methods. In addition, new tools for ISO 14001 implementation will be provided.

SH&E professionals continue to face the challenge of convincing management of the need to optimize safety, health, and environmental (operational risk) management systems and align them with an organization's business objectives. This, in part, requires the development of skills in facilitating and communicating effective risk assessments in an organization. Specifically, professionals should become familiar with the EMS LSS principles and incorporate them into standard business practices within their organization.

Risk Assessment: A Practical Guide to Assessing Operational Risks, First Edition.
Edited by Georgi Popov, Bruce K. Lyon, and Bruce Hollcroft.
© 2016 John Wiley & Sons, Inc. Published 2016 by John Wiley & Sons, Inc.

A major hurdle to the adoption of EMS is the perception that the EMS cost–benefit ratio is unfavorable. For this reason, SH&E professionals should learn to recognize business cost drivers and justify EMS implementation expenditures early in the process of environmental aspects' adoption and product development stages. In this chapter, an EMS model that incorporates LSS, risk assessment, environmental aspects' scores variation reduction, and residual risk reduction is presented, along with a case study that provides a practical demonstration of these concepts.

This chapter presents a value-added project for EMS implementation. Since the standard is so broad, this chapter addresses risk assessment methods, the Plan, Do, Check, Act (PDCA) cycle, a significant aspects' evaluation decision-making tree, LLS tool implementation, and residual risk reduction (Tague, 2004, p. 390).

18.3 ENVIRONMENTAL MANAGEMENT SYSTEMS (EMS)

The International Organization for Standardization's ISO 14000 is a family of standards that address various aspects of environmental management. For instance, ISO 14001:2004 and ISO 14004:2004 focus on EMS. At the time of this writing, ISO 14001:2004 is under review, and the final updated version is expected toward the end of 2015. It defines a set of environmental management requirements for EMS (International Organization for Standardization, a). The standard provides requirements for organizations to protect the environment by implementing pollution prevention (P2) strategies and improve their environmental performance.

One of the difficulties associated with EMS is to establish proper procedures for the determination and evaluation of environmental aspects due to their significant variation. An "environmental aspect" is defined by the US Environment Protection Agency (EPA) as an element of a facility's activities, products, or services that can "interact" with the environment. This chapter provides some methods for measuring and reducing environmental aspects' scores with the use of LSS principles combined with risk assessment tools, variation reduction, and sustainability methods.

The proposed EMS implementation addresses environmental, safety, and health needs in the risk assessment processes to prevent or minimize the work-related hazards and risks associated with the use, storage, maintenance, and disposal of environmental aspects. One of the goals is to educate business managers; engineers; chemical and equipment manufacturers; SH&E professionals; administrators; and workers to understand and implement EMS methods and apply this knowledge and skills to reduce environmental risk and prevent pollution.

18.4 ISO 14001 IMPLEMENTATION

With globalization, a growing number of companies are implementing the ISO 14001, EMS standard. Some of the benefits and incentives envisioned by management are as follows:

- Competitive advantage
- New business opportunities
- Improved environmental performance
- Integration with Health and Safety Management System (ANSI Z10-2012)
- Possible ISO 14001 registration.

A summary of the elements and steps required for a company to successfully implement the EMS requirements includes the following:

- Secure top management commitment early in the process.

- Form an EMS committee. Members of the committee should have a thorough understanding of the ISO 14001 standard.
- Involve affected functions and staff levels in the planning process.
- Develop self-assessment process.
- Identify environmental aspects and impacts.
- Set objectives for environmental risk reduction.
- Build on existing business practices.
- Align business objectives, Health and Safety (H&S) management system, and EMS.

It is essential to realize that an EMS implementation requires significant investment in personnel and time. As stated before, one challenge is to establish proper procedures for the determination and evaluation of the environmental aspects and their variations.

To demonstrate the application, a case study of a fictitious company *ABC Environmental Company* (Environmental & Ecological Services Industry, Primary SIC Codes: 8731-11) is used in the following sections. The methodology includes the development of new tools for ISO 14001 implementation. The objective is to reduce aspects' Significance Scores. The target is to achieve a statistically significant reduction of Significance Scores. Typical Six Sigma tools can be utilized to evaluate aspects' Significance Scores. Control charts and graphs may be developed to demonstrate the effectiveness of the proposed control measures.

An effective EMS implementation requires the involvement of all managers and employees, from top management to the field or production employees depending on the type of the organization. An EMS is a continual business cycle of planning, implementing, reviewing, and improving the processes and activities that any organization undertakes to meet its environmental obligations and continually improve its environmental performance. Like other management systems, EMS is based on the "PDCA" model, which emphasizes on the concept of continual improvement (shown in Figure 18.1).

The PDCA cycle steps and requirements outlined in the ISO 14001:2004 standard are presented in Table 18.1.

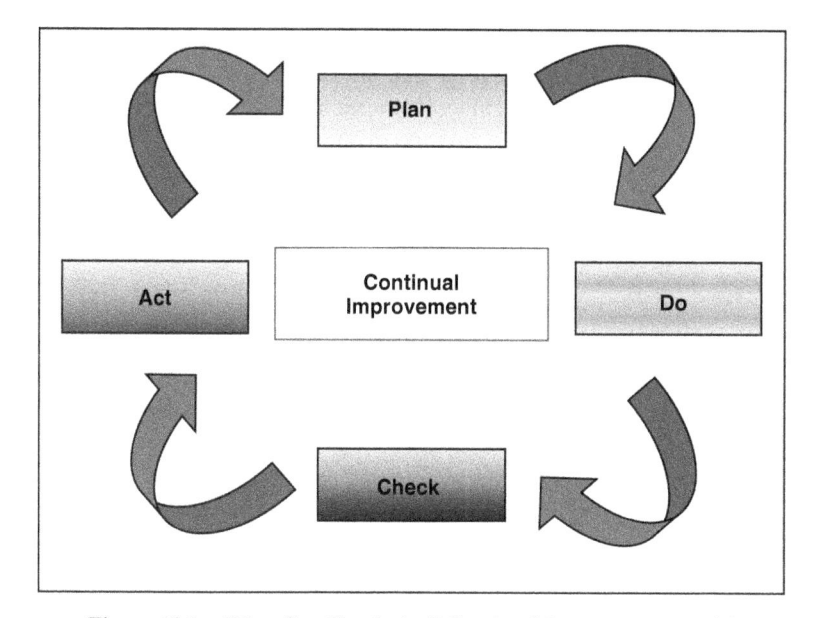

Figure 18.1 "Plan, Do, Check, Act" Continual Improvement Model

TABLE 18.1 Detailed PDCA Cycle Steps Based on ISO 14001:2004 Standard

PDCA Cycle	ISO 14001:2004 Standard
Plan	*4.2 Environmental Policy*
	4.3 Planning
	4.3.1 Environmental Aspects
	4.3.2 Legal and Other Requirements
	4.3.3 Objectives, Targets, and Program(s)
	ISO 14004 4.3 Planning/General guidance – Planning Add: Setting of objectives and targets and establishment of program(s) to achieve them (International Organization for Standardization, b)
Do	*4.4 Implementation and Operation*
	4.4.1 Resources, Roles, Responsibility, and Authority
	4.4.2 Competence, Training, and Awareness
	4.4.3 Communication
	4.4.4 Documentation
	4.4.5 Control of Documents
	4.4.6 Operational Control
	4.4.7 Emergency Preparedness and Response
Check	*4.5 Checking*
	4.5.1 Monitoring and Measurement
	4.5.2 Evaluation of Compliance
	4.5.3 Nonconformity, Corrective Action, and Preventive Action
	4.5.4 Control of Records
	4.5.5 Internal Audit
Act	*4.6 Management Review*

As indicated in Table 18.1, the required steps outlined in ISO 14001 include (4.2) Environmental Policy, (4.3) Planning, (4.4) Implementation and Operation, (4.5) Checking, and (4.6) Management Review. As with any management system endeavor, the planning aspect is vital to successful implementation. Due to some limitations, this chapter will focus only on the planning process step. However, the supporting interactive tool will include suggestions for documents control and statistical tools.

18.4.1 Environmental Policy and Planning

Before an organization begins an EMS implementation, a cross functional team (CFT) of qualified individuals likely to be involved in the process is selected and assembled. An effective leader should be selected by management to serve as the EMS management representative and other knowledgeable personnel to serve on the team or committee. The quality of the leader and team members selected directly reflects management's commitment to the EMS and determines its ultimate success. Team members involved in an EMS implementation must have a broad knowledge of organization's operations and environmental aspects, as well as a willingness to learn, and a commitment to continual improvement.

During the planning process, the organization should establish an environmental policy. A sample environmental policy is presented below:

ISO 14001 IMPLEMENTATION

ABC Environmental Company is committed to an environmental management system that will ensure continual quality improvement and pollution prevention in all aspects of our business. We are committed to compliance with all applicable environmental laws and/or regulations wherever we do business. We will continually search for and implement activities that will improve our abilities to properly utilize natural and man-made resources both economically and conservatively.

The planning process is critical to the fulfillment of the environmental policy and the application and maintenance of the EMS. Typically, planning will involve the following elements:

- Identification of environmental aspects and the determination of those that are significant
- Identification of applicable legal requirements and other requirements
- Setting of internal performance criteria
- Setting of objectives and targets and establishment of program(s) to achieve them.

Risk assessment (risk *identification*, *analysis*, and *evaluation*) is a critical component of the EMS planning process and is essential in assisting management in making decisions and determining the best use of available resources. Information generated by the environmental risk assessment process should be used in the establishment and improvement of other components of the EMS.

18.4.2 Environmental Aspects

The most essential part of the environmental risk assessment process is the *identification* and *evaluation* of environmental aspects. If significant environmental hazards remain unidentified, the risks derived from these hazards go unmanaged presenting a threat to the organization. According to ISO 14001 standard, every environmental aspect is required to be evaluated as part of the EMS (International Organization for Standardization, a). The US EPA provides the following definitions:

Environmental Aspect - an element of a facility's activities, products, or services that can or does interact with the environment. These interactions and their effects may be continuous in nature, periodic, or associated only with events, such as emergencies.

Environmental Impact - as any change to the environment, whether adverse or beneficial, resulting from a facility's activities, products, or services.

Significant Environmental Aspect - one that may produce a significant environmental impact.

The methodology described below defines the evaluation criteria used to determine the degree of significance of each aspect on the Environmental Aspects and Impacts List, upon the identification and inclusion of a new aspect to this list. The following *Environmental Aspect Significance Determination System* is considered a good practical example:

Purpose

The key to a successful registration for conformance with ISO 14001 is to precisely determine the organization's Environmental Aspects and Impacts. The EMS team should evaluate which aspects are the most significant and therefore require the most attention. For practical purposes, two definitions should be introduced before the initial evaluation takes place.

"3.3 – Environmental Aspect: Element of an organization's activities, products or services that can interact with the environment.
 # A significant environmental aspect is an environmental aspect that has or can have a significant environmental impact." and

"3.4 – Environmental Impact
Any change to the environment, whether adverse or beneficial, wholly or partially resulting from an organization's activities, products or services." (International Organization for Standardization, a)

There are three distinct requirements contained within Section 3 of ISO 14001:

1. The organization shall identify the environmental aspects of its activities, products and/or services. In other words the organization must understand how it interacts with the environment.
2. The organization shall identify the specific environmental aspects that can be controlled, and over which it can be expected to have influence.
3. Arrive at a list of significant environmental aspects based upon the individual environmental impact of each environmental aspect. The importance of this third step cannot be underestimated. The final list of significant environmental aspects will provide the basis for the Environmental Policy statement, and the Environmental Objective(s) and Targets. In other words the list of significant environmental aspects drives the entire content and scope of the operational portion of the EMS.

An example of a significant environmental aspect decision-making tree is presented in Figure 18.2.

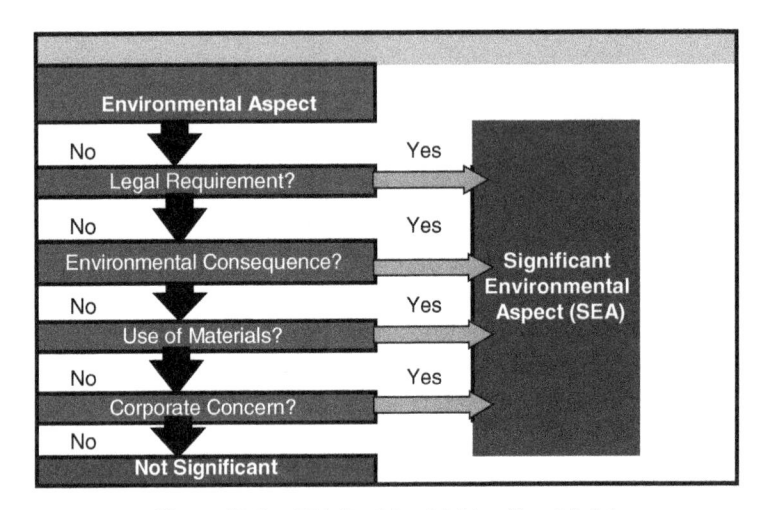

Figure 18.2 SEA Decision-Making Tree Model

This procedure defines the evaluation criteria used to determine the degree of significance of each aspect on the Environmental Aspects and Impacts List, upon the identification and inclusion of a new aspect to this list or upon updating an existing aspect. This procedure applies to all environmental aspects on the list. It is the responsibility of the process review team leader to modify the environmental aspects list according to the input received by the process review team.

All of the organization's activities, products, and/or services must be included within the environmental aspects review framework. The organization must assemble a list of all potential environmental aspects. The purpose of this activity is not to determine the importance of each aspect. The goal is to create a complete listing of how the organization interacts with the environment. The organization must include the effect of routine and nonroutine activities/situations in the list. Nonroutine activities/situations examples include the following:

➢ Emergency conditions
➢ Start-up
➢ Shutdown activities.

The procedure that is developed to produce the list of environmental aspects must recognize the concept of continual improvement. The procedure and the list of aspects can be influenced by future actions such as new legislation, changes in operations, etc. The Management Review (4.6) section mandates continual improvement because the EMS must be reviewed on a periodic basis to insure that it is still effective based upon the organization's activities, products, and/or services (International Organization for Standardization, a).

18.4.3 Identify Environmental Aspects

Environmental aspects can be identified based on the following:

- Those that ABC Environmental Company can control
- Those that over which it can influence by its management activities
- A review of activities and services conducted by and for ABC Environmental Company under normal conditions and nonroutine operations/activities
- A review of products produced for and used by ABC Environmental Company under normal conditions and nonroutine operations/activities.

Content requirements of environmental aspects:

- Understand how it interacts with the environment.
- Consider control versus influence.
- Arrive at a list of *SEA* based upon impacts.
- Develop a procedure to support the decision-making process.

The process of arriving at a list of SEA must be a reasonable and understandable procedure. Therefore, the following scoring procedure was developed. The procedure is intended to help the EMS team streamline the scoring process and evaluate the risks.

18.4.4 Identification Process

Normally, this process will be completed using a process review team that consists of representatives of different departments and disciplines throughout the company:

A. The process review team leader coordinates and schedules team meetings based on changes in process, timing for formal annual review, or any operational trigger that may impact the current significance of environmental aspects and impacts for the organization.
B. The process review team leader collects all relevant process activity forms, aspect lists, and other data that may need to be referenced by the team. He or she may act as a facilitator to assist process owners and operators in identifying the critical information.
C. At the formal, annual review, this team identifies aspects associated with all processes, activities, and services provided. At more frequent, informal meetings, only the process, activity, or service under revision will be reviewed.
D. The team accumulates and considers all relevant quantitative and qualitative data (e.g., hazardous materials and quantities used, air emissions, locations where the process activities and services are performed, potential for spills, legal risks (LRs) associated with the process, etc.) about each aspect under discussion.
E. The team identifies potential environmental impacts, both positive and negative (e.g., contamination of ground or surface water resulting from a chemical spill, impacts to species

habitat, reduced material consumption due to recycling efforts, etc.) associated with each aspect identified.
F. The team evaluates the degree of significance of each aspect using the evaluation criteria provided in the Environmental Aspect Significance Determination list.

Those aspects with extreme ratings on the list will be considered by management as the potential SEA upon which to develop or alter existing objectives and programs. The final list of environmental aspects, their impacts, and those selected as significant will be updated in the Environmental Aspects and Impacts List. An Excel environmental risk assessment matrix can be developed for convenience and data processing. This document can be used to develop environmental objectives and revise or refocus environmental management programs and the environmental policy where necessary.

There are three (3) major sections that make up the Environmental Aspects and Impacts List:

- Significance Scores without Controls
- Significance Scores with Controls
- Overall Significance Rating Chart.

Common to these three sections are the following:

- Location
- Department
- Index
- Aspect
- Impacts to Environmental Properties (IEP)
- Polarity Adjustment (PA)
- Impact Subtotal
- Frequency of use.

18.4.5 Location, Department, Index, and Aspect

Location and Department columns have been provided as an option to further define the aspect, if necessary. In some cases, the aspect might be significant enough in more than one location and/or department to warrant an additional score of that aspect in more than one line item. In most cases the aspect will only require a single line item listing, but an indication of a more specific location and/or department might be necessary to track that aspect in the event it changes ownership or location throughout its life cycle. The absence of an indicator in either field indicates "not applicable" or "all of the above." The Location and Department legends are shown below:

Example Locations

l: Laboratory
w: Warehouse building
f: Field (in general)
t: T building
a: Abatement building.

Example Departments

a: Abatement
r: Remediation
k: ABC supply

it: Information technology/securities
o: General overhead
ac: Accounting
hr: Human resources.

The Index column is a serialized alphanumeric designator to differentiate each line item as new aspects are added to the list. The letter designator is used to define in what process the aspect was discovered. It is possible that an aspect might be discovered in multiple processes and requires more than one line item listing and index designator. The number designator will be incremented with respect to the last number assigned within each letter upon the addition of a new aspect to the list. This legend defines those letters and their processes:

f: Field process
a: Administration process.

18.4.6 Impacts to Environmental Properties

Each of the 10 IEP (see Figure 18.3) will be rated using +1, −1, or 0, which are defined below:

+1: Means that aspect has a positive impact on the environment for that property.
−1: Means that aspect has a negative impact on the environment for that property.
0: Is used when an aspect has no impact on the environment for that property or when that property is not applicable for that aspect.

When rating the impacts in the Significance Scores without Controls section, consider the aspect's impact on that property in an unregulated environment, in and of itself, and outside of any controls an organization might have on it in a worse-case scenario. When rating the impacts in the Significance Scores with Controls section, consider the aspect's impact on that property under an organization's complete control within its current regulations and operational standards.

18.4.7 Impact Subtotal and Polarity Adjustment

These 10 impacts will then be summed together into an Impact Subtotal. That subtotal will then be used in the remaining Significance Score's calculation. In the event the subtotal equals zero, a polarity adjustment value will be substituted as a replacement to the Impact Subtotal and instead be used in the remaining Significance Score's calculation. Without this polarity adjustment, that calculation, which is a product of several factors, would use this unacceptable zero factor and generate a meaningless zero score.

In the Significance Scores without Controls section, the polarity adjustment will be a value of −1. In the Significance Scores with Controls section, the polarity adjustment will be a value of +1. An example of an environmental aspects' evaluation procedure is presented in Figure 18.3.

ISO 14004
4.3. Planning
General guidance — Planning
Identification of environmental aspects and the determination of those which are significant

				SIGNIFICANCE SCORES WITHOUT CONTROLS											
				IMPACTS TO ENVIRONMENTAL Properties											
Aspect ID #	Location	Department	ASPECT	Landfill Use	Air Pollution	Groundwater Contamination	Surface Water Contamination	Subsurface Soil Contamination	Surface Soil Contamination	Species Degradation	Habitat Degradation	Conservation of Resource	Quality of Life	Polarity Adjustment	Impact Subtotal
1	f	a	asbestos contaminated mat'l waste	−1	−1	−1	−1	−1	−1	−1	−1	−1	−1		−10
2	l	a	asbestos samples (bulk)	−1	−1	−1	−1	−1	−1	−1	−1	−1	−1		−10
3		a	lead samples (3rd party)	−1	−1	−1	−1	−1	−1	−1	−1	0	−1		−9

Figure 18.3 Environmental Aspects Evaluation Procedure Example

The SH&E professional and stakeholders will have to decide what kind of environmental aspects' risk assessment methodologies to use. Two different rating scales will be presented for the other possible variables to be added to environmental aspects' evaluation. The variables presented below will be the same for both sections:

- Significance Scores without Controls
- Significance Scores with Controls.

18.4.8 Impact Severity

Impact Severity is a rating that indicates how severe the aspect might impact the environment in the event of an uncontrolled full release. The EMS team may decide to use 1–3 rating such as 1 = low; 2 = medium; and 3 = high. It can also be determined using a scale of 1–5, shown in Table 18.2.

TABLE 18.2 Severity 1–5 Scale Description Example

Numerical Rating	Descriptive Word	Consequence
5	Catastrophic	Total system loss, chemical release with lasting environmental or public health impact
4	Critical	Major property damage and business downtime, chemical release with temporary environmental or public health impact
3	Marginal	Minor subsystem loss or damage, chemical release triggering external reporting requirements
2	Negligible	Nonserious equipment or facility damage, chemical release requiring only routine cleanup without reporting
1	Insignificant	Inconsequential with environmental losses or environmental chemical release

18.4.9 Impact Probability

Impact Probability is a rating that indicates how probable the aspect might impact the environment in the event of an uncontrolled full release. Again, the team may decide to use a typical S × P 1–3 rating such as 1 = low; 2 = medium; and 3 = high. It can also be determined using a scale of 1–5, as represented in Table 18.3.

TABLE 18.3 Impact Probability Example

Numerical Rating	Descriptive Word	Probability Descriptions
5	Frequent	Likely to occur repeatedly
4	Probable	Likely to occur several times
3	Occasional	Likely to occur sometime
2	Remote	Not likely to occur
1	Improbable	So unlikely, can assume occurrence will not be experienced

The above stated two ratings can be used to estimate Risk Level (RL) using Impact Severity × Impact Probability or S × P. In some cases, the team may decide to use typical S × P risk assessment matrix and avoid overcomplicating the calculations. Simplified methodology may be sufficient for small organizations. However, calculating only Risk Level may not be sufficient. To demonstrate why adding more variables is necessary, the author presents the following practical example.

Environmental remediation managers are familiar with asbestos floor tile removal process. Black mastic between the floor tiles and the floor is usually considered asbestos-containing material (ACM). Not to mention, it is difficult to remove the black mastic from a concrete floor. It is a very time-consuming process. Environmental remediation companies are using different mastic removers. The following mastic removers will be used to demonstrate environmental risks:

- Soy based
- Regular (*R*)
- Heavy duty (HD).

There are many advantages and disadvantages to consider in choosing one over the other two solutions. The use of mastic remover depends upon the type of job(s) that the company is working on. If the abatement crew is working in a hospital, school, or other occupied buildings, they generally use soy-based mastic remover. If the company is working in a building that is to be demolished, they may decide to use heavy duty mastic remover.

The heavy duty is far superior when it comes to being effective. The workers are able to work faster and reuse the mastic remover on other areas. The soy based is the slowest and least effective and can only be used in *one* application. The regular is somewhere in between the other two. Soy-based mastic remover is more environmentally friendly and heavy duty application usually leads to volatile organic compound (VOC) releases. Regular mastic remover also emits VOCs. If we only use traditional risk assessment methodology ($S \times P$) to calculate the Risk Level, we will see that there is a very minor difference between HD (RL = 6) and soy (RL = 4) mastic removers. A 1–5 rating scale was used for this example illustrated in Figure 18.4.

ISO 14001
Check
4.5.1. Monitoring and Measurement
SIGNIFICANCE SCORES WITHOUT CONTROLS

				IMPACTS TO ENVIRONMENTAL PROPERTIES														
Aspect ID #	Location	Department	ASPECT	Landfill Use	Air Pollution	Groundwater Contamination	Surface Water Contamination	Subsurface Soil Contamination	Surface Soil Contamination	Species Degradation	Habitat Degradation	Conservation of Resource	Quality of Life	Polarity Adjustment	Impact Subtotal	Impact Severity	Impact Probability	Risk Level S*P
43-1	w	a	mastic remover Soy	0	0	0	0	0	0	−1	−1	−1	1		−2	2	2	4
43-2	w	a	mastic remover R	−1	−1	−1	−1	0	−1	−1	−1	−1	−1		−9	2	2	4
43-3	w	a	mastic remover HD	−1	−1	−1	−1	−1	−1	−1	−1	−1	−1		−10	3	2	6

Figure 18.4 Environmental Aspects Evaluation Procedure Example

However, many environmental remediation managers have realized that Risk Level estimation does not present the "big picture" overview. For instance, using regular mastic remover in an occupied office building or a school may lead to legal fees and complaints from the occupants. Therefore, adding more categories/variables may be necessary. The variables presented below are just a few examples.

18.4.10 Frequency

Frequency is a rating that indicates the frequency of use of an aspect by an organization or the frequency of that aspect's presence on its property. It can be determined using a scale of 1–3 or 1–5. Both options are presented in Tables 18.4 and 18.5.

TABLE 18.4 Frequency 1–3 Rating Example

Numerical Rating	Description
1	Rarely
2	Frequently
3	Everyday use

TABLE 18.5 Frequency 1–5 Rating Example

Numerical Rating	Description
1	Rarely
2	Infrequently
3	Frequently
4	Continuously (nonroutine operations)
5	Everyday use (routine operations)

18.4.11 Legal Risks

Legal Risks is a rating that indicates the degree of risk and the impact probability of a lawsuit in the event of an uncontrolled full release of that aspect in the environment. It will be determined using a scale of 1–3 or 1–5. Both options are presented in Tables 18.6 and 18.7.

The following rating is to be used in the Significance Scores with Controls section only.

TABLE 18.6 Legal Risks 1–3 Rating Examples

Numerical Rating	Description
1	Violates environmental industry-specific regulations
2	Violates general regulations
3	No known risks

TABLE 18.7 Legal Risks 1–5 Rating Examples

Numerical Rating	Descriptions
5	Violates environmental industry-specific regulations
4	Violates general regulations
3	Violates voluntary standards
2	Violates company policy
1	No known risks

18.4.12 Current Controls

Current Controls is a rating that indicates the degree of existing controls for an environmental aspect. Again, the 1–3 or 1–5 rating scales can be used (Tables 18.8 and 18.9).

TABLE 18.8 Current Controls 1–3 Rating Example

Numerical Rating	Description
1	Acceptable
2	Needs improvement
3	Unacceptable

TABLE 18.9 Current Controls 1–5 Rating Example

Numerical Rating	Descriptions
5	Acceptable
4	Acceptable with additional considerations
3	Needs minor improvements
2	Needs major improvements
1	Unacceptable

18.4.13 Significance Score for Significance Scores without Controls Section

The Significance Score in this section is calculated using the following equation:

$$\text{Significance Score} = \sum \text{IEP} * (\text{PA} * S * P * F * \text{LR})$$

The sum of IEP multiplied by Impact Subtotal or polarity adjustment (PA) multiplied by Frequency of use (F) multiplied by Impact Severity (S) multiplied by Impact Probability (P) multiplied by Legal Risks (LR).

In addition, a Personnel Risk (PR) column can be added to align EMS with H&S Management System.

18.4.14 Personnel Risk

Again, the team will decide whether to use a 1–3 or 1–5 rating scale (Tables 18.10 and 18.11).

An example of Personnel Risk (PR) 1–5 rating is presented in Table 18.11.

In this case, the formula will be:

$$\text{Significance Score} = \sum \text{IEP} \times (\text{PA} \times S \times P \times F \times \text{LR} \times \text{PR})$$

TABLE 18.10 Personnel Risk 1–3 Rating Scale Example

Numerical Rating	Description
1	Catastrophic
2	Marginal
3	Insignificant

TABLE 18.11 Personnel Risk 1–5 Rating Scale Example

Numerical Rating	Descriptive Word	Consequence
5	Catastrophic	Death or permanent total disability
4	Critical	Permanent, partial, or temporary disability in excess of 3 months
3	Marginal	Minor injury or illness, lost workday incident
2	Negligible	Minor injury or illness, not a day away from work incident or minor medical treatment
1	Insignificant	First aid

An example of environmental aspects' evaluation procedure that includes Frequency of use, Impact Severity, Impact Probability, Legal Risks, and Personnel Risk is presented in Figure 18.5.

Figure 18.5 Environmental Aspects Risk Assessment Evaluation Procedure Example

Using the mastic remover example, the risk level scores can be compared. Evaluating environmental aspects using impact severity and impact probability only may not be sufficient and even misleading. Figure 18.6 provides a comparison of the Significance Scores with the additional variables and the Risk Level score discussed previously.

Figure 18.6 Mastic Remover Environmental Risk Assessment Example with Additional Variables

The Risk Level is not included in the calculations. It is presented here just for visualization and to demonstrate that soy-based mastic remover had equal Risk Level to the regular mastic remover. However, adding more variables to the equation leads to a disproportionate Significance Score.

ISO 14001 IMPLEMENTATION

The examples above show that Significance Scores without Controls are not acceptable. Therefore, environmental aspects will have to be controlled properly to avoid Personnel Risk and protect the environment.

18.4.15 Significance Scores with Controls Section

The Significance Score in this section is calculated using the following equation:

$$\text{Significance Score with Controls} = \sum \text{IEP} \times (\text{PA} \times \text{IS} \times \text{HoC})$$

Once the Hierarchy of Controls is established, Significance Scores could be evaluated again using the same methodology. However, we have to account for Hierarchy of Controls. If the control measures are acceptable and sufficient, the Impact Subtotal calculation with controls will be multiplied by 5. An example of Significance Scores with Controls is presented in Figure 18.7.

Location	Department	ASPECT	Landfill Use	Air Pollution	Groundwater Contamination	Surface Water Contamination	Subsurface Soil Contamination	Surface Soil Contamination	Species Degradation	Habitat Degradation	Conservation of Resource	Quality of Life	Polarity Adjustment	Impact Subtotal	Hierarchy of Controls	Significance Score WITH Controls
f	a	asbestos contaminated mat'l waste	−1	1	1	1	1	1	1	1	−1	1		6	5	30
l	a	asbestos samples (bulk)	−1	1	1	1	1	1	1	1	−1	1		6	5	30
	a	lead samples (3rd party)	−1	0	0	0	0	0	0	0	0	1	1	0	5	5

Figure 18.7 Environmental Aspects Evaluation Procedure Including Controls Example

18.4.16 Overall Significance Rating Chart

This last section can be used to calculate an overall significance rating or the difference between the scores without controls and with proper control measures. The scores for each aspect can be sorted and color coded to compare each aspect's rating. Once all scores have been established and calculations made, the entire list can be sorted by Significance Scores to reveal the most extreme ratings from highest to lowest.

This rating is calculated as follows:

The absolute value of the difference between Significance Scores with Controls and Significance Scores without Controls, which is illustrated in Figure 18.8

Location	Department	ASPECT	Impact Severity	Impact Probability	Frequency	Legal Risks	Personnel Risk	Product of Impact Factors w/Controls	Significance Score WITHOUT controls	Significance Score WITH controls
f	a	asbestos contaminated mat'l waste	5	1	2	2	2	40	−6000	1200
l	a	asbestos samples (bulk)	4	1	3	2	2	48	−4800	1440
l	a	lead samples (3rd party)	2	1	1	1	2	4	−162	20

Figure 18.8 Environmental Aspects Evaluation

The larger the value of the spread, the more extreme (higher) the rating, which generally translates into the following:

The more hazardous the aspect, typically the more controls ABC applies to that aspect. In some cases, multiple controls may have to be applied. The evaluation team may decide to use color coding to visualize the risks and shade the more extreme ratings in a dark shaded region.

The graphical representation is calculated by finding the minimum and maximum values of all Significance Scores available.

After the initial environmental aspects' evaluation, it becomes obvious that aspects without controls have enormous Significance Scores variation. One of the identified objectives was to reduce aspects' Significance Scores. The target was to statistically reduce Significance Scores compared to environmental aspects' scores without sufficient controls. Some of the tools will be discussed in the next section.

18.5 EMS AND IMPLEMENTATION OF LEAN SIX SIGMA PRACTICES

Combining Lean and the Six Sigma (LSS) methodologies has become popular during the last decades. Most of these LSS efforts have been focused on manufacturing process. A case study developed by the author based on lessons learned transferring *Lean* from the production floor to the environmental operations is presented along with a model that integrates LSS, and environmental remediation management.

CASE STUDY

Having experience in Environmental Aspects Scores Analyses, the team in charge of the overall effort decided to conduct a Lean pilot project to test possible conflicts between *Lean* goals and environmental performance.

The case study project begins with a *Process Cycle Efficiency* (PCE) evaluation. PCE is a widely used Lean tool that provides a calculation of the "Value-Added Time" [*work that customer would recognize as necessary to create the product or service they are to purchase*] divided by "Total Lead Time" [*how long the process takes from start to finish*] (iSixSigma). The "*service*" in this study is the removal of ACM floor tiles and black mastic. The *PCE formula* is shown below:

$$\text{Process Cycle Efficiency (PCE)} = \text{Value-Added Time}/\text{Total Lead Time}$$

Comparing the PCE of both the heavy duty mastic and the soy-based mastic remover method, it becomes apparent that the heavy duty process is more efficient. The soy-based mastic remover requires more Total Lead Time (172) compared to the heavy duty method (71). The resulting PCE indicates that the soy-based method, which requires more time (considered a waste factor in Lean), is less desirable from a Lean practice standpoint as illustrated in Figure 18.9.

However, from an environmental standpoint, the heavy duty mastic remover has a much higher environmental aspect Significance Score due to the harsh chemicals used. Therefore, EMS CFT should provide recommendations on when and how to use different environmental aspects.

One of the goals for LSS program is to reduce variation, and thus, new control methods were proposed for each environmental aspect to reduce the risk and demonstrate statistically significant reduction of Significance Scores. As a result, new tools and statistical methods using LSS were developed to solve the problem. The EMS LSS tool is based on a PDCA or coantinuous improvement approach described in the "Lean Six Sigma: Process Improvement Tools and Techniques" textbook (Summers, 2011). In addition, the risk assessment tool could be considered significantly expanded version of failure mode and effects analysis (FMEA) methodology (Tague, 2004, p. 236).

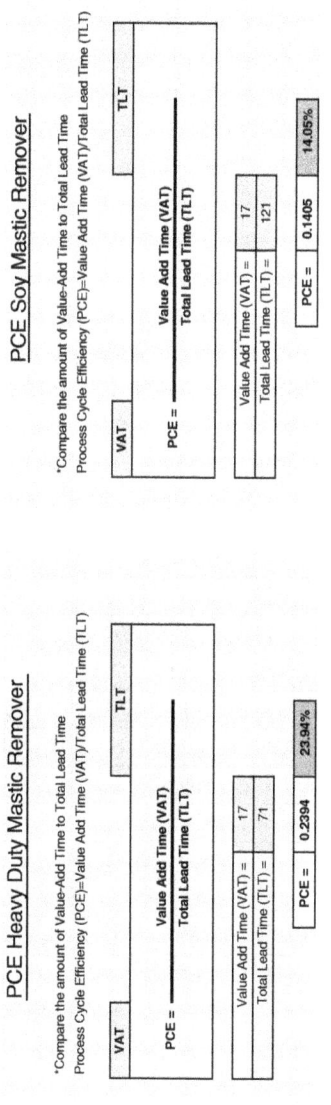

Figure 18.9 PCE for Heavy-Duty versus Soy-Based Mastic Remover

First, a histogram was prepared to demonstrate the enormous variation in the environmental aspects' scores as illustrated in Figure 18.10.

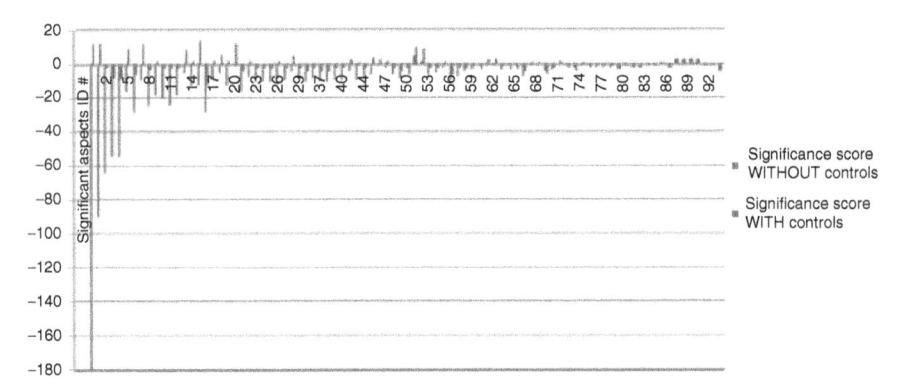

Figure 18.10 Significance Scores Histogram

Second, standard MS Excel tools were used to calculate and compare the mean, mode, median, range, standard deviation, variance, Kurtosis, and skewness (StatSoft, n.d.b) as presented in Table 18.12. Other tools, such as QI Macros are available (QIMacros, 2014).

TABLE 18.12 Significance Scores

	Significance Score Without Controls	Significance Score with Controls
Mean	−10.5319	−0.21277
Standard error	2.326167	0.487049
Median	−5.5	−2
Mode	−1	−2
Standard deviation	22.55302	4.722111
Sample variance	508.6388	22.29833
Kurtosis	35.94634	1.519692
Skewness	−5.42011	1.049506
Range	185	26
Minimum	−180	−12
Maximum	5	14
Sum	−990	−20
Count	94	94

It was determined that there was significant difference between the scores without controls and the scores after the process controls were implemented. Then, a t-test was used to determine whether or not the difference was statistically significant (StatSoft, n.d.a). The t-test results showed a statistically significant difference with a t critical of 1.66 and t-stat of −4.21.

The objective of such a project risk assessment is to reduce risk and demonstrate a statistically significant reduction of risk level scores. If the aspects' Significance Scores are substantially reduced, the project should be considered successful. In this case, the standard deviation was reduced by nearly 79%, with the mean and the median scores also significantly reduced.

The required resources for the project were not related to purchasing of new equipment and software. However, significant investment in time and personnel is required. The development of such a system required approximately 150 hours of EMS sponsor's time. In addition EMS team meetings and individual interviews resulted in an additional 550 hours of various team members time. It will be difficult to justify such expense if the organization is not fully committed to environmental, health, and safety excellence.

Other typical Six Sigma tools, like FMEA, can be used to integrate EMS, H&S management principles, and risk assessment of environmental remediation operations. One such example is included in the supplemental materials. Interactive EMS LSS tools included in this chapter are for educational purposes only. If the readers choose to implement such tools, they are encouraged to develop new tools or modify the supplemental materials.

18.6 CONCLUSIONS

Reduction of aspects' Significance Scores was achieved. This practical example demonstrated a statistically significant reduction of Significance Scores. The project was successful and accepted by the management. It reduced standard deviation by nearly 79%. The mean and the median scores were also significantly reduced.

SH&E professionals have to develop management skills and diversify their knowledge to overcome difficulties during such projects. After the initial analysis, it became clear that such significant investment in EMS Significance Scores variation reduction projects cannot be easily justified based on risk assessment alone. Future leaders in the safety profession will have to develop statistical skills and demonstrate knowledge in financial management. In order to defend such projects, SH&E professionals have to be familiar with a variety of risk management techniques, LSS tools, and financial management principles.

Being an expert in environmental compliance is not enough to complete successful complex projects. Complex projects require multidisciplinary knowledge and cross discipline management skills. Safety leaders have to become familiar with different organizational structures and a variety of stakeholders' interests to complete such projects. Professionals have to be prepared to deal with various levels of the organizational management and demonstrate competencies.

REVIEW QUESTIONS

1. Describe the purpose of Environmental Management Systems
2. Name applicable standards or guidelines, which address Environmental Management Systems and Occupational Safety and Health Management Systems.
3. State the primary purpose of an environmental aspects' risk assessment.
4. Describe the Significant Environmental Aspects (SEA) determination process.
5. Explain the importance of environmental aspects' evaluation utilizing without controls and with controls comparison.
6. Express your opinion about EMS and Lean Six Sigma integration.
7. Provide examples of risk assessment and Lean Six Sigma tools used for EMS projects.

REFERENCES

International Organization for Standardization. ISO 14001.2004. Environmental Management Systems – Requirements with Guidance for Use, n.d.a-a Retrieved from: http://www.iso.org/iso/iso_catalogue/catalogue_ics/catalogue_detail_ics.htm?csnumber=31807 (accessed December 28, 2015).

International Organization for Standardization. ISO 14004.2004. Environmental Management Systems – General Guidelines on Principles, Systems and Support Techniques, n.d.b-b Retrieved from: http://www.iso.org/iso/iso_catalogue/catalogue_ics/catalogue_detail_ics.htm?csnumber=31808&ICS1=13&ICS2=20&ICS3=10 (accessed December 28, 2015).

iSixSigma. Retrieved from: http://www.isixsigma.com/dictionary/process-cycle-efficiency-pce/ (accessed December 28, 2015).

QIMacros. Excelerating Lean Six Sigma, 2014. Retrieved from: https://www.qimacros.com/ (accessed December 28, 2015).

StatSoft. t-Test for Independent Samples, n.d.a-a Retrieved from: http://www.statsoft.com/Textbook/Basic-Statistics#t-test for independent samples (accessed December 28, 2015).

StatSoft. What are Basic Statistics?, n.d.b-b Retrieved from: http://www.statsoft.com/textbook/basic-statistics/?button=1 (accessed December 28, 2015).

Summers, D. C. S.. *Lean Six Sigma: Process Improvement Tools and Techniques*. Boston, MA: Prentice Hall, 2011.

Tague, N. R. *The Quality Toolbox*. Milwaukee, WI: ASQ Quality Press, 2004.

19

BUSINESS ASPECTS OF OPERATIONAL RISK ASSESSMENT

ELYCE BIDDLE
Department of Industrial and Management Systems Engineering, West Virginia University, Morgantown, WV, USA

19.1 OBJECTIVES

- Introduce business aspects
- Present Prevention through Design's role in making the business case
- Present case studies that support the business case

19.2 INTRODUCTION

The main objective of a safety, health, and environmental (SH&E) professional is to control the risk of hazards to the worker and environment to an acceptable or tolerable level. An initial risk assessment is necessary to reveal the existence of hazardous conditions that need control through elimination, reduction, or avoidance. Unfortunately, instituting a solution is not always inexpensive let alone "free" to the organization.

Although proof of preventing occupational injury, illness, or fatality alone has often driven industry to make changes, the lack of adoption of known effective solutions has clearly demonstrated that there were other reasons behind SH&E business decisions (Biddle, 2013). Organizations continually face increased global competition, rapidly changing technology, and decreased access to scarce resources. Under these conditions, SH&E efforts to ensure a safe and healthful work environment must compete with other organizational needs. Without compelling information about the value of SH&E efforts to the organization, management may view these programs and activities as a lower priority than projects that have established a clearer connection to their bottom line. The challenge for occupational safety

---

Risk Assessment: A Practical Guide to Assessing Operational Risks, First Edition.
Edited by Georgi Popov, Bruce K. Lyon, and Bruce Hollcroft.
© 2016 John Wiley & Sons, Inc. Published 2016 by John Wiley & Sons, Inc.

and health professionals is to describe the value of SH&E efforts in terms that are understood and accepted within the business community. A business case addresses that challenge.

A business case in occupational safety and health can be defined as a method that captures the effects of implementing programs or activities on employee health (injury or illness), risk management, and the business process. Capturing detailed cost and benefit data through the use of a business case provides for generating customary financial business metrics, such as net present value (NPV), internal rate of return (IRR), return on investment (ROI), and discounted payback period (DPP), which are meaningful to business management. It is important to note that these are not just terms used in the finance department, but they are terms universally understood by those making resource decisions, creating effective tools to demonstrate the value of your initiatives. Additionally, the business case provides a mechanism for capturing nonfinancial impacts, those that cannot be directly or easily monetized, such as changes in customer satisfaction, corporate social responsibility, product defects, presenteeism, and corporate reputation. In reality, the business case answers the question: What's in it for the company? Or in this instance ... Why should the company implement programs and practices that reduce the risk of adverse worker health effects?

The aim of this chapter is to describe methods of the Business Case Development Tool to systematically appraise both the costs and consequences of an action implemented at the worksite in an effort to reduce or eliminate the risk of adverse health effects to the worker – to derive the business value of your efforts. Examples include how these methods have been applied to small and large organizations and before and after the solutions have been implemented. Being able to develop a business case will assist in positioning environmental health and safety (EHS) as essential contributor to the business function through completing your primary objective of reducing health risk.

19.3 THE BUSINESS CASE DEVELOPMENT TOOL

There are a number of general business case guidelines, but few address the needs of the occupational safety and health field in general let alone tied to hazard evaluation and risk assessment. Additionally, there are a limited number of models that include strategic, descriptive, or analytic tools and instructional documentations to capture the benefits and costs to the business of implementing occupational EHS solutions (Biddle et al., 2011). Finally, an easy-to-use computerized tool to develop the business case did not exist.

Building on the collaborative efforts of NIOSH and ORC supported by the American Industrial Hygiene Association and the American Conference of Governmental Industrial Hygienists (AIHA, 2009) to determine the value of the industrial hygiene profession and using the wisdom contained in the ANSI/ASSE Z590.3-2011 and the ISO 31000 (ISO 31000:2009, 2009; ANSI/ASSE/ISO Z690), a Business Case Development Tool was designed. Specifically, this tool can be used to generate a business case to demonstrate the value to an employer of Prevention through Design (PtD) efforts to an employer integrated into the occupational safety and health management systems.

The Business Case Development Tool was designed to ease development of the business case by SH&E professionals for use in making decisions or for presentation to the organizational managers charged with making resource decisions. The tool can be used to select among alternative solutions or demonstrate the wisdom of a solution already selected. The information provided by the tool is expressed in the language understood by all management, not just those in occupational safety and health. Yet, the tool includes the critical components that define the SH&E profession – risk management.

The tool consists of a five-step process and individual descriptive or analytic tools or instruments to complete each of the steps. Figure 19.1 illustrates the major components of the tool, while Figure 19.2 provides an illustration of the selection of tools or instruments available for use in performing Step 1 when a specific Tool has been selected by the analyst.

THE BUSINESS CASE DEVELOPMENT TOOL

Business Case Developer				
About an Occupational Safety and Health Business Case				
Step 1	Describe current OSH condition		Show/hide modules	Description
Step 2	Conduct risk assessment		Show/hide modules	Description
Step 3	Identify changes from current OSH condition		Show/hide modules	Description
Step 4	Determine value		Show/hide modules	Description
Step 5	Develop recommendations report		Show/hide modules	Description
Instructions Manual				

Figure 19.1 Opening Page of the Business Case Developer Tool

Step1	Describe the Current Situation	
	Tool 1a	Identify problem and desired outcomes
	Tool 1b	Draft list of key business and EHS objectives
	Tool 1c	Draft business case project assumptions

Figure 19.2 The Business Case Developer Tool – Step 1

19.3.1 Steps of the Tool

The first step of the process represents the formative phase providing the background for determining the value of SH&E activities in current and future time frames. Assessing the current situation involves identifying the health protection problem(s), describing what actions are currently in place to address the problem(s), and determining the business objectives and their importance to the successful operation of the business. Understanding the current situation provides critical baseline information needed to identify interventions or solutions that could be implemented, continued, or revised to improve the current state of safety and health.

The second step of the development takes an in-depth look at the SH&E hazards and risks within the business process at the current time. More specifically, hazards are identified first so that the risks arising from those hazards can be evaluated and determined if they are tolerable or not. Hazards include all aspects of technology and activity that produce risk. Hazards can be physical, biological, chemical, mechanical, or psychosocial; risks can be focused on the health and safety of the worker or the business. In the end these are three main issues that must be addressed:

1. What can go wrong? – the hazard
2. How bad could it be? – the consequences of the hazard
3. How often might it happen? – the likelihood of the consequences.

The next step begins by identifying the solution(s) to hazard(s) recognized in Step 2. Consideration of PtD concepts, including the hierarchy of controls, is used to evaluate and select possible solutions for continued analysis. The business processes identified in Step 2 are revisited to determine what changes to those business processes result from the intervention or solution being considered. These changes again include both the risk of business loss or interruption and the risk of adverse worker health outcomes or environmental risks. A second risk analysis is performed considering the effect of implementing the solution(s) being considered. The relationship of hazard and consequences is evaluated using tools recommended in ANSI Z690.3/ISO/IEC 31010:2009. This step ends with providing a final risk measure – one that calculates the remaining business and SH&E risk, providing the decision maker a full understanding of the effect on risk of implementing the solution to mitigate or eliminate the hazard.

Step 4 shows the costs and benefits of the changes from the current state that were pinpointed in Step 3. The costs and benefits can be captured in dollars, percent, numbers, or simple narrative phrases. These costs and benefits are the basis for computing the metrics commonly used to evaluate business investments within a corporation or commercial enterprise setting. Financial metrics, the effects that an investment has on profit and financial condition of the company, use the costs and benefits associated with the solution or intervention being analyzed expressed in dollars. NPV, ROI, and DPP are some of the more meaningful metrics to business management. Additionally, those costs and benefits that are not expressed in dollars, such as contributions to business objectives, are summarized and presented in specific terms.

Step 5 is the last step, but by no means the least important step, in developing a business case. The final report integrates all the previous steps into a language and form that is understood and appreciated by those in the decision-making team. The report can be a lengthy written document that includes all the findings – from the hazards identified to changes in risk measures and financial and nonfinancial metrics – sources of the information used, and the members of the working team. It can also take the form of an executive summary or a PowerPoint presentation. The content and format is strongly dependent on the desires of those receiving the report.

In addition to the Business Case Development Tool being easy to use by analysts that are not trained in finance or economics, it is also flexible as it can be used under a variety of situations or conditions. It can be used in any of the stages of implementing solutions – preoperational, operational, postoperational, or postincident – defined in ANSI/ASSE Z590.3-2011.

19.4 BUSINESS CASE EXAMPLES

The following case examples help illustrate the flexibility of the tool and the benefits it provides when fully utilized.

19.4.1 Case Example One: Post Incident

In July 1999 three workers were killed in a crane collapse during the construction of a stadium pictured in Figure 19.3. A crane was lifting a section of the roof weighing over 450 t when it collapsed (pictured after the collapse in Figure 19.4). OSHA's investigation revealed that the crane's rated load was exceeded when the roof section was first lifted off the ground; workers were not kept clear of the suspended 450 tons load during the morning of the lift; and personnel platforms were used to lift workers during dangerous weather conditions. On the day of the collapse, winds gusted up to 26 miles per hour and workers indicated that it was too dangerous to operate that crane. Unfortunately, following a dispute over accelerating production to ensure meeting the opening day, the safety director left the company. It was decided that work would proceed as originally planned (OSHA, Abbott Laboratories, and Georgetown University, 2005).

Table 19.1 provides the financial information that was derived as part of the analysis. In the end, the project was over a year behind schedule and more than $620 million over budget.

Despite the importance and the impact that these costs provided when discussing the lessons learned from this case, use of the Business Case Development Tool could have added the financial metrics and nonfinancial losses associated with the project. Perhaps even more important would be the addition of the exploration of risk management, including defining the work processes in the construction project, which should have led to the adoption of PtD methodologies.

BUSINESS CASE EXAMPLES 413

Figure 19.3 Stadium Construction Before Collapse

Figure 19.4 Extensive Damage to Stadium After Collapse

TABLE 19.1 Costs Incurred as a Result of the Stadium Collapse

Cost Categories	Expenses (in Millions of Dollars)
Construction costs	413.9
Interest paid on bonds	330.8
Repairs	100.0
Jury awards	99.0
Total	**943.7**

19.4.2 Case Example Two: Regulatory Requirement

Tetrachloroethene or perchloroethylene (also known as PERC – CAS NO. 127-18-4) has long been recognized as an effective cleaning solvent and for the past 60 years has been the most commonly used solvent in the garment cleaning industry. As a volatile organic solvent, PERC poses serious health hazards if work and environmental exposure is not properly controlled. PERC can affect the body through respiratory and dermal exposure. Effects of chronic exposure to PERC include dizziness, impaired judgment and perception, dermatitis, damage to the liver and kidneys, depression of the central nervous system, and respiratory disease (NIOSH, 1997). Furthermore, the International Agency for Research on Cancer (IARC) classified PERC in group 2A, meaning that it is probably carcinogenic to humans, including esophageal cancer, lymphoma, cervical, and bladder cancer (Earnest et al., 1997; IARC, 1995). By February 2012, the Environmental Protection Agency (EPA) posted their final health assessment for PERC indicating that PERC is a "toxic chemical with both human health and environmental concerns" and is a "likely human carcinogen" (EPA, 2012a, 2012b).

Regulations restricting the use of PERC in the United States have been increasing since the 1990s but remain principally focused on reducing or eliminating its ozone depletion properties and environmental pollution contributions. Under the 1990 revisions of the Clean Air Act, in 1993 EPA issued technology-based national emission standards for hazardous air pollutants (NESHAP) that required operators to control PERC emissions at individual dry cleaners. As part of the EPA process, these standards underwent review that led to revisions accounting for the new developments in production practices, processes, and control technologies, with final standards going into effect in 2006. The regulations include a phaseout of PERC use at dry cleaners located in residential buildings by December 21, 2020, along with requirements that already have reduced PERC emissions at other dry cleaners. EPA's 2007 air toxics standards for the halogenated solvent cleaning industry also address PERC as it set limits for a group of toxics that include this solvent. EPA also set the maximum contaminant level for PERC under the Safe Drinking Water Act.

OSHA established mandatory permissible workplace exposure limits and provided guidance to reduce worker exposure, which includes recommendations for personal protective equipment. Although not specifically addressing PERC, OSHA standards that may apply when workers are exposed to PERC include Hazard Communication (29 CFR 1910.1200), General requirements for personal protective equipment (29 CFR 1910.132), and Respiratory Protection (29 CFR 1910.134).

As a result of the increasing attention to the health and environmental concerns surrounding PERC, including impending more stringent regulations, extensive research and development have provided acceptable PERC alternatives to the garment cleaning market. Principle alternatives include petroleum solvents, silicone-based solvents, liquid carbon dioxide, and wet cleaning (Sinsheimer et al., 2007). Each alternative has different physical properties that affect their SH&E hazards. These alternatives positioned the US garment cleaning industry of over 37,000 establishments (US Census Bureau, 2012) to make decisions about which of the solvent substitutes and processes should be selected.

Over the past two decades, numerous efforts by the US EPA, the Center for Neighborhood Technology (CNT), the Pollution Prevention Center, and the Toxics Use Reduction Institute (TURI) have been undertaken to assist garment cleaning industry in the decision-making process. The EPA and TURI performed evaluations of alternative chemicals and processes focusing on their cost and performance, including the health and safety effects (EPA, 1993; TURI, 1996; Ellenbecker & Geiser, 2011).

The CNT explored alternatives "in which cleaners can use new processes that not only are environmentally friendly to workers, garments and communities, but also allow the small "mom and pop" cleaner to continue to operate profitably" (Star & Ewing, 2000). The Pollution Prevention Center focused on determining the viability of professional wet cleaning in California (Sinsheimer et al., 2004, 2007).

Developing a business case began with investigating the financial advantage of each garment cleaning process. Table 19.2 presents a comparison of the initial cost of equipment and installation for five of the more common processes. Interestingly, these costs confirmed that professional wet cleaning was a viable option for exploration by those organizations who were interested in reducing the exposure to PERC in the profession of garment cleaning industry in the 1990s.

The differences in operational costs between wet and dry cleaning processes were collected in nine case studies. Funding and technical assistance to California companies willing to replace their existing dry cleaning process with wet cleaning began in 1995. Tables 19.3–19.5 provide the results from that work led by Peter Sinsheimer at the Pollution Prevention Center at Occidental College.

Operational costs coupled with the machine and installation costs were used to develop financial measures, which assist making decisions in the selection and adoption of chemical substitution and

TABLE 19.2 Comparison of Costs

Process	Machine ($)	Installation ($)
PERC dry cleaning	43,900	2,500–5,000
Hydrocarbon	61,000	5,000–6,000
GreenEarth (liquid silicone)	63,000	5,000–6,000
CO_2	140,000	50,000
Wet cleaning	40,000	2,000–2,500

Source: Data from Fong et al. (2006).

TABLE 19.3 Monthly Operating Expenses for Dry Cleaning Processes

Monthly Expense	Firm 1	Firm 2	Firm 3	Firm 4	Firm 5	Firm 6	Firm 7	Firm 8	Firm 9
Solvent	50	100	90	56	66	133	929	200	113
Detergent	53	16	21	39	50	5	133	13	83
Water	44	20	N/A	N/A	N/A	N/A	73	N/A	111
Electricity	89	143	187	93	300	244	156	642	364
Gas	278	466	221	267	488	552	438	488	435
Filters	60	25	11	23	N/A	40	882	40	7
Hazardous waste	100	54	40	54	35	100	100	117	50
Machine upkeep	147	239	83	132	172	111	158	298	119
Equipment	430	299	270	299	375	375	375	375	375
Regulatory fees	108	108	81	108	91	48	119	74	65
Total ($)	1359	1470	1004	1071	1577	1608	3365	2247	1723

Source: Data from Sinsheimer et al. (2004).

TABLE 19.4 Monthly Operating Expenses for Wet Cleaning Processes

Monthly Expense	Firm 1	Firm 2	Firm 3	Firm 4	Firm 5	Firm 6	Firm 7	Firm 8	Firm 9
Solvent	0	0	0	0	0	0	0	0	0
Detergent	121	246	82	167	63	60	0	300	400
Water	34	30	N/A	N/A	N/A	N/A	540	N/A	91
Electricity	50	115	132	75	120	144	140	324	292
Gas	266	510	144	255	353	387	334	353	408
Filters	0	0	0	0	N/A	0	762	0	0
Hazardous waste	0	0	0	0	0	0	0	0	0
Machine upkeep	24	24	24	24	83	24	0	24	24
Equipment	208	208	208	280	193	379	379	379	379
Regulatory fees	0	0	0	0	0	0	24	0	0
Total	$703	$1133	$590	$801	$811	$994	$2179	$1379	$1593

Source: Data from Sinsheimer et al. (2004).

redesigned equipment. Using data collected by the State of California's Air Resources Board (Fong et al., 2006) and a 5-year time frame, the NPV was derived comparing each process to wet cleaning and PERC (see Table 19.6).

The results demonstrate that the wet cleaning process is the best selection from a financial perspective – even without the probable reduction in occupational injuries and illnesses or improvements in

TABLE 19.5 Cost Reductions in Operating Expenses Using Wet Versus Dry Cleaning Processes

	Firm 1	Firm 2	Firm 3	Firm 4	Firm 5	Firm 6	Firm 7	Firm 8	Firm 9
Cost reduction ($)	656	337	414	270	765	614	1185	867	130
Average savings: $582 monthly										

Source: Data from Sinsheimer (2008).

TABLE 19.6 Net Present Value of Alternative Garment Cleaning Processes

Base Versus Comparison	Net Present Value ($)
PERC versus **wet clean**	29,061
Hydrocarbon versus **wet clean**	31,924
GreenEarth versus **wet clean**	53,099
CO_2 versus **wet clean**	170,911
PERC versus hydrocarbon	2,863
PERC versus GreenEarth	24,038
PERC versus CO_2	141,850

productivity that are typically included in NPV calculations and could have been easily added using the Business Developer Tool.

However, the effect on the environment was considered in nearly every wet cleaning study, regardless of the other issues considered. Furthermore, public presentation often referred to the wet cleaning process as being a "green" solution – pointing to the associated positive corporate social responsibility of making the change. Publication titles, such as *Fashioning a Greener Shade of Clean: Commercialization of Professional Wet Cleaning in the Garment Care Industry* and *The Viability of Professional Wet Cleaning as a Pollution Prevention Alternative to Perchloroethylene Dry Cleaning*, clearly demonstrated the link to environmental emphasis. The article "Being green while staying clean in Malibu (Colony Cleaners)" was published in *The Malibu Times* to publicize the conversion of one local cleaner to the wet cleaning process. In *Leading the Green Cleaning Wave,* Hesperian Cleaners was praised as the first in Alameda County to be a Bay Area Green Business following decision to "go green" when changing to wet cleaning. The concept of informing customers of the company commitment to being "green" transcends to naming the company – The Greener Cleaner, a professional garment cleaning shop in Chicago. A critical reason for adopting this PtD engineering control was improving the environment and the connection of that improvement to the company reputation.

19.4.3 Case Example Three: Operational

The University of California, Davis Agricultural Ergonomics Research Center, AgSafe (a nonprofit occupational safety and health organization), and NIOSH worked collaboratively in two separate efforts to reduce or eliminate ergonomic risk factors (RFs) for grape harvest workers in the wine grape industry in Northern California (Myers et al., 2002).

The initial effort conducted the risk assessment, which began with the description of the following work process, illustrated in Figure 19.5. The grape pickers rapidly move down a row, reaching to grasp and cut grape clusters and dropping them into plastic containers. These containers are pushed down the row with the worker's legs moving sideways. When the worker determines that the container or tub is full, he lifts, carries, and leans his body against the vine lifting the tub over his head to finally dump

Figure 19.5 Wine Grape Harvest Process Consists of Six Basic Tasks

TABLE 19.7 Risk Factor Comparison

	Large Tub	Small Tub
Lifting force	57 lb (season average)	46 lb (season average)
Sliding force	19–22 lb (terrain differences)	13–16 lb (terrain differences)
NIOSH lifting equation	3.4	2.4
Energy expenditure	47.7% of aerobic capacity	45% of aerobic capacity
Back injury probability	0.64	0.60

Source: Data from Myers et al. (2006).

the grapes into a larger container. The worker returns to the vine where the last grape was picked and the cycle begins again. The grape tub held an average of 57 lb, but weights of up to 80 lb have been recorded in the field. A robust risk assessment led to recommending a smaller-sized tub to minimize injuries from lifting and carrying cut grapes in the original study. Table 19.7 provides the changes in risk assessment associated with implementing this simple PtD solution (Myers et al., 2006).

In addition to the hazard identification and associated risk assessment, the team measured the reduction in pain and injury. The financial metrics measuring the ROI for this engineering control were calculated but not made available.

A second project was designed using the original team's expertise to determine if the smaller tubs were still being used by those who had originally adopted them and to identify the reasons for the continued use. Financial data was requested from the participating companies, but none was provided. However, local suppliers provided cost estimates for the small and large tub, $13 and $11, respectively. Tub handles were modified by adding grips at an additional cost of $.50 per tub, resulting in an additional expense of $2.50 per tub for each smaller purchased. Using the estimated maximum number of tubs purchased annually, use of the smaller tub required an investment of $1250. The companies did not track the injury and illness cases specifically attributable to lifting wine grape tubs, so a ROI metric could not be calculated. However, using the average cost of one back injury in California of $56,874, a total of 22,000 smaller tubs could be purchased without a negative ROI.

However, using the principles of the tool, it became clear that the decision to continue using the smaller tubs was based on the nonfinancial benefits. The companies in the study indicated that the most important business objectives included reducing worker turnover rates; improving worker morale; decreasing worker aches, pain, and injury; meeting harvest time frame; and improving or maintaining wine grape quality. The initial study highlighted the decreased worker aches, pain, and injury. Interviews and surveys included in the second study demonstrated that workers were very pleased with the smaller tubs, and as a result the labor turnover rate was substantially reduced – many workers remained in their jobs for over 15 years and one company had maintained the same workforce since adoption of the smaller tubs. With maintaining the same skilled tradesman during the harvest season, meeting the time frame for harvest is far more certain. Furthermore, the smaller tubs maintained the quality of the grape because there were fewer layers of grapes to cause crushing.

This analysis described the human risk reduction, provided some financial measures, and provided a qualitative assessment of business objectives. If the tool had been used to develop the business case, it could have added the unique ability to identify and quantify the reduction in business risk.

19.4.4 Case Example Four: Postoperational

Management of a large not-for-profit hospital planned to replace all existing soiled linen and trash collection receptacles with new containers of a single size and shape. This provided an opportunity for the risk management authority responsible for the safety and health of workers to explore alternative trash bags with the goal of reducing occupational injury and illnesses associated with lifting and carrying bags containing linen and trash. The following is a description of the first steps in an ongoing business case development project.

The decision was to evaluate the ergonomic advantages of LiteLift™ ergonomic bags compared to the current bags being used in linen collection and disposal. LiteLift™ bags are specially designed with a handle in the bottom of the bag, and as the employees tie the top of the bag, the tie becomes a second handle. Use of two handles allows the carriers to balance the load while lifting and carrying.

The project began by establishing a team of staff who had either interest in the project, were safety and health professionals, or had resource allocation authority or understanding. The risk assessment of the conventional linen bags began with the description of the following work process, illustrated in Figure 19.6.

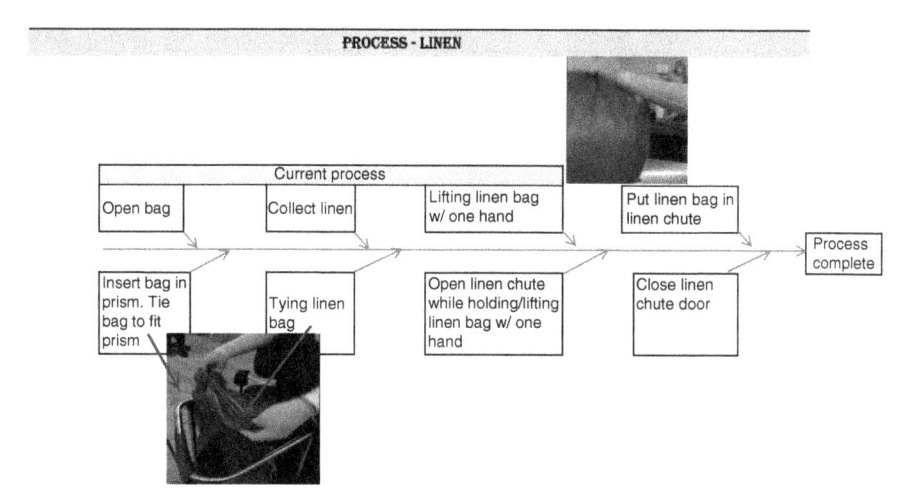

Figure 19.6 Current Process for the Collection Through Disposal of Soiled Linen

The next step in developing a business case is to identify the main safety and health hazards. Although the safety and health professionals track the hazards, completing the form found in the tool provided the opportunity to maintain specific records of this project. Three main ergonomic hazards were identified and recorded in the form below.

Figure 19.7 Hazards and Their Potential Effects Recorded

With the hazards identified, the next step involved determining the risk factor (RF) for each of the three potential effects. RF is the number resulting from multiplying the probability of the potential effect occurring by a number used to measure the severity of the expected loss in case of the occurrence of the potential effect. There are a large variety of risk assessment methods, but for this project, the ergonomic risk assessment of the current process was conducted utilizing the simple risk assessment matrix described in the PtD standard (ANSI/ASSE Z 590.3-2011). The risk to human health is not the only risk associated with workplace hazards. The risk to the continuity of business operations should also be considered. A similar risk assessment matrix was utilized to estimate that risk. Figures 19.8 and 19.9 present the risk assessment results for both types of risk.

H&S risk assessment matrix (RAM): numerical grading and scoring				
Assess Risk Associated with Potential Effects				
	Potential Effect(s) #	1 Lift St	2 Back	3 Cl
Severity ranking:	3	3	4	3
Probability ranking:	4	4	4	3
Total		12	16	9

Figure 19.8 Risk of Health Effects Is Determined and Entered in H&S RAM

Business Risk Assessment Matrix: Numerical Ratings			
Risk of business continuity loss from hazardous work environment			
Outcomes	Financial	Ethical	Legal
Intensity rating: 4	4	4	4
Likelihood rating: 4	4	4	3
Total	16	16	12

Figure 19.9 Risk of Business Continuity Loss Is Determined and Entered in B RAM

To present a 30,000 ft view of the current state, hazards and consequences are then presented utilizing a modified Bow-Tie risk assessment methodology shown in Figure 19.10. The RF numbers are transferred to the modified Bow-Tie risk assessment diagram.

As was the case in recording the hazards associated with the current process, consideration of PtD concepts, including the hierarchy of controls, should be discussed and documented even though the solution had already been selected. This works to ensure that the identified solution was selected using established criteria and no other solution would be preferred at this stage of analysis. The team provided their expertise and input into identifying the barriers and preventive actions that are included in the analysis. Figure 19.11 provides the PtD hierarchy of controls used to select the most effective options and examples of controls applied.

With the solution being agreed upon, the project continued with a pilot test using the LiteLift™ to lift, carry, and dispose of soiled linens. The same risk assessment methodology was utilized to evaluate hazards and consequences after the substitution of the bags. Hazards and consequences for the new bags are presented utilizing a modified Bow-Tie risk assessment methodology again shown in Figure 19.12.

Calculating the residual risk and risk reduction scores was the final step in the risk assessment. The calculations showed a 37% risk reduction. The results of this analysis are presented in Figure 19.13. However, management acceptance of a PtD solution is not always based on RA alone. As a result the more detailed financial/nonfinancial analysis is often necessary.

Completion of the pilot test and the risk assessment portion found in the tool led to numerous recommendations. It was determined by the team that smaller-sized bags would better control the weight of the bag. It was also determined that the unique handle on the lower end of the bag allows lifting with two hands and equal weight distribution. Therefore, they decided to modify their training to include two-handed lifting techniques using the LiteLift™ bags to further reduce the risk of injury or illness. For the manufacturer, the team recommended that a fill line, preferably yellow, should be added to the bag design. Having an easily visible fill line would allow the employees to avoid overfilling the bags and lifting heavy bags.

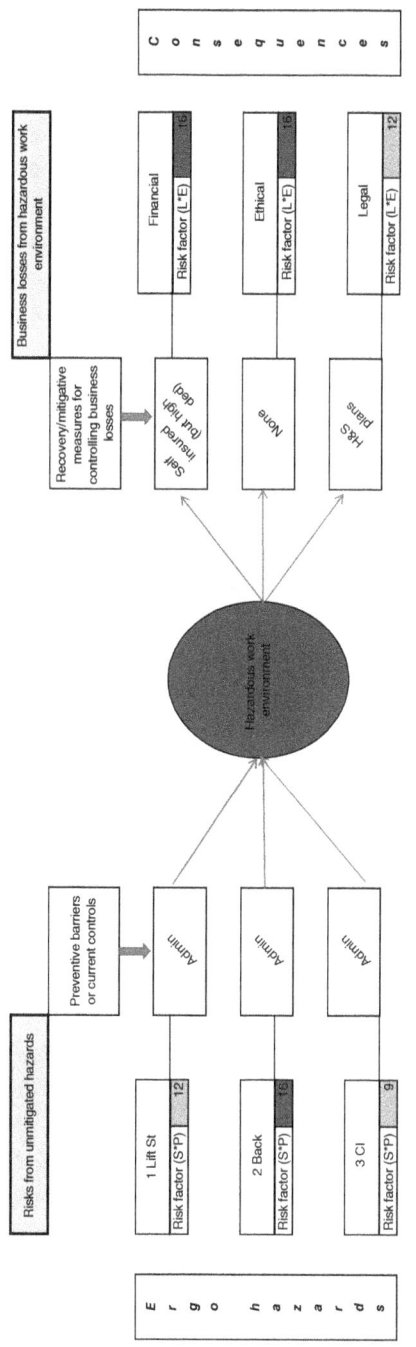

Figure 19.10 Modified Bow-Tie Diagram Provides Overview

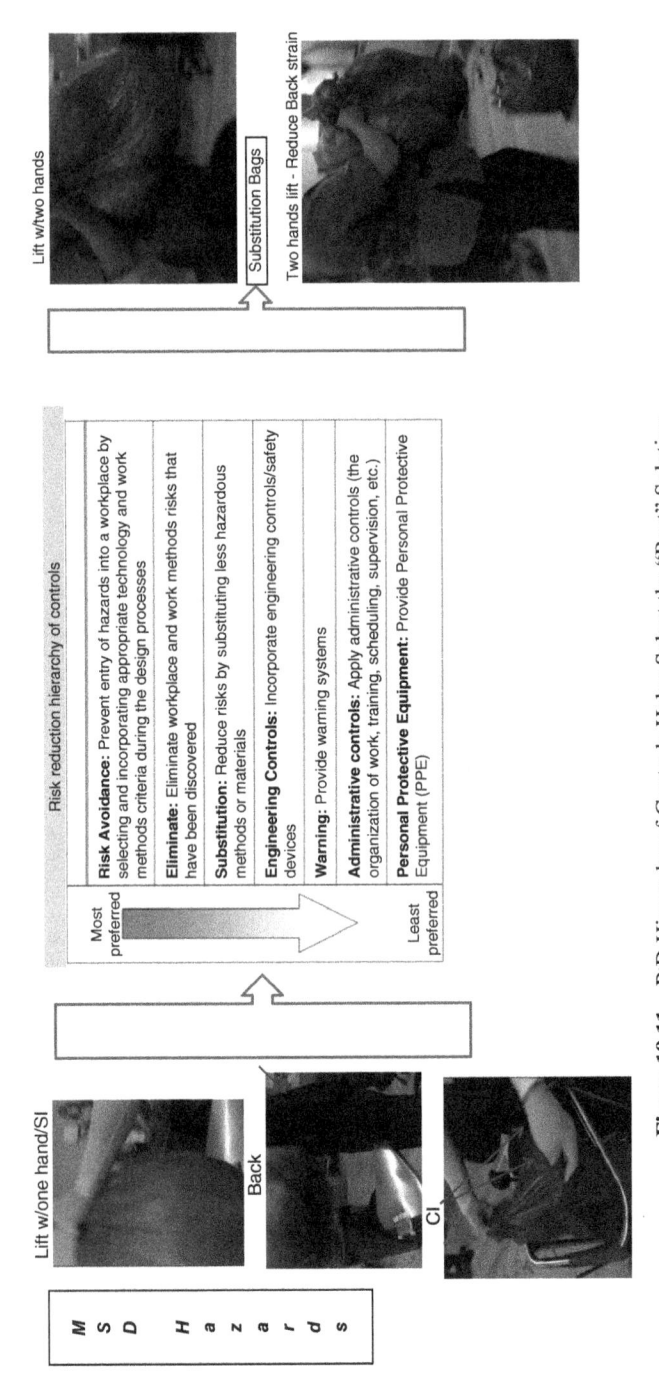

Figure 19.11 PtD Hierarchy of Controls Helps Select the "Best" Solution

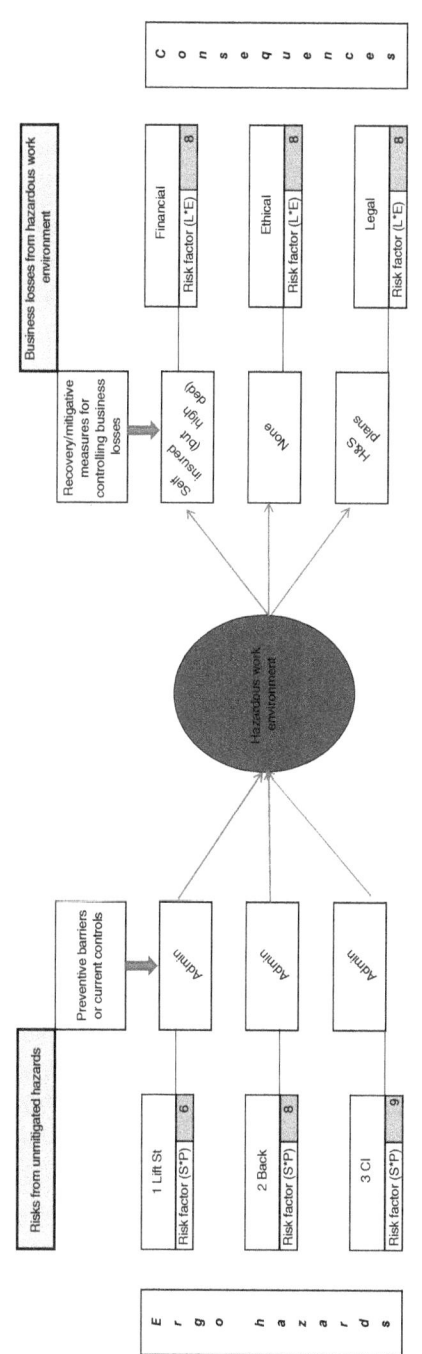

Figure 19.12 Modified Bow-Tie Diagram Overview of "New" Risk Assessment Associated with the Solution

HEALTH AND SAFETY RISK REDUCTION					
	Hazards	Risk Factor CS	Risk Factor FS	Risk Reduction (%)	Residual Risk(%)
	1 Lift St	12	6	50.00	50.00
	2 Back	16	8	50.00	50.00
	3 Cl	9	9	0.00	100.00
	Total	37	23	37.84	62.16
Is this an acceptable level of health and safety risk?			No		

Figure 19.13 Risk Reduction Recorded in the Tool

19.5 CONCLUSION

SH&E professionals agree that hazard identification and risk assessment are critical to protecting the worker. Ensuring the best outcomes for the health and safety of workers requires attention to selecting appropriate solutions, such as those grounded in PtD concepts.

However, demonstrating the business value that a solution contributes can be challenging for SH&E professionals who do not have the expertise or experience in such efforts. They must learn the world of business and "corporate speak" and make the case to executive management that SH&E solutions, activities, and programs are not only necessary but also good business investments. Whether you have a large corporate program, a small department, or single SH&E professional or technician, the Business Case Development Tool makes deriving the business case a simple, logical, and easy-to-understand connection between the risk assessment and the bottom line – financial or nonfinancial.

REVIEW QUESTIONS

1. Explain what is meant by a "business case" in safety, health, and environmental (SH&E) efforts.
2. Explain why developing a business case is important to SH&E.
3. Describe methods used to develop a business case.
4. Identify the steps used in developing a business case for SH&E.
5. List several risk assessment tools that are helpful in developing a business case.

REFERENCES

American Industrial Hygiene Association (2009). Strategy to Demonstrate the Value of Industrial Hygiene. Information available at: https://www.aiha.org/votp_new/products/index.html (accessed January 21, 2016).

ANSI/ASSE Z590.3-2011. Prevention Through Design: Guidelines for Addressing Occupational Hazards & Risks in Design & Redesign Processes, 2011. Information available at: http://www.asse.org/shoponline/products/Z590_3_2011.php (accessed December 30, 2015).

ANSI/ASSE/ISO Z690. Risk Management Standards Package, n.d. Information available at: http://www.asse.org/shoponline/products/EZ690-PKG.php (accessed December 30, 2015).

Biddle E., 2013. "Business cases: supporting PtD solutions." *Professional Safety* 58(3):56–64.

Biddle, E., Carande-Kulis, V., Woodhull, D., Newell, S., and Shroff, R. 2011. The Business Case for Occupational Safety, Health, and Environment and Beyond. Clark, S., Burke, R.J. and Cooper, C.L. (Eds.), *Occupational Health and Safety*. Farnham, England: Ashgate Publishing Ltd.

REFERENCES

Earnest, G.S., Spencer, A.B., Smith, S.S., et al. Control of Health and Safety Hazards in Commercial Dry Cleaners: Chemical Exposures, Fire Hazards and Ergonomic Risk Factors *(NIOSH Publication No. 97-150)*. Washington, DC: U.S. Department of Health and Human Services (DHHS), CDC, NIOSH, 1997.

Ellenbeck, M. & Geiser, K. 2011. "The origins of the Massachusetts toxics use reduction program." *Journal of Cleaner Production,* 19(5), 389–396.

EPA. *Multiprocess Wet Cleaning: Cost and Performance Comparison of Conventional Dry Cleaning and an Alternative Process (EPA Document No. 744-R-93-004)*. Washington, DC: Author, 1993.

EPA. Fact sheet on perchloroethylene, also known as tetrachloroethylene, 2012a. Retrieved from http://epa.gov/oppt/existingchemicals/pubs/perchloroethylene_fact_sheet.html (accessed January 21, 2016).

EPA. EPA releases final health assessment for tetrachloroethylene (perc)/public protections remain in place (Press release), 2012b. Retrieved from http://yosemite.epa.gov/opa/admpress.nesf/0/E99FD55271CE029F852579A000624956 (accessed January 21, 2016).

Fong, M., Chowdhury, H.R., Houghton, M., et al. (2006). *California Dry Cleaning Industry Technical Assessment Report*. Sacramento, CA: California EPA, Air Resources Board, Stationary Source Division, Emissions Assessment Branch.

International Agency for Research on Cancer (IARC). *Evaluation of Carcinogenic Risks to Humans: Dry Cleaning, Some Chlorinated Solvents and Other Industrial Chemicals*. IARC Monographs on the Evaluation of Carcinogenic Risks to Humans, 1995.

ISO 31000:2009. Risk Management – Principles and Guidelines, 2009. Information available at: http://www.iso.org/iso/home/standards/iso31000.htm (accessed December 30, 2015).

ISO/IEC 31010:2009. Risk Management – Risk Assessment Techniques, 2009. Information available at: http://www.iso.org/iso/home/standards/iso31000.htm (accessed December 30, 2015).

Myers, J. M., Miles, J. A., Tejada, D. G., Faucett, J., Janowitz, I. Weber, E., Smith, R., and Garcia, L., 2002. "Priority risk factors for back injury in agricultural field work: vineyard ergonomics." *Journal of Agromedicine,* 8(1), 39–54.

Myers, J. M., Miles, J. A., Faucett, J., Fathallah, F., Janowitz, I., Smith, R., and Weber, E. A., 2006."Smaller loads reduce risk of back injuries during wine grape harvest." *California Agriculture,* 60(1), 25–31.

Occupational Safety and Health Administration, Abbott Laboratories, and Georgetown University. The Business Case for Safety: Adding Value and Competitive Advantage, 2005. Presentation at The Center for Business and Public Policy and available at: https://www.osha.gov/dcsp/success_stories/compliance_assistance/abbott/abbott_casestudies/index.html (accessed December 30, 2015).

Sinsheimer, P.J., Grout, C., Namkoong, A., et al. *Fashioning a greener shade of clean: Commercialization of professional wet cleaning in the garment care industry*. Los Angeles, CA: Pollution Prevention Education & Research Center, 2004.

Sinsheimer, P.J., Grout, C., Namkoong, A., et al. (2007). The viability of professional wet cleaning as a pollution prevention alternative to perchloroethylene dry cleaning. *Journal of the Air & Waste Management Association,* 57, 172–178.

Star, A. & Ewing, S. *Real-World Wet Cleaning: A Study of Three Established Wet-Cleaning Shops*. Chicago, IL: Center for Neighborhood Technology, 2000.

Toxics Use Reduction Institute (TURI). *Garment Wet Cleaning: Utopia Cleaners, 1996 (TURI Technical Report No. 35)*. Lowell, MA: University of Massachusetts, Author, 1996.

U.S. Census Bureau. County business patterns, 2012. Retrieved from www.census.gov/econ/cbp/index.html (accessed January 21, 2016).

20

RISK ASSESSMENT: GLOBAL PERSPECTIVES

JIM WHITING

Consultant, Risk at Workplaces Pty Ltd, Indooroopilly, QLD, Australia

20.1 OBJECTIVES

- Using ISO 31000 for maturity assurance and conformity
- Global uptake of ISO 31000: International Risk Management Standard
- Global comparison of risk tolerability criteria
- Tolerability criteria for planning new industries/locations
- Investment to prevent a fatality
- Shifting the paradigm from absolute safety to risk management – ALARP and reasonably practicable
- Changing traditional language to risk-based language for more effective safety risk conversations

20.2 INTRODUCTION

In these days of business globalization, professional safety, health, and environmental (SH&E) practitioners often need to be aware of what are the international practices and standards around the world. Global organizations need assurance of the uniformity, consistency, and harmonization of their risk management standards, policies, and processes, in general, and their SH&E standards in particular.

In 2009, ISO, the international body in charge with achieving global standardization, finalized a standard risk management system (RMS) to achieve consistency and reliability in risk management by creating ISO 31000, a standard that is applicable to all forms of risk. It should be noted that in 2011, ISO 31000:2009 was nationally adopted in the United States as an American National Standard,

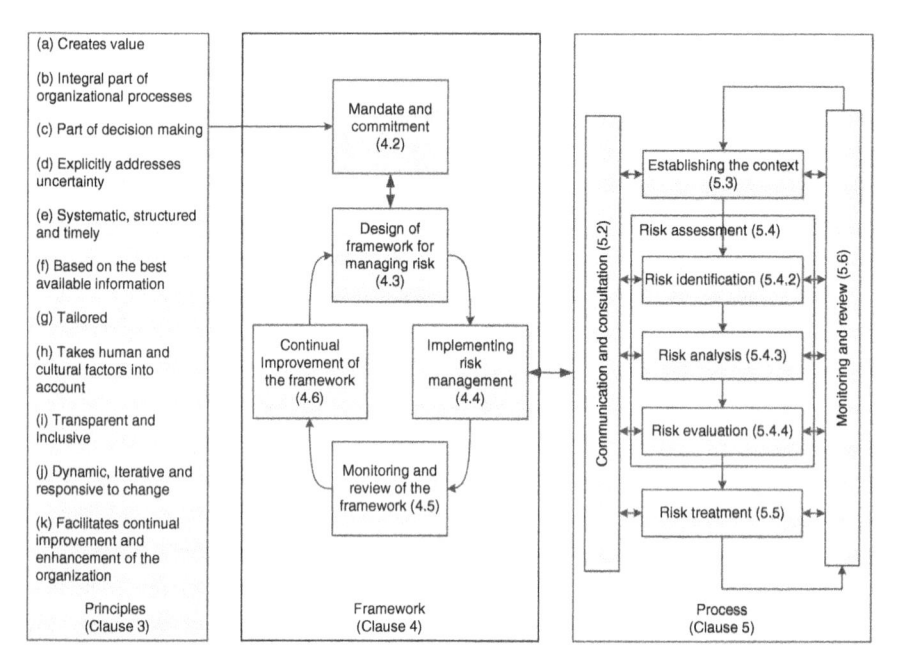

Figure 20.1 Relationships Between the Risk Management Principles, Framework, and Process. *Source:* Taken from ANSI/ASSE Z690.2-2011 (Courtesy of the American Society of Safety Engineers)

ANSI/ASSE Z690.2-2011, *Risk Management Principles and Guidelines*, and is identical to the ISO standard. The International Risk Management (RM) Standard consists of:

1. Principles – a framework and a structured process for effective RM shown in Figure 20.1
2. Consistently defined RM vocabulary and language
3. A set of performance criteria
4. One common overarching RM process shown in Figure 20.2
5. Guidance on how that process should be integrated into the decision-making processes of any organization.

20.3 USING ISO 31000 FOR MATURITY ASSURANCE AND CONFORMITY

The relatively new risk management standard ISO 31000:2009 is being applied worldwide and now allows an organization of any size and business activity to assess the maturity and adequacy of its RMS. Many organizations currently have at least informal risk management practices and processes, which include a number of fundamental components of an RMS as detailed in ISO 31000/ANSI/ASSE Z690.2, ANSI/AIHA/ASSE Z10-2012, ANSI/ASSE Z590.3-2011 Prevention through Design, and even the Occupational Safety and Health Administration's (OSHA) Process Safety Management standard, 1910.120 (briefly described in Chapter 6 – What-If Analysis). However, the means of gaining confidence and assurance that those practices and processes are adequate, mature, and effective are often lacking. Even if an organization has not already adopted a formal risk management process for particular types of risk or business circumstances, it can and should decide to carry out a regular critical review of its existing RM practices and processes in the light of this standard's requirements.

Many organizations have made the extra step to integrate their RMS for consistency in handling all risk domains. ISO 31000 becomes the overarching envelope that provides uniform management

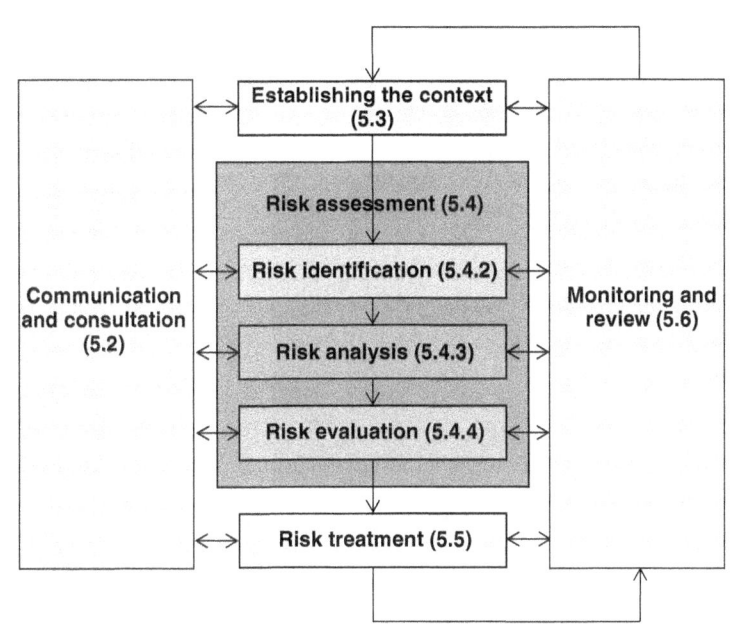

Figure 20.2 Risk Management Process. *Source:* Taken from ANSI/ASSE Z690.2-2011 (Courtesy of the American Society of Safety Engineers)

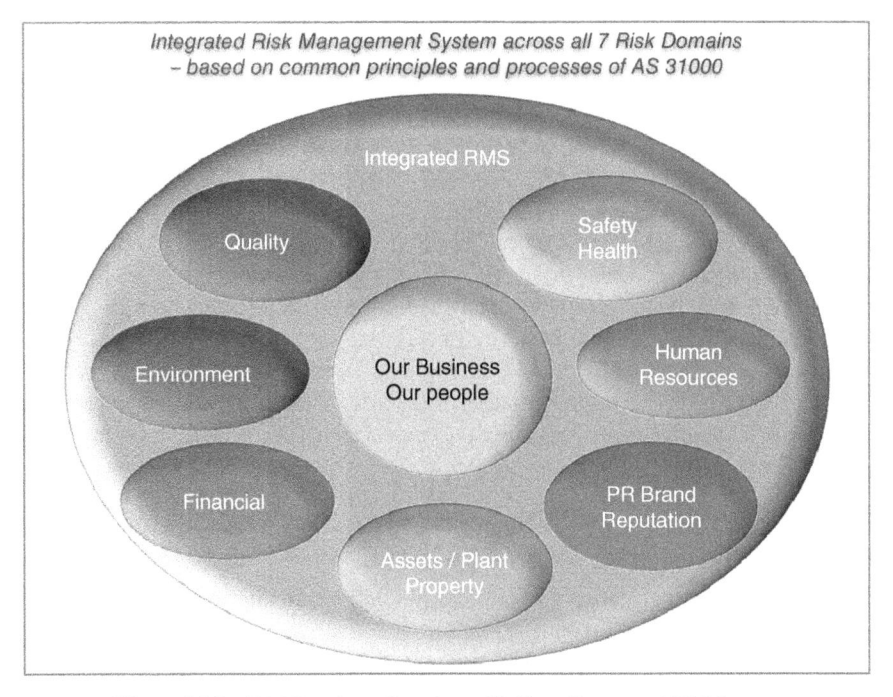

Figure 20.3 RM Envelope Creating a Uniform Integrated RM System

processes for all types of risks. Integrating RMS is often referred to as creating an Enterprise Risk Management (ERM) System or an Integrated Risk Management (IRM) System, as shown in Figure 20.3.

A process called *conformity assessment* consists of tools constructed and used to review, assess, and even audit how well the organization conforms and complies with the standard and hence provides a measure of maturity and adequacy of the organization's own system.

ISO 31000 is a *Principles and Guidelines* or *Performance* Standard rather than a *Specification* Standard. As such it is NOT normally intended to be used for formal external assessment for *certification* purposes as many organizations do with ISO 9001 (Quality Management Systems) and ISO 14001 (Environment Management Systems) and soon to come, ISO 45001 (OHS Management Systems). Nevertheless the RM Standard does provide an excellent basis for the construction of a maturity assessment tool.

Specification standards such as ISO 9001, ISO 14001, and ISO 45001 use terminology such as *shall* to specify conformance requirements, whereas ISO 31000 uses terminology such as *should*. In that way the RM Standard still details what an adequate and mature RMS should be. For example, the standard states in Clause 3(b) Principles that *"an organization* **should** *at all levels comply with the principle that Risk Management is an Integral Part of all Organizational Processes."*

Consequently the assessment tool includes questions requiring the organization to provide information and evidence that risk management is not a stand-alone activity separate from the main activities and processes of the organization. To demonstrate conformance, the organization needs to answer the questions how risk management is part of the responsibilities of each level of management and an integral part of all organizational processes, including strategic planning and all project and change management processes.

The standard covers each of the three main components (and the relationships between them) of a comprehensive, mature RMS. As shown in Figure 20.1, they are given as follows:

- Principles for managing risk (22 questions)
- Framework in which it should occur (79 questions)
- Risk management process shown in Figure 20.2 (164 questions).

A practical conformity assessment tool (Whiting, 2012) consists of three sets of comprehensive evidence-seeking questions designed for each detailed *"should" expectation* of the standard in its Principles, Framework, and Process. It can be used internally or externally as a first- or second- or even third-party audit tool. It can be tailored to any activity and risk domain. According to the answers to these questions, an overall rating of Maturity Levels 1–5 can be determined as in Figure 20.4.

Although a first-party conformity review, assessment, or audit can often be subjective in nature, the tool when used rigorously can yield rating measures of maturity and adequacy that do help improve objectivity and precision in evaluating the current status of a system. If performed diligently, first-party audits and self-assessments can provide

(1) feedback to everyone in the organization that will allow them to have confidence and assurance that the RMS is both being implemented and achieving effectiveness;
(2) a measure of the organization's evolutionary continuous improvement effort to assess the return on investment for sustaining that effort.

The ISO 31000 Standard and a conformity assessment tool (Whiting, 2012) can be applied throughout the life of an organization or a project and to a wide range of activities, including strategies, decision making, operations, processes, functions, projects, products, services, and assets.

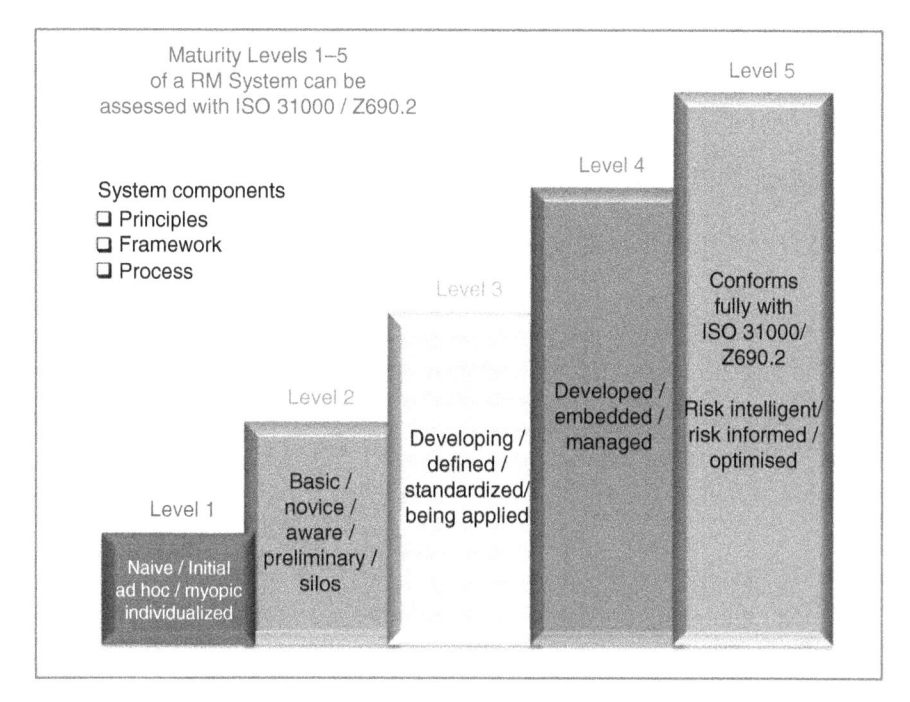

Figure 20.4 Maturity Levels Determined by Conformity Assessment

As a benchmark measurement at any given time, they provide an effective means of following change management and continuous improvement. They can be used by any public, private, or community enterprise, association, group, or individual. Therefore, this standard is not specific to any industry or sector. As well, this standard and conformity assessment tool can be applied to any type of risk domain, whatever its nature, and whether positive (upside risk) or negative (downside risk) consequences are being considered. Establishing internal and external benchmarking at a single organization, national or international level is possible.

20.4 GLOBAL UPTAKE OF ISO 31000: INTERNATIONAL RISK MANAGEMENT STANDARD

Since 2009, there has been widespread acceptance and adoption by many countries of the ISO 31000 Standard as their national RM Standards as illustrated in Figure 20.5. ISO 31000, as previously described, is not used for certification but does provide freedom for organizations to decide how deeply and fully they establish and adhere to its principles.

The real strength of ISO 31000 is that it encourages risk managers and their organizations to understand and make use of the relationships, commonalities, and differences between various risk management methods, standards, and best practices across all risk domains, rather than just SH&E concerns.

It is designed to be used to harmonize risk management processes in existing and future management system standards such as ISO 9001 QMS, ISO 14001 EMS, and ISO 45001 OHSMS. It provides a common approach in support of standards dealing with specific risks and/or sectors but does not replace those standards.

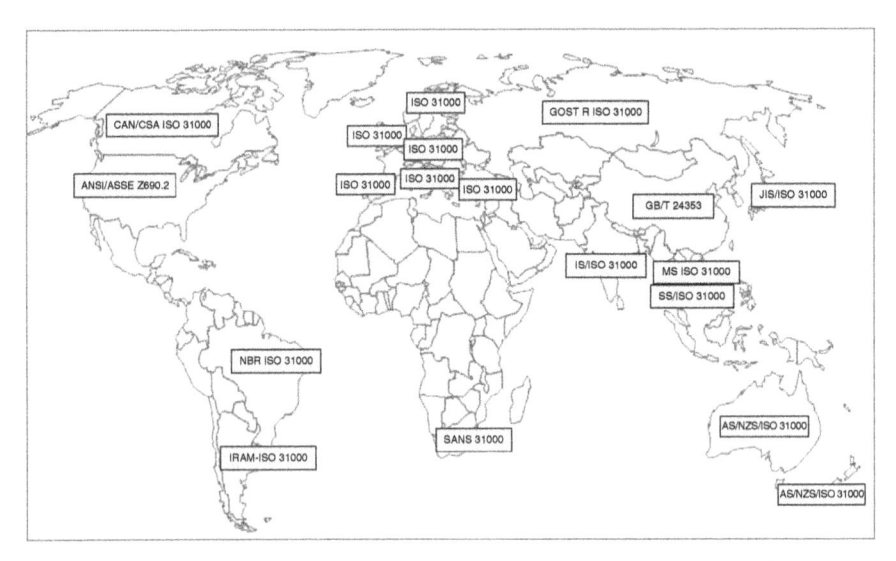

Figure 20.5 Global Adoption of ISO 31000 (Representative, Non-exhaustive List)

20.5 GLOBAL COMPARISON OF RISK TOLERANCE CRITERIA

A critical stage in the RM Process is *Evaluation* (see Chapters 2–4). At that stage in the process, the risk manager needs to evaluate the sizes of the risks calculated or estimated in the preceding *Analysis* stage by comparing them with predetermined criteria developed in the initial *Establish Context and Scope* stages. The criteria of most importance are the prior agreed risk acceptability levels and whether the risk level is being continually managed down so far as is reasonably practicable (SOFAIRP).

Risks to people can be represented in two ways. Both are a combination of the likelihood of an event happening (e.g., an accident at a major hazard installation) and the possible consequences – in terms of harm to people:

20.5.1 Individual Risk

Individual risk is the likelihood or probability or chance that a particular individual at a particular location under specific exposure circumstances will be harmed. It is usually described in numerical terms such as "*a 1 in 20,000,000 chance of being killed by lightning per annum (p.a.)*." But assessment of individual risk does not take account of the total number of people at risk from a particular event. Individual risk is usually expressed as the probability of fatality of an individual per year such as

1 in 1000 p.a., 1 chance in 1000 p.a., 10^{-3} p.a., 1E−03 p.a.

As an example if the fatality rate for a traffic risk is estimated as 10,000 driver fatalities p.a. in 100,000,000 drivers, then the individual risk is expressed as 10,000/100,000,000 p.a. = 1 in 10,000 p.a. or 10^{-4} p.a., or 1E−04 p.a.

If the risk of work fatalities for a generic or specific risk exposure is estimated for an organization of 20,000 exposed employees to be 4 p.a., then the individual risk is expressed as 4/20,000 p.a. = 1 in 5000 p.a. = 2 in 10,000 p.a. = 2 chances in 10,000 p.a. = 2×10^{-4} p.a. = 2E−04 p.a.

In reverse, if the individual risk for a given exposure is estimated as 2 chances in 10,000 p.a., then:

- For an organization of 20,000 exposed employees, there is a risk of 20,000 × 2/10,000 p.a. = 4 fatalities p.a.

- For an organization of 5000 employees (all other risk factors the same), there is a risk of $5000 \times 2/10{,}000 = 1$ fatality p.a.
- For an organization of 1000 employees (all other risk factors the same), there is a risk of $1000 \times 2/10{,}000 = 0.2$ fatality p.a. OR 1 fatality in 5 years.

20.5.2 Societal Risk

Societal risk is a way to estimate the chances of numbers of people being harmed from an incident. The likelihood of the primary event (an accident at a major hazard plant) is still a factor, but the consequences are assessed in terms of level of harm and numbers affected, to provide an idea of the scale of an accident in terms of numbers killed or harmed.

Societal risk can also be expressed as a *potential loss of life* (PLL), which is the number of fatalities that may be expected to occur each year, averaged over a long period. The number should be small: if 100 people are each exposed to a risk level of 10 in a million per year, the PLL is 0.001.

The PLL is a useful basis for cost/benefit analyses (CBA) of risk reduction measures, via the *implied cost of averting fatality* (ICAF): ICAF = cost of measure/(initial PLL minus the reduced PLL).

Such calculations are often controversial as they appear to require a value to be placed on life, but these calculations are commonly used internationally and may aid decision making in regard to adopting control measures for major hazards. For example, a *low ICAF* for a proposed risk reduction/treatment measure implies that the measure is *highly effective* because the cost is low compared to the risk reduction achieved. Conversely, a high ICAF implies a relatively ineffective risk reduction measure, indicating that the money should be diverted to an alternative.

20.6 TOLERABILITY CRITERION FOR INDIVIDUAL RISK

In the United Kingdom and the Netherlands, many decades ago, public safety risk criteria for new industrial hazardous activities have been set by government regulation. The starting point for determining these criteria was the statement *"that any additional risk from exposure to a new hazardous activity to a member of the public should not be significant when compared with risk in everyday life."* Following this approach, for new major hazard installations (say, a petrochemical plant or nuclear reactor), the maximum acceptable level for individual risk for a nearby resident has been taken arbitrarily as the risk level that increases the risk of death by **a maximum of 1%** compared with all other causes. For developed countries, the individual "natural death" risk for the population group of 10–14-year-olds is close to 10^{-4} p.a. or 1 in 10,000 p.a. and has been taken as the basic "background" fatality risk as shown in Figures 20.6 and 20.7. The maximum acceptable **individual risk** was hence established as

$$1\% \times 1 \text{ chance in } 10{,}000 \text{ p.a.} = 1/100 \times 1 \text{ in } 10{,}000 \text{ p.a.} = 1 \text{ chance in } 1{,}000{,}000 \text{ p.a.} = 10^{-6} \text{ p.a.}$$

In other words, the risk of a fatal accident to which an individual anonymous member of the public is exposed because of his/her continuous presence (365 days/year) in the neighborhood of a new hazardous activity shall be less than one in a million years. Risk exposure levels of less than 10^{-8} per year, or less than once in 100 million years, are considered to be very low/almost negligible. This individual risk level is consistent with low/almost negligible risk in other areas.

> *Note. The authors recommend that a very low risk NOT be classified as "negligible" or insignificant' as that implies that they can be neglected. No risk can ever be entirely neglected. Rather a more appropriate management approach is to recognize that even a very low risk can increase and become of concern if the risk factors change unfavourably. Hence it is recommended to use "very low" risk as the descriptor*

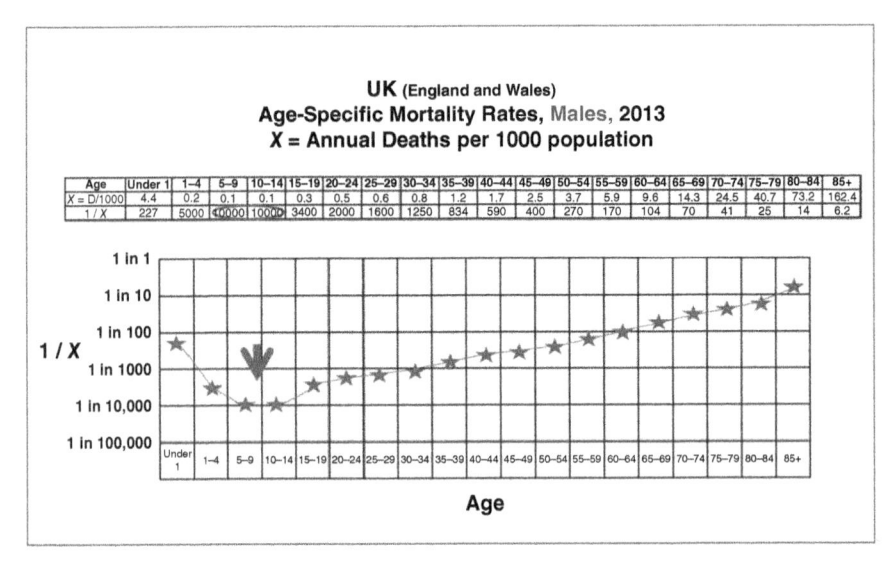

Figure 20.6 Fatality Rates in the United Kingdom by Age from All Exposures *Source:* UK Office for National Statistics, http://www.ons.gov.uk/ons/rel/vsob1/death-reg-sum-tables/2013/sty-mortality-rates-by-age.html

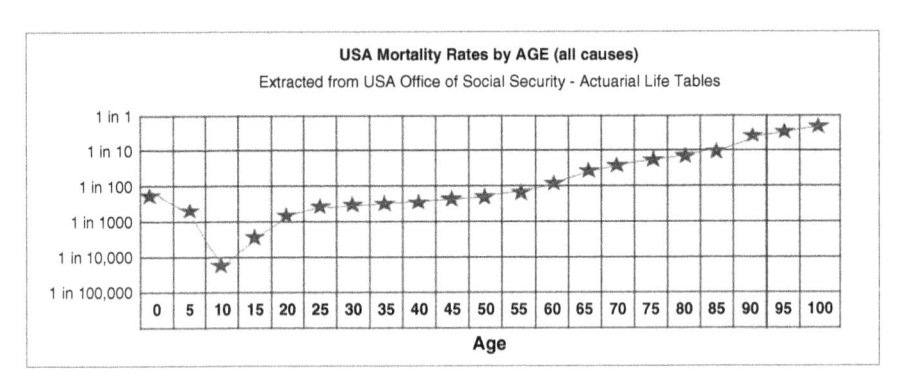

Figure 20.7 Fatality Rates in the United States by Age from All Exposures. *Source:* Extracted from Social Security Administration (2010)

only after considering if the risk factors would not change significantly and/or quickly then the monitoring frequency to revisit a risk assessment can be set as very low as well. [Ref: J. Loss Prev. Process Ind., 1991, Vol 4, January 1991]

Figures 20.6 and 20.7 provide examples of national mortality rates by age.

In the United States, the concept of 10^{-6} p.a. was originally an arbitrary number, finalized by the US Food and Drug Administration around 1977 as a screening level of *essentially zero* or *de minimis risk*. This concept was traced back to a 1961 proposal by two scientists from the National Cancer Institute (NCI) regarding methods to determine "safety" levels in carcinogenicity testing (Kelly, 1991).

The proposal for *de minimis risk* was contained in a 1973 notice and eventually adopted in 1977 in the Federal Register entitled "Compounds Used in Food-Processing Animals: Procedures for Determining Acceptability of Assay Methods Used for Assuring the Absence of Residues in Edible Products of Such Animals," commonly called the "Sensitivity of Method" regulations (USFDA, 1973). The term *de minimis* is an abbreviation of the legal concept, *de minimis non curat lex* – translated as

TABLE 20.1 Individual Fatality Risk Tolerability Criteria

Individual Fatality Risk Tolerability Criteria	Exposed Worker (p.a.)	Exposed Member of the Public (p.a.)
Max tolerable threshold	1 in 1000	1 in 10,000
	1×10^{-3}	1×10^{-4}
Broadly tolerable or acceptable levels	1 in 1,000,000	1 in 1,000,000
	1×10^{-6}	1×10^{-6}

the law does not concern itself with trifles. In other words, 10^{-6} p.a. was developed as a level of risk below, which was considered a "trifle" and not of regulatory concern.

A survey of worldwide risk tolerability criteria (AIChE, 2009, Appendix B) shows similarities among the criteria around the world. The data have been extracted from various publications and provide a benchmarking perspective.

The author's consideration of numerous worldwide, reasonably well-established, and widely accepted criteria for individual fatality risk tolerability criteria leads to the indicative levels in Table 20.1. Also see graphical representation in Figure 20.9.

CAUTION: It must be emphasized that these data can be used for internal assurance but cannot be used for regulatory compliance without checking with the appropriate regulatory authority in the home country.

20.7 TOLERABILITY CRITERIA FOR PLANNING NEW OPERATIONS

The foundation for choosing quantitative risk tolerability criteria is the following principles:

(1) The exposed persons such as nearby residents should not be involuntarily subject to a risk from a new exposure that is significant compared to the "background" risk associated with existing hazards.
(2) Individual and societal risk should be considered separately.

Land use planning departments and regulators of major hazard facilities in different countries have established quantitative risk criteria for new land use developments adjacent to existing land users according to specific sensitivities of exposed public persons. They are shown in Table 20.2.

TABLE 20.2 Fatality Risk Criteria for Land Use Planning and Locations of New Exposures

Location of Exposed Persons	Fatality Risk Criteria[a] (per million p.a.)
Sensitive, for example, in hospitals, schools child care, aged care	0.5
Residential including hotels, motels, resorts	1
Commercial	5
Sporting including open space, parks	10
Industrial	50

[a]NSW Department of Planning (2011).

They represent very low risks compared to other everyday risks associated with their existing land uses.

20.8 INVESTMENT TO PREVENT A FATALITY

An interesting global perspective in risk management is how different countries use measures related to human life values when calculating *CBA* for deciding to commit to spending on proposed new or changed risk controls for a particular risk, *for example, should all school buses be fitted with seat belts?*

Extract Quote: Washington Post August 25, 2011– http://www.washingtonpost.com/local/feds-reject-request-to-require-seat-belts-on-school-buses/2011/08/25/gIQATJhseJ_story.html

> *In Thursday's Federal Register, NHTSA cited its 2002 report to Congress, which said that shoulder-lap belts are effective in reducing school bus fatalities, but the addition of the belts "**would increase capital costs.**" NHTSA estimated equipping each bench-style set would cost between $375 and $600, a total of between $5485 and $7346 for each large bus.*
>
> *"The benefits would be achieved at a cost of **between $23 and $36 million per equivalent life saved**," NHTSA said. Rather than face a federal mandate, NHTSA said state and local governments should be left to decide whether to spend the money. Texas and California require school bus belts.*

The standard approach to CBA of risks to life is to convert them into equivalent costs. The monetary valuation of risks to life is often described as a *"value of life"* (SafetyNet, 2009). This phrase is convenient but inaccurate and also evokes strong emotional response. CBA evaluates small changes in risks for many people and does not attempt to value individual lives (*Refer to Chapter 19*).

The accumulation of risk to many people, which can be expected on average to result in the saving of one fatality, is better described as a "statistical fatality." For example, *a reduction in risk of 10^{-3} per year for each of 100 individuals over a period of 10 years would amount to a saving of one statistical fatality*. This distinction is important because it is much more reasonable to place a value on small changes in statistical risk than on individually identifiable lives. Presentation of this difficult and often emotive concept can be improved by using the term "**value of preventing a statistical fatality**" **(VPF).** This emphasizes that what is being valued is the reduction in risk to many lives rather than the actual lives that are at risk of being lost.

There are many different ways of considering $ values of saving lives at work and in societies generally. Some expressions used are given as follows:

- Value of a statistical life (VSL)
- VPF or statistical economic value for preventing a fatality
- Investment per equivalent life saved (IELS)
- Investment to prevent a fatality (IPF)
- Cost of a year of life saved (YOLS)
- Willingness to pay (WTP)
- Potential Loss of Life (PLL)
- Implied Cost of Averting Fatality (ICAF)

ICAF = cost of measure/(initial PLL–reduced PLL).

Such calculations are often controversial as they appear to require a value to be placed on life, but these calculations are commonly used internationally and may aid decision making in regard to

TABLE 20.3 Variations Between Countries in Value of a Statistical Life (VSL)

Number of Studies Averaged and Estimated Mean Value of a Statistical Life by Country (in thousands of 1995 US dollars)		
Country	Number of Values	Mean Value
Australia	1	2126
Austria	2	3253
Canada	5	3518
Denmark	1	3764
France	1	3435
Japan	1	8280
New Zealand	3	1625
South Korea	2	620
Sweden	4	3106
Switzerland	1	7525
Taiwan	2	956
United Kingdom	7	2281
United States	39	3472

Source: Reprinted with permission from Miller (2000).

TABLE 20.4 Some Examples of International VPFs/VoLS

Context	VPF or VoLS	Source
United States – School bus seat belts	$23 → $36 million per equivalent life saved	
Road fatalities on US roads	$5.8 million as the statistical economic value for preventing a human fatality	
New Zealand VPF accident compensation system	$3.4 million	Wren (2011)
OECD and EU countries	$1.5 → 5.4 million	OECD (2012)

adopting control measures for major hazards. For example, a low ICAF for a proposed risk reduction measure implies that the measure is highly effective because the cost is low compared to the risk reduction achieved. Conversely, a high ICAF implies a relatively ineffective risk reduction measure, indicating that the money should be diverted to an alternative.

A VSL is the $ amount that a group of people, say, a government, is willing to pay for a fatal risk reduction in the expectation of saving 1 anonymous life. These values can be estimated from what governments reflect in their social policies such as how much is invested in coronary care units, road safety engineering measures, vaccinations, etc.

These values are determined by the risk acceptance of the group (not always consciously) and are strongly influenced by social, cultural, and economic factors. Tables 20.3 and 20.4 show variations among countries according to these factors in the 1990s but would be quite different now as countries' economic circumstances have changed.

20.9 SHIFTING THE PARADIGM FROM ABSOLUTE SAFETY TO RISK MANAGEMENT

Internationally, more and more countries have shifted and are shifting their safety philosophies and practices from unrealistic *absolute safety* or even *zero risk* beliefs to more appropriate models of managing risk to *as low as reasonably practicable* (ALARP). The safety regulators in many countries have been traditionally *prescriptive* in describing absolute obligations to detail how safety risks were to be eliminated or prevented or stopped. In recent decades, legal and regulatory safety obligations are being described in *ALARP/performance-based frameworks* rather than prescriptive, unconditional *ensure without risk* models. National and international safety standards are now more and more written in terms of ALARP or the equivalent.

20.9.1 What Is Reasonably Practicable?

A risk-informed performance-based regulatory approach to safety laws defines *reasonably practicable*, in relation to a duty to ensure health and safety, as meaning that what was reasonably able to be done in relation to ensuring health and safety, taking into account and weighing up all relevant matters including the following:

(a) The **likelihood** of the hazard or the risk concerned occurring
(b) The **degree of harm** that might result from the hazard or the risk
(c) What the person concerned **knows**, or ought reasonably to know, about:
   (i) The hazard or the risk
   (ii) Ways of eliminating or minimizing the risk
(d) The **availability and suitability** of ways to eliminate or minimize the risk
(e) **After** assessing the extent of the risk and the available ways of eliminating or minimizing the risk, the **cost** associated with available ways of eliminating or minimizing the risk, including whether the cost is **grossly disproportionate** to the risk.

It is important to note some misunderstandings of CBA and ALARP. The concept of ALARP or SOFAIRP does NOT imply that **cost** alone nor **capacity to pay** for a reasonably practicable risk control, mitigation, or treatment measure can ever be legal justifications or defenses for not implementing the measure. Moral and legal factors will always emphasize the expectation that even if a risk is tolerable, the closer the risk level is to an agreed intolerable level, then the more investment in risk management is required to drive the risk lower. Hence the $ width of the triangle widens in the following figures. In terms of *individual risk*, ALARP or SOFAIRP tolerability principles are described in Figures 20.8 and 20.9 with the emphasis on "grossly disproportionate."

In terms of **societal or group risk**, ALARP or SOFAIRP tolerability principles are described in Figure 20.10 with a common expression of the "intolerable" frequency of incidents involving more than 1000 fatalities per incident (e.g., Bhopal or 2×500 passenger planes colliding in midair).

Note: As with individual risk, it is recommended that risk zones such as "negligible" and "insignificant" or "broadly tolerable" are not used. If a risk is to be tolerable, it needs to be below the "intolerable" threshold *and* has to be able to be shown to be ALARP regardless of how much it is below the threshold.

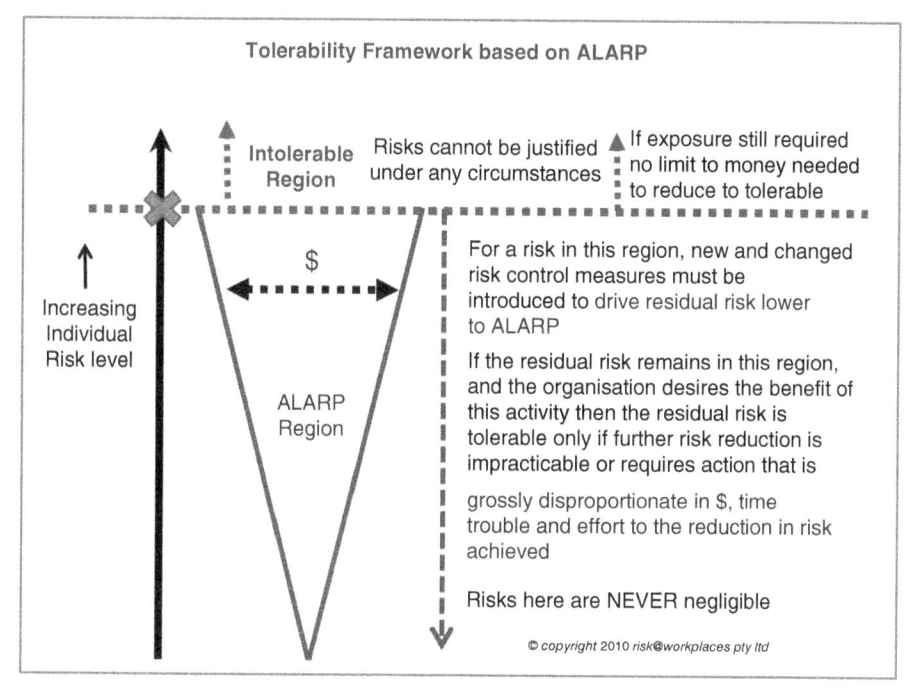

SHIFTING THE PARADIGM FROM ABSOLUTE SAFETY TO RISK MANAGEMENT

Figure 20.8 ALARP Risk Tolerability Framework for Individual Risk. *Source:* Adapted/extended from R2P2 – HSE (UK) and IEC/61508 Part 5 Annex B "Safety Systems"

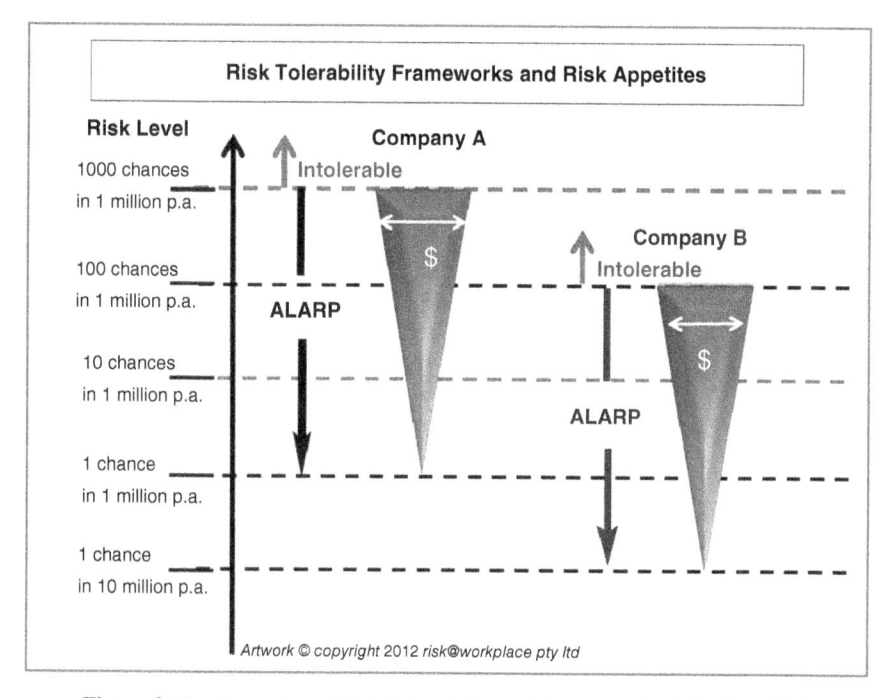

Figure 20.9 Examples of Risk Tolerability and Appetites for Individual Risk

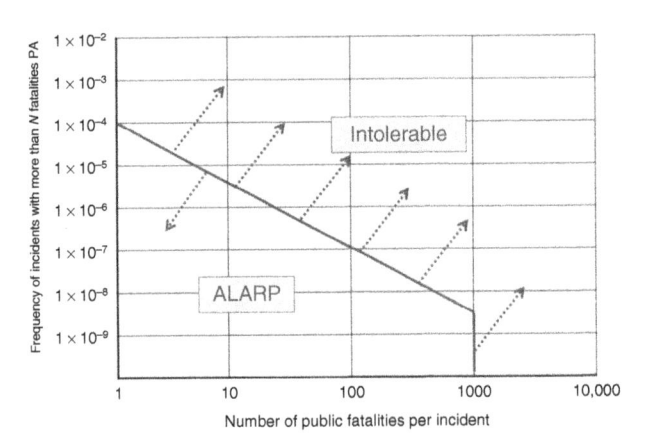

Figure 20.10 Indicative Societal or Group Risk Tolerability Framework (Generic Example for All Fatality Risks to Members of the Public)

20.10 MOVING TOWARD RISK-BASED LANGUAGE FOR MORE EFFECTIVE RISK CONVERSATIONS

If an SH&E practitioner wants to converse with others, internationally as well as nationally, then safety risk-based language can improve the quality and effectiveness of the conversation. Risk-based conversations can make safety discussions more realistic, objective, and solution focused as well as less argumentative. If risk-based conversations are to achieve this purpose, they need a common agreed language to express sometimes complex safety concepts.

The transition from compliance-based safety to risk-based safety in large part has occurred in many parts of the world and is beginning to take place in the United States. This requires *risk-centric* organizations and progressive risk professionals to use better terminology and language for risk-based conversations. For example, outdated terms such as loss control and loss prevention are now being replaced with safety risk management and risk control. *Appendix A* shows more commonly used risk-based language that is becoming more universal as preferred and recommended so as to clarify and reduce biases, misunderstandings, and misperceptions of the group and its individual members during safety-related discussions.

20.11 A CAUTIONARY CONCLUDING NOTE

Risk management is fundamentally about how to make risk-informed decisions when life and business circumstances are uncertain. Risk management is not about more *risk taking*, rather better *risk understanding* of the exposures we are currently not managing as well as we need to or could.

Safety is *managing risk to ALARP*, not *zero risk*. The *risk makers* and the *risk takers* need to be the *risk managers*.

REVIEW QUESTIONS

1. Explain how global practices influence national practices in operational risk management.

2. Provide examples of national standards around the world that have been adopted from the ISO 31000 Risk Management Standard series.

3. How can "Maturity Assurance and Conformity" of risk management practices be measured, and why is it important?

4. Describe what is meant by "Individual Risks" and "Societal Risks." Provide examples of each.

5. Describe criteria used to determine quantitative risk tolerability within an organization.

6. Define the difference between "Absolute Safety" and "Risk Management."

7. Provide examples of risk-based language and traditional safety-based language and explain how they differ.

REFERENCES

AIChE. Appendix B – Survey of Worldwide Risk Criteria. *Guidelines for Developing Quantitative Safety Risk Criteria*. Wiley. ISBN: 978-0-470-26140-8, 2009. http://onlinelibrary.wiley.com/doi/10.1002/9780470552940.app2/pdf (accessed December 30, 2015).

ANSI/AIHA/ASSE Z10-2012. *American National Standard – Occupational Health and Safety Management Systems*. Des Plaines, IL: American Society of Safety Engineers, 2012.

ANSI/ASSE Z590.3-2011. *Prevention Through Design: Guidelines for Addressing Occupational Hazards and Risks in Design and Redesign Processes*. Des Plaines, IL: American Society of Safety Engineers, 2011.

ANSI/ASSE Z690.2. *Risk Management Principles and Guidelines*. Des Plaines, IL: American Society of Safety Engineers, 2011.

Kelly, K. E. The Myth of 10^{-6} as a Definition of Acceptable Risk, *Paper Originally Presented at the 84th Annual Meeting Air & Waste Management Association Vancouver, B.C., Canada 16–21 June 1991*, 1991

Miller, T. R., 2000. "Variations between countries in VoLS – value of a statistical life." *Journal Transport Economics and Policy*, 34(Part 2 May), 169–188 http://scholar.google.com.au/scholar_url?url=http://www.researchgate.net/profile/Ted_Miller/publication/238369152_Variations_Between_Countries_in_Values_of_Statistical_Life/links/0c96052cde36d95d2c000000.pdf&hl=en&sa=X&scisig=AAGBfm3WreOfiy2C_5vSKZv89c3cmI9IfQ&nossl=1&oi=scholarr&ei=zvcxVa3JMMemmAXNm4DgAg&ved=0CB0QgAMoATAA (accessed December 30, 2015).

NSW Department of Planning, Australia, Hazardous Industry Planning Advisory Paper No. 4 – Risk Criteria for Land Use Safety Planning (HIPAP 4) (January 2011), 2011.

OECD. Recommended Value of a Statistical Life Numbers for Policy Analysis, pp. 125–136, 2012. http://www.oecd-ilibrary.org/environment/mortality-risk-valuation-in-environment-health-and-transport-policies/recommended-value-of-a-statistical-life-numbers-for-policy-analysis_9789264130807-9-en (accessed December 30, 2015).

SafetyNet. Cost–benefit analysis, 2009. Retrieved 18 April 2015 at: http://ec.europa.eu/transport/road_safety/specialist/knowledge/pdf/cost_benefit_analysis.pdf (accessed December 30, 2015).

Social Security Administration. Actuarial Life Table, 2010. Available at: www.ssa.gov/oact/STATS/table4c6.html.

U.S. Food and Drug Administration (USFDA). Compounds used in food-producing animals. Procedures for determining acceptability of assay methods used for assuring the absence of residues in edible products of such animals. Proposed rule. *Federal Register*, July 19, pp. 19226–19230, 1973.

Whiting, J. F. Using the New RM Standard ANSI Z690.2 to Assess the Maturity of your Risk Management System. ASSE PDC Safety 2012 – Denver, Session 662 Tuesday 5 June 2012, 2012. Available from: jim@workplaces.com.au (accessed December 30, 2015).

Wren, J. The Value of Preventable Injury Fatality (VPF) in New Zealand. Presentation for New Zealand Association of Economists Conference 30 June 2011, 2011.

APPENDIX 20.A

BETTER TERMINOLOGY AND LANGUAGE FOR RISK-BASED CONVERSATIONS

Traditional Safety Terminology	Preferred and Recommended Risk-Based Language
Loss control/loss prevention	**Safety risk management and risk control**: profits as well as losses – enabling positive outcomes as well as preventing negatives – maximizing the chances of gains, profits, benefits; safety is about a focus on *maximizing chances of gains not minimizing chances of losses*
Safety – as absence of harm, double negative	Safety – as presence of well-being, double positive
Safe acts/conditions	Standard, agreed acts/conditions
Unsafe acts, at-risk behaviors/at-risk conditions	Nonstandard, nonagreed behaviors/conditions
Safe	When risk is managed as low as reasonably practicable (**ALARP**)
Safer/safest	**Lower risk/lowest risk**
Event/scenario	If used interchangeably creates confusion. For example, the expression "The same *event* can lead to different consequences is valid but the same *scenario* can lead to different consequences is *not* valid." Reserve the term *event* for each discrete happening/action and *scenario* for all the events and circumstances needed to describe "How"/"When"/"Where"/"Who"/"What"
If safety is involved, money doesn't count!	Sounds like a good caring philosophy but it is an untrue, unbelievable statement, which corrodes credibility, trust, and respect. Better to use expressions such as "*when* a risk exceeds our agreed defined intolerable threshold level, and IF continued exposure to the risk is needed or desirable for legal, moral or commercial reasons there is no limit to time money effort needed to introduce measures that reduce the risk below the threshold." The reduced risk then also needs to be shown as always being managed to ALARP – not just at one point in time. *Tolerable* means *both* below intolerable and ALARP
Alertness, vigilance	**Situational awareness and mindfulness**
Violation, breach, failure, negligence, recklessness	Use nonjudgmental terms *such as* **variation, alternative,** ***deviation, and work-around*** so you will look for deeper underlying root causes of the variations
Shortcut	Smarter way of doing a job, which can be an approved variation but only after a formal authorization/approval process that must involve qualitative or semiquantitative risk assessments Always distinguish between: • Finding a shortcut (*smart*) and • Taking a shortcut without risk assessment (*dumb*)

APPENDIX

Traditional Safety Terminology	Preferred and Recommended Risk-Based Language
Safety measures, preventative measures, safeguards, barriers, layers of protection, mitigating factors, corrective actions	Use the single term **risk controls** for all of them
Causes of incidents and risks	All causes are missing or ineffective risk controls due to deeper underlying root causes based on systemic, physical, and work environment factors
Behavioral causes	Behaviors are consequences of deeper underlying root causes *not* seen as causes in themselves
Human error	Use the term **human factor** in preference to **human error** to emphasize that error is not a cause of an incident or a risk of an incident. It is a consequence of the underlying human factors/mismatches between a job's requirements and the person's capabilities and limitations. The mismatches are usually due to systemic, physical, and work environmental factors
Possible, probable, potential used interchangeably and hence confusingly	**Possible** = absolute, *yes*/*no*, black/white – it is or it isn't; has no range of values – cannot be used to express a level of likelihood • cannot use meaningless terms *quite possible* or *remotely possible*
	Probable = relative, not absolute – can use *likely, chances, odds*; always has a range of values – used to express a level of likelihood
	Potential = confusing – it can be used to express either *possible* or *probable*
Probability	**Likelihood, chances, odds** are risk terms preferred for nonquantitative users
Likelihood can be expressed as either a **frequency** or a **probability**	**Frequency** can be used *retrospectively* to indicate how often an actual incident has been occurring in the past *and also* it can be used *prospectively* to predict how often the risk of an incident may occur in the future
	Likelihood, chances, odds can be used *only* prospectively to express predictive estimate of how likely the risk will occur
	Often better to use the terms "chance" or "odds" *not* decimal 0.001 or unfamiliar exponential 1E−03 notation, for example: 1 chance in 100 ladder climbs 1 chance in 10,000 valve operations The odds are 1 in 1000 holes drilled
	Avoid using fractions of % – hard to interpret. For example, use 1 chance in 1000 rather than 0.1%
	Always question any assessor's perception that 1% or 1 chance in 100 is a small likelihood. It is a large likelihood
Exposure	How often and how long exposed (in financial RM it is $ quantum)
Frequency of exposure	How often, for example, exposed to noise daily (or yearly or every shift)
Duration of exposure	How long, for example, exposed to asbestos 3 h/shift (or 100 h p.a.)

INDEX

Acceptable risk, 5, 7, 20, 32, 60, 61, 84, 85, 146, 270, 277, 285, 383
   Levels, 1, 4, 6–9, 11, 13, 47, 55–57, 59, 74, 84–85, 91, 93, 94, 99, 105, 110, 125, 149, 150, 174, 209, 214, 215, 218, 222, 224, 274, 276, 280, 295, 344, 379, 383
Accident, 8, 12, 25, 50, 58, 62, 76, 78, 107, 108, 123, 132, 135, 149, 169, 195, 267, 274, 276, 279, 285, 293, 297, 305, 333, 335, 370, 378, 432, 433
Adams, Paul, 210, 214
Administrative controls, 26, 40, 60, 62, 79–82, 89, 109, 124, 137, 149, 171, 194, 272, 277, 285, 313, 358, 367
ALARA. *See* As low as reasonably acceptable (ALARA)
ALARP. *See* As low as reasonably practicable (ALARP)
ALARP Model, 61
American Conference of Governmental Industrial Hygienists (ACGIH), 186, 352, 410
American Industrial Hygiene Association (AIHA), 248, 250–252, 256–258, 261–262, 349, 350, 410
   Ergonomics committee, 349–350
American Institute of Chemical Engineers' (AIChE) Center for Chemical Process Safety, 16, 123
American National Standards Institute (ANSI), 4, 27, 28, 50, 122, 145, 164, 181, 209, 228, 263, 269, 323
American Petroleum Institute, 12
American Society of Safety Engineers, 2–3, 6, 28, 50, 105, 209, 222, 284
   Designing for safety, 239
ANSI/AIHA/ASSE Z10, 4, 12, 18, 20–21, 27–28, 31, 51, 52, 62, 63, 70–72, 74, 77, 90, 93, 94, 99, 100, 172, 198–199, 204, 210, 214, 217, 220, 248–249, 255, 263, 283, 294, 344, 347, 372, 390, 428

ANSI/ASSE A10.1-2011, 301–302
ANSI/ASSE Z244.12003 (R2014), 269, 283–285
ANSI/ASSE Z590.3, 2, 4–6, 20, 27, 29–30, 53–55, 61–63, 70–71, 74–75, 80–82, 85, 92, 95–99, 114, 122, 145, 147, 150–152, 155, 164, 166–167, 169, 187, 190, 204, 209, 210, 214–215, 217, 218, 220–224, 229–234, 244, 248, 251, 256, 258, 263, 324, 325, 329–331, 339, 344, 348, 410, 412, 428
   ANSI Z690.3 integration, 232, 233
   Hazard identification and analysis, 231
ANSI/ASSE Z690.1-2011, 6, 28, 50, 67–68, 76, 95, 228
ANSI/ASSE Z690.2, 6, 28–29, 50, 61, 88, 91–92, 96, 228–230, 234, 243, 244, 319, 323, 325, 371, 428, 431
ANSI/ASSE Z690.3-2011, 2, 6–7, 28, 29, 52, 53, 56–60, 62, 63, 68–71, 73, 76, 80, 85, 91–93, 96–99, 104, 109, 114, 122, 125, 146, 147, 149, 164, 166, 181, 228, 230, 234, 248, 324, 327, 330, 339, 375, 412
ANSI B11.0, 7–8, 20, 27, 30–31, 69–71, 82–83, 215, 269–275, 277, 279
   Risk assessment process, 275
ANSI B11.19, 30, 269, 271, 277
ANSI B11 standards, 269–270
ANSI B11.TR3-2000, 27, 31, 54, 269, 271, 278–280
   Two-factor risk model (4×4), 290
ANSI B11.TR7-2007, 19
ANSI-ITAA GEIA-STD-0010-2009, 19
ANSI/PMMI B155.1-2011, 20
As low as reasonably acceptable (ALARA), 32
As low as reasonably practicable (ALARP), 32, 33, 60, 61, 64, 85, 98, 100, 109, 110, 139, 206, 330, 427, 438–440, 442

Risk Assessment: A Practical Guide to Assessing Operational Risks, First Edition.
Edited by Georgi Popov, Bruce K. Lyon, and Bruce Hollcroft.
© 2016 John Wiley & Sons, Inc. Published 2016 by John Wiley & Sons, Inc.

ASSE Risk Assessment Institute, 3, 71, 88, 212
Asset, 33, 46, 59, 76, 86, 94, 96, 122, 148, 150, 153, 155, 213, 217, 369, 371, 373–375, 377–381, 384, 385, 430
Audit, 33, 50, 96, 212, 392, 430
Australian Government, 18
Aviation ground safety, 13–14

Barrier, 33, 50, 62, 80, 82, 148, 157, 169, 181–186, 194, 195, 206, 212, 214, 217, 271–272, 280, 287–290, 312, 330, 420, 443
Benefits, 60, 200, 204, 224–225, 339, 390, 410, 412, 442
    Financial, 203, 384, 436
    Nonfinancial, 202, 418
Bhopal Disaster, 25, 184, 193–195, 438
Bow tie analysis, 7, 114, 152, 164, 181–195
    Advantages and disadvantages, 183–184
    Case studies, 186–196
    Methodology, 182–185, 207, 220
Brainstorming, 6, 58, 97, 114, 122, 125–127, 129–132, 136, 139, 140, 142, 143, 147, 150, 157, 185, 186, 221, 234, 276, 297, 351, 353, 385
    Advantages and disadvantages, 125–126
    Methodology, 126
British Standards Institute (BSI), 9, 19, 27, 374
BSI IEC 61882:2001, 135–136
BS OHSAS 18001:2007, 9, 19, 52, 93, 295
Bureau of Labor Statistics (BLS), 76
    Census of Fatal Occupational Injuries, 296
    Work-related musculoskeletal disorders (WMSDs), 343
Bureau of Ocean Energy Management, Regulation and Enforcement (BOEMRE), 12
Business case, 183, 203, 243, 409–410, 424
    Cost/benefit data, 420
    Development/tools, 410–412
    Examples, 412–424
    In occupational safety and health, 409–410
Business continuity, 122, 192, 373–375, 420
Business Continuity Institute (BCI), 374
Business impact analysis/assessment (BIA), 375–376
Business losses impact descriptions, 188
Business risk assessment matrix (BRAM), 188–189, 192, 205, 420

Canadian Standards Association (CSA)
    CSA Standard Z1000-2006, 12–13, 18, 21
Carlson, Carl S., 35, 164, 166
Cartesian coordinate system, 55, 71
Causal factor, 33, 97, 108, 109, 124, 147, 330
Cause identification, 185, 186
Centers for Disease Control and Prevention, 375
Checklist, 6, 15, 16, 26, 58, 96, 99, 101, 102, 108, 114, 122, 124–128, 130–132, 140, 147, 150, 185, 220, 221, 250, 276, 292, 296, 297, 302–304, 313, 352, 354, 355
    Analysis method, 126–127
    Advantages and disadvantages, 127
    Process steps, 127
Christensen, Wayne C., 5, 210, 224, 229
Clemens, Patrick, 76, 146

Communication, 54, 58, 61–63, 70, 77, 81, 88, 94, 103, 194, 213–215, 220, 221, 229, 279, 292, 294, 325, 346, 351
Compliance, 24–25, 27, 28, 33, 50, 93, 139, 203, 210–213, 215–216, 221, 225, 237, 268–269, 311, 370, 393, 407, 440
Confined space entry, 24, 301, 308–311
Conformance, 33, 393, 430
Consequence, 2, 5, 7, 15, 26, 30, 33, 45, 46, 50, 51, 56, 59–61, 68, 79, 92, 96, 97, 102, 149, 181–184, 206, 222, 284, 312, 327, 375, 378, 411, 431
    Analysis, 59
    Identification, 136, 185
    Severity of, 44–45, 54, 59, 60, 68, 69, 71–74, 76–77, 137–138, 167, 174, 188, 204, 280, 380, 381, 398
Construction, 25, 99, 101, 102, 104, 109, 291–292, 296, 297
    Hazards, 301
    Pre-project planning process, 297–298
    Project risk assessment, 297–299
    Risk factors, 297
Context
    Establishing, 53, 55–57, 68–69
    Purpose and scope, 56, 348
Continual improvement, 33, 51–52, 63–64, 93–94, 391–392
Contractor, 33
Control, 34
    Assessment, 60, 80
    Influences, 185
    Reliability, 74
    Selection, 188
Corrective action, 34
Cost-benefit analysis/data, 57, 85, 203, 213, 224, 353, 362, 380, 383–384, 390, 433
Covey, Stephen, 88
Critical Control Point (CCP) Decision Tree, 34, 168–170, 327–329
Current state risk, 34, 80, 84

Deepwater Horizon incident, 50
Definitions, 32–46
Dekker, Sydney, 294
Deming, W. Edwards, 52, 93–94
Design/redesign, 3, 4, 34, 40, 87, 209, 210, 214, 215, 221, 227, 229, 339
Design reviews, 51, 213–218, 224
Design safety
    Challenges, 211–214
    Standards, 214–215, 224–225
Design safety checklists, 210, 211
Design safety review, 34, 122, 148, 149, 209–210, 214–216, 218–224
    Process, 218–220
    Techniques, 221
Detection of failure, 74, 75, 81, 167
Documentation, 31, 62–63, 85–86, 100, 206, 277, 392, 410
Documenting risk, 85–86

INDEX 447

Ebola epidemic, 375
EMS Lean Six Sigma (LSS), 389–390
   Case study, 404–407
   Implementation, 404
Energy release
   Haddon's theory, 148, 216–217
Energy types and hazards, 217
Engineering controls, 60, 80–82, 272, 277, 313, 358
EN ISO 12100-2010, 8–9
Enterprise risk management, 370, 430
Environmental aspects, 390–396, 398
   Significant environmental aspect (SEA) decision-making tree, 394–395
Environmental impact, 393–395
Environmental management system (EMS), 399–400
Environmental Protection Agency (EPA), 15, 41, 271, 400, 403, 414
   Risk Management Plan (RMP), 15–16, 24, 123–125, 146
Ergonomic risk assessment tool (ERAT), 218, 339, 344, 349, 351–352, 365–366
Ergonomics, 34, 64, 353–355, 428–429
   Assessment process, 336–338
   Definition, 34, 344
   Design, 344–345
   Design review, 217–218
   ERAT, 354–359
   Hazards, 249, 271, 345–346
   Hierarchy of controls, 358
   Improvement process, 350–354
   Risk assessment, 221, 349–350
   Risk assessment techniques, 349
   Risk factors, 346, 347, 355–356
   Risks, 340
Escalating factors, 182, 185
European Directive 2002/44/EC, 234–235
European Union, 8, 19, 234
   Hazard Statements (H Statements), 253
   Risk Phrases (R-Phrases), 253
Event, 34
Event tree analysis, 59, 77, 182, 297
Exposure, 34–35, 68, 74, 79, 96, 97, 167, 169, 373, 378, 443
   Action value, 235
   Assessment, 35, 248
   Frequency of, 74–75, 79
   Limit value, 235
   Rating methodologies, 251
   Time duration of, 74, 79
ExxonMobil's Operations Integrity Management System, 3, 19

Failure
   Detection, 74, 75, 81, 167
   Mode(s), 35, 97, 163, 166–167
Failure mode and effects analysis (FMEA), 75, 78, 163–164, 185, 205
   Application, 175–177
   Process, 172–175
   Purpose and use, 164–165
Failure mode, effects and criticality analysis (FMECA), 164
Fatality and Serious Incidents (FSI), 59, 62, 77, 293–294
Fault, 38, 59, 77, 146, 147, 149, 166, 182, 221, 283, 297, 304, 321, 330
Fault tree analysis, 59, 147, 182, 329–330
Federal Railroad Administration (FRA), 12,
Federal Transit Administration (FTA), 306
Field-level analysis, 101,
Financial benefits. *See* Benefits
Financial risk, 35
Fish bone diagram, 201, 325
Food and Drug Administration (FDA), 12, 169, 320, 323–326, 434
   Good Manufacturing Practices (GMP), 321, 323
Food processing
   Biological hazards, 321–322
   Chemical hazards, 322–323
   Physical hazards, 323
   Risk assessment techniques, 320–321
   Risks, 319–320
Frequency, 35, 400–401, 443
Frequency of exposure, 35, 42, 68, 74, 75, 79, 97, 167, 169
Fundamental methods, 99–100, 109–110
Future state risk, 35, 80

Global perspectives, 427

Haddon, William, Jr., 148, 216–217
Hand and Arm Vibration Evaluation, 234–236
Harm, 35
Hayes, Fred, 273
Hazard, 36, 94, 210, 373
Hazard analysis, 36, 94–112, 182, 210
   Formal methods, 103–112
   Informal methods, 99–103
Hazard Analysis and Critical Control Points (HACCP), 319–321, 324–325
   Appropriate level of protection (ALOP), 321
   Integration of PtD methods, 337–340
   Integration of risk assessment methods, 325–337
   Methodology, 324–325
Hazard analysis and risk assessment process, 53, 96–99, 220–224
Hazard and operability analysis (HAZOP), 122, 135–136, 185, 221, 297
Hazard area (zone), 36
Hazard-based efforts, 94–95
Hazard checklist, 108, 220, 296, 302
Hazard communication, 26, 94–95, 414
Hazard control, 75, 82, 83, 94, 167, 271, 272, 277, 282, 286, 301
Hazard control hierarchy, 82, 83, 271, 272, 286
Hazard descriptions, 115–117, 306
Hazard identification, 36, 58, 122, 146, 185, 186, 221
Hazard identification methods, 58
Hazardous, 37
Hazardous event/scenario, 37, 182
Hazard risk, 36
Hazard/risk avoidance, 36
Hazard/risk elimination, 37

Health and Safety Executive (HSE), 24, 49–50, 123, 235–236, 251, 253, 296
  Likelihood levels, 299
  Maintenance task analysis example, 304–305
  Risk assessment matrix, 299
  Severity levels, 298
Hierarchy of controls, 37, 60–63, 79–83, 209, 229, 353, 403
High level controls, 83
High-severity/low-probability events, 169
Howard, John, 213
Human error, 35, 107, 116, 138, 142, 146, 166, 221, 276, 294, 297, 306, 330, 346, 347, 373, 378, 443
Human factors, 37, 116, 146, 164, 221, 279, 344, 346, 373
Human Factors and Ergonomics Society, 344
Hunter, Thomas, 224

IEC/ISO 31010:2009. *See* ANSI/ASSE Z690.3-2011
Incident, 37
Incident investigation, 58, 103, 212, 293, 350
Individual risk, 432, 433, 438
Industrial hygiene (IH), 247–248
Industrial hygiene risk assessment, 247
  Advanced REACH Tool (ART), 254
  Case study, 256–260
  Concept, 248–249
  Control banding nanotool, 260
  Control of Substances Hazardous to Health (COSHH) Essentials Tool, 251, 253
  Dermal exposure, 260–261
  Exposure Rating Categorization Scheme, 252
  Exposure Rating Methodologies, 251
  Health effect and exposure methodology, 261
  Health risk prioritization, 255–256
  Health risk rating matrix, 252
  Health risk rating methodology, 250
  Identifying health risks, 249–250
  Modified HRR/IH FMEA method, 256, 259
  OSHA's calculation for mixtures, 254
  PtD alignment, 261–264
  Stoffenmanager, 254
Initial risk, 37, 80
The Institute, 370–371
International Ergonomics Association, 344
International Labor Office ILO-OSH 2001, 52
International Standards Organization (ISO), 228
International SEMATECH, 78–79, 139, 172–173
ISO 9001, 430–431
ISO 14001-2004, 52, 389–391, 430
  Implementation, 390–404
ISO 22000, 321
ISO 22301:2012, 373
ISO 31000:2009, 228, 410, 427–432
  Maturity conformance, 428–431
ISO 45001-2016, 28, 52, 93, 430, 431
ISO Guide 73:2009. *See* ANSI/ASSE Z690.1-2011

Job hazard analysis (JHA), 58, 93, 99, 104–109, 119, 153, 292, 301, 349
Job risk assessment, 93, 99, 100, 106, 109–110

Job safety analysis (JSA). *See* Job hazard analysis
Johnson, William G., 4, 122

Law of large numbers, 79
Lawrence Livermore National Laboratory (LLNL), 261
Layers of protection analysis (LOPA), 114, 194, 206
Lean Six Sigma (LSS), 404, 407
Legal risks, 400–402
Level of risk, 2, 4, 14, 21, 32, 33, 38, 41, 56, 59, 60, 68, 83, 85, 95, 97, 249, 285, 324, 330
Liberty Mutual Research Institute for Safety-2013, 343
Life cycle, 38, 147, 210, 215–216, 224
Likelihood, 35, 38, 50, 77, 79, 86, 94, 96, 97
Likelihood analysis, 59
Likelihood of business loss descriptions, 188, 205
Likelihood of occurrence, 72–73, 77–78
Loss, 25, 35, 36, 39, 45, 52, 76, 77, 94, 147, 150, 188, 194, 204–206, 212, 220–222, 236, 249, 274, 312, 350, 369, 371–373, 375, 377, 380–382, 385, 411, 419, 433, 440
Loss analysis, 107–108, 212, 221
Lower-level controls, 84

Machine safety
  Case study, 279–282
  Consensus standards, 268, 276
  Control systems, 273, 279
  Estimating risk, 278–279
  Hazards, 270–271
  Hazard control hierarchy, 271–273
  Maintenance and service, 282–284
  OSHA standards, 268
  Risk assessment, 274–278
  Safeguarding, 271–273, 277–278
  Selecting machines for assessment, 274
Main, Bruce W., 50, 68, 69, 71, 110, 137, 210, 213, 216, 273
Management of change (MOC), 13, 26, 27, 51, 214, 294–296
  Standards requiring, 294–295
Management review, 46, 214, 392, 395
Mandated assessments, 123
Manual material handling, 211, 300
Manuele, Fred A., 23, 24, 36, 50, 60, 69, 71, 75, 76, 85, 96–100, 146, 167, 169, 209, 210, 214, 224, 228, 229, 293, 294
Michaels, David, 11
MIL-P-1629, 164
MIL-STD-882E, 9–11, 31, 33, 54, 74, 146, 149, 150, 152, 269, 305–308
Mine Safety and Health Administration (MSHA), 101–102
Mishap, 38
Mitigation Measure, 38, 192–193, 195
Monitoring, 63–64

National Academy of Sciences (NAS), 262, 263
National Aeronautics and Space Administration (NASA), 12, 164, 320
  Space Shuttle Columbia explosion, 62

INDEX

National Fire Protection Association (NFPA), 13, 125, 216
  NFPA 654, 125
  NFPA 70E, 13, 31
National Institute for Occupational Safety and Health (NIOSH), 29, 213, 229, 344
  Noise, 236
  Prevention through Design (PtD) Initiative, 213, 229, 244
  Stress, 345–346
National Safety Council (NSC), 14, 29, 105, 107, 214, 229, 354
  Institute for Safety through Design, 214, 229
Noise Measurements, 236–237
Non-routine activities, 101

Occupational exposure limits, 240
Occupational health and safety (OHS), 4, 9, 12, 13, 18–21, 27–28, 39, 51, 52, 64, 71, 93, 210, 213, 214, 217, 237, 248, 253, 263, 269, 294, 295, 344, 347, 372
Occupational health and safety management systems (OHSMS), 27, 39, 51–52, 71, 93, 210, 214, 217, 248, 263, 295, 344, 347, 372,
Occupational Safety and Health Administration (OSHA), 11, 14–16, 24, 52, 93, 99, 104–106, 123–125, 131, 137–139, 146, 172, 216, 254, 259, 267–270, 293, 345, 412, 414, 428
  Confined space, 311
  Consensus standards referenced by, 276
  Construction, 301
  Control of Hazardous Energy, 268, 269, 283, 292
  Ergonomics, 345
  Machine safeguarding, 270–271, 273
  Noise, 236
  Particulate-total and respirable, 237
  Reporting fatalities and multiple hospitalization incidents, 293
Occupational Size-Selective Criteria and Particles Size Sampling, 237–238
Occupy Movement–Occupy Port of Oakland, 57
Operational risk, 39, 370–381
  Business aspects, 409
Operational risk management system (ORMS), 39, 51–52, 61, 91–93, 371
Organization, 39, 371
Organizational culture, 68, 69, 85, 212, 214, 277, 294
Organizational risk, 369–372
  Assessment, 369, 373–375
  Assessment process, 376
  Case study, 385–386
  Definitions, 372–373
  External sources, 370
  Internal sources, 370
  Management, 370–372
  Management questions, 377
  Maximum acceptable outage, 375
  Recovery time objectives, 375
  Risk categorization, 378–379
  Risk matrix, 380
  Risk profile, 379
  Risk register, 386
OSHA Voluntary Protection Program (VPP), 52, 62, 93, 99

Perceived risk, 96, 274, 279, 351
Personal protective equipment (PPE), 80–81
  OSHA 1910.132 standard, 25, 100
  Hazard analysis, 99–100, 103, 118
Personal risk. *See* Individual risk
Plan, do, check, act (PDCA). *See* Continual improvement
Preliminary hazard analysis, 145, 147, 150–151, 153–155, 185–187, 189–191, 221, 292, 325
  Application, 153–156
Preliminary hazard list (PHL), 147
Pre-operational, 212
Pre-startup safety review (PSSR), 26
Pre-task hazard analysis, 99, 101–102, 104–106, 292, 301–303
Prevention controls/measures, 192–193
Prevention effectiveness (PE), 74–75, 81, 167
Prevention through design (PtD), 4–5, 7, 39, 210, 212, 213, 229, 277, 319, 410
  Business process, 243
  Case study, 234–243
  Concept, 229
Preventive action, 39
Probability, 39, 50, 97
Probability of occurrence, 74–75
  Descriptions, 188
Process, 39
Process hazard analysis, 121–125, 146
Process life-cycle, 4, 35, 38, 97, 210
Process Safety Management (PSM), 14, 25–26, 123–125, 146, 172, 428
Procurement, 212
Project-oriented tasks, 291–293
  Construction, 296
  Error-traps, 294
  Fall hazard analysis, 311–316
  Maintenance and service, 304–315
  Operating hazard analysis (OHA) method, 305–308
  Pre-entry hazard analysis, 308–310
  Safe work methods, 299–301
Protection controls, 79, 83, 84, 149, 183
Protection factor (PF), 83–84, 280
Protective device, 40
Protective measures, 40
  EN ISO 12100-2010, 8–9
  EPA, 15–16, 24
  European Union, 8, 49
  Federal Railroad Administration, 11
  Fire protection, 13, 18, 19
  ISO 14121-1, 19
  Japan Government, 18
  MIL-STD-882E-2012, 9–11, 31–32
  NFPA 70E, 31
  NIOSH, 18
  OSHA, 11, 14–15, 19, 24, 100
  Safety professionals, 1–17

Qualitative
  risk assessment, 40, 72
Quantitative
  risk assessment, 40, 73

Rausand, Marvin, 146
Raw risk, 40, 80
Reactive measures, 182–183
Reasonable foreseeable misuse, 40
Reason, James, 182, 294
Recordkeeping. *See* Documentation
Redundancy, 142
Reliability, 16, 42, 68, 74, 80, 81, 114, 163, 164, 172, 279, 285, 321, 370, 383, 427
Residual risk, 40, 60, 80, 240, 243, 277
RIMS, the Risk Management Society, 370–371
Risk, 41
Risk acceptance, 41
Risk actions, 70–71
Risk analysis, 41, 50, 59–60, 67–68
Risk assessment, 50, 94–96
   Definition, 2, 41, 210, 230
   Process, 42, 53, 96–99, 239–243
   Provisions for, 4–16, 18–21, 24, 31–32, 49, 100, 227–238, 344
   Purpose, 23–24, 52–53
   Team, 57–58, 107, 352, 376
   Techniques/methodologies, 6–7, 14–16, 114
   Triggers, 50, 87–88
Risk assessment code (RAC), 74
Risk assessment matrix, 43, 53–55, 60, 71, 75, 90, 110, 185, 186, 281, 298–299, 308
Risk avoidance, 42, 61, 81
Risk-based decision making, 11
Risk-based efforts, 95–96
Risk-centric, 42
Risk control, 9, 29, 31, 53, 57, 61, 79, 81, 91, 94, 98, 103, 106, 107, 109, 147–149, 169, 190, 216, 222, 225, 285, 302, 304, 312–313, 355, 358, 367, 381, 383–385, 436, 438, 440, 442, 443
Risk criteria, 42, 54, 56–57, 67–69
Risk description, 42, 167–168
Risk elimination, 81
Risk evaluation, 42, 50, 60–61, 67–68
Risk factors, 42, 70–71, 74, 204–205
Risk financing, 61
Risk identification, 42, 50, 58
Risk index, 307
Risk indicators, 346
Risk level, 42, 54, 70, 74–75, 196
Risk management, 42
Risk management framework, 50
Risk management plan, 43
Risk management process, 92, 238–239, 323, 326, 372, 429
Risk matrix. *See* Risk assessment matrix
Risk perception, 440
Risk priority matrix, 138
Risk priority number, 43, 75, 167, 204, 240, 281–282, 327
Risk professional, 43
Risk profile, 43, 389
Risk rank, 43, 86, 98, 135, 146, 152, 164, 180, 190, 351
Risk reduction, 43, 79–84, 220, 240, 256
Risk reduction measures, 43, 259, 277, 280, 285, 295

Risk Reduction Program, 12
Risk register, 43, 86, 381, 384, 386
Risk retention, 43, 61
Risk score/scoring, 54–55, 74–75, 137–139
Risk scoring system, 68–71, 74–75, 152–153, 167
   Three and four-factor systems, 69, 75, 167–172
   Two-factor system, 68–69, 74, 167, 276–277
   Variables, 70–71
Risk source, 43
Risk tolerance, 44
Risk treatment, 44, 61
Risk values, 70–72
Rostykus, Walt, 218
Rumsfeld, Donald, 247

Safe, 44, 210
Safeguard/Safeguarding, 44
Safety, 44
Safety design reviews, 209–225
Safety function, 44
Safety professionals, 44
Safety task analysis card (STAC), 104
Safety work procedures, 44
Semi-quantitative risk assessment, 72–73
Serious incidents (injuries), 50, 59, 62, 75, 77, 293–294, 378
Severity
   Descriptions, 152, 171, 176, 187–188, 191, 193
   Levels, 36, 54, 59, 75–77, 84, 152, 153, 171, 174, 187, 191, 204, 222, 256, 258–259, 276, 278, 280, 281, 298, 319, 327, 351
Severity of consequence, 30, 44–45, 74–77, 188, 380
Singapore Standard SS, 19
Six sigma, 233
   Define, measure, analyze, improve, control (DMAIC) logic, 233, 234
Societal risk, 433, 435
Society of Fire Protection Engineers (SFPE), 13
Sony cyber-attack, 375
Stakeholder, 45
Standard, 45
Step back 5 x 5, 103
Stephans, Roger, 149
Stop, look, analyze, and manage (SLAM), 101–102
Strategic risk, 45
Structured what-if analysis (SWIFT), 131–133, 140–143
Substitution, 81
Supplier, 45, 212
Swiss Cheese Model of Defenses, 182
System, 45
System safety
   Fundamental tenets, 149–150
   Process, 9–11

Take 5 for safety, 102–103
Task, 45
Task hazard analysis, 292, 301–303, 305
Threat, 45–46, 373

Tolerable risk, 46
Triggers, 46

Unacceptable risk, 44, 61, 75, 84–85, 101, 167, 210, 285, 382
User, 46

Value of statistical life (VSL), 436, 437
Vulnerability, 46, 74, 373

Walline, Dave, 79, 88, 212
Warning, 46, 81

What-if analysis, 121–122, 125, 127–130, 139–140
  Advantages and disadvantages, 128
  Process steps, 125–136
What-if/checklist analysis, 130–131
Whiting, James, 106, 109
Whole system risk, 97
Work-related musculoskeletal disorders (WMSDs), 343, 345–346
World Health Organization (WHO), 320
Worst credible consequence, 46, 76
Worst conceivable consequence, 46, 76